The Microskills Hierarchy
A Pyramid for Building Cultural Intentionality

NINTH EDITION

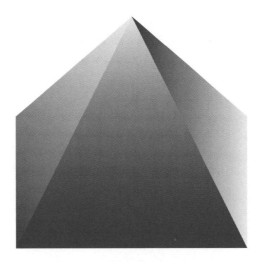

Intentional Interviewing and Counseling

Facilitating Client Development in a Multicultural Society

Allen E. Ivey, Ed.D., ABPP

Distinguished University Professor (Emeritus)
University of Massachusetts, Amherst
Consultant: Microtraining/Alexander Street Press

Mary Bradford Ivey, Ed.D., NBCC

Amherst, Massachusetts Schools
Consultant: Microtraining/Alexander Street Press

Carlos P. Zalaquett, Ph.D., M.A., Lic., LMHC

Professor, The Pennsylvania State University

CENGAGE
Learning

Australia • Brazil • Mexico • Singapore • United Kingdom • United States

**Intentional Interviewing and Counseling:
Facilitating Client Development in a
Multicultural Society, 9th Edition**
Allen E. Ivey, Mary Bradford Ivey,
Carlos P. Zalaquett

Product Director: Marta Lee-Perriard

Product Manager: Julie Martinez

Content Developer: Elizabeth Momb

Product Assistant: Kimiya Hojjat

Marketing Manager: Jennifer Levunduski

Content Project Manager: Rita Jaramillo

Digital Content Specialist: Michelle Wilson

Art Director: Vernon Boes

Manufacturing Planner: Judy Inouye

Production Service: MPS Limited

Text and Cover Designer: Lisa Delgado

Cover Image: OlgaYakovenko/Masterfile

Compositor: MPS Limited

For product information and technology assistance, contact us at
Cengage Learning Customer & Sales Support, 1-800-354-9706.

For permission to use material from this text or product,
submit all requests online at **www.cengage.com/permissions**.
Further permissions questions can be e-mailed to
permissionrequest@cengage.com.

Library of Congress Control Number: 2016936348

Student Edition:

ISBN: 978-1-305-86578-5

Loose-leaf Edition:

ISBN: 978-1-337-27776-1

Cengage Learning
20 Channel Center Street
Boston, MA 02210
USA

Cengage Learning is a leading provider of customized learning solutions
with employees residing in nearly 40 different countries and sales in more
than 125 countries around the world. Find your local representative at
www.cengage.com.

Cengage Learning products are represented in Canada by Nelson
Education, Ltd.

To learn more about Cengage Learning Solutions, visit **www.cengage.com**.

Purchase any of our products at your local college store or at our preferred
online store **www.cengagebrain.com**.

Printed in the United States of America
Print Number: 03   Print Year: 2017

Love is listening.

Paul Tillick, Ph.D., Licentiate of Theology
University Professor, Harvard University
Most influential theologian of the last century

To the multicultural scholars who have changed the nature and practice of counseling and psychotherapy

Patricia Arredondo, Ed.D. President, Arredondo Advisory Group, author of the Multicultural Competencies and Guidelines, past president of the American Counseling Association, National Latina/o Psychological Association, APA Society for Counseling Psychology

Eduardo Duran, Ph.D. Private practice, consultant, author of *Native American Postcolonial Psychology* and *The Soul Wound* describing historical trauma of Native Americans, professor of psychology in several graduate settings, continues to teach and lecture in community settings all over the world

Thomas Parham, Ph.D. Vice chancellor, University of California, Irvine, past president of the Association for Multicultural Counseling and Development and the Association of Black Psychologists (Distinguished Psychologist), 100 Black Men of America Wimberly Award

Paul Pedersen, Ph.D. Professor emeritus Syracuse University, first White scholar to introduce multicultural issues to the helping fields, author of 40 books, American Psychological Association Award for Distinguished Contributions to the International Advancement of Psychology

Derald Wing Sue, Ph.D. Professor, Columbia University, originator of the Multicultural Competencies, nationally and internationally known for writing on microaggressions, past president of the Society for Counseling Psychology President's Committee on Race

ABOUT THE AUTHORS

Allen E. Ivey is Distinguished University Professor (Emeritus), University of Massachusetts, Amherst. He is the founder of Microtraining Associates, an educational publishing firm, and now serves with Microtraining/Alexander Street Press as a consultant. Allen is a Diplomate in Counseling Psychology and a Fellow of the American Counseling Association. He is past president and Fellow of the Society for Counseling Psychology. He is also a Fellow of the American Counseling Association (where he made the first presentation on neuroscience and counseling), Society for the Psychological Study of Culture, Ethnicity, and Race, and the Asian American Psychological Association. He has keynoted conferences in 25 countries, but is most proud of being named a Multicultural Elder at the National Multicultural Conference and Summit. Allen is author or coauthor of more than 40 books and 200 articles and chapters, translated into 25 languages. He is the originator of the microskills approach, which is fundamental to this book.

Courtesy of Allen E. Ivey

Mary Bradford Ivey is a consultant with Microtraining/Alexander Street Press and a former school counselor. She has served as a visiting professor at the University of Massachusetts, Amherst; Keene State College University of Hawai'i; and Flinders University, South Australia. Mary is the author or coauthor of many articles and of 16 books, translated into multiple languages. She is a Nationally Certified Counselor (NCC) and has held a certificate in school counseling. She is also known for her work in promoting and explaining developmental counseling in the United States and internationally, with a special background to the prevention of bullying. Her elementary counseling program was named one of the 10 best in the nation at the Christa McAuliffe Conference. She is one of the first 15 honored Fellows of the American Counseling Association for her extensive contributions to the multicultural and social justice field, as well as her well-known video demonstrations and writing.

Courtesy of Mary Bradford Ivey

Carlos P. Zalaquett is a professor in the Department of Educational Psychology, Counseling, and Special Education at the Pennsylvania State University, and a licensed mental health counselor in the State of Florida. He is also vice president for the United States and Canada of the Society of Interamerican Psychology, president of the Pennsylvania Mental Health Counselors Association, and past president of the Florida Mental Health Counseling Association, the Suncoast Mental Health Counselors Association (SMHCA), and the Florida Behavioral Health Alliance. Carlos is the author or coauthor of more than 50 scholarly publications and five books, including the Spanish version of *Basic Attending Skills*. He has received many awards, including the University of South Florida's Latinos Association's Faculty of the Year, the Tampa Hispanic Heritage's Man of Education Award, and the SMHCA Emeritus Award. His current research uses a neuroscience-based framework to compare brain activity and self-reported decision making. This cutting-edge research integrates mind, brain, and body in the exploration of human responses central to counseling and psychotherapy. He is an internationally recognized expert on mental health, counseling, psychotherapy, diversity, and education and has conducted workshops and lectures in 11 countries.

Courtesy of Carlos Zalaquett

CONTENTS

LIST OF BOXES

PREFACE

Welcome to the ninth edition of *Intentional Interviewing and Counseling: Facilitating Client Development in a Multicultural Society*, the original, most researched system in the basics of skilled counseling and psychotherapy. You will find a completely updated and rewritten revision, based on the latest research, and made even more user friendly through restructuring and a new organization.

The microskills approach has become the standard for interviewing, counseling, and psychotherapy skills training throughout the world. Based on more than 500 data-based studies, used in well over 1,000 universities and training programs throughout the world, the culturally sensitive microskills approach is now available in 20 translations. The emphasis is on clarity and providing the critical background for competence in virtually all counseling and psychotherapy theories.

Easy to teach and learn from, students will find that the content, transcripts, case illustrations, and exercises help ensure that they can immediately take to the "real world" the concepts presented in the textbook.

An alternative version of this text is available. *Essentials of Intentional Interviewing* (3rd ed.) covers the skills and strategies of interviewing, counseling, and psychotherapy in a briefer form, with less attention to theory, research, and supplementary concepts.

The Microskills Tradition and Basic Competencies

The backbone of this book continues the original emphasis on competencies. What counts is that students first develop a foundation by becoming competent in listening and empathic skills. This is followed by step-by-step movement through the microskills hierarchy, through which the major aspects of a successful interview are introduced. Students who work with this book will be able to

- Engage in the basic skills of the counseling or psychotherapy session: listening, influencing, and structuring an effective session with individual and multicultural sensitivity.

- Conduct a full session using only listening skills by the time they are halfway through this book.

- Master a basic structure of the session that can be applied to many different theories:

  1. Develop an *empathic relationship* with the client.
  2. Draw out the client's *story*, giving special attention to strengths and resources.
  3. Set clear *goals* with the client.
  4. Enable the client to *restory* and think differently about concerns, issues, and challenges.
  5. Help the client move to *action* outside the session.

- Observe counseling and therapy in action through the many interview example transcripts throughout the book. We consider this a central part of learning the application of skills and theories with many diverse clients.

- Integrate ethics, multicultural issues, and positive psychology/wellness into counseling practice.

- Analyze with considerable precision their own natural style of helping and, equally or perhaps more important, how their counseling style is received by clients.

- Become able to integrate basic aspects of neuroscience into the session. Develop a client-centric approach, full of genuine desire to help others and advance our communities and societies.

Empathy and empathic communication have become even more central to the microskills framework. While they have always been there, they are now a centerpiece, associated with each and every skill. Students will be able to evaluate each intervention for its quality of empathic understanding and whether or not it facilitates the interview process. Every transcript in this text includes process discussions that illustrate the various levels of empathy. Students will be able to evaluate on the spot how their interviewing leads affect the client.

The Portfolio of Competencies is emphasized in each chapter. Students have found that a well-organized portfolio is helpful in obtaining good practicum and internship sites and, at times, professional positions as well. Students may complain about the workload, but if they develop a solid portfolio of competencies, use the interactive website to reinforce learning, and engage in serious practice of skills and concepts, it will become clear how much they have learned. The portfolio concept and the authors' videos increase course satisfaction and ratings.

New Competency Features in This Ninth Edition

The coming decade will bring an increasing integration of mental and physical health services as we move to new, more sophisticated and complete systems to help clients and patients. Innovations in team practice are bringing counselors and psychotherapists together more closely with physicians, nurses, and human service workers. Furthermore, neuroscience, neurobiology, and brain research are leading to awareness that body and mind are one. Actions in the counseling session affect not only thoughts, feelings, and behaviors but also what occurs in the brain and body. Many exciting new opportunities await both students and instructors.

This ninth edition of *Intentional Interviewing and Counseling* seeks to prepare students for culturally intentional and flexible interviewing, counseling, and psychotherapy. The following features have been added or strengthened as we prepare for this new future.

- *Listening lights up the brain.* The power and importance of attending behavior and empathy are now further validated by neuroscience research showing that specific parts of the brain are activated during empathic listening.

- *Crisis counseling, suicide assessment and prevention*, and a transcript of *cognitive behavioral therapy* are given increased attention. Students can take the learning from earlier chapters to develop beginning competence in these critical aspects of practice. The CBT transcript shows the specifics of work with automatic thoughts and demonstrates clearly how students can use this strategy.

- Included is a newly integrated chapter on the *action influencing skills* (Chapter 12). The skills of self-disclosure, feedback, logical consequences, directives/instruction, and psychoeducation are now presented together through data and transcripts of a four-interview case study with a single client, who makes progress and becomes able to free herself with the counselor to discuss deeper, more critical relationship issues.

- Our emphasis on *multicultural and social justice* has once again been enlarged. With this edition, we introduce Eduardo Duran's concept of the Soul Wound and the historical and intergenerational issue of cultural and individual trauma. New to this edition are specific session recommendations to help clients who have encountered racism, sexism, bullying, and the many forms of harassment and oppression.

- The critical issue of recognizing *stress* and its dangerous impact on the brain and body is emphasized throughout, while also noting that appropriate levels of stress can be positive and necessary for learning, change, and building resilience to master more serious and challenging stress. Research in wellness and neuroscience has revealed the importance of positive psychology and therapeutic lifestyle changes (TLCs) as a supplement to stress management and all theoretical approaches.

- The fifth stage of the interview—action—has been given increased attention with the *action plan*. The action plan is a systematic, comprehensive approach to homework and generalization from the interview to the "real world." Albert Ellis gave us the term *homework*, which for some clients feels like school. The action plan is more systematic, with an emphasis on collaboration and client decision as to how to take the interview into daily life.

- *Self-actualization, intentionality*, and *resilience* are clarified and given increased emphasis as goals for the interview. Resilience, especially, has become more central as an action goal to enable clients to adapt and grow as they experience stress. A new section focuses on what we would like to see for our clients as a result of the counseling session. Of course we want to facilitate their reaching their own desired ends, but we also seek to encourage the development of resilience skills to better cope with future stresses and challenges.

- Increased attention and emphasis is given to *transcripts* in most chapters, showing how the skills are used in the interview and their impact on client conversation, leading to personal growth. We see how empathy is demonstrated and rated in the session. The Client Change Scale illustrates how the client is learning and progressing the session. At times, reading key transcripts aloud will bring the interview even more to the here and now.

- *Increased integration of cutting-edge neuroscience with counseling skills.* Counseling and psychotherapy change the brain and build new neural networks in both client and counselor through neural plasticity and neurogenesis. Special attention is paid to portions of the brain (with new illustrations) that are affected in the helping process. Neuroscience research stresses a positive wellness orientation to facilitate neural development, along with positive mental health. An updated neuroscience/neurobiology appendix with additional practical implications is also included. Students will find that virtually all of what we do in the helping fields is supported by neuroscience research.

- One of the most important changes in this edition is a refined and more precise definition of *empathy*. Drawing from neuroscience, paraphrasing is now associated with cognitive empathy, reflection of feeling with affective empathy, and mentalizing (understanding the client's world more holistically) with the summary.

- *CourseMate*, our optional online package, a popular and effective interactive ancillary, has been updated. The many case studies and interactive video-based exercises provide practice and further information leading to competence. Downloadable forms and feedback sheets make it easier for students to develop a Portfolio of Competence. Students who seriously use these resources report that they understand the session better and perform better on examinations.

Supplementary Materials

This text is accompanied by several supporting products for both instructors and students.

MindTap

MindTap for *Intentional Interviewing and Counseling: Facilitating Client Development in a Multicultural Society*, Ninth Edition, engages and empowers students to produce their best work—consistently. By seamlessly integrating course material with videos, activities, apps, and much more, MindTap creates a unique learning path that fosters increased comprehension and efficiency.

For students:

- MindTap delivers real-world relevance with activities and assignments that help students build critical thinking and analytic skills that will transfer to other courses and their professional lives.
- MindTap helps students stay organized and efficient with a single destination that reflects what's important to the instructor, along with the tools students need to master the content.
- MindTap empowers and motivates students with information that shows where they stand at all times—both individually and compared to the highest performers in class.

Additionally, for instructors, MindTap allows you to:

- Control what content students see and when they see it with a learning path that can be used as is or matched to your syllabus exactly.
- Create a unique learning path of relevant readings, multimedia, and activities that move students up the learning taxonomy from basic knowledge and comprehension to analysis, application, and critical thinking.
- Integrate your own content into the MindTap Reader, using your own documents or pulling from sources such as RSS feeds, YouTube videos, websites, Googledocs, and more.
- Use powerful analytics and reports that provide a snapshot of class progress, time in course, engagement, and completion.

Online Instructor's Manual

The Instructor's Manual (IM) contains a variety of resources to aid instructors in preparing and presenting text material in a manner that meets their personal preferences and course needs. It presents chapter-by-chapter suggestions and resources to enhance and facilitate learning.

Online Test Bank

For assessment support, the updated test bank includes true/false, multiple-choice, matching, short answer, and essay questions for each chapter.

Cengage Learning Testing powered by Cognero

Cognero is a flexible, online system that allows you to author, edit, and manage test bank content as well as create multiple test versions in an instant. You can deliver tests from your school's learning management system, your classroom, or wherever you want.

Online PowerPoint

These vibrant Microsoft® PowerPoint® lecture slides for each chapter assist you with your lecture by providing concept coverage using images, figures, and tables directly from the textbook.

Acknowledgments

Our Thanks to Our Students

National and international students have been important over the years in the development of this book. We invite students to continue this collaboration. Weijun Zhang, a former student of Allen, is now the leading coach and management consultant in China. He wrote many of the National and International Perspectives on Counseling Skills boxes, which enrich our understanding of multicultural issues. Amanda Russo, a student at Western Kentucky University, allowed us to share some of her thoughts about the importance of practicing microskills. We give special attention to Nelida Zamora and SeriaShia Chatters, both former students of Carlos. Nelida worked with us closely in the development of two sets of videos, *Basic Influencing Skills* (3rd ed.) and *Basic Stress Management Skills* for Alexander Street Press/Microtraining Associates. Nelida Zamora also gave permission to use a transcript of her demonstration session with Allen in Chapters 9 and 10. SeriaShia Chatters helped develop the DVD sets and book videos, important in making the nature of helping skills clear. She is now a faculty member at The Pennsylvania State University. Our graduate students at the University of South Florida volunteered their time to participate in the videos that are on the supplemental website. We are especially appreciative of the quality work of Kerry Conca, Megan Hartnett, Jonathan Hopkins, Stephanie Konter, Floret Miller, Callie Nettles, and Krystal Snell.

Our Thanks to Our Colleagues

Machiko Fukuhara, president of the Japanese Microcounseling Association and president of the International Council on Psychology, Inc., has been central in Mary and Allen's life, work, and writing for many years. Thomas Daniels, a distinguished Canadian professor, has also been with us as stimulating coauthor, friend, and provocateur. These two have been central in the development of microcounseling and its expansion internationally.

James Lanier has been a good friend and influential colleague. He is the person who helped us move from a problem-oriented language to one that is more positive and hopeful. Robert Marx developed the Relapse Prevention form of Chapter 14. Mary and Allen's two-hour meeting with Viktor Frankl in Vienna clarified the centrality of meaning in counseling, along with specifics for treatment. William Matthews was especially helpful in formulating the five-stage interview structure. Lia and Zig Kapelis of Flinders University and Adelaide University are thanked for their support and participation while Allen and Mary served twice as visiting professors in South Australia.

David Rathman, Chief Executive Officer of Aboriginal Affairs, South Australia, has constantly supported and challenged this book, and his influence shows in many ways. Matthew Rigney, also of Aboriginal Affairs, was instrumental in introducing us to new ways of thinking. These two people first showed us that Western individualistic ways of thinking are incomplete, and therefore they were critical in bringing us early to an understanding of multicultural issues.

The skills and concepts of this book rely on the work of many different individuals over the past 30 years, notably Eugene Oetting, Dean Miller, Cheryl Normington, Richard Haase, Max Uhlemann, and Weston Morrill at Colorado State University, who were there at the inception of the microtraining framework. The following people have been personally and professionally helpful in the growth of microcounseling and microtraining over the years: Bertil Bratt, Norma Gluckstern-Packard, Jeanne Phillips, John Moreland, Jerry Authier, David Evans, Margaret Hearn, Lynn Simek-Morgan, Dwight Allen, Paul and Anne Pedersen, Patricia Arredondo, Lanette Shizuru, Steve Rollin, Bruce Oldershaw, Oscar Gonçalves, Koji Tamase, Elizabeth and Thad Robey, Owen Hargie, Courtland Lee, Robert Manthei, Mark Pope, Kathryn Quirk, Azara Santiago-Rivera, Sandra Rigazio-DiGilio, and Derald Wing Sue.

Fran and Maurie Howe have reviewed seemingly endless revisions of this book over the years. Their swift and accurate feedback has been significant in our search for authenticity, rigor, and meaning in the theory and practice of counseling and psychotherapy.

Jenifer Zalaquett has been especially important throughout this process. She not only navigates the paperwork but is instrumental in holding the whole project together.

Julie Martinez has now worked with us as consulting editor through six editions of this book. At this point, we almost feel that she is a coauthor. Elizabeth Momb, our action editor, has been a blessing and her expertise and patience are "over the top." It is always a pleasure to work with the rest of the group at Cengage Learning, notably Rita Jaramillo, Vernon Boes, and Kimiya Hojjat. Our manuscript editor, Peggy Tropp, has become a valuable adviser to us and has been a joy with her understanding support. We would also like to acknowledge the efforts of our project manager, Lynn Lustberg of MPS Limited.

We are grateful to the many thoughtful reviewers for their valuable suggestions and comments for this new edition. They shared ideas and encouraged the changes that you see here, and they also pushed for more clarity and a practical action orientation.

Again, we ask you to send in reactions, suggestions, and ideas. Please use the form at the back of this book to send us your comments. Feel free to contact us also by email. We appreciate the time that you as a reader are willing to spend with us.

Allen E. Ivey, Ed.D., ABPP
Mary Bradford Ivey, Ed.D., NCC, LMHC
Carlos Zalaquett, Ph.D., MA, LMHC,
Licensiado en Psicología
email: allenivey@gmail.com
mary.b.ivey@gmail.com
cpz1@psu.edu

To the Student: Demystifying the Helping Process

Demystify: make less mysterious or remove the mystery from.
—Webster's Online Dictionary

Demystify: to make something easier to understand.
—Cambridge Advanced Learner's Dictionary & Thesaurus

What makes counseling and psychotherapy work? The actual nature of what is happening in the session remained mysterious until 1938, when Carl Rogers, founder of person-centered counseling, began to provide answers. As the first to demystify, he used the newly invented wire recorder to record live counseling sessions. He soon found that what therapists said they did in the interview was not what actually happened. Among his important discoveries was that an empathic relationship between counselor and client is fundamental to success.

As audio technology progressed, recording and analyzing interviews became common. Nonetheless, questions remained. Among them were "What are the key *behaviors* facilitating client growth?" Finding the central components of this interpersonal relationship called therapy remained elusive. "What is listening?" "Is nonverbal behavior an important aspect of successful therapy?" "How best can we structure an effective session and treatment plan?"

Demystifying the behaviors of a successful counseling session. Until the microskills approach came along, the counseling and psychotherapy field had not yet identified the specific actions and behaviors of effective interviewing. With colleagues at Colorado State University, Allen obtained a grant from the Kettering Foundation to research the interview in depth. For the first time, the group was able to video record using 2-inch-wide videotape (compare that to your smartphone—the world changes). Until this point, no one had examined how verbal skills are related to nonverbal behavior.

Attending, the first behavioral skill. The importance of listening (later termed attending behavior) came to the CSU group almost by chance. To test our new technology, we videotaped Rhonda, our secretary, in a demonstration session. She totally failed to attend to the student she was interviewing—looked away, had awkward verbal hesitations, and shifted her body uncomfortably. She frequently changed the topic, seldom following the interviewee. When we reviewed the video, we identified attending behavior dimensions for the first time: appropriate eye contact, comfortable body language and facial expression, a pleasant and smooth vocal tone, and verbal following—staying with the client's topic. Three of the key elements of listening and communicating empathy turned out to be nonverbal, a major discovery for our highly verbal profession.

When Rhonda and Allen viewed the videotape, she noticed the same behaviors we listed above. After a short discussion period, Rhonda went back for another video session and listened effectively, and even looked like a counselor. All that happened in a half-hour!

Taking microskill learning home. The next level of demystification came when Rhonda returned after the weekend. "I went home, I attended to my husband, and we had a beautiful weekend!" We had not expected that learned interview behavior would generalize to real life. We became aware of the importance of teaching communication microskills to clients and patients. Children, couples, families, management trainees, psychiatric patients, refugees, and many others have now been taught specifics of communication via the skills taught in this book. Think of microskill teaching as an effective counseling and therapy change strategy in itself.

Demystify your own helping style through video. This book, *Intentional Counseling and Interviewing,* asks you to look at yourself on video as you practice counseling skills. The majority of you now have smartphones, computers, or small cameras that provide the opportunity to see yourself as others see you. Go through practice sessions with classmates and friends and obtain valuable feedback.

The microskills demystification goes viral. Allen's first book was translated into multiple languages and has become a regular part of the curriculum in counseling, social work, psychology, and other departments in the United States and abroad. Working for her doctoral degree at the University of Massachusetts, Mary soon joined Allen and was the first person to teach listening skills to managers. Carlos, trained in the microskills as part of his graduate program in Chile, soon taught them as part of the first course on counseling skills in South America and has translated the skills into Spanish.

What about multicultural issues? About a year after the identification of key skills of listening, Allen was enthusiastically teaching a workshop. He talked of attending behavior, including the importance of eye contact, but then a beginning counselor from Alaska challenged him and described her experience with Native Inuits. She pointed out that traditional people could see direct eye contact and close face-to-face interaction as uncaring or even hostile. One can still attend, but we need to consider the natural nonverbal and verbal communication style of each culture. This led us to give central attention to multicultural issues, as you will see throughout this text.

And now, the demystification of neuroscience and neurobiology. Our most recent venture has been into this newly relevant field. Research in neuroscience has further demystified the helping process. Not too surprising is the discovery that almost all of what has been done in our field is validated by neuroscience: *counseling changes the brain (and the body)*. Neurobiology has become relevant as we learn the impact of stress and trauma on mental *and* physical health. Appendix IV provides a detailed basic discussion with many illustrations. We recommend referring there for more specifics as you read and discover neuroscience within the chapters.

The National Institute of Mental Health is leaving the pathology model of the present *Diagnostic and Statistical Manual of Mental Disorders* in favor of brain-based assessment and treatment. Research on what happens in therapy is changing rapidly. The holistic brain/body approach now includes exploration of how personal interaction even changes DNA and gene functioning, depression as a biological disease, and how social conditions affect human development. Such findings are leading to a new holistic approach suggesting new strategies for facilitating physical and mental health.

Many clients will come to you with some knowledge of the brain, because of extensive coverage of new findings in the media. Even with a beginning knowledge of key brain processes, you can now explain the importance of focusing on stress management, increasing emotional regulation, and using counseling and therapy collaboratively to build resilience and developmental growth. Whether we come from a traditional psychoanalytic, a cognitive behavioral, or an environmentally oriented approach, how we affect the client's brain and body will be clarified by neuroscience, neurobiology, and related fields.

To be continued. The learning process of demystification constantly brings something new and exciting. You may want to visit a rather basic YouTube introduction to neuroscience and counseling by Allen and Mary, using the search terms *allen ivey* or *spark lecture*. More generally, the search terms *neuroscience* and *neurobiology* will lead you in fascinating directions on YouTube and elsewhere. At the conclusion of this book, you will find many more specific suggestions for exploring the helping process on YouTube and the Internet.

SECTION I

The Foundations of Counseling and Psychotherapy

Section I presents the conceptual underpinnings of effective counseling and therapy. Building on this foundation, the first half of the book focuses on listening skills and the structuring of an effective session. Later chapters will discuss influencing skills and strategies, designed to provide you with many possibilities for empowering your clients to take charge, find meaningful goals, and change their lives. The book concludes with integrative applications and illustrates how these skills can be applied to multiple theories of counseling and therapy, and how you can integrate this learning with your own way of being to advance your own personal style of helping.

Chapter 1, Intentional Interviewing, Counseling, and Psychotherapy, offers an overview and a road map of what this book can do for you. We begin by defining interviewing, counseling, and psychotherapy. Counseling is best considered both a science and an art. We present the central skills of counseling, but it is you who will make this knowledge live in the interview and therapy session. We also ask you to record an interview before you start to identify your natural helping skills. You are not taking this course by chance; something has led you here, with unique abilities, oriented to helping others. You will be asked to reflect on what brings you to the helping field: What do you want to do to help others grow and develop?

Chapter 2, Ethics, Multicultural Competence, Positive Psychology, and Therapeutic Lifestyle Changes, presents crucial aspects of all counseling and psychotherapy. Ethics—the professional standards that all major helping professions observe and practice—provides counselors and psychotherapists with guidelines on issues such as competence, informed consent, confidentiality, power, and social justice. Multicultural competence focuses on cultural awareness, knowledge, skills, and action to meet the highly diverse clients we are likely to meet. You will be asked to examine yourself as a multicultural individual. Positive psychology, wellness, and therapeutic lifestyle changes (TLCs) enable clients to identify their strengths and resources to build resilience. This approach significantly facilitates resolving client life issues, focusing on what they "can do" rather than what they "can't do."

Chapter 3, Attending Behavior and Empathy Skills, presents the most basic fundamentals of counseling and psychotherapy. Without the listening and attending skills, an empathic relationship cannot occur. Many beginning helpers inappropriately strive to solve the client's issues and challenges in the first 5 minutes of the session by giving premature advice and suggestions. Please set one early goal for yourself: *Allow your clients to talk*. Observe closely how they are behaving, verbally and nonverbally. Your clients may have spent several years developing their concerns, issues, and life challenges before consulting you. Listen first, last, and always.

Chapter 4, Observation Skills, builds on attending behavior and gives you the further opportunity to practice observing your clients' verbal and nonverbal behavior. You are also asked to observe your own nonverbal reactions in the session. Clients often come in with a "hangdog" and "down" body posture. Between your observation and listening skills, you can anticipate that they will later have more positive body language, as well as a new story and a better view of self. You can help their body to stand up straight and their eyes to shine.

Begin this book with a commitment to yourself and your own natural communication expertise. Through the microskills approach, you can enhance your natural style with new skills and strategies that will expand your alternatives for facilitating client growth and development.

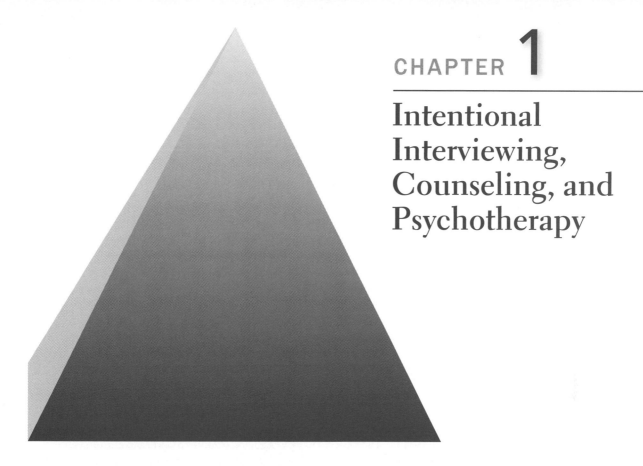

CHAPTER 1

Intentional Interviewing, Counseling, and Psychotherapy

We humans are social beings. We come into the world as the result of others' actions. We survive here in dependence on others. Whether we like it or not, there is hardly a moment of our lives when we do not benefit from others' activities. For this reason it is hardly surprising that most of our happiness arises in the context of our relationships with others.

—The Dalai Lama

Chapter Goals and Competency Objectives

Each chapter of this book is organized around the counseling framework of awareness, knowledge, skills, and action. Awareness and knowledge of counseling and therapy are obviously basic, but they are not fully meaningful until manifested in skills taken into action. Furthermore, action implies taking awareness, knowledge, and skills beyond the textbook to the real world.

Awareness and Knowledge

▲ Define and discuss similarities and differences among interviewing, counseling, and psychotherapy, and review who actually conducts most of the helping sessions. This may be surprising and rewarding.

▲ Explore the session as both science and art. We ask you to reflect on yourself as a potential helper. While science undergirds what is said here, you as an independent artist will find your own integration of knowledge and skills.

▲ Identify **intentionality** and **cultural intentionality** as bases for increasing your flexibility to reach a wide variety of clients facing an endless array of concerns, issues, and challenges.

▲ Examine key goals of counseling and psychotherapy: self-actualization, resilience, and resolution of client issues.

▲ Consider the place of cutting-edge neuroscience in your own work and for the future of counseling and psychotherapy.

▲ Identify the locations where counseling and psychotherapy are practiced.

Skills and Action

▲ The microskills hierarchy provides a foundation for interview action, not only in a beginning form but also in conjunction with multiple theories and approaches to counseling and therapy, including person-centered, cognitive behavioral therapy (CBT), crisis counseling, and many others.

▲ As your first practical exercise, record a counseling session demonstrating your natural style of communicating and helping. This provides a baseline so that later you can examine how your counseling style may have changed and grown during your time with this book. Practicing and developing a Portfolio of Competencies provide a summary and journal of your experience.

Introduction: Interviewing, Counseling, and Psychotherapy

Sienna, 16 years old, is 8 months pregnant with her first child. She says, "I wonder when I'll be able to see Freddy [baby's father] again. Mom works hard to keep him away from me. I mean, I want him involved; he wants to be with me, and the baby. But my mom wants me home. His mom said she's looking for a two-bedroom apartment so we could possibly live there, but I know my mom will never go for it. She wants me to stay with her until I graduate from high school and, well, to be honest, so that this never happens again [she points to her belly]."

I listen carefully to her story and later respond, "I'm glad to hear that Freddy wants to be involved in the care of the child and maintain a relationship with you. What are your goals with him? What happens when you talk with your mom about him?"

"I don't know. We don't really talk much anymore," she says as she slumps down in her chair and picks away at her purple nail polish. I reflect her sad feelings, but as I do so, she brightens up just a bit as she recalls that most of the time she does get along with her mother fairly well.

She then describes her life before Freddy, focusing mainly on the crowd she hung around, a group of girls whom she says were wild, mean, and tough. Her mood returns to melancholy, and she seems anxious and discouraged. At the same time, the session has gone smoothly and we seem to have a good relationship. I say, "I sense that you have a good picture of what you are facing. Well, it seems that there's a lot to talk about. How do you feel about continuing our conversation before sitting down with your mom?"

Surprisingly, she says, "No. Let's talk next week with her. I think she might come. The baby is coming soon and, well, it'll be harder then." As we close the session, I ask her, "As you look back on our talk together, what comes to mind?" Sienna responds, "Well, I feel a bit more hopeful and I guess you're going to help me talk about some important issues with my mom, and I didn't think I could do that."

This was the first step in a series of five sessions. As the story evolved, we invited Freddy for a session. He turned out to be employed and was anxious to meet his responsibilities, although finances remained a considerable challenge. A meeting with both mothers followed, and a workable action plan for all families was generated. I helped Sienna find a school with a special program for pregnant teens.

This case exemplifies the reality of helping. We often face complex issues with no clear positive ending. If we can develop a relationship and listen to the story carefully, clearer goals develop, and solutions usually follow.

Reflective Exercise Love is listening

Famed theologian Paul Tillich says "Love is listening." *Listening, love, caring,* and *relationship* are all closely related. These four words could be said to be the center of the helping process.

- What relevance do these words have in the meeting with Sienna?
- What are your reactions and thoughts about the centrality of these words?
- How might the science and art of counseling and therapy speak to this issue?

Defining Interviewing, Counseling, and Psychotherapy

The terms *interviewing, counseling,* and *psychotherapy* are used interchangeably in this text. The overlap is considerable (see Figure 1.1), and at times interviewing will touch briefly on counseling and psychotherapy. Both counselors and psychotherapists typically draw on the interview in the early phases of their work. You cannot become a successful counselor or therapist unless you have solid interviewing skills.

Interviewing is the basic process used for gathering data, providing information and advice to clients, and suggesting workable alternatives for resolving concerns. Interviewers can be found in many settings, including employment offices, schools, and hospitals. Professionals in many areas also use these skills—for example, in medicine, business, law, community development, library work, and many government offices.

Closely related to interviewing, **coaching** is "partnering with clients in a thought-provoking and creative process that inspires them to maximize their personal and professional

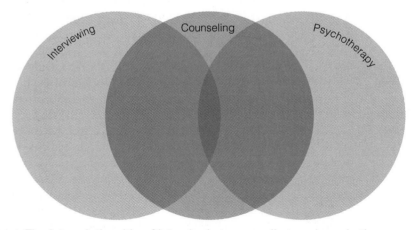

FIGURE 1.1 The interrelationship of interviewing, counseling, and psychotherapy.

potential" (International Coach Federation, 2015). Professional coaches are now hired in increasing numbers. The skills presented in this book are also basic to coaching (see Moore, 2015). You will find many aspects of the newer coaching movement closely related to counseling's history, but it is now recognized as a separate profession, although many counselors also become formal coaches.

Counseling is a more intensive and personal process. Although counselors and therapists interview to gain client information, counseling is more about listening to and understanding a client's life challenges and developing strategies for change and growth. Counseling is most often associated with the professional fields of counseling, human relations, clinical and counseling psychology, pastoral counseling, and social work and is also part of the role of medical personnel and psychiatrists.

Psychotherapy focuses on more deep-seated difficulties, which often require more time for resolution. Historically, psychotherapy was the province of psychiatrists, but they are limited in number, and today they mostly offer short sessions and treat with medications. This means that professionals other than psychiatrists conduct most talking therapy. Table 1.1 shows a total of 24,050 psychiatrists along with more than 1 million other helping professionals. Thus, it is only logical that other specialists, such as clinical and counseling psychologists, clinical mental health counselors, and clinical social workers, now provide most psychotherapy. All these psychiatrists and other professionals interview clients to obtain basic facts and information as they begin therapy, and they often provide counseling as part of the therapeutic process. The skills and concepts of intentional interviewing and counseling are equally necessary for the successful conduct of longer-term psychotherapy.

Importance of Attending Sessions. For counseling and therapy to work, clients need to attend their sessions. Do they?

One study, which examined 23,034 clients attending a total of 218,331 psychotherapy sessions, found that about 21% of clients did not return after the first session. About 50% finished at four sessions, another 25% completed their work in 5 to 10 sessions, and most had completed therapy by the 35th session (Carlstedt, 2011). An analysis of 650 studies that included more than 83,000 clients found that nearly 20% of all the clients in the studies ended their treatment early. Dropout rates were highest among the youngest participants (Swift & Greenberg, 2012). In addition, Sue and Sue (2013) note that close to 50% of clients from minority groups drop out after the first session. This suggests that although some clients find that a few sessions provide the help and information they need, many others fail

---

TABLE 1.1   Numbers of Helping Professionals

School Counselors	135,080	Marriage and Family Counselors	30,150
College Counselors	72,050	Clinical, Counseling, and School Psychologists	104,730
Mental Health Counselors	120,010	Child, Family and School Social Workers	286,520
Vocational Rehabilitation Counselors	103,890	Health Care Social Workers	145,920
Substance Abuse and Behavioral Disorder Counselors	85,180	Psychiatrists	24,060
Human Service Workers—Social and Human Service Assistants	354,800	Professional Coaches identified by the International Coaching Federation	30,000

U.S. Department of Labor. (2015). *Occupational Outlook Handbook*. http://www.bls.gov/ooh/home.htm. The Labor Department updates these data regularly.

to find counseling and therapy valuable. Dropouts represent opportunities missed for both client and counselor. Effective use of the skills and interventions presented in this book can help you help others by creating a stronger relationship with clients and reducing dropouts.

The Science and Art of Counseling and Therapy

Is therapy effective? Yes! (Lambert, 2013). Counseling and psychotherapy now have a solid research and evidence base that enables us to identify many qualities and skills that lead to effectiveness. This research focuses on the factors that contribute to establishing an effective working relationship with the client (empirically supported relationships), as well as the effect of specific interventions (empirically supported treatments) (Norcross, 2011). Science has demonstrated that the specifics of listening skills are identifiable and are central to competent helping.

But an evidence-based approach by itself is not enough. Counseling is both a science and an art. You as counselor are similar to an artist whose skills and knowledge produce beautiful paintings out of color, canvas, and personal experience. You are the listener who will provide color and meaning to the interpersonal relationship we call helping.

Like the artist or the skilled athlete, you bring a natural talent to share with others and the flexibility to respond to surprises and change direction when necessary. Theories, skills, and strategies remain essential, but you are the one who puts them together and can uniquely facilitate the development and growth of others.

Neuroscience and neurobiology have now added a new scientific dimension to counseling and therapy practice. Throughout this text, we will be sharing findings from these fields that have useful implications for daily practice. Each conversation we have with a client has the potential for affecting not only the mind but also the brain, which in turn can change the body. As we examine the mind/brain/body connections, you will discover that the vast majority of what counselors and therapists do now has additional importance and validation.

Also essential is to keep the client at the center of our attention. The centricity of clients is important because they are the largest contributors to their own change and improvement. Furthermore, they are the ones who can provide you with feedback about your work, if you dare to ask. We encourage you to seek feedback from your clients regarding the quality of the working relationship, the specific interventions, and progress toward reaching their goals. Adapting your work guided by the feedback you receive can increase your effectiveness (Duncan, Miller, Hubble, & Wampold, 2010). Effective counseling and therapy include the client, the therapist, the relationship, the treatment, the feedback, and the contextual factors surrounding this process.

Reflective Exercise Where is your place in the helping field?

- Do you see yourself emphasizing interviewing, counseling, or psychotherapy?
- Given the many possibilities for service, which of the professions listed in Table 1.1 appeals to you most at this time?
- Would you rather work in a school, a community mental health clinic, a hospital, a business, or private practice?
- What are your thoughts as to your responsibility in meeting the overall mental health needs of minority or economically disadvantaged clients?
- Would you like to work to improve the overall mental health and well-being of your society?

BOX 1.1   National and International Perspectives on Counseling Skills

Problems, Concerns, Issues, and Challenges—How Shall We Talk About Client Stories?

James Lanier, University of Illinois, Springfield

There are different ways of listening to client stories. Counseling and therapy historically have tended to focus on client problems. The word *problem* implies difficulty and the necessity of eliminating or solving the problem. Problem may imply deficit. Traditional diagnosis such as that found in the *The Diagnostic and Statistical Manual of Mental Disorders* (5th ed.; *DSM–5*; American Psychiatric Association, 2013) carries the idea of problem a bit further, using the word *disorder* with such terms as *panic disorder, conduct disorder, obsessive-compulsive disorder*, and many other highly specific disorders. The way we use these words often defines how clients see themselves.

I'm not fond of problem-oriented language, particularly that word *disorder*. I often work with African American youth. If I asked them, "What's your problem?" they likely would reply, "I don't have a problem, but I do have a concern." The word *concern* suggests something we all have all the time. The word also suggests that we can deal with it—often from a more positive standpoint. Defining *concerns* as *problems* or *disorders* leads to placing the blame and responsibility for resolution almost solely on the individual.

Recently, there has been increasing and particular concern about that word *disorder*. More and more, professionals are realizing that the way people respond to their experiences is very often a *logical response to extremely challenging situations*. Thus, the concept of posttraumatic stress *disorder* (PTSD) is now often referred to as a *stress reaction*. Posttraumatic

stress reaction (PTSR) has become an alternative name, thus normalizing the client's response. Still others prefer to *avoid naming* at all and seek to work with the thoughts, emotions, and behaviors of the stressed clients.

Finding a more positive way to discuss client concerns and stories is relevant to all your clients, regardless of their background. *Issue* is another term that can be used instead of *problem*. This further removes the pathology from the person and tends to put the person in a situational context. It may be a more empowering word for some clients. Carrying this idea further, *challenge* may be defined as a call to our strengths. All of these terms represent *an opportunity for change*.

Remember, if you listen carefully to most stories, what at first seems "abnormal" often will gradually become more understandable as you discover that the client has presented a "normal" response to an insane situation.

As you work with clients, please consider that change, restorying, and action are more possible if we help clients maintain awareness of already existing personal strengths and external resources. Supporting positive stories helps clients realize the positive assets they already have, thus enabling them to resolve their issues more smoothly and effectively, and with more pride—specifically, they become more actualized. Then you can help them restory with a *can do* resilient self-image. Out of this will come action, generalizing new ideas and new behaviors to the real world.

The next section extends science and art to cultural intentionality, collaborating with clients from different backgrounds to achieve growth and become more flexible and intentional themselves. Please take a moment first to review Box 1.1, which explores how traditional counseling too often focuses only on problems. James Lanier suggests positive ways to draw out clients' stories and focus more on strengths.

Cultural Intentionality: The Flexible, Aware, and Skilled Counselor

All interviewing and counseling is multicultural.
Each client comes to the session embodying multiple voices from the past.
—Paul Pedersen

The culturally intentional counselor acts with a sense of purpose (intention), skill, and respect for the diversity of clients. There are many ways to facilitate client development. *Cultural intentionality* is a central goal of this text. We ask you to be yourself but also to realize that to reach a wide variety of clients, you need to be flexible, constantly changing behavior and learning new ways of being with the uniqueness of each client.

First, let us describe the word *intentionality*. Intentionality speaks to the importance of being in the moment and responding flexibly to the ever changing situations and needs of clients. Beginning students are often eager to find the "right" answer for the client. In fact, they are so eager that they often give quick patch-up advice that is inappropriate. Even experienced counselors can become encapsulated into one way of thinking.

In short, flexibility—the ability to move in the moment and change style—is basic to the art form of helping. But this needs to be based on solid knowledge, awareness, and skills that are then turned into culturally intentional action. For example, your own personal issues or cultural factors such as ethnicity, race, gender, lifestyle, socioeconomic background, or religious orientation may have biased your response and session plan for Sienna.

The words *cultural intentionality* speak to the fact that the interview occurs in a cultural context, and we need to be aware of diversity and difference. Culturally intentional counseling and psychotherapy are concerned not with which single response is correct but with an awareness that different people from varying backgrounds respond uniquely. We can define **cultural intentionality** as follows:

> Cultural intentionality is acting with a sense of capability and flexibly deciding from among a range of alternative actions. The culturally intentional individual has more than one action, thought, or behavior to choose from in responding to changing life situations and diverse clients. The culturally intentional counselor or therapist remembers a basic rule of helping: *If a helping lead or skill doesn't work—try another approach!*

Multiculturalism, also referred to as diversity or cross-cultural issues, is now defined quite broadly. Once it referred only to the major racial groups, but now the definition has expanded in multiple ways. The story is that we are all multicultural. If you are White, male, heterosexual, from Alabama, a Methodist, and able-bodied, you have a distinct cultural background. Just change Alabama to Connecticut or California, and you are different culturally. Similarly, change the color, gender, sexual orientation, religion, or physical ability, and your cultural background changes your worldview and behavior. Multiculturalism means just that—many cultures.

We are all multicultural beings. Culture is like air: We breathe it without thinking about it, but it is essential for our being. Culture is not "out there"; rather, it is found inside everyone, markedly affecting our view of the world. Continually learn about and be ready to discuss cultural difference**.**

Discussed in detail in Chapter 2, **multicultural competence** is imperative in counseling and psychotherapy. We live in a multicultural world, where every client you encounter will be different from the last and different from you. Without a basic understanding of and sensitivity to a client's uniqueness, you may fail to establish a relationship and true grasp of the client's issues. Throughout this book, you will examine the multicultural issues and opportunities we all experience.

Reflective Exercise Developing your own culturally intentional style

- What is your family and cultural background, and how does that affect the person you are?
- How has each new experience or setting changed the way you think?
- Has this led to increasing flexibility and awareness of the many possibilities that are yours?
- Can you listen and learn from those who may differ sharply from you?

Resilience and Self-Actualization

> When we tackle obstacles, we find hidden reserves of courage and resilience we did not know we had. And it is only when we are faced with failure do [sic] we realise that these resources were always there within us. We only need to find them and move on with our lives.
>
> —A. P. J. Abdul Kalam
> 11th President of India

Many, even most, of our clients come to us feeling that are not functioning effectively and are focused on what's wrong with them. They are *stressed*. Clients may feel *stuck*, *overwhelmed*, and *unable to act*. Frequently, they will be unable to make a career or life decision. Often they will have a *negative self-concept*, or they may be depressed or full of anger. This focus on the negative is what we want to combat as we emphasize developing client intentionality, resilience, and self-actualization.

We cannot expect to solve all our clients' issues and challenges in a few sessions, but in the short time we have with them, we can make a difference. First think of what cultural intentionality and flexibility mean for you as a counselor or therapist. Clients will benefit and become stronger as they feel heard and respected and they discover new ways to resolve their concerns. Resolving specific immediate issues, such as choosing a college major, making a career change, deciding whether to break up a long-term relationship, or handling mild depression after a significant loss, will help them feel empowered and facilitate further action.

Resilience is a short- and long-term goal of effective counseling and therapy. We seek to help clients "bounce back" and recover when they encounter serious life challenges, including the traumatic. We do not want just to resolve issues and concerns, we also want to help our clients handle future difficulties, become more competent, and respect themselves more. When our clients adapt and learn from stressful threat, adversity, tragedy, and trauma, they are building their strengths and resilience.

The development of client cultural intentionality is another way to talk about resilience. As counselors, we want to be flexible and move with changing and surprising events, but clients need the same abilities. Helping a client resolve an issue is a good contribution to increasing client resilience. You have helped the client move from stuckness to action, from indecision to decision, or from muddling around to clarity of vision. Pointing out to clients who change that they are demonstrating resilience and ability is even better, as it facilitates longer-term success. Counseling's ultimate goal is to teach self-healing—the capacity to use what is learned in counseling to resolve other issues in the future. This is the ultimate demonstration of achieved resiliency.

Self-actualization as a goal of counseling and therapy was central to the world of both Carl Rogers and Abraham Maslow. Closely related to cultural intentionality and resilience, self-actualization is defined as

> the curative force in psychotherapy—*man's tendency to actualize himself, to become his potentialities* . . . to express and activate all the capacities of the organism. (Rogers, 1961, pp. 350–351)

> intrinsic growth of what is already in the organism, or more accurately of what is the organism itself. . . . self-actualization is growth-motivated rather than deficiency-motivated. (Maslow, Frager, & Fadiman, 1987, p. 66)

Regardless of the situation in which our clients find themselves, we ultimately want them to feel good about themselves, in the hope of good results (i.e. resilience). Both Rogers and Maslow had immense faith in the ability of individuals to overcome challenges and take charge of their lives.

Self-actualization also means that all of us exist in relationship to others. Counseling and psychotherapy sessions are indeed for the individual client, but both Rogers and Maslow also gave central importance to **being in relation** to others. They were fully aware that clients and their cultural backgrounds were unique. Increasingly, professionals are talking of building *resilient self-actualization in relation to others and their cultural/environmental context (CEC)*.

Reflective Exercise What are the goals of counseling and therapy?

- Self-actualization is a challenging concept. What does it mean to you?
- What experience and supports have led you to become more yourself, what you really are and want to be?
- How have you bounced back (resilience) from major challenges you have faced?
- What personal qualities or social supports helped you grow?
- What does this say to your own approach to counseling and psychotherapy?

Let us now turn to the skills and strategies that are aimed at developing cultural intentionality, resilience, and self-actualization and that, above all, provide the foundation for establishing an effective working relationship with your clients.

The Microskills Hierarchy: The Listening and Action Skills of the Helping Process

Counseling and psychotherapy require a relationship with the client; we seek to help clients work through issues by drawing out and listening to their stories.

Microskills identify the behavioral foundations of intentional counseling and psychotherapy. They are the specific communication skills that provide ways for you to reach many types of clients. They will clarify the "how" of all theories of counseling and therapy. You master these skills one by one and then learn to integrate them into a well-formed session.

Effective use of microskills enables you to anticipate how clients may respond to your interventions. With practice you will be able to match the microskills to the developmental and idiosyncratic characteristics of each client. We will offer you ways to adapt your communication to work with children and adult clients at every stage of development. And if clients do not respond as you expect, you will be able to shift to skills and strategies that match their needs.

The **microskills hierarchy** (see Figure 1.2) summarizes the successive steps of intentional counseling and psychotherapy. The skills rest on a base of ethics, multicultural competence, neuroscience, positive psychology, and resilience (Chapter 2). On this foundation rests attending and observation skills (Chapters 3 and 4), which are key to successful use of all the other aspects of the helping interview.

You next will move up the microskills pyramid to the empathic basic listening skills of **questioning, observation, encouraging, paraphrasing, summarizing**, and **reflecting feelings** (Chapters 5–7). Unless you have developed skills of listening and respect, the upper reaches of the pyramid are meaningless and potentially damaging. Develop your own style of being with clients, but always respect the importance of listening to client stories and issues.

Once you have basic competence, you will be able to conduct a complete session using only listening skills. The **five-stage** structure provides a framework for integrating the microskills into a complete counseling session. The **empathic relationship–story and strength–goals–restory–action** framework provides an overall system for you to use and

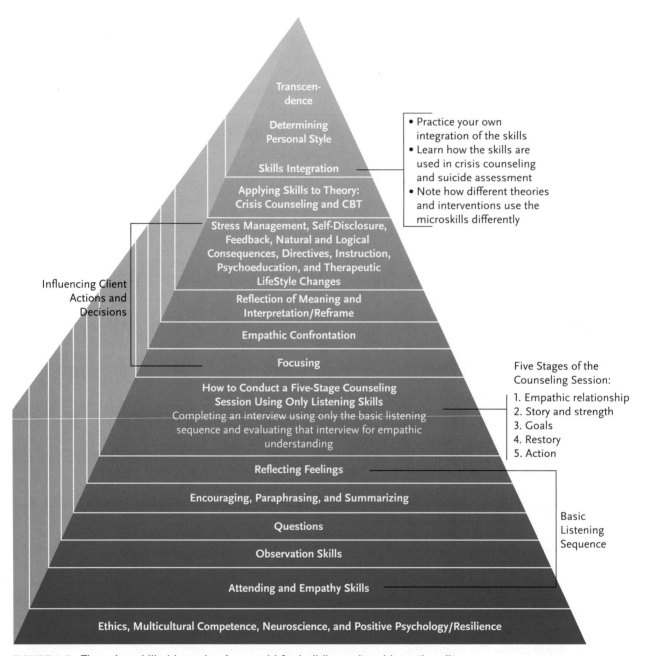

FIGURE 1.2 The microskills hierarchy: A pyramid for building cultural intentionality.

serves as a checklist for all your meetings with clients. You will use this framework when you practice completing a full session using only listening skills (Chapter 8).

Next you will encounter the influencing skills to help clients explore personal and interpersonal conflicts. **Focusing** will help you and your clients to see personal, cultural, and contextual issues related to their concerns (Chapter 9). **Empathic confrontation** (Chapter 10) is considered critical for client growth and change. Interpretation/reframing and **reflection of meaning** in Chapter 11 are important influencing skills to help clients think about themselves and their situations in new ways.

Concrete action strategies (Chapter 12) include an array of influencing skills that offer tested methods for change and for building resilience. Here you will find first the skills of self-disclosure and feedback, then strategies of **logical consequences**, along with some basics of **decision counseling**. This is followed by specific examples of the best ways to provide information and direction for the client, emphasizing stress management, **psychoeducation**, and how to use **therapeutic lifestyle changes** to provide more self-direction for clients.

By the time you reach Chapter 13, you will be prepared to build competence in multiple theories of counseling and psychotherapy. You will find that microskills can be organized into different patterns utilized by different approaches. Crisis counseling and cognitive behavioral counseling are presented with transcript of action alternatives.

At the apex of the microskills pyramid are integration of skills, developing your own personal style of counseling and therapy, and transcendence (Chapter 14). Competence in skills, strategies, and the five stages are not sufficient; you will eventually have to determine your own approach to the practice of counseling and psychotherapy. Counselors and psychotherapists are an independent lot; the vast majority of helpers prefer to develop their own style and their own blend of skills and theories.

Transcendence speaks about your capacity to go beyond yourself and successfully apply your newly mastered skills to help others. The ultimate test of your capacities is the benefit they can afford to others. Your growth is wonderful; helping others growth is even better!

As you gain a sense of your own expertise and power, you will learn that each client has a totally unique response to you and your natural style. While many clients may work well with you, others will require you to adapt intentionally to them and their individual, multicultural style.

The Microskills Teaching and Learning Framework

The model for learning microskills is practice oriented and follows this step-by-step progression:

1. *Introduction.* Focus on a single skill or strategy and identify it as a vital part of the helping process.

2. *Awareness, Knowledge, and Skills.* Read about the single skill or strategy and/or hear a lecture on the main points of effective usage from your instructor. Cognitive understanding is vital for skill development. However, understanding is not competence, nor does it show that you can actually engage in an effective interview, counseling, or therapy.

3. *Observation.* View the skill in operation via a transcript and process analysis—or better yet, watch a live demonstration or view a videotaped presentation.

4. *Multiple Applications.* Review different applications of the skills, variations according to diversity and other cultural dimensions, and additional ways in which the skill or strategy can be used.

5. *Action: Key Points and Practice.* The main issues of the chapter are summarized. Ideally, use video or audio recording for skill practice; however, role-play practice with observers and feedback sheets is also effective. Seek immediate feedback from your practice session. Use the Feedback Sheets. How did those who watched the session describe your interaction?

6. *Portfolio of Competencies and Personal Reflection.* Here you develop a summary of your interviewing, counseling, and psychotherapy abilities. Questions will ask you to summarize the meaning of the chapter for practice now and in the future.

BOX 1.2	Research and Related Evidence That You Can Use

Microskills' Evidence Base

More than 450 microskills research studies have been conducted (Ivey & Daniels, 2016). The model has been tested nationally and internationally in more than 1,000 clinical and teaching programs. Microcounseling was the first systematic video-based counseling model to identify specific observable counseling skills. It was also the first skills training program that emphasized multicultural issues. Some of the most valuable research findings include the following:

• *You can expect results from microskills training.* Several critical reviews have found microtraining an effective framework for teaching skills to a wide variety of people, ranging from beginning interviewers and counselors to experienced professionals who need to relate to clients more effectively. Teaching your clients many of the microskills will facilitate their personal growth and ability to communicate with their families or coworkers. The formal term for including teaching in your interviews is **psychoeducation**. Most chapters of this book will contain discussion of psychoeducational microskills in the activities section.

• *Practice is essential.* Practice the skills to mastery if the skills are to be maintained and used after training. *Use it or lose it!* Complete practice exercises, and generalize what you learn to real life. Whenever possible, audio or video record your practice sessions.

• *Multicultural differences are real.* People from different cultural groups (e.g., ethnicity/race, gender) have different patterns of skill usage. Learn about people different from you, and use skills in a culturally appropriate manner.

• *Different counseling theories have varying patterns of skill usage.* Expect person-centered counselors to focus almost exclusively on listening skills whereas cognitive behaviorists use more influencing skills. Microskills expertise will help you define your own theory and integrate it with your natural style.

• *If you use a specific microskill, then you can expect a client to respond in anticipated ways.* You can anticipate how the client will respond to your use of each microskill, but each client is unique. Cultural intentionality prepares you for the unexpected and teaches you to flex with another way of responding or another microskill.

• *Neuroscience and brain research now support clinical and research experience with the microskills approach.* Throughout this book, we will provide data from neuroscience and brain research. This research explains and clarifies much of what counseling and psychotherapy have always done and, at the same time, increases the quality and precision of our practice.

Most chapters will follow a model similar to this, although some are organized differently to meet the needs of the particularly skill(s) area. The early chapters provide the foundation and basic understanding and are organized independently, while later chapters typically follow this model. From time to time, you will see *Multiple Applications* as a supplement to some chapters.

We offer this learning framework because you can "go through" the skills quickly and understand them, but practicing them to full mastery is what makes for real expertise. We have seen many students "buzz" through the skills, but end with little in the way of actual competence. Teaching these skills to clients has also proven to be an effective counseling and therapeutic technique (Ivey & Daniels, 2016; Ivey, Ivey, Zalaquett, & Daniels, In press 2016).

The microskills of this book are key to developing an empathic relationship, drawing out the client's stories and issues, ensuring that change and growth will be the result of your conversations with your clients, and encouraging clients to develop self-healing—the capacity to apply what they have learned with you to other situations. Box 1.2 summarizes more than 450 data-based studies on the microskills framework, now used in thousands of settings around the world.

Counseling and Psychotherapy Theory and the Microskills

All counseling theories use the microskills, but in varying patterns with differing goals (see Table 1.2). Mastery of the skills will facilitate your ability to work with many theoretical alternatives. The microskills framework can also be considered a theory in itself, in which

TABLE 1.2   Microskills Patterns of Differing Approaches to the Interview

MICROSKILL LEAD	Decisional counseling	Person-centered	Logotherapy	Multicultural and feminist therapy	Crisis counseling	Cognitive behavioral therapy	Brief counseling	Motivational interviewing	Counseling/coaching	Psychodynamic	Gestalt	Business problem solving	Medical diagnostic interview
BASIC LISTENING SKILLS													
Open question	●	○	●	◐	◐	◐	●	●	●●	◐	●	◐	◐
Closed question	◐	○	●	◐	◐	●	◐	◐	◐	○	◐	◐	◐
Encourager	●	◐	●	◐	◐	◐	●	●	◐	◐	◐	◐	◐
Paraphrase	●	●	●	◐	◐	●	●	●	◐	○	◐	◐	◐
Reflection of feeling	●	●	◐	◐	◐	◐	●	◐	◐	◐	○	◐	◐
Summarization	◐	◐	●	◐	◐	◐	◐	●	◐	◐	○	◐	◐
INFLUENCING SKILLS													
Reflection of meaning	◐	●	●	●	○	○	○	◐	◐	◐	○	○	○
Interpretation/reframe	◐	○	◐	◐	◐	○	◐	●	○	●	●	◐	◐
Logical consequences	◐	○	◐	◐	●	◐	◐	◐	◐	○	○	◐	◐
Self-disclosure	◐	◐	◐	◐	◐	○	◐	◐	○	○	◐	○	○
Feedback	◐	◐	◐	◐	○	◐	◐	●	●	○	◐	◐	○
Instruction/ psychoeducation	●	○	○	◐	●	●	◐	◐	○	○	○	●	◐
Directive	◐	○	◐	◐	◐	●	○	◐	◐	○	●	●	●
CONFRONTATION (Combined skill)	◐	◐	◐	●	○	◐	◐	●	◐	◐	●	◐	◐
FOCUS													
Client	●	●●	●	◐	◐	●	●	●	●	◐	●	◐	◐
Main theme/issue	●	○	◐	◐	●	◐	●	◐	◐	○	◐	●	●
Others	◐	○	◐	◐	◐	◐	◐	◐	○	◐	◐	○	○
Family	◐	○	◐	◐	◐	◐	◐	◐	○	◐	○	○	○
Mutuality	◐	◐	◐	◐	○	◐	◐	●	○	○	○	○	○
Counselor/therapist	○	◐	◐	◐	○	○	○	○	○	○	○	○	○
Cultural/environmental/ contextual	◐	○	◐	●●	●	◐	◐	◐	◐	○	○	◐	○
ISSUE OF MEANING (Topics, key words likely to be attended to and reinforced)	Problem solving	Self-actualization, relationship	Values, meaning vision for life	How CEC impacts client	Immediate action, meeting challenges	Thoughts, behavior	Problem solving	Change	Strengths and goals	Unconscious motivation	Here-and-now behavior	Problem solving	Diagnosis of illness
COUNSELOR ACTION AND TALK TIME	Medium	Low	Medium	Medium	High	High	Medium	Medium	Medium	Low	High	High	High

LEGEND

● Frequent use of skill   ◐ Common use of skill   ○ Occasional use of skill counselor and client work together to enable the construction of new stories, accompanied by changes in thought and action.

In short, if you become competent in these thoroughly researched skills, tested in multiple clinical and counseling settings around the world, you will have developed a level of proficiency that will take you in many directions—not only in the helping fields, but also in business, medicine, governmental work, and many other settings.

Counseling and psychotherapy will be increasingly informed by research in neuroscience, and you will want to keep abreast of new developments. Relevant studies and implications for practice will be presented throughout this book. Neuroscience and brain research will lead us to major changes in the ways we think about counseling and psychotherapy in the next 10 years, enabling us to use the skills of this book in new ways with considerably more awareness and precision.

Neuroscience and Neurobiology: Implications of Cutting-Edge Science for the Future of Counseling and Psychotherapy

Psychotherapy is a biological treatment, a brain therapy. It produces detectable physical changes in our brain, much as learning does.

—Eric Kandel, Nobel Prize Winner

Our interaction with clients changes their brain (and ours). In a not too distant future, counseling will be regarded as ideal for nurturing nature.

—Óscar Gonçalves

No longer can we separate the body from the mind or the individual from his or her environment and culture. Counseling and psychotherapy are moving closer to medicine, neurology, and cognitive science. Counselors once argued against the "medical model." Influenced by preventive medicine, accountability, and neuroscience research, however, physicians are increasingly aware that what happens in the body is deeply influenced by the mind. And the counseling and therapy field has led this change in consciousness through effective helping skills and strategies. For example, many medical schools have adopted some version of the listening skills of this book as an essential part of education. Listening develops a relationship that encourages a client or patient to become more resilient and improve both mind and body. Furthermore, neuroscience information has become a constant, so expect your clients to be informed and ask questions about how counseling affects the brain.

Appendix IV contains a brief but comprehensive summary of key information on neuroscience/neurobiology. Please refer to this for more detailed information as you read through the text. You can see Allen and Mary speak with PowerPoints on elementary basics of neuroscience by inserting "allenivey" in the YouTube search field.

Brain Plasticity

Whether in interviewing, counseling, or psychotherapy, the conversation changes the brain through the development of new neural networks. This is an example of *brain plasticity and neurogenesis*. Throughout our lives, we are adding and losing many millions of neurons, synapses, and neural connections. Effective counseling and therapy develop useful new neurons and neural connections in the brain. Both your own and your client's brain functioning can be measured through a variety of brain-imaging techniques, most notably functional magnetic resonance imaging (fMRI) (Hölzel et al., 2011; Logothetis, 2008; Welvaert & Rosseel, 2014).

"Neuroplasticity can result in the wholesale remodeling of neural networks . . . a brain can rewire itself" (Schwartz & Begley, 2003, p. 16). If we are indeed affecting the brain in all our sessions, then perhaps neuroscience can help us understand a bit more of what is happening between counselors and clients.

But there is also negative plasticity—the loss of neural networks and neurons—associated with stress, whether hunger, poverty, bullying, or many types of trauma.

The Brain and Stress

Some 80% of medical issues involve the brain and stress (Ratey & Manning, 2014). You will find that, in one way or another, the vast majority of your work in counseling and psychotherapy includes stress as an underlying issue. Resolving stressors is critical in many styles of treatment. The evidence is clear that stress management and therapeutic lifestyle changes (see Chapter 2) are effective routes toward both mental and physical health and are necessary regardless of your counseling style or chosen theoretical approach.

Stress and stressful events leave a marked imprint on the brain. We need some stress for learning and for physical growth. Some people have compared the brain to a muscle: If it doesn't get exercise, it atrophies. But, like a muscle, it can be overstressed, which can result in damage and loss of neurons. Figure 1.3 shows the brain under severe stress (aversive condition) compared with the relative absence of stress (neutral condition).

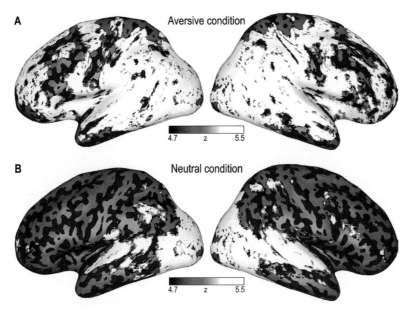

FIGURE 1.3 The brain under aversive stress.

Hermans, E., van Marle, H., Ossewaarde, L., Henckens, A., Qin, S., Kesteren, M., Schoots, V., Cousijn, H., Rijpkema, M., Oostenveld, R., & Fernández, G. (2012). Stress-related noradrenergic activity prompts large-scale neural network configuration. *Science, 334*, 1151–1153. Reprinted with permission from AAAS.

At another level, you will find stress involved in virtually all of the issues clients face. Admission to college, career choice, financial issues, and coping with racial or gender harassment are what many would term "normal" concerns. But these do not feel normal to our stressed and worried clients.

Listening to client stories is our first avenue to establishing an empathic relationship and understanding the client's world. Neuroscience's research on the brain has become a practical influence on our understanding of physical and mental health. Neuroscientists and even genetic researchers emphasize the importance of prevention and the key role of stress management. In this new paradigm, counseling skills and strategies remain first-line necessities. We need to listen empathically to clients' stories, join them in their view of the world, and work with them on an egalitarian basis toward growth, development, and action. But we now can do this with a solid scientific base, as neuroscience findings reveal that many of the traditional approaches are effective and appropriate.

As you read the following chapters, a few brain basics will help you locate and understand how the brain relates to the counseling process. Throughout this book, we will provide further information concerning areas of neuroscience/neurobiology that inform and clarify our practice. Many of you will find a new vocabulary here, but this language will become far more commonplace and important in daily use in the decades to come. Again, we suggest that you periodically explore Appendix IV on neuroscience/neurobiology as you encounter terms or concepts that you feel need further explanation. Do not worry about always learning the meaning of new words at this time. Gradually, over time, they will become part of your counseling vocabulary.

There are a number of other reasons why neuroscience and study of the brain will become more critical in the future years. Here are just a few:

1. The National Institute of Mental Health plans to institute a brain-based approach to counseling and therapy within the next 10 to 20 years, well within the time you will be practicing. This will replace the Diagnostic and Statistical Manual of Mental Disorders (DSM) with a totally new approach, one that is more amenable to the goals of the counseling process. It will be helpful if you take a few moments to visit your browser and search for "NIMH Research Domain Criteria."

2. This NIMH approach integrates neuroscience and neurobiology with medicine, counseling and therapy, developmental psychology, multicultural issues, and multiple sciences. Old clinical categories will be reevaluated and possibly eliminated. Diagnosis will be multidimensional and will lead to integrated treatment recommendations involving physicians, counselors and therapists, and the human services professions.

3. The media are full of new developments in neuroscience. Many of your clients will know this field through general reading and expect you to be knowledgeable as well. It is best that you develop a basic understanding, perhaps even including availability of pictures of the brain and body in your work setting.

4. You will find, like us, that knowledge of neuroscience and the brain enables you to be a more skilled practitioner. Currently, all three authors of this book constantly think of how the client's brain (and our own) is influenced by the interaction. Interview transcripts in this book will at times discuss how the client's brain is likely reacting to helping interventions. We find that, as a result of neuroscience/neurobiology, our work is more precise and successful. Neuroscience will improve your practice.

5. Neuroscience is stimulating and interesting, with constant new developments. Once one learns some basic vocabulary, it becomes fun and enjoyable.

Reflective Exercise How does neuroscience speak to you?

The idea of the brain being central in counseling practice is relatively new. In fact, the first conference presentation on the topic was when Allen spoke to the American Counseling Association in Hawai`i in 2006, and this counseling text was the first to give special attention to neuroscience. There has been an immense gain in awareness and acceptance since that time.

Some of you have taken biology or psychology courses (particularly social psychology) that speak with sophistication about the brain and its structures. For others, this will be a new topic with new vocabulary.

Here are some questions to consider:

- How do you react to the introduction presented above?
- Do you agree that neuroscience and the brain are relevant to your work as a counselor or psychotherapist?
- Will you seek to expand your knowledge of neuroscience and neurobiology?

Office, Community, Phone, and Internet: Where Do We Meet Clients?

> Regardless of physical setting, you as a person can light up the room, street corner, even the Internet. Smiling and a warm, friendly voice make up for many challenging situations. It is the *how* you are, rather than *where* you are.
>
> —Mary Bradford Ivey

First let's recognize that interviewing and counseling occur in many places other than a formal office. There are street counselors who work with youth organizations, homeless shelters, and the schools, as well as those who work for the courts, who go out into the community and get to know groups of clients. Counseling, interviewing, and therapy can be very informal, taking place in clients' homes, a neighborhood coffee shop or nearby park, or while they play basketball or just hang out on a street corner. The "office" may not exist, or it may be merely a cubicle in a public agency where the counselor can make phone calls, receive mail, and work at a computer, but not necessarily the place where he or she will meet and talk with clients. The office is really a metaphor for your physical bearing and dress—smiling, culturally appropriate eye contact, a relaxed and friendly nonverbal style.

As a school counselor, Mary Bradford Ivey learned early on that if she wanted to counsel recent immigrant Cambodian families, home visits were essential. She sat on the floor as the family did. She attended cultural events, ate and cooked Cambodian food, and attended weddings. She brought the Cambodian priest into the school to bless the opening ceremonies. She provided translators for the parents so they could communicate with the teachers. She worked with school and community officials to advocate for the special needs of these immigrants. The place of counseling and developing your reputation as a helper varies widely. Maintaining a pleasant office is important, but not enough.

Another approach, also used by Mary, is to consider the clientele likely to come to your setting. Working in a school setting, she sought to display objects and artwork representing various races and ethnicities. The brightness of the artwork worked well with children, and many parents commented favorably on seeing their culture represented. But most important, make sure that nothing in your office can be considered objectionable by any of those whom you serve.

Phone, Skype, and Internet Counseling

Historically, the emphasis has been on keeping the boundaries between counselor and client as clear and separate as possible, but this seems to be changing. Where once the therapist was opaque and psychologically unseen, you as a person have become more important. As the centrality of the fifth stage of the interview (action and follow-up) is recognized as increasingly critical to client change, many counselors are now using smartphones so that clients can follow up with them or ask questions. With smoking, alcohol, or drug cessation, being available can make a significant difference. However, this is also fraught with practical and ethical issues. You and the client both lose privacy, nonverbal communication will be missed, and confidentiality may be endangered. Skype and other visual phone services partially answer these questions, but they are still not the same as a face-to-face relationship.

Enter "online counseling" in your search engine, and a number of services will appear. Following is a composite result from Allen Ivey's visits to several online services:

Meeting life's challenges is difficult.

[Internet counseling center] enables you to talk with real-life professional counselors 24/7 in full confidentiality.

Choose a counselor.

First session is free.

Easy payments arranged.

We expect that more and more of these services will appear on the Internet, particularly now that a degree of face-to-face interaction is available on the phone and online. Some of these services may be quite helpful to clients at a reasonable cost, but others may be risky. Now search for "coaching services," and you will find an amazing array of possibilities. However, view all these sites with some attention to how ethical and professional standards can be met on the Internet.

Distance Credentialed Counselor (DCC) is a national credential currently offered by the Center for Credentialing and Education (CCE). Holders of this credential adhere to the National Board for Certified Counselors' Code of Ethics and the Ethical Requirements for the Practice of Internet Counseling. These professionals adapt their counseling services for delivery to clients via technology-assisted methods, including telecounseling (telephone), secure email communication, chat, videoconferencing, and other appropriate software (http://www.cce-global.org/DCC). Professional associations are still working on these new approaches to counseling delivery. As with all types of counseling and therapy, ethics is the first concern. In Chapter 2 (and expanded in Appendix II), we present some beginning issues in ethics. You will also find Internet links to the professional ethics of the main national helping organizations.

Your Natural Helping Style: Establishing Your Baseline

At the beginning of this chapter, you were asked to give your own response to Sienna's multiple issues. Your response reflects you and your worldview. Your use of microskills and the five-stage structure must feel authentic. If you adopt a response simply because it has been recommended, it will likely be ineffective for both you and your client. Not all parts of the microtraining framework are appropriate for everyone. You have a natural style of communicating, and these concepts should supplement your style and who you are. Learn these new skills, strategies, and concepts, but be yourself and make your own authentic decisions for practice.

Also develop awareness of the natural style of the clients with whom you work, particularly if they are culturally different from you. What is this experience like for you? What do you notice about your client's style of communication? Use your observations to expand your competence and later add new methods and information to your natural style.

An Important Audio or Video Exercise

We believe the following is one of the most central exercises in the book:

Find someone who is willing to role-play a client with a decision that needs to be made, a concern, an issue, or an opportunity. Interview or practice counseling that "client" for at least 15 minutes, using your own natural style.

Read pages 28–30 of Chapter 2 and follow the ethical guidelines as you work with a volunteer client. Ask the client, "May I record this session?" Also inform the client that the video or audio recorder may be turned off at any time. Common sense demands ethical practice and respect for the client.

Video, with feedback from colleagues and/or clients, is the preferred way to examine your counseling style. Many of you will have either a small video recorder or a video-capable camera or phone. Video makes it possible for us to discover what we are really doing, not just what we think we are doing. Feedback from others helps us evaluate both our strengths and the areas where we might benefit from further development and growth.

The volunteer client can select almost any topic for the session. A friend or classmate discussing a school or job problem may be appropriate. A useful topic might be some type of interpersonal conflict, such as concern over family tensions, or a decision about a new job opportunity.

When you have finished, ask your client to fill out the Client Feedback Form (Box 1.3). In practice sessions, always seek immediate feedback from clients, classmates, and colleagues. We suggest that you use the Client Feedback Form for this purpose, with your own adaptations and changes, throughout your practice sessions.

You may also find it helpful to continue using this form, or some adaptation of it, in your work in the helping profession. Professional counselors and therapists seldom offer their clients an opportunity to provide them with feedback. In the interest of a more egalitarian session, consider this type of feedback from your clients as a regular part of your practice. We ourselves have learned valuable and surprising things through feedback, particularly when we may have missed something.

Please transcribe the audio or video for later study and analysis. You'll want to compare your first performance with practice sessions and, ideally, with another, more detailed analytic transcript and self-evaluation at the end of this course of study.

You can photocopy the Client Feedback Form here. Occasionally, adding specific items for individual clients may enable them to write things that they find difficult to put into words.

Self-Assessment

Review your audio or video recording and ask yourself and the volunteer client the following questions. Include your thoughts about these questions in your Portfolio of Competencies.

1. We build on strengths. What did you do right in this session? What did the client notice as helpful?

BOX 1.3	Client Feedback Form

To Be Completed by the Volunteer Client

DATE _____

(NAME OF INTERVIEWER) _____   (NAME OF PERSON COMPLETING FORM) _____

Instructions: Rate each statement on a 7-point scale, with 1 representing "strongly disagree (SD)," 7 representing "strongly agree (SA)," and N as the midpoint "neutral." You and your instructor may wish to change and adapt this form to meet the needs of varying clients, agencies, and situations.

	Strongly Disagree SD			Neutral N		Strongly Agree SA	
1. (Awareness) The session helped you understand the issue, opportunity, or problem more fully.	1	2	3	4	5	6	7
2. (Awareness) The interviewer listened to you. You felt heard.	1	2	3	4	5	6	7
3. (Knowledge) You gained a better understanding of yourself today.	1	2	3	4	5	6	7
4. (Knowledge) You learned about different ways to address your issue, opportunity, or problem.	1	2	3	4	5	6	7
5. (Skills) This interview helped you identify specific strengths and resources you have to help you work through your concerns and issues.	1	2	3	4	5	6	7
6. (Skills) The interview allowed you to identify specific areas in need of further development to cope more effectively with your concerns and issues.	1	2	3	4	5	6	7
7. (Action) You will take action and do something in terms of changing your thinking, feeling, or behavior after this session.	1	2	3	4	5	6	7
8. (Action) You will create a plan of action to facilitate change after this session.	1	2	3	4	5	6	7

What did you find helpful? What did the interviewer do that was right? Be specific—for example, not "You did great," but rather, "You listened to me carefully when I talked about _____."

What, if anything, did the interviewer miss that you would have liked to explore today or in another session? What might you have liked to have happen that didn't?

Use this space or the other side for additional comments or suggestions.

2. What stands out for you from the Client Feedback Form?

3. What was the essence of the client's story? How did you help the client bring out his or her narrative/issues/concerns?

4. How did you demonstrate intentionality? When something you said did not go as anticipated, what did you do next?

5. How did you experience the session? How authentic and genuine did you feel?

6. Name just one thing on which you would like to improve in the next session you have. What actions will you take?

Key Points: The Art of Applying and Taking Action As You Work Through This Book

Welcome to the fascinating field of counseling and psychotherapy! You are being introduced to the basics of the individual counseling session, but the same skills are essential in group and family work. These therapeutic skills are essential whether you find yourself in a school or university, a community mental health clinic or hospital, or private practice. The microskills framework has been taught throughout the world of business, law, and medicine and used by UNESCO and others with disaster survivors, refugees, and AIDS workers. These skills are also basic to interpersonal communication—they can make a difference not only in client lives, but also in your own relationships with others.

This first chapter frames the entire book. The following key points are what we particularly want you to remember. The first competency practice exercise in this chapter asked you to examine yourself and identify your strengths as a helper. In the end, *you* are the person who counts, and we hope that you will develop your counseling skills based on your natural expertise and social skills. We hope that you enjoy the journey.

Following is a summary of awareness, knowledge, skills, and actions that you may want to take home:

Interviewing, Counseling, and Psychotherapy. These are interrelated processes that sometimes overlap. Interviewing may be considered the most basic; it is often associated with information gathering and providing necessary data to help clients resolve issues. Counseling focuses on normal developmental concerns or adjustment issues, whereas psychotherapy emphasizes treatment of more deep-seated issues. But the overlap is considerable, and we will see therapists engaged in counseling and counselors active in psychotherapy. The concept of coaching has not yet received full attention within the counseling field, but its positive orientation and its emphasis on co-discovery of client values and goals is highly consistent with the traditions of counseling.

The field of counseling and therapy is now well supported by empirical research. Therapy works! More recently, neuroscience findings have added to our understanding, solidifying counseling as a scientific undertaking. Nonetheless, it is you, the counselor or therapist, who effectively integrate the many aspects of research and theory, creatively apply these findings to the client, and seek their feedback to improve effectiveness.

Counseling Is Both Science and Art. The evidence base for our effectiveness in helping others is strong. Nonetheless, it is counselors and psychotherapists who take the science, the evidence base, and the microskills into concrete practice. Placing various oils on a painting is an art form based on specifics skills. However, it is the artist who arranges the elements into a creative whole. The challenge for you, as an artist, is to take the scientific side of counseling and psychology and facilitate the growth of your clients.

Cultural Intentionality. The culturally intentional counselor or therapist acts with a sense of capability and flexibly in deciding from among a range of alternative actions. *If a helping lead or skill doesn't work—try another approach!*

Resilience and Self-Actualization. A major objective of counseling and psychotherapy is enabling clients to find their own direction and enhance their potential. Self-actualization requires resilience and the ability to rebound from the inevitable stresses and challenges we all face.

Microskills and the Microskills Hierarchy. Microskills are single communication skill units (for example, questioning or reflection of feelings). They are taught one at a time to ensure mastery of basic counseling and therapy competencies.

The microskills hierarchy organizes microskills into a systematic framework for the eventual integration of skills in a natural fashion. The microskills rest on a foundation of multicultural competence, ethics, positive psychology/resilience, and neuroscience. The attending and listening skills are followed by focusing, confrontation, influencing skills, integration of skills, and your own analysis of your personal style of interviewing, counseling, and psychotherapy.

All counseling theories use the microskills with varying patterns and goals. Mastery of the skills facilitates a capacity to work with many theoretical alternatives.

Neuroscience/Neurobiology and Stress. Newer research in these areas indicates that virtually all of counseling and psychotherapy is valid and on the right track. Particularly important is neuroplasticity, or "rewiring" of the brain. Successful therapy may be expected to help clients develop new neural connections. Neuroscience and neurobiology will lead us to a deeper understanding of the importance of helping clients deal with stress.

Stress, which can be helpful for learning and action, is a real concern when one is overstressed. Dangerous cortisol can build up and injure neurons and neural networks. In addition to "normal" stressors such as divorce, separation, failing an exam, or even choosing a college, stress underlies most diagnostic categories in our field. Thus stress management becomes a central issue in virtually all counseling and therapy.

Places Where We Meet Clients. Interviewing and counseling occur in many places other than a formal office. Many services are also offered via the Internet.

Your Natural Helping Style. Microskills are useful only if they harmonize with your own natural style. Audio or video record a session with a friend or classmate; make and save a transcript. Later, as you learn more about session analysis in your counseling practice, continually examine and study your behavior. You'll want to compare this first interview with your performance in another recording some months from now.

Portfolio of Competencies and Personal Reflection

Additional resources can be found by going to CengageBrain.com and logging into the MindTap course created by your professor. There you will find a variety of study tools and useful resources that include quizzes, videos, interactive counseling and psychotherapy exercises, case studies, the Portfolio of Competencies, and more.

Developing a Portfolio of Competencies: Your Initial Video or Audio Recording

We recommend that you develop a Portfolio of Competencies as a journal of your path through this course and your reflections on your place in this field. This portfolio is a way of putting together what you have learned and your counseling practice. Students have used this portfolio as they apply for practicums or internships.

Your first recorded practice session will provide a critical foundation on which to build. We recommend developing a transcript of that first session. You can later compare your first interview with other practice sessions as you progress. Your self-assessment and personal reflection will help you assess progress chapter by chapter. This transcript can serve as a baseline as you learn and evaluate your skills and actions in counseling and therapy.

Assessing Your Level of Competence: Awareness, Knowledge, Skills, and Action

We speak in terms of four levels of competence in counseling skills. The first three are awareness and knowledge, basic competence, and intentional competence. *Awareness* refers to self-awareness and your ability to be genuine while *knowledge* refers to your understanding of counseling concepts. *Basic competence* asks you to practice the skills or strategies, showing that you know what they are and how to use them in the session. *Intentional competence* speaks to action and occurs when you can use a skill and anticipate how the client will respond. You can also demonstrate the flexibility to change your skill usage and/or personal style in accordance with the client's immediate and long-term needs.

The fourth level that you may want to achieve is *psychoeducational teaching competence*. Many of the skills and strategies of this book can be taught to clients as part of counseling and therapy. In addition, you may be asked to conduct more formal presentations in which you teach other counselors, volunteer peer counselors, or others who may benefit from listening skills training, such as businesspeople, clergy, or community service workers.

Please take a moment now to start the process of competency assessment for this chapter using the following checklist. As you review the items below, ask yourself, "Can I do this?" Check those dimensions that you currently feel able to do. Those that remain unchecked can serve as future goals. Do not expect to attain intentional competence on every dimension as you work through this book. You will find, however, that you will improve your competencies with repetition and practice.

Awareness and Knowledge. Can you define and discuss the following concepts?

❑ Distinctions and similarities among interviewing, counseling, and psychotherapy

❑ Balance of science and art that makes sense to you as appropriate for counseling

❑ Meaning and importance of cultural intentionality in counseling practice

❑ Resilience and self-actualization as potential goals for clients

❑ The microskills hierarchy and its relevance to practice

❑ The potential value of neuroscience for the practice of counseling and psychotherapy

❑ The places where counseling and psychotherapy are practiced

Basic Competence. We have asked you to take ideas from the chapter and actually try them out in your own life and/or the real world.

❑ Finding a volunteer client, conducting a session, obtaining client feedback, and evaluating your own natural style of helping

Intentional competence and **psychoeducational teaching competence** will be reviewed in later chapters.

Personal Reflection on This Introductory Chapter

This chapter has presented the foundations of this book. Take time to write down your reflections about interviewing and counseling. Consider adding your responses to the following questions to your Portfolio of Competencies. These ideas are the building blocks upon which you can begin the process of developing your own style and theory.

What stood out from this chapter? What are your major questions or concerns? What is your view of counseling and psychotherapy at this point?

How have personal relationships strengthened you? What positive assets and resources do you bring to counseling and therapy?

What did you discover in your first videotaped interview? What did you notice about your natural helping style? What did you do right? Where might you seek to improve?

Where do you see yourself in the helping field? What do you envision doing, whether in counseling and therapy or in some other field of endeavor?

How might you use ideas in this chapter to begin the process of establishing your own style and theory?

Keep a journal of your path through this course and your reflections on its meaning to you.

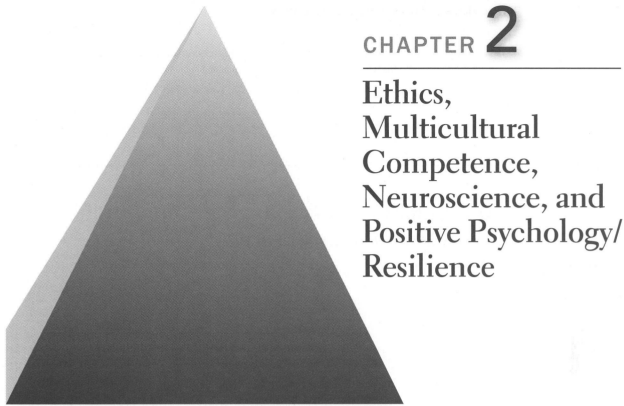

CHAPTER 2

Ethics, Multicultural Competence, Neuroscience, and Positive Psychology/ Resilience

Ethics, Multicultural Competence, Neuroscience, and Positive Psychology/Resilience

I am (and you also)
Derived from family
Embedded in a community
Not isolated from prevailing values
Though having unique experiences
In certain roles and statuses
Taught, socialized, gendered, and sanctioned
Yet with freedom to change myself and society.

Chapter Goals and Competency Objectives

Awareness and Knowledge

▲ Develop an understanding of the basics of ethical counseling and therapy practice.

▲ Examine your identity as a multicultural being, how dimensions of diversity and privilege may affect the session, and the central importance of multicultural competence.

▲ Identify multicultural strengths in clients as a path toward wellness and resilience.

▲ Identify wellness as an ethical practice.

Skills and Action

▲ Use the RESPECTFUL model for encountering multicultural difference.

▲ Understand the meaning and depth of microaggressions and have beginning skills in working with clients who experience this treatment.

▲ Define and apply positive psychology and wellness as a basis for fostering and building client resilience.

▲ Use therapeutic lifestyle changes (TLC) as positive wellness strategies in the session for physical and mental health.

Introduction: Ethics and the Counseling and Psychotherapy Process

Ethics is nothing else than reverence for life.

—Albert Schweitzer

Action indeed is the sole medium of expression for ethics.

—Jane Addams

Albert Schweitzer was awarded the Nobel Peace Prize in 1952. A renowned philosopher and musician, he earned his medical degree and started practice in a hut in Africa. Jane Addams founded Hull House in Chicago in 1889, which resulted in the formation of social work as a profession. She is again gaining attention for her pioneering work with the poor and community interventions. Both represent the ideals of ethical practice, taking philosophy into concrete action.

Ethics are thoughtful professional lists of do's and don'ts for our profession. Morals are the way we apply ethics through our commitment to excellence, reverence for others, and willingness to take action to engage life for ourselves and others.

Ethics and Morals	Anticipated Client Response
Ethics are rules, typically prescribed by social systems and, in counseling, as professional standards. They define how things are to be done. Morals are individual principles we live by that define our beliefs about right and wrong. A moral approach to interviewing and counseling allows us to apply ethical principles respectfully to our clients and ourselves.	Following professional ethics results in client trust and provides us with guidelines for action in complex situations. Morals represent our individual efforts and actions to follow ethical principles. A moral approach to interviewing and counseling helps us to remember that our personal actions count, both inside and outside the session. Furthermore, a moral approach to the session may ask you to help clients examine their own moral and ethical decisions.

Ethical codes can be summarized as follows: "Do no harm to your clients; treat them responsibly with full awareness of the social context of helping." As interviewers and counselors, we are morally responsible for our clients and for society as well. At times these responsibilities conflict, and you may need to seek guidance from documented ethical codes, your supervisor, or other professionals.

Effective practice is not only scientific, it is also ethical. Professional helping organizations such as the American Association of Marriage and Family Therapy (AAMFT), the American Counseling Association (ACA), the American Psychological Association (APA), and the National Association of Social Workers stress the importance of ethics in the helping relationship. Appendix II lists the websites for these professions' ethical standards.

The following section begins with some basic issues of professional ethics, followed by practical implications for your work in practice sessions with volunteer clients.

Ethics and Responsibility: A Summary

The core of ethical responsibility is to *do good and to do nothing to harm the client or society*. The bulk of ethical responsibility lies with you. A person who comes for help is vulnerable and open to destructive action by the counselor. The following are basic guidelines for you to consider as you review ethical standards in more detail. It is important and highly recommended that you review the more detailed summary in Appendix I and study the ethics links applying to the most relevant area of your future practice

1. *Maintain Confidentiality.* Counseling and psychotherapy rest on trust between counselor and client. You as therapist are indeed in a powerful relationship and the more trust you build, the more power you have. This book asks you to practice many basic strategies of counseling and therapy. It is essential that you maintain the confidence of your volunteer client. But, at the same time, you do not as a student have legal confidentiality. Those with whom you work should be aware of your student status.

   Confidentiality is designed to protect clients (not counselors), and only the courts, in the final analysis, can provide a guarantee of confidentiality.

2. *Recognize your limitations.* Maintain an egalitarian atmosphere with your volunteer "clients," classmates, or co-workers. Share with them before you begin the constraints of the situation, the task you wish to work through with them. Inform them that they are free to stop the process at any time. Do not use the interview as a place to delve into the life of another human being. The interview is for helping others, not examining them.

3. *Seek consultation.* As you practice the exercises presented throughout this text, remain in consultation with your professor, workshop leader, or mentor. Counseling and psychotherapy are often very private—it is important that you constantly obtain supervision and consultation in your work. You may also find it helpful to discuss your own growth as a helper with other students. At the same time, be very careful in discussing what you have learned about your clients.

4. *Be aware of individual and cultural differences.* This point will be stressed throughout this book. An emphasis on cultural issues can lead at times to stereotyping an individual. At the same time, an overemphasis on individuality may miss background multicultural issues.

5. Remember both the Golden Rule and the Platinum Rule in counseling and therapy:

   Treat the client as you would like to be treated.
   Treat clients the way they want to be treated.

   Put yourself in the place of the client. Every person deserves to be treated with respect, dignity, kindness, and honesty.

6. Give special attention to ethical treatment of children and their rights.

Ethics and Your Practice Sessions with Microskills

Practice interviews with volunteer clients are what will make this book come alive. Here you have the opportunity to test out various skills and strategies and to develop increasing competence. It is essential to record your sessions as often as possible, ideally for each skill set and strategy. Video recording with feedback is the most powerful method. Equipment is usually available on college campuses, but you may also have your own camera available in your computer, cell phone, or other devices. Many of you will have cameras capable of good resolution video.

As you work with volunteers, obtaining their informed consent is essential. The formal guidelines of the American Counseling Association (2014) offer useful guidance.

A.2. Informed Consent in the Counseling Relationship
A.2.a. Informed Consent
Clients have the freedom to choose whether to enter into or remain in a counseling relationship and need adequate information about the counseling process and the counselor. Counselors have an obligation to review in writing and verbally with clients the rights and responsibilities of both counselors and clients. Informed consent is an ongoing part of the counseling process, and counselors appropriately document discussions of informed consent throughout the counseling relationship.

When you work with children, the ethical issues around informed consent become especially important. Depending on state laws and practices, it is often necessary to obtain written parental permission before interviewing a child or before sharing information about the interview with others. The child and family should know exactly how any information is to be shared, and interviewing records should be available to them for their comments and evaluation. An essential part of informed consent is stating that both child and parents have the right to withdraw their permission at any point. Needless to say, these same principles apply to all clients—the main difference is parental awareness and consent.

When you enter into role-plays and practice sessions, inform your volunteer "clients" about their rights, your own background, and what they can expect from the session. For example, you might say:

I'm taking a practicum course, and I appreciate your being willing to help me. I am a beginner, so only talk about things that you want to talk about. I would like to [audio or video] record the interview, but I'll stop immediately if you become uncomfortable and delete it as soon as possible. I may share the recording in a practicum class or I may produce a written transcript of this session, removing anything that could identify you personally. I'll share any written material with you before passing it in to the instructor. Do you have any questions?

You can use this statement as a starting point, after adapting it to the specifics of your situation, and eventually develop your own approach to this critical issue. The sample practice contract in Box 2.1 may be helpful as you begin. You will want to adapt this to meet your own institutions requirements.

Awareness, Knowledge, and Skills of Ethics, Multicultural Competence, Positive Psychology, and Therapeutic Lifestyle Changes
Multicultural Competence

Every session has a cultural context that underlies the way clients and counselors think, feel, and behave.

—Carlos Zalaquett

Multicultural competence is imperative in the interview process. Awareness of our clients' multicultural background enables us to understand their uniqueness more fully. We live in a multicultural world where every client you encounter will be different from the last and different

BOX 2.1	Sample Practice Contract

The following is a sample contract for you to adapt for practice sessions with volunteer clients. When you counsel a minor, the form must be signed by a parent as appropriate under Health Insurance Portability and Accountability Act (HIPAA) standards (see Appendix II).

Dear Friend,

I am a student in interviewing skills at [insert name of class and college/university]. I am required to practice counseling skills with volunteers. I appreciate your willingness to work with me on my class assignments.

You may choose to talk about topics of real concern to you, or you may prefer to role-play an issue that does not necessarily relate to you. Please let me know before we start whether you are talking about yourself or role-playing.

Here are some dimensions of our work together:

Confidentiality. As a student, I cannot offer any form of legal confidentiality. However, anything you say to me in the practice session will remain confidential, except for certain exceptions that state law requires me to report. Even as a student, I must report (1) a serious issue of harm to you; (2) indications of child abuse or neglect; (3) other special conditions as required by our state [insert as appropriate].

Audio and/or video recording. I will be recording our sessions for my personal listening and learning. If you become uncomfortable at any time, we can turn off the recorder. The recording(s) may be shared with my supervisor [insert name and phone number of professor or supervisor] and/or students in my class. You'll find that recording does not affect our practice session so long as you and I are comfortable. Without additional permission, recordings and any written transcripts are destroyed at the end of the course.

Boundaries of competence. I am an inexperienced interviewer; I cannot do formal counseling. This practice session helps me learn interview skills. I need feedback from you about my performance and what you find helpful. I may give you a form that asks you to evaluate how helpful I was.

_____
VOLUNTEER CLIENT

_____
INTERVIEWER

_____
DATE

from you in some major way. Without a basic understanding of and sensitivity to a client's uniqueness, the interviewer will fail to establish a relationship and true grasp of a client's issues.

You may anticipate the client's response to your exhibiting multicultural competence.

Multicultural Competence	Anticipated Client Response
Your competence in multiculturalism is based on your level of *awareness*, *knowledge*, *skills*, and *action*. Self and other awareness and knowledge are critical, but one must also have the skills and the ability to act.	Anticipate that both you and your clients will appreciate, gain respect, and learn from increasing knowledge in intersecting identities, the nature of privilege, and multicultural competence. You, the interviewer, will have a solid foundation for a lifetime of personal and professional growth. You will be challenged to consider implications of social justice for your practice.

Critical to *awareness* is that interviewing and counseling have now become global phenomena. The early history of interviewing, counseling, and therapy is populated primarily by famous White male European and American figures, such as Sigmund Freud, Carl Jung, Carl Rogers, Viktor Frankl, Albert Ellis, and Aaron Beck. While their contributions are legion, they all give at best only minor attention to cultural difference or to women. The rise of the multicultural movement in the United States can be traced to the Civil Rights Act, followed by the growth of awareness in activists from groups such as African Americans, those who are of mixed race, women, the disabled, war veterans, and individuals who may identify as lesbian, gay, bisexual, transgendered, queer, intersex, or asexual (LGBTQIA). All these in different ways have identified and named oppression as a root cause of human distress. Counseling was slow to respond to these movements, but gradually our field has become a central force in what is termed "psychological liberation."

The Association for Multicultural Counseling and Development (AMCD) Multicultural and Social Justice Counseling Competencies (Ratts, Singh, Nassar-McMillan, Butler, & McCullough, 2015) offer helping professionals a framework for incorporating multicultural and social justice competencies into counseling theories, practices, and research.

The following section focuses on identifying some key dimensions of multiculturalism and asks you to examine your own awareness and understanding. This is followed by some of the major challenges we face in working with multicultural awareness, knowledge, skills, and action.

RESPECTFUL Interviewing and Counseling

> All interviewing and counseling are multicultural. The client brings many voices from the past and present to any counseling situation.
>
> —Paul Pedersen

The **RESPECTFUL model** (D'Andrea & Daniels, 2001, 2015) enables us to discover the multiple voices that clients bring to us. In addition, it provides a way for you to identify the past and present voices that affect your own thoughts, feelings, and behaviors. This framework is a basic *awareness* and *knowledge* opportunity. Please review the list in Box 2.2 and identify your multicultural self. It is possible that you have not thought of yourself as a multicultural being. As you consider the issues of multiculturality, we ask that you also examine your beliefs and attitudes toward those who are similar to and multiculturally different from you.

As you review your multicultural identity, what stands out for you among these voices? What might be surprising? Then look for strengths in each dimension that support you as an individual. What is most meaningful or salient in the way you think about yourself? There are certain givens in life, such as being a man or woman of a certain race or ethnicity, that affect how we see ourselves and others view us. But other dimensions can be as important or more important in our identity. For some of you, it may be spiritual or religious values or where you lived when you were growing up; for others, it may be your education or being raised in a lower-income situation. If you happen to be an older person or one who has been affected by physical or mental disability, that could be the most salient factor when you think of yourself.

Do not view the RESPECTFUL model as just a list of difficult issues and concerns. Rather, look at this list as a source of information about your own and your clients' resilience. Clients can draw amazing strength from their religious or spiritual background, the positive pride associated with racial/ethnic identity, their family background, or the community in which they grew up.

Nonetheless, there is also the possibility of **cultural and historical trauma** in each of the RESPECTFUL dimensions. Large historical events and daily microaggressions in the form of insults to one's color, ethnicity, gender, sexual orientation, religious beliefs, or disability result in personal and group trauma, frustration, anger, hopelessness, and depression. Bullying is frequently part of these insults.

Multiculturalism needs to consider present and past histories of oppression, but we need also to look for positive strengths and what enables each of us to build resilience. These strengths can be used by to work more comfortably with difficult and challenging situations.

Intersections among multicultural factors are also critical. For example, consider the biracial family (e.g., a child who is both of Chinese and White parents or African descent and Latina/o). Both children and parents are deeply affected, and categorizing an individual

BOX 2.2	The RESPECTFUL Model

Identify yourself on each of the dimensions of the RESPECTFUL model, and identify strengths and positives than can be associated with each. Then ask yourself how you might work with those who are different from you.

The 10 Dimensions	Identify yourself as a multicultural being.	What personal and group strengths can you develop for each multicultural dimension?	How effective will you be with individuals who differ from you on each dimension?
R Religion/spirituality			
E Economic/social class background			
S Sexual identity			
P Personal style and education			
E Ethnic/racial identity			
C Chronological/lifespan status and challenges			
T Trauma/crisis (may be single trauma or repeated racism, sexism, bullying, etc.)			
F Family background and history (single- or two-parent, extended family, etc.)			
U Unique physical characteristics (including disabilities, false standards of appearance, skills and abilities)			
L Location of residence, language differences			

into just one multicultural category is inappropriate. Or think of the Catholic lesbian woman who may be economically advantaged (or disadvantaged). Or the South Asian gay male with a Ph.D. For many clients, sorting out the impact of their sometimes conflicting multiculturality may be a major issue in counseling.

You can add the RESPECTFUL model to your *skills* and *action* by helping clients extend their understanding of themselves as cultural beings and building cultural health. At times, providing them with a handout on the RESPECTFUL framework and discussing with them the meaning they take out of looking at themselves as a person of many cultures.

The Soul Wound and Historical Trauma

> The medicine is already within the pain and suffering. You just have to look deeply and quietly. Then you realize that it has been there the whole time.
> —Saying from the Native American oral tradition

> The Native idea of historical trauma involves the understanding that the trauma occurred in the soul or spirit.
>
> —Edward Duran

The **intergeneration transmission of trauma** was first identified by Israeli researchers examining the lives of second- and third-generation survivors of the Nazi Holocaust. Children and grandchildren of survivors exhibited increased depression, psychiatric issues, and even suicide (Shoshan, 1989; Soloman, Kotter, & Mikulincer, 1988). The trauma of severe abusive treatment can persist over generations. Furthermore, there is now clear neurobiological evidence that epigenetics changes in the genome can be transferred from one generation to the next—and onward from that point to future generations (Kellermann, 2013).

Other groups react similarly to historical loss—African Americans (slavery and continued societal devaluation), Japanese Americans (forced relocation to U.S. labot camps during World War II), and Latina/o Americans (Spanish colonization and a history of mistreatment in the United States). Research has shown that this type of historical loss also affects the mental health of Native American youth, and very likely others whose cultures have suffered trauma. With greater awareness of the trauma, Native American adolescents are more anxious and become aware of cultural loss, loss of people, and cultural mistreatment (Amenta, Whitbeck, & Habecker, 2015).

The **Soul Wound** that occurs with and from historical trauma has been most clearly defined by Duran (2006) as he outlines the experiences of Native Americans over the generations. Between 1870 and 1900, at least 80% were killed and their lands lost as they were moved to reservations. As with Holocaust survivors, the traumatic wound does not disappear, but remains "in the soul." Box 2.3 describes how the soul wound develops.

| BOX 2.3 | A Story of How the Soul Wound Develops |

Africa's South Sudan has been the site of some of the world's most horrific wars, torture, and rape. An estimated 2 million people have been killed and another 4 million displaced.

Talia Aligo (not her real name) is an honors high school senior in Atlanta and soon will be attending college on a large scholarship. Her family was lucky enough to escape to the United States and save all their lives. She is a Muslim and, with a dark complexion, has always encountered racism and prejudice from White students, but she has remained incredibly resilient and has triumphed. Her college admission essay detailed the hurts experienced since her arrival in the United States.

After 9/11, Life became more challenging for Talia. The comments and slurs became louder and more frightening, almost wherever she went. She did not wear her hijab head covering outside anymore, as she knew that this would only result in more difficulties for her. But she did have a photo of herself taken wearing the hijab, which she shared with a friend. The "friend" passed the photo on, and the school situation became worse. One of her "friends" said to her directly, "All Muslims are terrorists."

Talia has learned to deal with fear. She is a living example of the Soul Wound—a displaced person, not accepted but reviled in her adopted land. The treatment that nondominant people experience in this culture obviously builds anger and frustration, but we miss the very real fear that they encounter daily. What will someone say to me? Will I receive decent treatment and service in a restaurant or store? Am I in danger of being attacked? Am I safe? People of Color, those with disability, individuals and families in or near poverty, lesbian, gay, bisexual, transgendered, queer, intersex, and asexual (LGBTQIA) people and others face this same fear. We could (and should) also think of the young child or teen who is bullied—that experience is not so very different.

When you counsel or conduct therapy with vulnerable individuals, there is always the possibility of their having experienced dealing with prejudice and being bullied—anger/frustration and fear are very likely to be part of their lives. Research indicates that over time, these overt and covert insults, microaggressions, and direct insults result in a trauma response to unfamiliar and even familiar social interactions. This narrows their world and can cause persistent anxiety and depression. It may be necessary to conduct an assessment for trauma, develop an intervention that addresses not only presenting issues but also other concerns such as empowering their cultural identity, and use communication skills that are sensitive to trauma issues.

Listening to their stories, of course, is the first thing to do, but we also need to empower these clients by building resilience and focusing on their existing personal and family strengths. A critical part of this will be facilitating understanding of cultural pride, cultural health, and cultural identity.

The trauma continues. Native Americans still experience continued racism and even denial of their existence by many governmental institutions. They have lower incomes, a higher rate of alcoholism, abuse in the family, high suicide rates, and hopelessness.

African Americans have their own version of the Soul Wound, a result of continued individual and institutional racism since the time of slavery and "Jim Crow" segregation. Latinas/os have been disparaged by politicians and face further threat, which ultimately can lead to hopelessness and fear. Asian Americans, too, still struggle with harassment and microaggressions.

Psychological liberation from historical trauma occurs when clients discover that what they saw as a personal issue is not just "their problem." With the counselor's help, clients begin to see that external and historical racism, sexism, heterosexism, or other form of oppression are the underlying causes of many of their concerns (Ivey & Zalaquett, 2009). For example, veterans returning from Vietnam were often hospitalized for what we now call posttraumatic stress disorder (PTSD). But at the time there was no such term, so they were given typical diagnoses such as depression, mania, or schizophrenia. The Veterans Administration asked therapists to search for malingers who were faking their symptoms just to obtain benefits.

Traumatized war veterans also carry the burden of the Soul Wound. Figley (1995) was among the first to point out that families and children of war trauma survivors had many similar issues to their injured father's. It was veterans themselves, gathering in discussion groups without the "benefit" of professionals, who discovered the root underlying cause of their issues—the trauma of the Vietnam War. Through this group work, many relieved themselves of psychological issues and moved on to health. As psychiatrists observed this phenomenon, they came up with the label for PTSD. We believe that this term is inaccurate, because external stressors are the real cause of the internalized issues. Consider deleting the word *disorder* and using just *posttraumatic stress (PTS)*. By labeling stress symptomology as a "disorder," psychiatry pathologized what, in truth, is a logical result of living in an insane environment.

Similarly, the helping professions have all too often failed to see that the issues that clients bring to us are deeply involved with societal dysfunction, harassment, and oppression. The multicultural movement shows that we need to examine external and historical causes of personal concerns, whether clients are People of Color, those affected by a disability, those with gender or sexual orientation concerns, or children and adolescents who are bullied.

The stress of internalized racism was studied in 656 African Americans, Asian Americans, Latinas/os, and Native Americans (Campó & Carter, 2015). Personal racial harassment, microaggressions, and discrimination from institutions such as businesses and banks can lead to anxiety and depression, as well as loss of self-worth—the Soul Wound. The authors point out that White society has the resources and power to make and enforce decisions, define "normal" behavior, and define reality.

Surrounded by White majority society, People of Color can easily appropriate beliefs about themselves that are negative to their self-image at a conscious or unconscious level. White cultural standards of beauty, damaging stereotypes, denial of the history of racism and oppression, and devaluing of one's own cultural group tend to result in emotional reactions of shame, anger, and fear (also see Forsyth & Carter, 2012).

Reflective Exercise Soul wound

What is the meaning of this for you, the counselor or therapist?

Awareness of this issue is critical, and when it comes up in the session, you need to be ready to understand and encourage storytelling, which may include elements of harassment, exclusion, and bullying. All these lead to anxiety and fear.

Despite these challenging issues, it is also important to focus on positives, strengths, and resources for resilience in clients, their families, and their culture. Two real interviews illustrating how to develop resilience resources in the face of cultural oppression are presented in Chapters 9 and 10.

Privilege as a Multicultural Interviewing Issue

Power from unearned privilege can look like strength when it is in fact permission to escape or to dominate.

—Peggy McIntosh

Privilege is power given to people through cultural assumptions and stereotypes. Just being in some RESPECTFUL categories offers an individual immediate privilege. McIntosh (1988) comments on the "invisibility of Whiteness." European Americans tend to be unaware of the advantages they have because of the color of their skin. The idea of special privilege has been extended to include men, those of middle- or upper-class economic status, and others in our society who have power and privilege. For those with White privilege comes increasing awareness that White people are a minority in the global population. Within the United States, there are presently more infants and preschoolers of so-called "minorities" than Whites, so that during your work life in helping, Whites will become the new minority.

Income inequality and rigid class structures are now recognized as a central multicultural issue, perhaps even more important to many than other RESPECTFUL issues. In most nations, a small group holds the bulk of the wealth and believes that others "lower" than they are at fault for their own condition—they just need to study and work harder. Sociologists speak of education and society producing social reproduction—the fact that over generations, class and income levels change very little. As a counselor, you will see students accumulate immense amounts of debt—and many of these same students fail to complete their education. Connections and personal influence are the way that most people find internships and later good jobs.

In short, societal structures are such that moving from one social class to another is quite challenging. This is not true just in the United States. Researchers have found that the children of politicians, physicians, lawyers, and top businesspeople fill the universities of not only the United States but also France, Germany, Great Britain, and even socialist Sweden. Moreover, even under full communism in Russia, the elite followed the same pattern as in capitalist countries. These classic findings (Bourdieu & Passeron, 1990) remain true today, not only for education but in virtually all other areas of privilege.

What does this type of privilege mean for you as you counsel with clients and others who face these obstacles? Listening to stories is obviously important, but a more activist teaching, consulting, coaching role is needed. Consider becoming a mentor and helping your clients understand and cope with a challenging system. This is an area for social justice action.

Quite a few Whites, males, heterosexuals, middle-class people, and others currently enjoy the convenience of not being aware of their privileged state and become angry at any challenge. The physically able see themselves as "normal," with little awareness that they are only "temporarily able" until old age or a trauma occurs. Out of privilege comes stereotyping of less dominant groups, thus further reinforcing the privileged status. Research has found that rich people have less empathy and generosity (Goleman, 2013), although we are all aware of the generosity and positive influence of Microsoft's Bill Gates and Facebook's Mark Zuckerberg.

Thus, we need to avoid stereotyping anyone or any group's cultural identities. To say that all White rich males are insensitive or that all those who have experienced serious trauma are deeply troubled is just another form of stereotyping. The well-off often seek to help others, and most trauma survivors are resilient. Look for individual uniqueness, strength, and openness to change. Multicultural awareness enriches uniqueness only when it allows us to become more aware of how much each person differs from the others.

You, the interviewer, face challenges. For example, if you are a middle-class European American heterosexual male and the client is a working-class female of a different ethnicity or race, it can become more difficult to gain trust and rapport. If you are a young Person of Color and the client is older, White, and of a markedly different spiritual orientation, again it will take time to develop a relationship and working alliance.

Remember that the issues the client brings to you are the ones he or she currently sees as most important. Although these concerns often relate to multicultural identity, it is generally best to keep them in your awareness and only discuss them in the session if it seems potentially helpful to the client. However, there are areas where a much stronger stand needs to be taken. For example, if you are working with a woman who is trying to please her husband and accepts being beaten now and then, naming this as an issue related to trauma and sexism is often essential. However, this still must be done carefully to ensure safety, even to the point of taking the client to a safe house. Respectful naming is the act by which we help clients identify biased or racist actions or situations. Helping clients rename and reframe their situation and live life in new ways is one way we can help rewire brain networks in more positive ways.

Clients may be talking about an academic issue and, in this process, may mention frustration with university facilities that don't meet their physical needs. They still need emotional and cognitive support to resolve their immediate issues, but they can also benefit from awareness of societal privilege that works against them. And you, as a counselor, have a social justice responsibility to work with the campus and community to increase awareness. This may lead to a discussion of disability rights and/or what occurs in clients' daily lives around handling their issues.

Political Correctness: How Can We Respect Differences?

Political correctness (PC) is a term used to describe language that is calculated to provide a minimum of offense, particularly to the racial, cultural, or other identity groups being described. Conservative, liberal, and other commentators have denounced the existence of PC. The term and its usage are hotly contested.

Political correctness originated specifically to encourage people to use proper and respectful names for those whose color or culture is different from their own. The idea has since been extended to issues such as personal safety and denial of speaking rights to visiting lecturers whose opinions differ from those of the majority. The term has caused loud outcries in media and political circles. We are not going to take a stand on these issues, but we do believe that respect and the ability to listen empathically to the "other side" may be more helpful in the long run. Respect for others is the issue, and denouncing political correctness has become an excuse for some to increase their harassment of others.

Given this controversy, what is the appropriate way to name and discuss cultural diversity? We argue that interviewers and counselors should use language empathically, and we urge that you use whatever terms the client prefers. Let the client define the name that is to be used, as it is respect for the client's point of view that counts in the issue of naming.

A woman is unlikely to enjoy being called a girl or a lady, but you may find some who use these terms. Some people in their 70s resent being called elderly or old, whereas others embrace and prefer this language. At the same time, you may find that the client is using language in a way that is self-deprecating. A woman struggling for her identity may use the word *girl* in a way that indicates a lack of self-confidence. The older person may benefit from a more positive view of the language of aging. A young person struggling with sexual identity may find the word *gay* or *lesbian* difficult to deal with at first. You can help clients by exploring names and social identifiers in a more positive fashion.

Race and ethnicity are often central topics in discussions of multicultural issues. *African American* is considered the preferred term, but some clients prefer *Afro-Canadian* or *Black*. Others feel more comfortable being called *Haitian*, *Puerto Rican*, or *Nigerian*. A person from a *Latina/o* background may well prefer that term, but others might more comfortable with *Chicano*, *Mexican*, *Mexican American*, *Cuban*, *Puerto Rican*, *Chilean*, or *Salvadorian*. Some *American Indians* prefer *Native American*, but most prefer to be called by the name of their tribe or nation, such as *Lakota*, *Navajo*, or *Swinomish*. Some Caucasians would rather be called *British Australians*, *Irish Americans*, *Ukrainian Canadians*, or *Pakistani English*. These people are racially White but also have an ethnic background.

Knowledge of the language of nationalism and regional characteristics are also useful. American, Irish, Brazilian, or New Zealander (or "Kiwi") may be the most salient self-identification. Yankee is a word of pride to those from New England and a word of derision for many Southerners. Midwesterners, those in Outback Australia, and Scots, Cornish, and Welsh in Great Britain often identify more with their region than with their nationality. Many in Great Britain resent the more powerful region called the Home Counties. And we must recognize that the Canadian culture of Alberta is very different from the culture of Ontario, Quebec, or the Maritime Provinces.

Awareness, Knowledge, Skills, and Action for Multicultural Competence

The history of our field has shown the gradual development of the meaning and necessity of multicultural competence. This training is now a requirement for medical licensure in several states, has become a standard in many helping professions, and is a core component of the United Nations' Convention on the Rights of Children (United Nations, 2015). In addition, it has had increasing influence in business management and other fields.

Let us examine the multicultural guidelines and competencies in more detail.

Awareness: Be Aware of Your Own Assumptions, Values, and Biases

Awareness of yourself as a cultural being is a vital beginning to authenticity. Unless you see yourself as a cultural being, you will have difficulty developing awareness of others. We will not elaborate further in this area, as the previous discussion of the RESPECTFUL model focuses on awareness.

The competency guidelines also speak to how contextual issues beyond a person's control affect the way the person discusses issues and problems. Oppression, discrimination, sexism, racism, and failure to recognize and take disability into account may deeply affect clients without their conscious awareness. Is the problem "in the individual" or "in the environment"? For example, you may need to help clients become aware that issues such as tension, headaches, and high blood pressure may be results of the stress caused by harassment and oppression. Many issues are not just client problems but also problems of a larger society.

> **Reflective Exercise**
>
> Have you experienced cultural trauma and microaggressions, and how has this affected your sense of trust in others, both individuals and groups?
>
> Perhaps most challenging, where might you have biases of favoritism toward others or toward your own group? Or perhaps you may discover some unconscious biases against others. How might that affect your interviewing practice?
>
> End this self-examination with stories of strength and resilience that come from your life experience. What are you proud of? Build on what you can do, rather than what you can't do.

Knowledge: Understand the Worldview of the Culturally Different Client

> These [racial] assaults to black dignity and black hope are incessant and cumulative. Any single one may be gross. In fact, the major vehicle for racism in this country is offenses done to blacks by whites in this sort of gratuitous never- ending way. These offenses are **microaggressions**. Almost all black-white racial interactions are characterized by white put-downs, done in automatic, pre- conscious, or unconscious fashion. These mini-disasters accumulate. It is the sum total of multiple microaggressions by whites to blacks that has pervasive effect on the stability and peace of this world.
>
> —Chester Pierce, 1974, p. 515

Worldview is formally defined as the way you and your client interpret humanity and the world. People of different historical, religious, and cultural backgrounds worldwide often have vastly different philosophic views on the meaning of life, right and wrong, and personal responsibility versus control by fate. Because of varying multicultural backgrounds, we have different worldviews in the way we see and think about people. Often central to differences in worldview are microaggressions, which mount over time, resulting in damage not only to the psyche, but also to the body.

Pierce in 1974 focused on the African American experience. Derald Wing Sue (2010; Sue & Sue, 2016) has done much to bring Pierce's early work to national attention and to extend this issue of microaggressions to all dimensions of the multicultural cube. His work is cited in current media with strong support from many minority groups, but it is also widely attacked by the media as divisive and merely "politically correct." Sue has brought the *name* microaggressions to the nation's attention; enabling your clients to name their experience in this way is part of the route toward psychological liberation.

Many people still deny the power of the harassment and bullying of microaggressions. It is best that you be prepared to watch for instances of microaggressions, help clients name them for what they are, and provide counseling for emotional support, cognitive understanding, and deciding when and how to respond to these painful events. Enable clients to use these situations as an opportunity to develop increased resilience.

A classic study found that 50% of minority clients did not return to counseling after the first session (cited in Sue & Sue, 2016). This book seeks to address cultural intentionality by providing you with ideas for multiple responses to your clients. If your first response doesn't work, be ready with another.

BOX 2.4   Stories of Microaggressions

Jenny Galbraith earned one of Harvard University's highest honors when she was named a Ledecky Undergraduate Fellow. Nonetheless, here are her words describing the African American Harvard experience (Galbraith, 2015).

> The professor is lecturing about "racial tensions" without naming Michael Brown, Tamir Rice, Freddy Gray, Rekia Boyd, Sandra Rice. . . . I want to stand up and scream about how the things he is talking about tear bodies apart.
>
> I wonder how many students in that very White classroom are feeling what I feel at that moment. I look to my left and my right and see students jotting down notes, continuing on to the next notion of cost-benefit analysis. I send an innocuous text to a friend—I think I just want to feel less alone, that feeling when a moment hits you deeper than it hits those around you. . . .

I am thinking of the time when my uncle's neighbor called the cops on him because he dared to walk in his own backyard. Because he dared to exist in that space that he literally owned. Because he dared to exist at all. Reflection is causing me pain. I want to tell my friend I understand.

This is about what it means not to fit into Harvard's mold, what it means to know that any moment might twist your stomach into knots. There's no easy way to fix this, for Harvard, for America.

The following provide further examples of the small hurts of microaggressions that pile up over time and become traumatic (Binkley & Whack, 2015).

Sheryce Holloway is tired of White people at Virginia Commonwealth University asking if they can touch her hair or if she knows the latest dance moves.

At Chicago's Loyola University, Dominick Hall says that groups of White guys stop talking when he walks by, and sometimes people grip their bags a little tighter.

Katina Roc said she will never forget the day two years ago when she sat down in class at West Virginia University and a White student a few seats away collected his things and moved away.

Derald Wing Sue, Professor at Columbia University, speaks perfect English, but he is sometimes asked, "You speak perfect English, but where are you from?" He says, "Portland, Oregon." The all-too-frequent response is, "No, I mean where are you from?" expecting to hear China.

Carlos Zalaquett, full Professor at Penn State University, still finds clerks following him around the store if he is not wearing a suit. Recently, he paid for an item in the back section of a well-known big-box store. With his receipt in his hand, he was stopped at the exit and quizzed by three employees who doubted that he had already paid. Carlos expects these incidents and is prepared to deal with them as they happen. He uses his resilience to teach those who would deny him equal treatment.

Many male African Americans cannot get a taxi unless someone else, perhaps their wife, hails it for them—even if they are wearing an expensive business suit.

Traditional approaches to counseling theory and skills may be inappropriate and/or ineffective with some groups. We also need to give special attention to how socioeconomic factors, racism, sexism, heterosexism, and other oppressive forces may influence a client's worldview. Box 2.4 illustrates how multiculturalism is an intricate part of our social tapestry, and Box 2.5 reminds us that multiculturalism belongs to all of us.

Despite the recent presidency of Barack Obama, racial disparities remain. Racial minorities still are more likely to drop out of school and at all levels tend to be more dissatisfied with the educational system. While college attendance has nearly doubled, recent court decisions have resulted in fewer minorities at "top" state universities. Beyond schooling, People of Color encounter more poverty, violence, income disparities, and a variety of other discriminatory situations. Although middle- and upper-class minorities are somewhat protected, they all-too-frequently suffer the same mistreatment. You will find that many minority people have learned to wear a suit while White people are allowed to be more casual.

BOX 2.5 **National and International Perspectives on Counseling Skills**

Multiculturalism Belongs to All of Us

Mark Pope, Cherokee Nation and Past President of the American Counseling Association

Multiculturalism is a movement that has changed the soul of our profession. It represents a reintegration of our social work roots with our interests and work in individual psychology.

Now, I know that there are some of you out there who are tired of culture and discussions about culture. You are the more conservative elements of us, and you have just had it with multicultural this and multicultural that. And, further, you don't want to hear about the "truth" one more time.

There is another group of you that can't get enough of all this talk about culture, context, and environmental influences. You are part of the more progressive and liberal elements of the profession. You may be a member of a "minority group" or you have become a committed ally. You may see the world in terms of oppressor and oppressed. I'll admit it is more complex than these brief paragraphs allow, but I think you get my point.

Here are some things that perhaps can join us together for the future:

1. We are all committed to the helping professions and the dignity and value of each individual.

2. The more we understand that we are part of multiple cultures, the more we can understand the multicultural frame of reference and enhance individuality.

3. Multicultural means just that—many cultures. Racial and ethnic issues have tended to predominate, but diversity also includes gender, sexual orientation, age, geographic location, physical ability, religion/spirituality, socioeconomic status, and other factors.

4. Each of us is a multicultural being and thus all interviewing and counseling involve multicultural issues. It is not a competition as to which multicultural dimension is the most important. It is time to think of a "win/win" approach.

5. We need to address our own issues of prejudice—racism, sexism, ageism, heterosexism, ableism, classism, and others. Without looking at yourself, you cannot see and appreciate the multicultural differences you will encounter.

6. That said, we must always remember that the race issue in Western society is central. Yes, I know that we have made "great progress," but each progressive step we make reminds me how very far we have to go.

All of us have a legacy of prejudice that we need to work against for the liberation of all, including ourselves. This requires constantly examining honestly and at times this self-examination can be challenging, even painful. You are going to make mistakes as you grow multiculturally; but see these errors as an opportunity to grow further.

Avoid saying, "Oh, I'm not prejudiced." We need a little discomfort to move on. If we realize that we have a joint goal in facilitating client development and continue to grow, our lifetime work will make a significant difference in the world.

Skills and Action to Cope with the Results of Discrimination and Build Cultural Health

> Culture and unique life experiences wire the brain. A brain that has been raised with pain, poverty, and discrimination is very different from one raised in privilege.
> —Allen Ivey

What we see, hear, feel, smell, touch, and taste go to distinct areas of the brain. These cultural/environmental/contextual data are integrated together to produce what we call cognition and emotion, the way we know the world—our worldview. New environmental stimuli combine with memories of past experience in the hippocampus. The students and professors in Box 2.4 all experienced instances of microaggressions, which resulted in increased amounts of damaging cortisol in the brain. If these happen often enough, neural brain networks related to anger and fear will be strengthened and changes in the brain can be permanent.

To maintain any sense of balance in the face of these hurts, the person needs to have developed a base of resilience and faith, trust, and pride in their family's cultural background. This is **cultural health**. What is called the **attentional network** is key to how we attend to the world and then integrate internal and external perceptions (Ivey, 2016). People with varying cultures and life experience have different integrations of their lives. Out of this comes a "brain map" in memory which guides us in the future. Solid memories of strength enable coping with these situations, but they still hurt. Counseling seeks to strengthen these positive connections.

The issue of action in the interview to help those who experience microaggressions and prejudice is illustrated in two interviews Allen held with Nelida Zamora, a Cuban American (Chapters 9 and 10). Nelida encountered a painful microaggression in her first graduate counseling class. She had asked her professor a few questions with her Cuban American accent. After class she was asked, "Where are you from?" Miami did not satisfy the fellow student, so it was asked again, "No, where are you really from?" This seemingly small issue was burned into her memory, resulting in less self-confidence. Notice the parallels with Derald Wing Sue's continuing Columbia University experiences.

In these action sessions, Allen built on Nelida's strengths in her family and culture. As you view these interviews, you may find some counseling hints to help clients deal with microaggressions. In short, we need to move beyond knowledge and awareness and take our skills into the session to work with challenging issues.

Chapter 9 on Focusing provides specific guidelines that may be helpful in dealing with internalized racism, sexism, bullying, family abuse, and other hurtful experiences.

If you encounter many good things in life, the brain map will take a positive turn. Good input generally results in good output. On the other hand, less effective and damaging output comes from being raised in a family, community, and region (plus media) providing only prejudiced information. Negative beliefs about self or others easily become embodied and hard to change. Those who live with difficult life experiences may tend to interpret daily life in negative terms, experiencing the Soul Wound and internalized oppression. They also may feel anger and fear toward those more fortunate. Another way to put this is "garbage in—garbage out." If people are fed the garbage of life, what can we expect?

What do we do with instances of microaggressions and harassment in counseling? Seek to move the perceptual frame and interpretations of life issues and concerns, and build resilience and build strength through skilled use of communication skills such as those presented here. With individual clients, first watch for signs and stories that represent microaggressions. Draw out the story sensitively, and be willing to self-disclose and share your support appropriately. Enlightened use of the many available theoretical alternatives is important, particularly multicultural counseling and therapy (MCT) and social justice advocacy. Often just sitting with a client is not enough. Watch for teachers and other influential persons who may be problematic. Seek help from community leaders.

Help clients name the issue and identify contextual/environmental factors. Educate clients to understand their goals, expectations, and legal rights—and provide tools to address the situation. Apply advocacy skills and exercise institutional intervention skills on behalf of clients if needed. Use the guidelines offered by the Multicultural and Social Justice Counseling Competencies (MSJCC) (Ratts, Singh, Nassar-McMillan, Butler, & McCullough, 2015) in practice. At the same time, we must not impose our beliefs on clients. The client needs to be ready to act.

In the following section, we discuss positive psychology as a route to developing resilience. Search for positive stories of strength; help the client remember the resources he or she has from family and friends; identify what the client has done right. All these and more can be useful in facilitating the ability to bounce back and cope with present and future challenges. In addition, therapeutic lifestyle changes (TLCs) provide a number of actions that clients can take to improve mental health. Among those discussed later in this chapter are exercise, drawing on spiritual and religious resources, building cultural identity through cultural health, relaxation and meditation training, and many others, including the counseling relationship itself.

Positive Psychology and Therapeutic Lifestyle Changes: Building Client Resilience

> Your mind is a powerful thing. When you fill it with positive thoughts, your life will start to change.
>
> —Anonymous

> Optimists literally don't give up as easily and this links to greater success in life.
>
> —Elaine Fox

If you help clients recognize their strengths and resources, you can expect them to use these positives as a basis for resolving their issues. There is a second story behind the first story of client difficulties and concerns. Where have they done things right and succeeded? Of course, our first task is to draw out the worrisome story and issues that brought them to the interview or counseling session. Searching for the positive psychology/wellness story does not deny client concerns and serious challenges. But our aim is constantly to watch and listen for strengths that will eventually be part of the solution.

Resilience and Optimism

Positive psychology's central aim is to encourage and develop optimism and resilience. *Optimism* is defined in various dictionaries with many affirmative words—among them, *hope, confidence,* and *cheerfulness.* It also includes a trust that things will work out and get better, a sense of personal power, and a belief in the future. Optimism is a key dimension of resilience and the ability to recover and learn from one's difficulties and challenges.

Resilience is the ability to bounce back from setbacks, temporary failure, and early or late trauma of many types. People who are more optimistic have an increased ability to eliminate, reduce, or manage stressors and negative emotions. Furthermore, they are better able to approach and face their difficulties. Optimists tend to live healthier lives, suffer less from physical illness, and, of course, feel better about themselves and their abilities (Kim, Park, & Peterson, 2011; Nes & Segerstrom, 2006).

Out of this research has come an effective six-point scale to measure optimism (see Box 2.6). We suggest that you use this scale to assess your own level of optimism. At times you may want to share this scale with clients, or ask them questions to discover their level of optimism. One of our goals in counseling and therapy is to increase resilience, and helping clients become more optimistic and hopeful is part of this process.

Using the scale in Box 2.6, you can develop an optimism score by adding items 2, 3, 4 and a pessimism score by adding 1, 5, and 6. These six items have proven both reliable and predictive. We suggest that you use them as an indication of what a positive attitude toward life means. A positive attitude appears critical for the development of resilience.

BOX 2.6   A Six-Point Optimism Scale

Please say how much you agree or disagree with the following statements: 1 = Strongly disagree, 2 = Somewhat disagree, 3 = Slightly disagree, 4 = Slightly agree, 5 = Somewhat agree, 6 = Strongly agree.

1. _____ If something can go wrong for me, it will.

2. _____ I'm always optimistic about my future.

3. _____ In uncertain times, I usually expect the best.

4. _____ Overall, I expect more good things to happen to me than bad.

5. _____ I hardly ever expect things to go my way.

6. _____ I rarely count on good things happening to me.

Scheier, M., Carver, C., & Bridges, M. (1994). Distinguishing optimism from neuroticism (and trait anxiety, self-mastery, and self-esteem): A reevaluation of the Life Orientation Test. *Journal of Personality and Social Psychology, 67*(6), 1063–1078. Copyright © 1994 by the American Psychological Association. Reproduced with permission.

Positive Psychology and Resilience	Anticipated Client Response
Help clients discover and rediscover their strengths. Find strengths and positive assets in clients and in their support system. Identify multiple dimensions of wellness. In addition to listening, actively encourage clients to learn new actions that will increase their resilience.	Clients who are aware of their strengths and resources can face their difficulties and resolve issues from a positive foundation. They become resilient and can bounce back from obstacles and defeat.

Building Resilience Through Strengths and Resources

> Positive psychology brings together a long tradition of emphasis on positives within counseling, human services, psychology, and social work.
> —Martin Seligman

In recent years, the field of counseling has developed an extensive body of knowledge and research supporting the importance of positive psychology, a strength-based approach. Psychology has overemphasized the disease model and all too often places a self-defeating and almost total focus on difficulties, ignoring the client's own strengths in the resolution of issues. Nonetheless, the positive psychology movement is well aware that happiness is not possible in the midst of excessive stress.

In addition to drawing out the problematic negative stories, we also need to search for stories of strength and success. Seek out and listen for times when clients have succeeded in overcoming obstacles. Listen for and be "curious about their competencies—the heroic stories that reflect their part in surmounting obstacles, initiating action, and maintaining positive change" (Duncan, Miller, & Sparks, 2004, p. 53). In the microskills, we call this the "positive asset search." Find something that the client is doing right, and discover the client's resources and supports. Use these strengths and positive assets as a foundation for personal growth and resolving issues.

Encouraging and teaching clients to becoming fully engaged in life is basic to positive psychology. Your client may have "retired" from life to be safe and secure, but this leads easily to depression. In terms of life satisfaction, engagement was found to be more central than happiness. Find areas in which your clients have been involved, activities that they care about. Often they have stopped exercising, meditating, or playing tennis or golf. They may have cut themselves off from friends, church, or other groups that provide interest and support. How much time do they spend on passively watching TV or other screen time, taking them away from daily life?

"Life—be in it" is a classic Australian saying. Australia is known as one of the happiest countries in the world.

Therapeutic lifestyle changes (TLC), discussed in the following section, are a key route to identifying and encouraging an engaged lifestyle. Examples of lifestyle changes include increased exercise, better nutrition, meditation, and helping others, which in turn helps us feel better about ourselves.

At the same time, when we view engagement from a broad multicultural perspective, many clients, regardless of race or ethnicity, are hungry, abused, or suffering from trauma. Perhaps their unemployment benefits are about to run out, or a family member may be seriously ill. These clients may consider it a luxury to find the time to study better nutrition, to exercise, or, particularly, to meditate. But you can help the engagement process and lead the client to greater resilience by encouraging a short walk—even around the block. You can provide information on how to make wiser food decisions. Meditation may make no sense to some, but deep breathing, visual imaging of family strengths, or short relaxation training can help alleviate stress. A client might be reminded to draw on spiritual traditions or help with a church project. Helping others builds compassion and changes the way the brain functions (Fowler & Christakis, 2010; Sepella, 2013).

Martin Seligman was surprised to find in his research that meaning was the most essential aspect of a satisfying life. This would not surprise Viktor Frankl, famous survivor of the Nazi concentration camp at Auschwitz, who stated that those of us who find a *why*, a meaning for our lives, can live with virtually any *how*. Meaning and a life vision carry many people through the most difficult times in their lives. (For more information on Frankl, meaning, and a positive life vision, see Chapter 11.)

TLCs for Stress Management, Building Mental and Physical Health, Brain Reserve, and Resilience*

> It is unethical for a physician (or counselor) to meet with a patient and fail to prescribe exercise.
>
> —John Ratey, M.D.

Physical Exercise. The preceding quote from John Ratey, who teaches exercise science at Harvard Medical School, is significant for those of us who interview, counsel, or engage in psychotherapy. We have many complex theories and methods, but the helping fields have forgotten to make the simple prescription of exercise central to the helping session. Key to stress management and behavioral health is getting blood flowing to the brain and body. Exercise increases brain volume; it is also valuable in obesity prevention, may help prevent cancer, and may slow the onset of Alzheimer's disease.

Two studies offer examples of how exercise affects mental health. Duke University found that after one year, those with major depressive disorder who did regular exercise received as much benefit as they would have gained from medications. In addition, regular exercisers were less likely to relapse and return into depression. Another research study found that women with generalized anxiety disorder who did two weeks of resistance training or aerobic exercise decreased their worry symptoms by 60% (Herring, Jacob, Suveg, Dishman, & O'Connor, 2012).

---

*By permission of Allen E. Ivey © 2013, 2016. Permission is granted for duplication with the request that this copyright information remain on the document.

TABLE 2.1   Therapeutic Lifestyle Changes: Physical Exercise

Facilitative	Destructive
Enhances sleep Increases dopamine, gray matter Reduces depression, anxiety Increases lifespan	Increases wakefulness Obesity Ill health Reduces lifespan Increases likelihood of heart disease (4 hours TV = 80% increase in heart death)

Potential activities in the session. Be sure to check on the physical activities of each client. Watch for changes that indicate less exercise. At appropriate moments, point out that exercise has positive benefits for physical and mental health. Follow up regularly to see if an exercise plan has been implemented. Refer to a physician in cases where there is any history of significant illness or present signs of ill health. Recommendations for clients who are beginning to exercise for the first time require a visit to a health provider.

Nutrition, Weight, and Supplements. Relatively new to our field is the importance of helping clients deal with the central health and brain issue of nutrition. *Avoid the whites* (pasta, sugar, salt) and *snack only on healthy food.* Vegan, vegetarian, and Mediterranean diets have proven to be effective. Fat activates genes that cause apoptosis (cell death). Obesity facilitates diabetes and other illnesses, including cancer, and it encourages the development of Alzheimer's disease. Supplements can be valuable, but encourage clients to consult their physician before taking supplements.

Potential activities in the session. As this can be an area of sensitivity for some clients, issues of eating need to be approached with care. Those who are concerned about prejudice against weight need special attention and support. Intake forms before interviewing or counseling begins can ask for information on eating style and nutrition, thus opening the way for you to bring up the topic later. What can you share with your clients in a way that they might listen? Basically, we suggest initiating a brief discussion of one of the issues mentioned here and observing client response. Too much information may overwhelm the client and result in no action. Those who want to think seriously about better nutrition will likely need referral to a nutritionist, as most medical personnel do not have sufficient time or interest to explore this issue. David Katz, M.D., of Yale, who was nominated in 2009 for

TABLE 2.2   Therapeutic Lifestyle Changes: Nutrition

Facilitative	Destructive
Organic food Healthy snacks Low-fat, complex carbs Olive oil Richly colored fruits/vegetables Wild salmon, fish oil, flaxseed Walnuts and other nuts Pure water	Junk food causing inflammation Sugar/pasta/white bread Palm oil, cottonseed oil, etc. "Dirty dozen" fruits and vegetables (with most pesticide residue) Meat hormones and antibiotics Processed foods/snacks Walnuts and other nuts BHP plastic water bottles

TABLE 2.3 Therapeutic Lifestyle Changes: Social Relations

Facilitative	Destructive
Love, sex	Isolation, living alone
Joyful relationships	Negativism, criticism
Extended lifespan	Brain cell death
Higher levels of oxytocin ("love hormone")	Anger, fear, stress
Helping others	Ignoring or harming others

U.S. Surgeon General, talks of "forks, fingers, and feet." ("Fingers" refer to the importance of not smoking.) The "3-F" framework is a possible way to start the conversation.

Social Relations. Being with people in a positive way makes a significant difference in wellness. Interpersonal relationships, of course, are often the central issue in interviewing and counseling. We want our clients to engage socially as fully possible, as this not only builds mental health but also builds the brain and body. Love and close relationships build health. People with negative or ambivalent relationships have been found to have shorter protective telomeres, thus predicting age-related disease (Uchino, Bowen, Carlisle, & Birmingham, 2012).

Potential activities in the session. Naturally, this is where careful listening skills become essential, but in addition to negative stories, we need to search for the positive stories and strengths that can lead to better relationships. There is a long history in counseling and psychotherapy of searching for the "problem," identifying weaknesses that need to be corrected, and diagnosing and labeling before searching out what the client "can do" rather than what he or she "can't do." Of course, clients don't come to us looking for what they did right. Rather, they want solutions to alleviate their issues—and as soon as possible. Thus, we need to draw out the problematic story and the issues behind it, but simultaneously listen for strengths and supports, then use these as a foundation for positive change.

Albert Ellis, the famed originator of rational emotive behavioral therapy (REBT), is also considered by many to be the founder of the cognitive behavioral therapy (CBT) tradition. As a teen, he found himself awkward, with little in the way of social skills. He decided that he wanted a girlfriend, so he started asking girls on dates, both those he knew and those he didn't. He was virtually always turned down, but he kept taking the risk. He did not let failure get him down, and eventually he succeeded. Many of your clients have difficulty taking social risks, meeting new people and relating to them. One of your tasks is to listen to their stories and, working with them, to find new ways of relating. One of these is role-playing social tasks that your client finds challenging.

Cognitive Challenge. Take a course, learn a language, learn to play an instrument—basically do something different for growth and the creation of new neural networks. Uncertainty can be growth producing, challenging the assumptions that we have worked so hard to accumulate while young. With a brain already full of well-connected pathways, adult learners should "jiggle their synapses a bit" by confronting thoughts that are contrary to their own.

Potential activities in the session. Check out the degree of stimulating and cognitive involvement of your clients. Generally, we want to encourage and support more. The destructive factors all too easily lead to depression. However, an overly active schedule, both

TABLE 2.4   Therapeutic Lifestyle Changes: Cognitive Challenge

Facilitative	Destructive
Any cognitive challenge	TV, too much screen time
Change of any type	Repetitious routine tasks
Learning a musical instrument	Being alone
Learning a new language or skill	Vegetating, sitting too much
Playing card games and word games	Boredom
Playing brain games	Taking the easy way out

cognitively and socially, can lead to anxiety and sleeplessness. Our goal is to find cognitive/emotional balance.

Sleep. A full rest is critical for brain functioning and development of new neural networks. Lack of sleep is one of the indications for depression or anxiety.

Potential activities in the session. Your office intake form needs to include questions about how well clients sleep. Your task in the session is to follow through and learn more. If clients are not sleeping well, you need to make frequent contact with them and follow their sleep patterns. However, counseling around sleep issues belongs primarily to experts and medical professionals. If you sense serious problems, such as sleep apnea or continued inability to sleep, referral to medical professionals and possibly a formal sleep study is essential.

Meditation and Relaxation. If you meditate 10–20 minutes or more each day, it will make a significant difference and calm you throughout the day. Evidence from the University of Wisconsin (Kabat-Zinn & Davidson, 2012) is clear—meditation makes a positive difference in your brain, even increasing gray matter. A relaxed focus in extended prayer, the lighting of candles, saying the Rosary, or attending healing services may function similarly to meditation and help the immune system. You can download Jon Kabat-Zinn's *Guided Mindfulness Meditation* at the Apple Store. An app called *GPS for the Soul* provides an excellent introduction to meditation and daily exercises to help us slow down and "be here now."

Potential activities in the session. The first thing to do is assess what areas of stress management, meditation, and relaxation may be relevant to your client. Then contract with the client for what is most meaningful. You can teach many basic forms of meditation as you talk with a client in almost any situation.

TABLE 2.5   Therapeutic Lifestyle Changes: Sleep

Facilitative	Destructive
7-9 hours of sleep	Sleep deprivation
Reading, meditating, quiet, no TV	Screen time
Increases metabolism, hormones	Sleepy at school/job
Consolidates learning/memory	Parts of brain turn off
Positive mood, increase in motivation	Increased risk of accidents
Critical for physical health	Loss of emotional control
	Increased eating

TABLE 2.6   Therapeutic Lifestyle Changes: Meditation and Relaxation

Facilitative	Destructive
20 minutes meditation (or more) 10 minutes systematic muscle relaxation Yoga Tai Chi Prayer Spiritual meditation Listening to soft or culturally appropriate music Focus on a mountain, hill, or stream Deep breathing	Lack of awareness of what stress does to mind and body Failing to take time for oneself Overinvolvement in too many activities Continued worry and anxiety Focus on money, achievement, status Always needing to be busy

TABLE 2.7   Therapeutic Lifestyle Changes: Cultural Health/Cultural Identity

Facilitative	Destructive
Loud and proud about one's multicultural identity, its full RESPECTFUL dimensions Awareness of family strengths Awareness of positives in one's school, community, region, or nation, but also able to identify and seek to correct weaknesses Free and able to change self and work toward involvement, action, and change	Denial of one's culture and/or lack of awareness that one has a culture Alienated from family Only focuses on negatives in the community, region, or nation, or ignores or denies weaknesses or limitations Passive and oblivious to responsibility for change in the world

Multicultural Pride and Cultural Identity. Our personal identity as multicultural beings affects both our mental and physical health. The harassment that comes with racism, ethnic prejudice, and lack of opportunity deeply affects People of Color. Regardless of our race or ethnicity, we still face issues of religious discrimination and favoritism, economic injustice, ableism, sexism, heterosexism, and other forms of oppression.

You were asked earlier in this chapter to identify your multicultural identity through the RESPECTFUL model. A positive psychology/wellness approach recognizes that multiple forms of oppression do exist, but finding personal and cultural strengths can increase one's self-respect and strengthen identity.

Potential activities in the session. If clients show interest, introduce them to the RESPECTFUL model and ask them for examples of strengths and positive assets that will lead to resilience. All aspects of the RESPECTFUL model have heroes who have made a difference. Who are the cultural heroes and models? What have been their strengths for action, success, or survival that the client can draw on? Look for strength stories in each RESPECTFUL area. What is the client's own personal commitment to the values that he or she has discussed, and how might they be acted upon?

While the seven factors discussed here are fundamental to mental and physical health, there are other TLCs that may be as or more useful to many of your clients. They are summarized briefly in Box 2.7. As always, TLCs should be considered in light of the possible need for referral to other sources of assistance. At the same time, you can make a significant difference in helping clients to think through their own mental and physical health plans.

BOX 2.7   Additional Therapeutic Lifestyle Changes

Drugs and Alcohol

Drugs can harm or destroy brain cells. Drugs are especially dangerous for the critical ages 13–18. Teen marijuana use increases the risk of psychosis by 8%–10% and causes cognitive decline. Research shows that one or two glasses of alcohol for men and one for women help the brain and heart and delay Alzheimer's. But alcohol is also associated with increases in breast and other cancers. Moderation is the key.

Medication and Supplements

Interactions are often missed by professionals. More than one antidepressant is no better than one, and exercise and meditation are often as good as meds. Look for USP or DSVP verification on supplements, and check with a physician. For those over 45, a daily baby aspirin is highly recommended. Watch blood pressure, body mass index (BMI), and dental hygiene. Dangerous inflammation can often not be seen, so try to control with appropriate diet. There are strong data linking omega-3 with mental and physical health.

Positive Thinking/Optimism/Happiness

Positive thoughts and emotions rest primarily in the executive frontal cortex, while negative emotions of sadness, anger, fear, disgust, and surprise lie in the deeper limbic parts of the brain. Research shows that positive thinking and therapy can build new neural structures and actually reduce the power and influence of negative neural nets.

Beliefs, Values, and Spirituality

Researchers have found that those with a strong spiritual orientation recovered more quickly and left the hospital sooner. This has been replicated in several studies. Research on cognitive therapy found that using spiritual imagery with depressed clients was highly effective. Imagery and metaphor can affect several areas of the brain simultaneously. A strong positive faith or belief system can make a significant difference in one's life. One can visualize positive examples of strength models of faith, such as Jesus, Moses, Buddha, Mohammed, Viktor Frankl, Gandhi, or Martin Luther King. Visualizing positive strengths of family members, such as grandparents, can remind one of basic value systems. Your belief system and spiritual orientation contribute to healthy, positive thinking and an optimistic attitude.

Take a Nature Break, Rather Than a Coffee Break

The *Wall Street Journal* (Wang, 2012) has summarized research showing that memory, attention, and mood increase by 20% if one goes for a 10-minute quiet walk in nature, or even spends time viewing nature scenes in a quiet room. Japanese research validates the value of nature; even looking at or visualizing the color green seems to be helpful. Small breaks can be an important part of stress management—and even improve studying and grades.

No Smoking

Smoking narrows blood vessels, reducing blood flow to the brain. It is our primary addictive drug and most dangerous drug in terms of early death. An issue is that smoking often relaxes and increases one's ability to concentrate. Successful no-smoking counseling typically requires a committed client and close follow-up by phone or email—continuous support and reminders appear to help.

Control Screen Time

The many types of screens we use are changing our brains, resulting in short-term attention and increasing both hyperactivity and obesity. The constant flicker both activates and tires the brain. A study by the National Heart, Lung, and Blood Institute (2013) revealed that children ages 8–18 average 4.5 hours a day watching TV, 1.5 hours on a computer, and 1 hour playing video games. As a result, they go to bed later and have more difficulty getting up.

Art, Music, Dance, Literature

What is the client's passion? Returning to and becoming involved with the arts can be life changing. The relaxation and here-and-now emotions of music and dance bring satisfaction and a needed break in routine. Many of us find peace and oneness through participating in or viewing the arts.

Relaxing and Having Fun

This will increase dopamine release to the nucleus accumbens. Dance, tennis, enjoying a sunset, you name it—it is good for the brain.

More Education

It is clear that the farther you go in the educational system, the less your chance of Alzheimer's and the better your health. Ensure that child raising provides richness and encourages reading. Continuing education as we age helps develop new neural networks. Take courses; learn a musical instrument or a new language. However, education is often oriented to the privileged. Those from economically disadvantaged backgrounds receive less. Even when they go to college, they drop out more frequently, often with high levels of debt. Support networks are needed. Work for better and more just schooling.

Money and Privilege

Not having to worry about funds, being able to get the best medical care, and buying the best food are not options available to all. Most of us who read this page, regardless of race/ethnicity, enjoy some form of privilege. Regardless of background, many benefit from being educated and reasonably affluent. White privilege brings consequential benefits to this group. Coming into the world with privilege also brings responsibility.

Helping Others and Social Justice Action

Working for justice and volunteering make a positive difference to both helper and helpee. Stress hormones are reduced, and telomeres lengthen. Fetal development depends on a healthy mother. Poverty, abuse, and trauma produce damaging cortisol, destroying brain cells permanently. Racial and oppressive harassment of all types causes stress and raises cortisol. Neuroscience and genetic/epigenetic research clearly shows that environmental conditions are key to health and that social justice action will enable children and adults to function more fully, effectively, and joyfully.

Joy, humor, zest for living, and KEEP IT SIMPLE

Client: Doctor, I can't sleep. I keep thinking someone is under my bed and I get up several times a night to look.

Psychiatrist: I can fix that. Four years of therapy twice a week.

Client: How much will it cost?

Psychiatrist: $100. An hour.

Client: I can't afford it, sorry.

Two years later, they meet on the street.

Psychiatrist: Nice to see you. How are things going?

Client: Great, no more problems. I told my neighbor, and he came over and cut off the legs of the bed.

We often make things too complex with fancy theorizing. TLCs are a shortcut to health that will change our practice. Learned optimism heals and can change a life. Focus on strengths and what clients *can do* rather than on what they can't. The positive approach builds cognitive reserves that will enable them to meet challenges more effectively. Keep it simple, use humor and laughter, and increase zest for life.

Action: Key Points and Practice of Ethics, Multicultural Competence, Positive Psychology, and Therapeutic Lifestyle Changes

Ethics and the Counseling and Psychotherapy Process. Helping professionals must practice within boundaries of their competence, based on education, training, supervised experience, state and national credentials, and appropriate professional experience. Their main goal is to do good and avoid harm. Their actions are regulated by their helping profession's code of ethics. Confidentiality provides the basis for trust and relationship building. Informed consent requires telling clients of their rights. When recording sessions, we need permission from the client. Helping professionals are asked to work outside the interview to improve society and are called upon to act on social justice issues.

Multicultural Competence. Interviewing and counseling are global phenomena used with individuals from many different cultures and customs. It is an ethical imperative that interviewers and counselors be multiculturally competent and continually increase their awareness, knowledge, skills, and action in multicultural areas.

The RESPECTFUL Model. The RESPECTFUL model lists 10 key multicultural dimensions, thus showing that cultural issues will inevitably be part of the interviewing and client relationship. Privilege is close related to race and ethnicity. We need to be aware that being White, male, and economically advantaged often puts a person in a privileged group. The interviewer needs to be client centered rather than directed by "politically correct" terminology. This is an issue of respect, and clients need to say what is comfortable and appropriate for them.

The Soul Wound and Historical Trauma. Soul Wound occurs with and from historical trauma. Traumatic wounds like the ones suffered by Native Americans do not disappear; they remain "in the soul" and are passed from generation to generation. The concept of the Soul Wound has relevance to all minority groups, including young people and women, who may experience bullying and/or abuse.

Privilege as a Multicultural Interviewing Issue. Privilege is power given to people through cultural assumptions and stereotypes. A privileged perception could make client stories invisible and shortchange the counseling process. Avoid stereotyping anyone or any group's cultural identities. Look for individual uniqueness, strength, and openness to change.

Microaggressions. "Racial microaggressions are brief and commonplace daily verbal, behavioral, or environmental indignities, whether intentional or unintentional, that communicate hostile, derogatory, or negative racial slights and insults toward people of color" (Sue, 2010). Microaggressors are often unaware that they are harming another person. Repeated racial harassment (or bullying) can literally result in posttraumatic stress. What seems small at first is damaging through repetition.

Awareness, Knowledge, Skills, and Action for Multicultural Competence. Awareness, knowledge, and skills are meaningless unless we act. Given the diversification of our society, developing all four is a must, but it is a lifelong process of continuing learning. Awareness of yourself as a cultural being is a vital beginning. Learn about your own and other worldviews. Use a culturally and diversity sensitive approach to interviewing and counseling. Adapt your strategies in a culturally respectful manner.

Political Correctness. Political correctness (PC) is a term used to describe language that is calculated to provide a minimum of offense, particularly to the racial, cultural, or other identity groups being described.

Bias and Prejudice Wound Clients' Soul and Affect Their Brain. The way we see the world depends on past learning. There is no "immaculate perception." Attentional and salience networks relates to how we attend to the world and then integrate internal and external perceptions. Brain structures such the insula, thalamus, hippocampus, and medial prefrontal cortex are involved in this process. Their task is to first integrate and pass on messages and then make meaning of external perceptions, based on past learning and experience. Bias and prejudice are almost locked in the brain. Prejudice is not just a way of thinking; it becomes a way of being and reinforces itself by interpreting people different from ourselves in negative ways. We should actively seek to unlock ourselves from biases.

Positive Psychology: Building Client Resilience. Positive psychology promotes the development of optimism and resilience. Optimism can be measured with scales such as the Six-Point Optimism Scale. Search for positive assets, which are the resources and strengths that clients bring with them. Clients aware of their strengths and resources are more resilient and can face challenges in a more effective manner.

Therapeutic Lifestyle Changes: Specifics on Which to Build Wellness for Lifetime Resilience. Most of us can benefit from the therapeutic lifestyle changes (TLCs) for managing stress and building mental and physical health, brain reserve, and resilience. An effective plan to implement the TLCs can improve the quality of your life and work. You can use the TLCs in every interview to help clients increase well-being and effective coping. As with your own health program, it does little good to change all the TLCs at once. The client can select those that are most immediately meaningful.

Practice and Feedback: Individual, Group, and Microsupervision

Additional resources can be found by going to CengageBrain.com and logging into the MindTap course created by your professor. There you will find a variety of study tools and useful resources that include quizzes, videos, interactive counseling and psychotherapy exercises, case studies, the Portfolio of Competencies, and more.

Intentional counseling and psychotherapy are achieved through practice and experience. They will be enhanced by your own self-awareness, emotional competence, and ability to observe yourself, thus learning and growing in skills.

The competency practice exercises that follow are designed to provide you with learning opportunities in three areas:

1. *Individual practice.* A short series of exercises gives you an opportunity to practice the concepts.

2. *Group Practice and Microsupervision.* Practice alone can be helpful, but working with others in role-playing sessions or discussions is where the most useful learning occurs. Here you can obtain precise feedback on your counseling style. And if video or audio recordings are used with these practice sessions, you'll find that seeing yourself as others see you is a powerful experience.

3. *Self-assessment.* You are the person who will use the skills. We'd like you to look at yourself as a counselor or therapist through the feedback for these exercises.

Individual Practice

Exercise 2.1 Review an Ethical Code
In Appendix II, select the organization that is most relevant to your interests, visit the suggested website, and examine its ethical code in detail. Also select the code of one other helping profession, and look for similarities and differences between the two codes. Website addresses are correct at the time of printing but can change. For a key word web search, use the name of the professional association and the words *ethics* or *ethical code*.

Exercise 2.2 An Exploration in Social Justice
Anyone who is reading this text has already made a commitment to explore social justice in relation to his or her own life. Simply being or thinking of being a counselor and entering the helping fields represents a social justice orientation. What do you see as the role of social justice in the helping profession? How do these issues relate to your own professional and personal life? Are you willing to include social justice issues both in your interviews and in activities in your community?

Exercise 2.3 Self-Awareness Practice and Taking the RESPECTFUL Model to the Interview

Work through the RESPECTFUL model to increase your own understanding of yourself as a multicultural being. Then engage in an interview with a volunteer client exploring the model. What does each of you learn?

❏ Identify yourself on each component of the model. Then write answers to the questions provided there. In addition, list where you stand within each area of privilege. What are your areas of privilege? Less privilege?

❏ Have you experienced cultural trauma and microaggressions, and how has this affected your sense of trust in others, both individuals and groups?

❏ Perhaps most challenging, where might you have biases of favoritism toward others or toward your own group? How does this affect those who are different from you in interviewing practice?

❏ End this self-examination with stories of strength and resilience that come from your life experience. What are you proud of? Build on what you can do, rather than what you can't do.

Exercise 2.4 Personal and Volunteer Client Therapeutic Lifestyle Assessment

Conduct a review of the TLC areas, and make a list of your strengths and areas where you might gain benefit from working on an area for growth. Do not select more than two areas. We suggest that you give particular attention to physical exercise, but only if you are really interested.

Go through the same exercise with a volunteer client.

Group Practice and Microsupervision

Exercise 2.5 Informed Consent

Box 2.1 presents a sample informed consent form, or practice contract. In a small group, develop your own informed consent form that is appropriate for your practice sessions, school, or agency.

Portfolio of Competencies and Personal Reflection

Determining your own style and theory can be best accomplished on a base of competence. Each chapter closes with a self-assessment of competencies and a personal reflection exercise asking your thoughts and feelings about what has been discussed. By the time you finish this book, you will have a substantial record of your competencies and a good written record as you move toward determining your own style and theory.

Use the following as a checklist to evaluate your present level of mastery. Check those dimensions that you currently feel able to do. Those that remain unchecked can serve as future goals. Do not expect to attain intentional competence on every dimension as you work through this book. You will find, however, that you will improve your competencies with repetition and practice.

Assessing Your Level of Competence: Awareness, Knowledge, Skills, and Action

Awareness and Knowledge. Can you define and discuss the following concepts?

❑ Key aspects of ethics as they relate to counseling and psychotherapy: competence, informed consent, confidentiality, power, and social justice

❑ The four dimensions of multicultural competence: becoming aware of your own assumptions, values, and biases; understanding the worldview of the culturally different client; developing appropriate intervention strategies and techniques; and acting guided by all of these

❑ Positive psychology and wellness

❑ Contextual factors of the wellness model

❑ At least six TLCs

Basic Competence. Here you are asked to perform the basic skills in a more practical context, such as an evaluation or an actual counseling session. This initial level of competence can be built on and improved throughout your use of this text.

❑ Write an informed consent form.

❑ Define yourself as a multicultural being.

❑ Evaluate your own wellness using the TLCs.

❑ Take another person through a wellness assessment.

Intentional competence and **psychoeducational teaching competence** will be reviewed in later chapters.

Personal Reflection on Ethics, Multicultural Competence, Positive Psychology, and Therapeutic Lifestyle Changes

Reflecting on yourself as a future counselor or psychotherapist as part of your Portfolio of Competencies can be a helpful way to review what you have learned, evaluate your understanding, and think ahead to the future. Here are some questions that you may wish to consider.

What stood out for you personally in the section on ethics? What one thing did you consider most memorable for your practice? Some people consider ideas of social justice and action in the community a controversial topic. What are your thoughts?

How comfortable are you with ideas of diversity and working with people different from you? Can you recognize yourself as a multicultural person with many dimensions of diversity? What are your thoughts about children's rights, and how would this concept influence your work?

Wellness and positive psychology have been stressed as a useful part of the counseling and psychotherapy interview. At the same time, relatively little attention has been given so far to the very real problems that clients bring to us. While many difficult issues will be covered throughout this text, what are your personal thoughts at this moment on wellness and positive psychology? How comfortable are you with this approach. What do you think about the therapeutic lifestyle changes? Which TLCs do you need to work on?

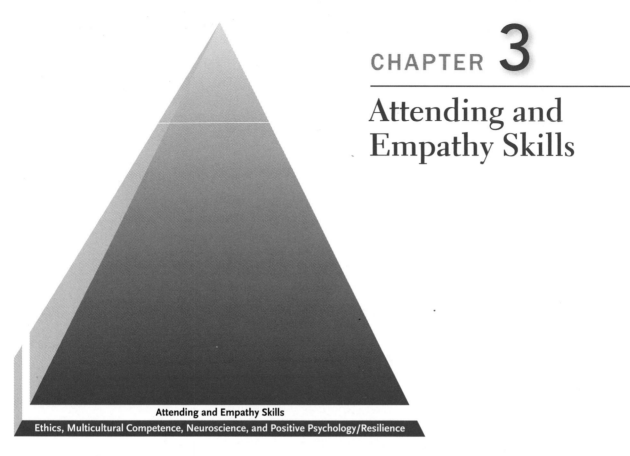

CHAPTER 3

Attending and Empathy Skills

Attending and Empathy Skills

Ethics, Multicultural Competence, Neuroscience, and Positive Psychology/Resilience

When someone really hears you without passing judgment on you, without taking responsibility for you, without trying to mold you, it feels good. When I have been listened to, when I have been heard, I am able to re-perceive my world in a new way and go on. It is astonishing how elements that seem insoluble become soluble when someone listens. How confusions that seem irremediable become relatively clear flowing streams when one is heard.

—Carl Rogers

Chapter Goals and Competency Objectives

Awareness and Knowledge

▲ Develop a solid understanding of how attending behavior, attention, and selective attention form the basis for all counseling and therapy.

▲ Understand how basics of neuroscience explain and expand the importance of attention and empathy.

▲ Learn how teaching microskills of listening is useful therapeutic strategy.

Skills and Action

▲ Increase your skill in listening to clients, and communicate that interest.

▲ Establish an empathic relationship with your clients.

▲ Adapt your attending patterns to the needs of varying individual and cultural styles of listening and talking.

▲ Develop recovery skills that you can use when you are lost or confused in the session. Even the most advanced professional doesn't always know what is happening. When you don't know what to do, attend!

Introduction: Attending Behavior: The Foundational Skill of Listening

This chapter on Attending Behavior and the next on Observation Skills are complementary and may be read together. Attending focuses on the counselor's verbal and nonverbal behavior, whereas observation skills focus on specifics of clients' nonverbal and verbal behaviors.

Attending behavior, essential to an empathic relationship, is defined as supporting your client with individually and culturally appropriate verbal following, visuals, vocal quality, and body language/facial expression. Listening is the core skill of attending behavior and is central to developing a relationship and making real contact with our clients.

What Are the Specific Behaviors of Listening?

Listening is more than hearing or seeing. You can have perfect vision and hearing but be an ineffective listener. How can we identify effective listening more precisely? The following exercise may help you to define listening in terms of clearly observable behaviors.

One of the best ways to identify and define listening skills is to experience the opposite—poor listening. Please stop for just a moment and think back to what was going on when someone failed to listen to you or you felt that someone ignored you or distorted what you said. How did you feel inside? What was the other person doing that showed he or she was not listening? The situations you remember illustrate the importance of being heard and the frustration you feel when someone does not listen to you.

A more active and powerful way to define and clarify listening is to find a partner to role-play a session in which one of you plays the part of a poor listener. The poor listener should feel free to exaggerate in order to identify concrete behaviors of the ineffective counselor. The "client" ideally should continue to talk, even if the counselor appears not to listen. If no partner is available, think again of a specific time when you felt that you were not heard. First ask the "client" how he or she felt "inside" or emotionally when the counselor did not listen. Then think together about what were the specific and observable *behaviors* that indicated lack of listening? Later, compare your thoughts with the ideas presented in this chapter.

An exaggerated role-play is often humorous. However, on reflection, your strongest memory of poor listening may be feelings of disappointment and even anger. This exercise demonstrates clearly that attending and listening behaviors make a significant difference to the session. These are the ways in which you communicate empathy and understanding to the client. They are the behavioral roots of the working alliance and a good counseling relationship.

It is the *observable behaviors* that affect the client immediately. Examples of poor listening and other ineffective interviewing behaviors are numerous—and instructive. If you are to be effective and competent, do the opposite of the ineffective counselor: Attend and listen!

Neuroscience brain imaging has proven the importance of attending in another way. When a person attends to a stimulus, such as the client's story, many areas of the brain of both counselor and client become involved (Posner, 2004). Specific areas of the brain show activity. In effect, attending and listening "light up" the brain in effective interviewing, counseling, and psychotherapy. Without attention, nothing will happen.

Now let us turn to further discussion of how skills and competence can "light up" the session.

Attending Behavior: The Skills of Listening

Attending behavior will have predictable results in client conversation. When you use each of the microskills, you can anticipate how the client is likely to respond. These predictions are never 100% perfect, but research has shown that we can generally expect specific results from various types of helping interventions (Daniels, 2010). If your first attempt at listening is not accepted by the client, you can intentionally flex and change the focus of your attention or try another approach to show that you are hearing the client.

Attending Behavior	Anticipated Client Response
Support your client with individually and culturally appropriate visuals, vocal quality, verbal tracking, and body language, including facial expression.	Clients will talk more freely and respond openly, particularly about topics to which attention is given. Depending on the individual client and culture, anticipate fewer breaks in eye contact, a smoother vocal tone, a more complete story (with fewer topic jumps), and a more comfortable body language.

We can't learn all the possible qualities and skills of effective listening immediately. It is best to learn the counseling and therapy behaviors step by step. Attending and observation are the places to start.

Awareness, Knowledge, and Skills of Attending Behavior and Empathy Skills

Attention is the connective force of conversations and of empathic understanding. We are deeply touched when it is present and usually know when someone is not attending to us. The way one attends deeply affects what is talked about in the session. Also important is to observe clients' reactions. Learning what to do and what not to do will help determine what might be better and more effective in helping that client.

Attending behavior is the first and most critical skill of listening. It is a necessary part of all interviewing, counseling, and psychotherapy. Sometimes listening carefully is enough to produce change.

To communicate that you are indeed listening or attending to the client, you need the following "3 V's + B":*

1. **Visual/eye contact.** Look at people when you speak to them.

2. **Vocal qualities.** Communicate warmth and interest with your voice. Think of how many ways you can say, "I am really interested in what you have to say," just by altering your vocal tone and speech rate. Try that now, and note the importance of changes in *behavior*.

3. **Verbal tracking.** Track the client's story. Don't change the subject; stay with the client's topic.

---

*We thank Norma Gluckstern Packard for the 3 V's + B acronym.

4. Body language/facial expression. Be yourself—authenticity is essential to building trust. To show interest, face clients squarely, lean slightly forward with an expressive face, and use encouraging gestures. Especially critical, smile to show warmth and interest in the client.

The 3 V's + B reduce counselor talk time and provide your clients with an opportunity to tell their stories with as much detail as needed. As you listen, you will be able to observe your clients' verbal and nonverbal behavior. Note their patterns of eye contact, their changing vocal tone, their body language, and topics to which your clients attend and those that they avoid. Also note the individual and cultural differences in attending.

These attending behavior concepts were first introduced to the helping field by Ivey, Normington, Miller, Morrill, and Haase (1968). Cultural variations in microskills usage were first identified as central to the model by Allen, when he worked with native Inuits in the central Canadian Arctic. He found that sitting side by side with them was more appropriate than direct eye contact (body language, facial expression, and visuals vary among cultures) and that developing a solid relationship was as important as staying on the verbal topic. Nonetheless, smiling, listening, and a respectful and understanding vocal tone are things that "fit" virtually all cultures and individuals. Through this, Allen became much closer to the Inuits he was teaching. In short, *attending behavior and listening are essential for human communication, but we need to be prepared for and expect individual and multicultural differences.*

A common tendency of the beginning counselor or psychotherapist is to try to solve the client's difficulties in the first 5 minutes. Think about it: Clients most likely developed their concerns over a period of time. It is critical that you slow down, relax, and attend to client stories and look for themes in their narratives. Use the 3 V's + B to more fully understand the client's concerns and build rapport.

Visual: Patterns of Eye Contact

Not only do you want to look at clients but you'll also want to observe breaks in eye contact, both by yourself and by the client. Clients often tend to look away when thinking carefully or discussing topics that particularly distress them. You may find yourself avoiding eye contact while discussing certain topics. There are counselors who say their clients talk about "nothing but sex" and others who say their clients never bring it up. Through eye-contact breaks or visual fixation, vocal tone, and body shifts, counselors indicate to their clients whether the current topic is comfortable for them.

Cultural differences in eye contact abound. Direct eye contact is considered a sign of interest in European North American middle-class culture. However, even here people often maintain more eye contact while listening and less while talking. Furthermore, if a client from any culture is uncomfortable talking about a topic, it likely better to avoid too much direct eye contact.

Research indicates that some *traditional* African Americans in the United States may have reverse patterns; that is, they may look more when talking and slightly less when listening. Among some traditional Native American and Latin groups, eye contact by the young is a sign of disrespect. Imagine the problems this may cause the teacher or counselor who says to a youth, "Look at me when I am talking to you!" when this directly contradicts the individual's basic cultural values. Some cultural groups (for instance, certain traditional Native American, Inuit, or Aboriginal Australian groups) generally avoid eye contact, especially when talking about serious subjects. This itself is a sign of respect.

Persons with disabilities represent a cultural group that receives insufficient attention. Box 3.1 provides an overview of some key issues.

BOX 3.1   Attending Behavior and People with Disabilities

Attending behaviors (visuals, vocals, verbals, and body language) all may require modification if you are working with people who are disabled. It is your role to learn their unique ways of thinking and being, for these clients will vary extensively in the way they deal with their issues. Focus on the person, not on the disability. For example, think of a person with hearing loss rather than "a hearing impaired client," a person with AIDS rather than "an AIDS victim," a person with a physical disability rather than "a physically handicapped individual." So-called handicaps are often societal and environmental rather than personal.

People Who Have Limited Vision or Blindness
Eye contact is so central to the sighted that initially you may find it very demanding to work with clients who are blind or partially sighted. Some may not face you directly when they speak. Expect clients with limited vision to be more aware of your vocal tone. People who are blind from birth may have unique patterns of body language. At times, it may be helpful to teach them the skills of attending nonverbally even if they cannot see their listener. This may help them communicate more easily with the sighted.

People Who Have Hearing Loss or Deafness
An important beginning is to realize that some people who are deaf do not consider themselves impaired in any way. Many of this group were born deaf and have their own language (signing) and their own culture, a culture that often excludes those who hear. You are unlikely to work with this type of client unless you are skilled in sign language and are trusted among the deaf community.

You may counsel a deaf person through an interpreter. This experience will not prove to be positive without some training in the use of an interpreter in addition to a basic understanding of deaf culture. Too many counselors speak to the interpreter instead of to the client and use phrases such as "Tell him. . . ." This certainly will cut off the client. Also, eye contact is vital, whether in direct counseling with a deaf client or while using an interpreter.

Those who have moderate to severe hearing loss will benefit if you extensively paraphrase their words to ensure that you have heard them correctly. Speak in a natural way, but not fast. Speaking more loudly is often ineffective, as ear mechanisms often do not equalize for loud sounds as they once did. In turn, teaching those with hearing loss to paraphrase what others say to them can be helpful to them in communicating with others.

People with Physical Disabilities
First, each person is unique. We suggest that you consider yourself one of the many who are temporarily able. Age and life experience will bring most of you some physical challenges. For older individuals, the issues discussed here may become the norm rather than the exception. Approach all clients with humility and respect.

Consider the differences among the following: a person who uses a wheelchair, an individual with cerebral palsy, one who has Parkinson's disease, one who has lost a limb, a client who is physically disfigured by a serious burn. They all may have the common problem of lack of societal understanding and support, but you must work with each individual from her or his own perspective. These clients' body language and speaking style will vary. What is important is to attend to each one as a complete person.

Vocal Qualities: Tone and Speech Rate

Your voice is an instrument that communicates much of the feeling you have about yourself or about the client and what the client is talking about. A comfortable "prosodic" tone tends to make clients feel more relaxed with you. Changes in pitch and volume, speech breaks and hesitations, and speech rate can convey your emotional reactions to the client. Clearing the throat, by you or your client, may indicate that words are not coming easily. If clients are stressed, you'll observe that in their vocal tone and body movements. And if the topic is uncomfortable for you or you pick up on a client's stress, your vocal tone or speech rate may change as well. Aucouturier and collaborators (2016) demonstrated that digital changes made to verbal communications led speakers to change their emotions in the direction shown in the digitally modified speech.

Keep in mind that different people are likely to respond to your voice differently. Think of the radio and television voices that you like and dislike. This strategy can be useful for many types of clients, ranging from the depressed to those who have impulse or have anger control issues. It can be useful for social skills training for the shy or the overly aggressive. It also becomes a way that all of us can improve emotional regulation.

Verbal underlining is another useful concept. As you consider the way you tell a story, you may find yourself giving louder volume and increased vocal emphasis to certain words and short phrases. Clients do the same. The key words a person underlines via volume and emphasis are often concepts of particular importance. At the same time, expect some important things to be said softly. When talking about critical issues, especially those that are difficult to talk about, expect a lower speech volume. In these cases, it is wise for you to match your vocal tone to the client's.

Accent is a particularly good example of how people will react differently to the same voice. What are your reactions to the following accents: Australian, BBC English, Canadian, French, Pakistani, New England (U.S.), Southern (U.S.)? Obviously we need to avoid stereotyping people because their accents are different from ours.

Exercise Tone of voice

Try the following exercise with a group of three or more people.

Ask the members of the group to close their eyes while you speak to them. Talk in your normal tone of voice on any subject of interest. As you talk to the group, ask them to notice your vocal qualities. How do they react to your tone, your volume, your speech rate, and perhaps even your regional or ethnic accent? Continue talking for 2 or 3 minutes. Then ask the group to give you feedback on your voice. Summarize what you learn here. If you don't have a group easily available, spend some time noting the vocal tone/style of various people around you. What do you find most engaging? Do some types of speech cause you to move away from the speaker?

This exercise often reveals a point that is central to the entire concept of attending. People differ in their reactions to the same stimulus. Some people find one voice interesting whereas others find that same voice boring; still others may consider it warm and caring. As vocal tone is so important in communicating emotion, this exercise and others like it reveal again and again that people differ, and that what is successful with one person or client may not work with another.

---

* This exercise was developed by Robert Marx, School of Management, University of Massachusetts.

Body Language: Attentive and Authentic

The anthropologist Edward Hall once examined film clips of Southwestern Native Americans and European North Americans and found more than 20 different variations in the way they walked. Just as cultural differences in eye contact exist, body language patterns also differ.

A comfortable conversational distance for many North Americans is slightly more than arm's length, and the English prefer even greater distances. Many Latinas/os prefer half that distance, and some people from the Middle East may talk practically eyeball to eyeball. As a result, the slightly forward lean we recommend for attending is not appropriate all the time.

What determines a comfortable interpersonal distance is influenced by multiple factors. Hargie, Dickson, and Tourish (2004, p. 45) point out the following:

Gender: Women tend to feel more comfortable with closer distances than men.

Personality: Introverts need more distance than extraverts.

Age: Children and the young tend to adopt closer distances.

Topic of conversation: Difficult topics such as sexual worries or personal misbehavior may lead a person to more distance.

Personal relationships: Harmonious friends or couples tend to be closer. When disagreements occur, observe how harmony disappears. (This is also a clue when you find a client suddenly crossing the arms, looking away, or fidgeting.)

Ability: Each person is unique. We cannot place people with physical disability in any one group. Consider the differences among the following: a person who uses a wheelchair, an individual with cerebral palsy, one who has Parkinson's disease, one who has lost a limb, or a client who is physically disfigured by a serious burn. Their body language and speaking style will vary. Attend to each client as a unique and complete person. Ensure that your working space makes necessary physical accommodations.

A person may move forward when interested and away when bored or frightened. As you talk, notice people's movements in relation to you. How do you affect them? Note your own behavior patterns in the session. When do you markedly change body posture? A natural, authentic, relaxed body style is likely to be most effective, but be prepared to adapt and be flexible according to the individual client.

Your authentic personhood is a vital presence in the helping relationship. Whether you use visuals, vocal qualities, verbal tracking, or attentive body language, be a real person in a real relationship. Practice the skills, be aware, and be respectful of individual and cultural differences. Box 3.2 demonstrates the impact of our attending behavior on people from different cultures, and Box 3.3 presents relevant research evidence regarding the use of counseling skills.

BOX 3.2   National and International Perspectives on Counseling Skills

Use with Care—Culturally Incorrect Attending Can Be Rude
Weijun Zhang, Management Consultant, Shanghai, China

The visiting counselor from North America got his first exposure to cross-cultural counseling differences at one of the counseling centers in Shanghai. His client was a female college student. I was invited to serve as an interpreter. As the session went on I noticed that the client seemed increasingly uncomfortable. What had happened? Since I was translating, I took the liberty of modifying what was said to fit each other's culture, and I had confidence in my ability to do so. I could not figure out what was wrong until the session was over and I reviewed the videotape with the counselor and some of my colleagues. The counselor had noticed the same problem and wanted to understand what was going on. What we found amazed us all.

First, the counselor's way of looking at the client—his eye contact—was improper. When two Chinese talk to one another, we use much less eye contact, especially when it is with a person of the opposite sex. The counselor's gaze at the Chinese woman could have been considered rude or seductive in Chinese culture.

Although his nods were acceptable, they were too frequent by Chinese standards. The student client,

probably believing one good nod deserved another, nodded in harmony with the counselor. That unusual head bobbing must have contributed to the student's discomfort.

The counselor would mutter "uh-huh" when there was a pause in the woman's speech. While "uh-huh" is a good minimal encouragement in North America, it happens to convey a kind of arrogance in China. A self-respecting Chinese would say *er* (oh), or *shi* (yes) to show he or she is listening. How could the woman feel comfortable when she thought she was being slighted?

He shook her hand and touched her shoulder. I told our respected visiting counselor afterward, "If you don't care about the details, simply remember this rule of thumb: in China, a man is not supposed to touch any part of a woman's body unless she seems to be above 65 years old and displays difficulty in moving around."

"Though I have worked in the field for more than 20 years, I am still a lay person here in a different culture," the counselor commented as we finished our discussion.

BOX 3.3	**Research and Related Evidence That You Can Use**

Attending Behavior

We begin with two classic studies:

A review of the attending literature concluded that smiling, orienting the body to face the client, leaning forward, using appropriate gestures, and establishing a medium distance of about 55 inches between helper and client are useful nonverbal behaviors (Hill, 2014; Hill & O'Brien, 1999). While the precision of this study is to be admired, it shows little sensitivity to cultural differences. You will still find this naïve approach in the media and even in professional textbooks. Cultural differences do exist for culturally appropriate attending behavior.

A second classic study is more relevant. Just because you think you are listening and being empathic does not mean that the client sees you that way. This study found that White counselors' perception of their expressed empathy and listening was not in accord with the perception of African Americans, who saw them as less effective (Steward, Neil, Jo, Hill & Baden, 1998). The way a counselor can "be with" a client tends to vary among cultures. Nwachuku and Ivey (1992) tested a program of culture-specific training and found that variations to meet cultural differences were essential.

Implications for your practice: Be mindful of your attending behavior and maintain awareness of multicultural difference. The way you are with a client may be received differently than you think.

Communication studies have given considerable attention to listening, with comparable results to what is discussed in this chapter (Ivey & Daniels, 2016). Bodie, Vickery, Cannava, and Jones (2015) examined attending and listening with communication studies students and found results similar to those of the original microskills counseling research (Ivey et al., 1968). Both studies (and many others) have found that systematic training in listening skills makes a significant difference in verbal and nonverbal behavior. The Bodie group study had national impact as it was the focus of a major article on listening in the *Wall Street Journal*—one example of how attending and listening are now reaching far beyond our early work. However, this work and the *Wall Street Journal* article still failed to recognize multicultural difference.

Researchers have documented the importance of communication skills training for physicians. Training improves physicians' communication, self-efficacy, confidence, and satisfaction with the training programs (Reiss, 2015). Furthermore, communication skills training has a positive effect on patient outcomes, such as satisfaction and perception that the physician understood their disease. There was moderate evidence that empathic listening improved health outcomes (Kelley, Kraft-Todd, Schapira, Kossowsky, & Riess, 2014; Riess, 2015).

Implications for your practice: Research shows that attending behavior "works," but remains sensitive to individual and cultural differences.

Verbals: Following the Client or Changing the Topic

Verbal tracking is staying with your client's topic to encourage full elaboration of the narrative. Just as people make sudden shifts in nonverbal communication, they change topics when they aren't comfortable. In middle-class U.S. communication, direct tracking is appropriate, but in some Asian cultures such direct verbal follow-up may be considered rude and intrusive.

Verbal tracking is especially helpful to both the beginning interviewer and the experienced counselor who is lost or puzzled about what to say next in response to a client. *Relax*; you don't need to introduce a new topic. Ask a question or make a brief comment regarding whatever the client has said in the immediate or near past. Build on the client's topics, and you will come to know the client very well over time.

The Central Role of Selective Attention. The normal human brain is wired to attend to stimuli and focus on what may be essential to accomplish the tasks at hand, while other potentially useful information falls into the background. Selective attention is central to interviewing, counseling, and psychotherapy. The thalamus is seen as the "switching station" that sends and exchanges specific messages with various brain regions, the brain stem, and spinal cord, enhancing body response to stimuli (see Appendix IV).

Clients tend to talk about what counselors are willing to hear. In any session, your client will present multiple possibilities for discussion. Even though the topic is career choice, a sidetrack into family issues and personal relationships may be necessary before returning to the purpose of the counseling session. But some counselors may not be as interested in career work, and most of their career clients end up talking about themselves and their personal history and end up in long-term therapy. How you selectively attend may determine the length of the session and whether or not the client returns.

A famous training film (Shostrum, 1966) shows three eminent counselors (Albert Ellis, Fritz Perls, and Carl Rogers) all counseling the same client, Gloria. Gloria changes the way she talks and responds very differently as she works with each counselor. Research on verbal behavior in the film revealed that Gloria tended to match the language of the three different counselors (Meara, Pepinsky, Shannon, & Murray, 1981; Meara, Shannon, & Pepinsky, 1979). Each expert indicated, by his nonverbal and verbal behavior, what he wanted Gloria to talk about!

Should clients match your language and chosen topic for discussion, or should you, the counselor, learn to match your language and style to that of the client? Most likely, both approaches are relevant, but in the beginning, you want to draw out client stories from their own language perspective, not yours. What do you consider most significant in the session? Are there topics with which you are less comfortable? Some counselors are excellent at helping clients talk about vocational issues but shy away from interpersonal conflict and sexuality. Others may find their clients constantly talking about interpersonal issues, excluding critical practical issues such as getting a job.

Observe the selective attention patterns of both you and your clients. What do your clients focus on? What topics do they seem to avoid? Now ask yourself the same questions. Are you particularly interested in certain thoughts and behaviors while perhaps missing other critical issues?

The Value of Redirecting Attention. There are times when it is inappropriate to attend to the here and now of client statements. For example, a client may talk insistently about the same topic over and over again. In such cases, what seems to work best is paraphrasing and/or summarizing the client's story so far, using his or her own words as much as possible (see Chapter 6). This can be followed by questions as you search for relevant details or a deliberate topic jump to more positive experiences and memories. But remember that clients who have been traumatized (such as by hospitalization, breakup of a long-term relationship, accident, or burglary) may need to tell their story several times.

A depressed client may want to give the most complete description of how and why the world is wrong and continue on with more negatives in their lives. We need to hear that client's story, but we also need to selectively attend and not pay attention only to the negative. Clients grow from strengths. Redirect the conversation to focus on positive assets when you observe a strength, a wellness habit (running, spirituality, music), or a resource outside the individual who might be helpful.

The most skilled counselors and psychotherapists use attending skills to open and close client talk, thus making the most effective use of limited time in the interview.

The Usefulness of Silence. Sometimes the most useful thing you can do as a helper is to support your client silently. As a counselor, particularly as a beginner, you may find it hard to sit and wait for clients to think through what they want to say. Your client may be in tears, and you may want to give immediate support. However, sometimes the best support may be simply being with the person and not saying a word. Consider offering a tissue, as even this small gesture shows you care. In general, it's always good to have a box or two of tissues for clients to take even without asking or being offered. Of course, don't follow the silence too long, search for a natural break, and attend appropriately.

There is much more happening in the brain than just silence. It turns out that the auditory cortex remains active when you are attending or listening to silence. Your brain remains highly sensitive, as revealed by functional magnetic resonance imaging (fMRI) and an increasing array of technologies, including computed tomography (CT), positron emission tomography (PET), electroencephalography (EEG), and diffusion-weighted magnetic resonance imaging (DW-MRI). Brain imaging has become a central area of research, with profound implications for counseling and therapy practice.

For a beginning counselor, silence can be frightening. After all, doesn't counseling mean talking about issues and solving problems verbally? When you feel uncomfortable with silence, look at your client with a supportive facial expression. If the client appears comfortable, draw from her or his body language and join in the silence. If the client seems disquieted by the silence, rely on your attending skills. Ask a question or make a comment about something relevant mentioned earlier in the session.

Talk Time. Finally, remember the obvious: *Clients can't talk while you do.* Review your sessions for talk time. Who talks more, you or your client? With most adult clients, the percentage of client talk time should generally be more than that of the counselor. With less verbal clients or young children, the counselor may need to talk slightly more or tell stories to encourage conversation. A 7-year-old child dealing with parental divorce may not say a word about the divorce initially. But when you read a children's book on feelings about divorce, he or she may start to ask questions and talk more freely. Play therapy may also help children tell their story and feelings about the divorce.

Training as Treatment: Social Skills, Psychoeducation, and Attending Behavior

Social skills training involves psychoeducational methods to teach clients an array of interpersonal skills and behaviors. These skills and behaviors include listening, assertiveness, dating, drug refusal skills, mediation, and job interviewing procedures. Virtually all interpersonal actions can be taught through social skills training.

Training as treatment is a term that summarizes the method and goal of social skills training. The microtraining format of selecting specific skill dimensions for education has become basic to most psychoeducational social skills programs. Teaching listening skills can be most helpful to many clients. Consider the following steps: (1) negotiate a skill area for learning with the client; (2) discuss the specific, concrete behaviors involved in the skill, sometimes presenting them in written form as well; (3) practice the skill with the client in a role-play in the individual or group counseling session; and (4) plan for generalization of the skill to daily life.

Shortly after the first work in identifying counseling and psychotherapy microskills, Allen Ivey was working with a first-year college student who suffered a mild depression and complained about the lack of friends. Allen asked the student what he talked about with those in his dormitory. The student responded by continuing his list of complaints and worries. With further probing, the student acknowledged that he spent most of his time with others talking about himself and his difficulties. It was easy to see that potential friends would avoid him. We all tend to move away from those who talk negatively and stay away from those who talk only about themselves and fail to listen to us.

On the spot, Allen talked to the student about attending behavior and its possible rewards. Emphasizing the importance of gaining trust and respect from others by listening, he presented the 3 V's + B. Allen suggested that the student might profit from actively listening to those around him rather than talking only about himself. The student expressed interest in learning these skills, and a practice session was initiated there in the session. First, negative attending was practiced, and the student was able to see how his lack of listening might contribute to his isolation in the dormitory. Then positive attending was practiced, and the student discovered that he could listen.

Allen and the student discussed specifics of selecting someone with whom to try these skills. When the student returned the following week, he had a big smile and reported that he had found his first friend at the university. Moreover, he discovered an important side effect: "I feel less sad and depressed. First, I don't feel so alone and helpless. The second thing I noticed was that when I am attending to someone else, I am not thinking about myself and then I feel better." When you are attending to someone else, it becomes much more difficult to think negatively about yourself. Instruction in attending behavior is one of the foundations of social skills training.

Many types of clients can benefit from learning and practicing these skills. Gearhart and Bodie (2011) have shown that teaching active listening and empathic skills builds closer relationships in a variety of populations. Early work in treatment settings demonstrated the value of teaching attending behavior to hospitalized patients (Donk, 1972; Ivey, 1973). A study of adult schizophrenics showed that teaching social skills with special attention to attending behavior was successful and that patients maintained the skills over a 2-month period (Hunter, 1984). Allen Ivey found that teaching attending and other microskills to veterans at a VA hospital was sufficient to enable them to return to their families and communities. Van der Molen (1984, 2006) used attending behavior and other microskills in a highly successful psychoeducational program in which he taught people who were shy (also known as the "avoidant personality") to become more socially outgoing. As just one other example, children diagnosed with attention deficit disorder (ADD) who receive skills training were less disruptive (Pfiffner & McBurnett, 1997).

Implications for your practice: Many clients can benefit from training and education in listening skills, particularly those who may be moderately depressed or lacking in social skills.

Empathy: Awareness, Knowledge, and Skills

Carl Rogers (1957, 1961) brought the importance of empathy to our attention. He made it clear that it is vital to listen carefully, enter the world of the client, and communicate that we understand the client's world as the client sees and experiences it. Putting yourself "in another person's shoes" or viewing the world "through someone else's eyes and ears" is another way to describe empathy. The following quotation has been used by Rogers himself to define empathy.

> This is not laying trips on people. . . . You only listen and say back the other person's thing, step by step, just as that person seems to have it at that moment. You never mix into it any of your own things or ideas, never lay on the other person anything that the person did not express. . . . To show that you understand exactly, make a sentence or two which gets exactly at the personal meaning the person wanted to put across. This might be in your own words, usually, but use that person's own words for the touchy main things. (Gendlin & Hendricks, n.d.)

Again, please recall the importance of empathy to the relationship, the "working alliance"; it is central to the 30% of *common factors* that make for successful interviewing, counseling, and psychotherapy (Miller, Duncan, & Hubble, 2005). When you provide an empathic response, you can anticipate how clients are likely to respond. Note below another description of empathy and the predictions that you can make.

Empathy	Anticipated Client Response
Experiencing the client's world and story as if you were that client; understanding his or her key issues and saying them back accurately, without adding your own thoughts, feelings, or meanings. This requires attending and observation skills plus using the important key words of the client, but distilling and shortening the main ideas.	Clients will feel understood and engage in more depth in exploring their issues. Empathy is best assessed by clients' reaction to a statement and their ability to continue discussion in more depth and, eventually, with better self-understanding.

Carl Rogers's thinking resulted in extensive work by Charles Truax (1961), who is recognized as the first person to measure levels of empathic understanding. He developed a 9-point scale for rating empathic understanding (Truax, 1961). Robert Carkhuff (1969), who originally partnered with Truax, developed a 5-point scale. These scales have been widely used in research and have practical applications for the session.

Many others have followed and elaborated on Rogers's influential definition of empathy (see Carkhuff, 2000; Egan, 2010; Ivey, D'Andrea, & Ivey, 2012). A common current practice is to describe three types of empathic understanding. This is the convention that we will use in this book. Chapter interview transcripts will be evaluated on the following scale.

Subtractive empathy: Counselor responses give back to the client less than what the client stated, and perhaps even distort what has been said. In this case, the listening or influencing skills are used inappropriately.

Basic empathy: Counselor responses are roughly interchangeable with those of the client. The counselor is able to say back accurately what the client has said. Skilled intentional competence with the basic listening sequence (see Chapter 8) demonstrates basic empathy. You will find this the most common counselor comment level in interviews. Rogers pointed out that listening in itself is necessary and sufficient to produce client change.

Additive empathy: Counselor responses that add something beyond what the client has said often are additive. This may be adding a link to something the client has said earlier, or it may be a congruent idea or frame of reference that helps the client see a new perspective. Effective use of the influencing skills of the second half of this book is typically additive. Feedback and your own self-disclosure, used thoughtfully, can be additive.

The three anchor points above are often expanded to classify and rate the quality of empathy shown in the session. You can use empathy rating in your practice with microskills. And later in your professional work, it is wise to check whether you have maintained interest in your clients and are fully empathic.

Client: I don't know what to do. I've gone over this problem again and again. My husband just doesn't seem to understand that I don't really care any longer. He just keeps trying in the same boring way—but it doesn't seem worth bothering with him anymore.

Level 1 Empathy: (subtractive) That's not a very good way to talk. I think you ought to consider his feelings, too.

(slightly subtractive) Seems like you've just about given up on him. You don't want to try any more. (interpreting the negative)

Level 2 Empathy: (basic empathy or interchangeable response) You're discouraged and confused. You've worked over the issues with your husband, but he just doesn't seem to understand. At the moment, you feel he's not worth bothering with. You don't really care. (Hearing the client accurately is the place to start all empathic understanding. Level 2 is always central.)

Level 3 Empathy: (slightly additive) You've gone over the problem with him again and again to the point that you don't really care right now. You've tried hard. What does this mean to you? (The question adds the possibility of the client's thinking in new ways, but the client still is in charge of the conversation.)

(additive and perhaps transformational) I sense your hurt and confusion and that right now you really don't care any more. Given what you've told me, your thoughts and feelings make a lot of sense to me. At the same time, you've had a reason for trying so hard. You've talked about some deep feelings of caring for him in the past. How do you put that together right now with what you are feeling? (A summary with a mild self-disclosure. The question helps the client develop her own integration and meanings of the issue at the moment.)

In the first half of this book, we recommend that you aim for interchangeable responses. What is essential for empathic understanding is careful listening and hearing the client accurately. This by itself often helps the client to clarify and resolve many issues. At the same time, be aware that slightly subtractive empathy may be an opening to better understanding. You may see your helping lead as interchangeable, but the client may hear it differently. Use unpredicted and surprising client responses as an opportunity to understand the client more fully. *It's not the errors you make; it is your ability to repair them and move on that counts!*

Many other dimensions of empathic understanding will be explored throughout this book. For the moment, recall the following points as central.

1. Aim to understand clients' experience and worldview as they present their story, thoughts, and emotions to you in a nonjudgmental supportive fashion.
2. Seek to communicate that understanding to the client, but avoid mixing "your own thing" in with what you say.
3. The above is the surest route to reaching that critical Level 3 of interchangeable empathic responding.

Neuroscience and Empathy

Historically, counseling and therapy have advocated and shown the importance of empathy, but empathy has always been a somewhat vague and sometimes controversial concept. Well-established data from neuroscience have changed our thinking. Empathy is identifiable through fMRI and other key technologies. Key to this process are the **mirror neurons**, which fire when humans or animals act *and* when they observe actions by another. Many believe that mirror neurons are one of the most significant discoveries in recent science.

One of the earliest studies measuring the importance of mirror neurons and empathy asked closely attached partners to watch each other (through a one-way mirror) receive a mild shock. It was found that the brain of the shocked partner fired in two areas—one representing physical pain and the other emotional. Most important, the observing partner's emotional pain centers fired simultaneously (Singer et al., 2004). Note that research consistently shows that the mirror neurons of children, adolescents, and adults diagnosed with conduct or antisocial personality disorder *do not activate* (Decety & Jackson, 2004). In fact, there is evidence that many with this diagnosis show pleasure when observing others in pain.

These basic findings have been replicated many times in different ways. For example, Marci, Ham, Moran, and Orr (2007) found that skin conductance of client and patient pairs was high and parallel when they both indicated that they felt a communication of empathic understanding. But this can work two ways in communication. Verbal communication is a joint activity, and an fMRI study found that this "neural coupling" disappears when story comprehension is not effective. When listening skills are not successfully implemented (i.e., subtractive), empathy falls apart.

Questions of interchangeable, subtractive, and additive empathy were explored by Oliveira-Silva and Gonçalves (2011). Forty participants watched actors respond to emotionally laden video stories. They found that participants who demonstrated higher levels of additive empathy had an increased heart rate, while those whose observations were interchangeable or subtractive had no "change of heart."

An important Japanese fMRI study examined the neural correlates of active listening (Kawamichi et al., 2015). Subjects described emotional experiences from their own lives on videotape. Each video clip of life episodes was responded to later by confederates playing the role of listeners, either with or without active listening. The subjects' brains were scanned while they heard positive and negative listening to their statements. fMRI results showed that active listening provides a reward and actually does "light up the brain" (see Figure 3.1). More specifically, the reward system of the *ventral stratium* was activated, as was the *medial PFC (mPFC)* related to cognitive empathy and the *right anterior insula* (affective empathy and emotional appraisal). Research by Eres, Decety, Louis, and Molenberghs (2015) suggests that amount of gray matter may actually be increased by empathic listening.

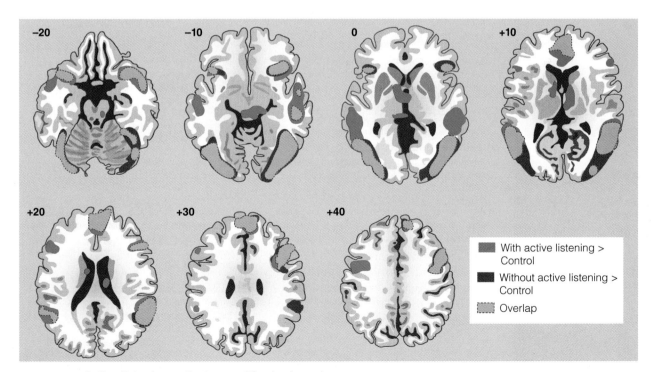

FIGURE 3.1 Active listening activates positive brain regions.

In short, your empathic being and ability to listen and be with clients are a vital part of helping your clients grow and change. Listening and empathy are not just abstract concepts—they are clearly measureable and make a difference in other people's lives (Stephens, Silbert, & Hasson, 2010).

Observe: Attending Behavior and Empathy in Action

The session presented below illustrates the importance of empathic attending skills and of using the skills with awareness of cultural and gender differences.

Azara, a 45-year-old Puerto Rican manager, was not promoted, although she thinks her work is of high quality. She is weary of being passed over and seeing less competent individuals take the position she feels she deserves.

The first session segment is designed as a particularly ineffective interview in order to provide a sharp contrast with the more positive effort that follows. In both cases, the counselor, Allen, has the task of developing a relationship and drawing out the client's story. Note how disruptive visual contact, vocal qualities, failure to maintain verbal tracking, and poor body language can lead to a poor session.

Negative Example

Counselor and Client Conversation	Process Comments
1. *Allen:* Hi, Azara, you wanted to talk about something today.	Allen fails to greet Azara warmly. He just starts and does nothing to develop rapport and a relationship, which is especially important in a cross-cultural session. He remains seated in his chair behind a desk. (The nonverbal situation is already subtractive.)
2. *Azara:* Yes, I do. I've come to you because there's been an incident at my job a couple of days ago. And I'm kind of upset about it.	Hesitantly, Azara sits down. She is upset, and immediately moves ahead with her issues regardless of what Allen does. She is clearly ready to start the session.
3. *Allen:* What is your job?	Allen's voice is aggressive. He ignores Azara's upset feelings and asks a closed question. An appropriate vocal tone communicates warmth and is critical in any relationship.
4. *Azara:* Well right now I'm an assistant manager for a company and I've worked at this company for 15 years.	Azara keeps trying. Allen looks down while she talks. Subtractive nonverbals.
5. *Allen:* So after 15 years you're still an assistant. When I was in business, I didn't take that long to get a promotion. Let me tell you about what I did to get ahead . . . [he goes on at length about himself].	The focus is taken away while Allen, the counselor, talks about himself. With this long response, he has more talk time than the client, Azara. The evaluative "put-down" is an example of how counselors inappropriately use their power and is, of course, totally subtractive.
6. *Azara:* (looking very puzzled) Yeah, I'm still an assistant after 15 years. But what I want to talk to you about is I was passed over for a promotion.	Is there an issue of discrimination here? By ignoring cultural issues, Allen will eventually lose this relationship. (Allen has NO idea what is going on. It could be cultural, it could not. This is definitely an important factor to consider, but he is not establishing an empathetic relationship. Regardless of who the client is, Allen would lose the relationship because of his inability to attend and listen!)

Counselor and Client Conversation	Process Comments
7. *Allen:* Could you tell me a little bit more about some of the things you might have been doing wrong?	Still looking out the window, he returns to Azara with an open question, but he continues to ignore the main issue and topic-jumps with an emphasis on the negative.
8. *Azara:* Well, I don't think I did anything wrong. I've gotten very good feedback from . . .	Azara starts defending herself here, but Allen interrupts. Changing topics and interruption are clear signs of the failure of empathic communication.
9. *Allen:* Well, they don't usually pass people up for promotions unless they're not performing up to standards.	The counselor supplies his interpretation, a subtractive negative evaluation without any data. He is not drawing out her story or really seeking to define her concerns.

Allen does not seem to listen to Azara. Furthermore, he confronts Azara inappropriately, and it is very unlikely that she will return for another session. Her European American counselor just doesn't "get it." This may seem extreme, but to many people who receive a performance review evaluating their work, there are parallels to the real world of business and being considered a minority. Let us hope that there are no counselors like Allen in the preceding example.

But let's give Allen another chance. What differences do you note in this second session?

Positive Example

1. *Allen:* Hi, Azara. Nice to see you. Please come in and sit where it looks comfortable.	Allen stands up, smiles, faces the client directly, and shakes hands. First impressions are always important. Allen provided positive, facilitative nonverbals.
2. *Azara:* Thank you, nice to see you too.	She sits down and smiles in return, but appears tense.
3. *Allen:* Thanks for coming in.	The counselor likes to honor the client's willingness to come to the session. It is a small attempt to equalize the power relationship that exists in counseling.
4. *Azara:* Thanks. I'm hopeful that you can help me.	Azara relaxes a little. The relationship has begun and, hopefully, will be enhance throughout the session. When we listen to a client, the brain "lights up."
5. *Allen:* Azara, I looked at your file before you came in and I see that you'd like to talk about a problem on the job. Is that right?	Looking at forms in the session is very likely to be subtractive. If you must look at files, share what you are looking at with your client. Read ahead as appropriate to your setting.
6. *Azara:* Yes, that's right.	Her mouth is a little tense and she sits back.

Even in this brief period of time, Allen has conveyed to Azara a genuine warmth and readiness to hear her story. The session continues with Allen discussing confidentiality and what to expect in the interview.

Next, Allen changes the subject to the here-and-now relationship,

7. *Allen:* Azara, Before we start, do you have any questions to ask me?	Allen closes the folder as he finishes necessary structuring of the session. He has found that with most clients it helps the relationship if he opens himself up and allows the clients to take control of the session for a moment. (Potentially additive statement, but we need to see what happens to determine if that is so.)

(continued)

Counselor and Client Conversation	Process Comments
8. *Azara:* It's good to know that our discussion goes no further. I do think that one of the reasons that I am not getting the promotions that I deserve is that I'm Latina. What are your thoughts about that?	She appears relieved and decides to be fully open. She is very direct, almost confrontational, and could put many of us "on the spot."
9. *Allen:* Discrimination is a very challenging area. I've worked with several People of Color who have had the same thing happen to them. This, plus some teaching in San Juan, Puerto Rico, gives me some understanding of what you are facing and your background.	Allen self-discloses here. What do you think of this? Would you discuss your own background?
10. *Azara:* That helps. I didn't know you had worked in my hometown. Do you speak Spanish?	Where possible, seek to learn the language of your clients—at a minimum, some key words.
11. *Allen:* Sad to say, I don't except for a very few words. But I've counseled several Latina and Latino clients, and things seemed to work OK. If you find me missing something or failing to understand, please let me know.	Be open about your experience, but do not overstate your competence.
12. *Azara:* Yes, I expected that, but I appreciate your sharing where you are. Let's move ahead and see how it goes.	She shows slight nonverbal signs of disappointment, but accepts "what is." The general warmth and support plus appropriate smiling makes acceptance more possible.

Allen also spends time discussing important cultural concerns with Azara, ensuring that she feels comfortable with him and that he understands important aspects of her culture. The session picks back up with Azara discussing the problem she is facing at work. Notice the difference in information gathered between the positive and negative examples.

13. *Allen:* So there is a concern about unfairness on the job? Would you like to tell me more?	Allen returns now to the job issue.
14. *Azara:* Okay, well, a few days ago I found out that I was passed over for a promotion at my job. And I've been with this company for 15 years. I was really pretty upset when I first found out, because the person who got the job, first of all is a male, he's only been with the company for 5 years. And you know I think I'm much better qualified than he is for this position. I've gotten really good evaluations from my supervisor, I have a great working relationship with my colleagues. . . . I was completely shocked to find out that I didn't get this promotion. 'Cause I was actually encouraged to apply for this job. And, you know, I didn't get it. This is . . . I'm just really, really angry.	Azara says a lot in this comment and we as counselors sometimes have difficulty in hearing it all. This is where the skills of paraphrasing and summarizing (Chapter 6) can be most important. The task of these skills is to repeat what the client has said, but in a more succinct form.
15. *Allen:* 15 years compared to 5, and you are really, really angry. And what I've heard makes you angry is that you've had a good record, you were even asked to apply for this job, and finally this White male who hasn't been there that long gets the job. Have I heard you correctly?	The counselor's summary of what has been said indicates that he has been engaging in verbal as well as nonverbal attending. "Have I heard you correctly?" is termed a "checkout" in the microskills framework. If you are accurate, you the client will often say "yes" or even "exactly!" This represents a Level 2, interchangeable response.
16. *Azara:* Yes, you heard me. . . . Well, I think it's discrimination. Now the problem I'm having (pause) . . . I think it's discrimination, but now I have to decide what I'm going to do, if I'm going to file a complaint. Will that upset my colleagues, will that get my boss, my supervisor, upset with me? I'm really worried about the consequences. I don't want to lose my job, but I think it's discrimination.	Having been heard, the client moves on.

Counselor and Client Conversation	Process Comments
17. *Allen:* Azara, it's a tough decision to make. If you file for discrimination, you set yourself up for a lot of hassles; if you don't file, then you're stuck with your anger and frustration. Could you tell me a bit more about that dilemma you are feeling?	Here Allen paraphrases the main ideas and reflects Azara's feelings as well. This is followed by an open question about the dilemma. Level 2 with some elements of additive empathy as he encourages her to tell more of her story.
18. *Azara:* Well, it's like I'm stuck, I don't know what to do. On the one hand, I think it's important to file the complaint because I think it will show the company that they really need to think about diversity in the workforce, and I'm kind of tired of being the only Latina working in this company for as long as I have, when you know they need to do something different. So I'm torn between that and being afraid of losing my job.	Azara summarizes key aspects of her conflict. The discrepancies or incongruity between herself and the company could be summarized this way: the responsibility to file a discrimination suit because it appears that the company is consistently being unfair versus the fear of losing her job if she takes this on.
19. *Allen:* So you're angry, afraid, frustrated. A lot of stuff comes together for you all at once.	Allen is sitting upright, forward trunk lean, supportive vocal tone, while he reflects her emotions and her dilemma. Appropriate nonverbals are always central to maintaining an empathic relationship.
20. *Azara:* Yes, that's right. And I don't know what to do about that.	The client provides her own "checkout" and speaks of her puzzlement.
21. *Allen:* One thing I heard you saying that I'd like to understand a little bit more, you had good evaluations, you say you have good relationships, success, a reasonable rate of promotion, at least raises along the line. I'd just like to hear at this point about examples of something specific that's gone right in the past. Something you're proud of. Because when a person talks to me about difficulties, it kind of makes them feel a little embarrassed, and I'd like to understand some of your strengths. I've got a general understanding of your concern, and we will come back to that. Could you tell me a little bit about some of your strengths too?	Now that the issues are clearer, Allen turns to the positive asset search. What are Azara's strengths that we can draw on as we work on these concerns? Note that Allen has avoided using the word "problem" as that is a defeatist, negative view of client issues. You will find that most counseling training books use a problem-centered language. This is a clear example of Level 3, additive empathy.

The session continues from here with Allen and Azara exploring her strengths.

Here we see a much stronger focus on Azara as a person with individual needs and feelings. A relationship has been established, and it is now possible to discuss multicultural issues as appropriate to the moment. Through attending and listening, we see her story and concerns more fully. A positive asset search for strengths has been initiated.

Attending and Empathy in Challenging Situations

Don't be fooled by the apparent simplicity of the attending skills. Some beginning interviewers and counselors may think that these skills are obvious and come naturally. They may be anxious to move to the "hard stuff." The more we work with beginning and experienced counselors, the more we realize how difficult is to master these skills. Cognitive learning through reading and study does not mean one has the skills and is *really* able to listen to clients empathically. Effective listening takes time, commitment, and intentional and deliberate practice.

You also may have wondered how attending behavior can be useful if you plan to work with challenging clients in schools, community mental health centers, or hospitals. The following examples from our personal experience illustrates the depth and breadth of attending.

Mary: Attending is natural to me, and the basic listening sequence has always been central to my work with children, but even with all my experience with children, sometimes I am at a loss as to what to do next. After some analysis, I found that if I moved back to my foundation in attending skills and focused carefully on visuals, vocals, verbal following, and body language, I could regain contact with even the most troubled child. Similarly, in challenging situations with parents, I have at times found myself returning to a focus on attending behavior, later adding the basic listening sequence and other skills. Conscious attending has helped me many times in involved situations. Attending is not a simple set of skills.

I train older students to work as school peer mediators and another group to be peer tutors for younger children. I have found that using the exercise at the beginning of this chapter on poor attending and then contrasting it with good attending works well as an introductory exercise. I then teach attending skills and the basic listening sequence to my student groups.

Allen: One of my most powerful experiences occurred when I first worked at the Veterans Administration with schizophrenic patients who talked in a stream of consciousness "word salad." I found that if I maintained good attending skills and focused on the exact words they were saying, they were soon able to talk in a more normal, linear fashion. I also found that teaching communication skills with video and video feedback to some troubled patients was effective. Sometimes attending was sufficient treatment by itself to move them out of the hospital. Depressed psychiatric inpatients in particular responded well to social skills training. However, I did find that highly distressed patients could learn only one of the four central dimensions at a time, as too many things were confusing for them. Thus, I would start with visual/eye contact and later move on to other attending skills.

Like Mary, when the going gets rough, I find that it helps me to return to basic attending skills and a very serious effort to follow what the client is saying as precisely as possible. In short, when in doubt, attend. It often works!

Practice, practice, practice: also known as "use it or lose it!"

The Samurai Effect, Magic, and the Importance of Practice to Mastery

Practice isn't the thing you do once you're good. It's the thing you do that makes you good.

—Malcolm Gladwell

Japanese masters of the sword learn their skills through a complex set of highly detailed training exercises. The process of masterful sword work is broken down into specific components that are studied carefully, one at a time. Extensive and intensive practice is basic to a samurai. In this process of mastery, the naturally skilled person often suffers and finds handling the sword awkward at times. The skilled individual may even find performance worsening during the practice of single skills. Being aware of what one is doing can interfere with coordination and smoothness in the early stages.

Once the individual skills have been practiced and learned to perfection, the samurai retire to a mountaintop to meditate. They deliberately forget what they have learned. When they return, they find the distinct skills have been naturally integrated into their style or way of being. The samurai then seldom have to think about skills at all; they have become samurai masters.

What is samurai magic, you may ask? Intentional practice!

Once upon a time, it was believed that giftedness was inherited. Thus, many of us have been taught that Mozart and Beethoven had a magical gift. Baseball fans still believe that Ted Williams and Joe DiMaggio "had it in their genes." It is a bit different from that. The "magic" of a solely genetic predisposition to giftedness is now recognized as a scientific error, but that error is still promoted in the popular media. Natural talent is there, but it needs to be developed and nurtured with careful practice. Expertise across all fields depends on persistence, practice, and the search for excellence (Ericsson, Charness, Feltovich, & Hoffman, 2006).

The neuroscience of "giftedness" has been detailed by David Shenk in his book *The Genius in All of Us* (2010). He finds that whether one is a master musician or a superstar athlete, natural talent may be there, but the real test is many hours and often years of detailed practice. We now know that Mozart, with many natural talents, was bathed in music by his demanding father, who was one of the first to focus on a detailed study of techniques and skills. From the age of 3, Mozart received intensive instruction, and his greatness magnified over time. Ted Williams carried his bat to school and practiced until dark.

Intentional practice is the magic! This means that you need to recognize and enhance your natural talents, but greatness only happens with extensive practice. Practice is the breakfast of champions. Skipping practice means mediocre performance.

Here is what Shenk (2010, pp. 53–54) found that relates directly to you and your commitment to excellence in interviewing, counseling, and psychotherapy:

1. *Practice changes your body.* Both the brain and the body change with practice.

2. *Skills are specific.* Each skill must be practiced completely if they are to be integrated in superior performance.

3. *The brain drives the brawn.* Changes in the brain are evident in scans. Areas of the brain relating to finger exercises or arm movements show brain growth in those areas. Expect the same in your brain as you truly master communication skills.

4. *Practice style is crucial.* One can understand attending behavior intellectually, but actually practicing the specific skills of attending is what will make the difference. One pass through is seldom enough.

5. *Short-term intensity cannot replace long-term commitment.* If Ted Williams did not continue to practice, his skills would have gradually been lost. You will want to take what you learn about counseling skills and use it regularly.

6. *Practice provides a continuous feedback loop*, which leads to even more improvement. In addition, feedback from colleagues on your counseling style and skills is especially beneficial.

We are asking you to focus on your natural gifts in communication and then add to them through practice and sharpening of new skills. You may find a temporary and sometimes frustrating decrease in competence, just as can happen with samurai, athletes, and musicians. Some of you may experience some discomfort in practicing the skill of attending. Others may find attending so "easy" that you fail to become fully competent in this most basic of listening skills (many experienced professionals still can't listen effectively to their clients).

Learn the skills of this book, but allow yourself time for integrating these ideas into your own natural authentic being. It does not take magic to make a superstar, but it does require systematic and intentional practice to achieve full competence in interviewing, counseling, and psychotherapy. Make your own magic.

Action: Key Points and Practice of Attending Behavior and Empathy Skills

Central Goals of Listening. Listening is central, but it is more than hearing or seeing. When we use attending behavior, we reduce counselor talk time to provide clients with ample opportunity to examine issues and tell their stories. Selective attention may be used to facilitate more useful client conversation. Attending with individual and cultural sensitivity is always a must. Observation skills will enable you to stay more closely in tune with your clients.

Four Aspects of Attending. Attending behavior consists of four simple but critical dimensions (3 V's + B), but all need to be modified to meet individual and cultural differences.

1. *Visual/eye contact.*
2. *Vocal qualities:* Your vocal tone and speech rate indicate much of how you feel about another person.
3. *Verbal tracking:* Don't change the subject. Keep to the topic initiated by the client. If selectively attend to an aspect of the story or a different topic, realize the purpose of your change.
4. *Body language/facial expression:* Be attentive and genuine. Face clients naturally, lean slightly forward, have an expressive face, and use facilitative, encouraging gestures.

Attending Behavior. Attending is easiest if you focus your attention *on the client rather than on yourself.* Again, your ability to be empathic and observe what is occurring in the client is central. Note what the client is talking about, ask questions, and make comments that relate to your client's topics. For example:

Client:	I'm so confused. I can't decide between a major in chemistry, psychology, or language.
Counselor:	(nonattending) Tell me about your hobbies. What do you like to do? or What are your grades?
Counselor:	(attending) Tell me more, or You feel confused? or Could you tell me a little about how each subject interests you? or Opportunities in chemistry are promising now. Could you explore that field a bit more? or How would you like to go about making your decision?

Listening and Individual and Multicultural Differences. Attending and empathy are vital in all human interactions, be they counseling, medical interviews, business decision meetings, or your behavior with friends and family. It is also important to note that different individuals and cultural groups may have different patterns of listening to you. For example, some may find the direct gaze rude and intrusive, particularly if they are dealing with difficult material.

Attending Behavior Research. Even if you believe you are listening and being empathic, your client may not see it that way. Be mindful of your attending behavior, and maintain awareness of multicultural difference. Systematic training in listening skills makes a significant difference in verbal and nonverbal behavior. Communication skills training improves professionals' communication, self-efficacy, confidence, and satisfaction. Furthermore, communication skills training has a positive effect on patient outcomes, such as satisfaction and perception that the provider understood their issue.

Empathy. Empathy is the ability to enter the world of the client and to communicate that we understand his or her world as the client sees and experiences it. Attending and listening behaviors are the ways in which you communicate empathy and understanding to the client. Empathy can be substractive, basic, or additive.

The Neuroscience of Active Listening and Empathy. Appropriate listening lights up the brain. In the brains of individuals who experience positive listening to their statements, the reward system of the ventral striatum and the medial PFC (mPFC), related to cognitive empathy, and the right anterior insula (affective empathy and emotional appraisal) are activated.

Training as Treatment. Teaching listening skills can be helpful to many clients. Consider the following steps: (1) negotiate a skill area for learning with the client; (2) discuss the specific and concrete behaviors involved in the skill, sometimes presenting them in written form as well; (3) practice the skill with the client in a role-play in the session or group counseling session; and (4) plan for generalization of the skill to daily life.

Practice Is of the Essence. Expertise across all fields depends on persistence, intentional practice, and the search for excellence. Awareness, knowledge, and skills can be effective only when you act.

Practice and Feedback: Individual, Group, and Microsupervision

Additional resources can be found by going to CengageBrain.com and logging into the MindTap course created by your professor. There you will find a variety of study tools and useful resources that include quizzes, videos, interactive counseling and psychotherapy exercises, case studies, the Portfolio of Competencies, and more.

Intentional interviewing, counseling and psychotherapy are achieved through practice and experience. Reading and understanding are at best a beginning. Some find the ideas here relatively easy and think that they can perform the skills. But what makes one competent in basic skills is practice, practice, practice.

Individual Practice

Exercise 3.1 Deliberate Attending and Nonattending
During a conversation with an acquaintance, deliberately attend and listen more carefully than you usually do. Maintain appropriate eye contact with an open, attentive posture. Use a supportive vocal tone, and focus carefully on what the other person is saying. Observe what happens and how conversations can change if you really seek to listen.

You may wish to contrast deliberate attending with nonattending. What happens when your eye contact wanders, your vocal tone shows disinterest, your body becomes more rigid, or you constantly change the topic? What did you learn from this experience?

Follow this up by sitting back at a meeting, at a party, or during a general social conversation in someone's living room. Who attends most effectively? What is the person who gives lesser attention doing? Where does the eye contact of the group flow? Consider other aspects of the 3 V's + B in your observations.

Exercise 3.2 Observing Verbal and Nonverbal Patterns

Observe 10 minutes of a counseling session, a television interview, or any two people talking. You will find these on YouTube easily—search for interview, counseling, or psychotherapy.

Visual/eye contact patterns. Do people maintain eye contact more while talking or while listening? Does the "client" break eye contact more often while discussing certain subjects than others? Can you observe changes in pupil dilation as an expression of interest?

Vocal qualities. Note speech rate and changes in intonation or volume. Give special attention to speech "hitches" or hesitations.

Verbal tracking. Does the counselor stay on the topic, or does he or she topic-jump. What are the patterns of selective attention?

Body language. What do you observe in the body language, gestures, and facial expression of counselor and client?

Group Practice and Microsupervision

Exercise 3.3 Using Attending Skills

The following instructions are suggested for use in video-based practice session in this and later chapters. The details here will not be repeated, but can be referred to as needed. Using the Feedback Form (see Box 3.4) can provide enough structure for a successful practice session without the benefit of recording equipment.

Step 1: Divide into practice groups. Get acquainted with each other informally before you go further.

Step 2: Select a group leader. The leader's task is to ensure that the group follows the specific steps of the practice session. It often proves helpful if the least experienced group member serves as leader first. Group members then tend to be supportive rather than competitive.

Step 3: Assign roles for the first practice session/microsupervision.

❑ *Client.* The first role-play client will be cooperative and present a story, talk freely, and not give the counselor a difficult time.

❑ *Counselor.* The counselor will demonstrate a natural style of attending behavior, practicing the basic skills.

❑ *Observer 1.* The first observer will fill out the feedback form (Box 3.4) detailing the counselor's attending behavior. Observation of these practice sessions could also be called "microsupervision" in that you are helping the counselor understand his or her behavior in the brief session. Later, when you are working as a professional helper, it is vital that you continue to share your work with colleagues through verbal report or

BOX 3.4   Feedback Form: Attending Behavior

_____ DATE
_____

NAME OF COUNSELOR   (NAME OF PERSON COMPLETING FORM)

Instructions: Provide written feedback that is specific and observable, nonjudgmental, and supportive.

1. _Visual/eye contact._ Facilitative? Staring? Avoiding? Sensitive to the individual client? At what points, if any, did the counselor break contact? Facilitatively? Disruptively?

2. _Vocal qualities._ Vocal tone? Speech rate? Volume? Accent? Points at which these changed in response to client actions? Number of major changes or speech hesitations?

3. _Verbal tracking and selective attention._ Was the client able to tell the story? Stay on topic? Number of major topic jumps? Did shifts seem to indicate counselor interest patterns? Did the counselor demonstrate selective attention in pursuing one issue rather than another? Did the client have the majority of the talk time?

4. _Attentive body language._ Leaning? Gestures? Facial expression? At what points, if any, did the counselor shift position or show a marked change in body language? Number of facilitative body language movements? Was the session authentic?

5. _Specific positive aspects of the session._

6. _Empathic communication._ Rate the quality of counselor responses in the session as subtractive, interchangeable, or additive.

7. _Discussion question._ What areas of diversity do the counselor and client represent? How does this affect the session? Keep in mind that all clients have a RESPECTFUL cultural background.

audio/videotape. Supervision is a vital part of effective interviewing, counseling, and psychotherapy.

❏ *Observer 2.* The second observer will time the session, start and stop any equipment, and fill out a second observation sheet as time permits.

Step 4: Plan and select a meaningful topic for the practice session. A suggested topic for the attending practice session is "Why I want to be an interviewer, counselor, or therapist." The client talks about her or his desire to join the helping profession, or at least to consider it as a future career.

Other possible topics include the following:

❏ A job you have liked and a job that you didn't (or don't) enjoy.

❏ A positive experience you have had that led to new learning about yourself.

While the counselor and the client plan, the two observers preview the feedback sheets.

Step 5: Record a 3-minute practice session using attending skills. The counselor practices the skills of attending, the client talks about the current work setting or other selected topic, and the two observers fill out the feedback sheets. Try not go beyond 3 minutes, but find a comfortable place before stopping.

Step 6: Review the practice session and provide feedback to the counselor for 12 minutes. Feedback has been called the "breakfast of champions," so give special attention here. Note the suggestions for feedback in Box 3.5. As a first step, the role-play client gives her or his impressions of the session and completes the Client Feedback Form from Chapter 1. This is followed by counselor self-assessment and comments by the two observers.

| **BOX 3.5** | **Guidelines for Effective Feedback** |

To see ourselves as others see us.
To hear how others hear us.
And to be touched as we touch others . . .
These are the goals of effective feedback.

Feedback is one of the skill units of the basic attending and influencing skills developed in this book; it is discussed in more detail in Chapter 12. But even before we study this skill, it is important to understand some basic guidelines.

The person receiving the feedback is in charge. Let the counselor in the practice sessions determine how much or how little feedback is wanted.

Feedback includes strengths, particularly in the early phases of the program. If the counselor requests negative feedback, add positive dimensions as well. People grow from strength, not from weakness.

Feedback is most helpful when it is concrete and specific. Not "Your attending skills were good" but "You maintained eye contact throughout,

except for breaking it once when the client seemed uncomfortable." Make your feedback factual, specific, and observable.

Corrective feedback should be relatively nonjudgmental. Feedback often turns into evaluation. Stick to the facts and specifics, though the word *relatively* recognizes that judgment inevitably will appear in many different types of feedback. Avoid the words *good* and *bad* and their variations.

Feedback should be lean and precise. It does little good to suggest that a person change 15 things. Select one to three things the counselor actually might be able to change in a short time. You'll have opportunities to make other suggestions later.

Check how your feedback was received. The client response indicates whether you were heard and how useful your feedback was. "How do you react to that?" "Does that sound close?" "What does that feedback mean to you?"

Finally, as you review the audio or video recording of the session, start and stop the recording periodically. Replay key interactions. Only in this way can you fully profit from the recording media. Just sitting and watching television is not enough; use media actively. Written feedback, if carefully presented, is an invaluable part of a program of counseling skill development.

Step 7: Rotate roles. Everyone should have a chance to serve as counselor, client, and observer. Divide your time equally!

Some general reminders. It is not necessary to compress a complete interview into 3 minutes. Behave as if you expected the session to last a longer time, and the timer can break in after 3 minutes. The purpose of the role-play sessions is to observe skills in action. Thus, you should attempt to practice skills, not resolve concerns or issues. Clients have often taken years to develop their interests and concerns, so do not expect to solve one of these problems or obtain the full story in a 3-minute role-play session.

Portfolio of Competencies and Personal Reflection

Assessing Your Level of Competence: Awareness, Knowledge, Skills, and Action

Use the following as a checklist to evaluate your present level of mastery. Check those dimensions that you currently feel able to do. Those that remain unchecked can serve as future goals. Do not expect to attain intentional competence on every dimension as you work through this book. You will find, however, that you will improve your competencies with repetition and practice.

Awareness and Knowledge. Can you define and discuss the following concepts?

❑ Identify and count the 3 V's + B as your observe the session.

❑ Define subtractive, interchangeable, and addictive empathy.

❑ Observe movement harmonics and movement dysynchrony.

❑ Outline key elements of observation of nonverbal communication.

Basic Competence. These are fundamentals required before moving on to the next skill area in this book. Can you:

❑ Demonstrate culturally appropriate visuals/eye contact, vocal qualities, verbal following, and body language in a role-played session.

❑ Increase client talk time while reducing your own.

❑ Stay on a client's topic without introducing any new topics of your own.

❑ Hear a client accurately so that you demonstrate interchangeable empathy.

❑ Mirror nonverbal patterns of the client. The counselor mirrors body position, eye contact patterns, facial expression, and vocal qualities.

Intentional Competence. In the early stages of this book, strive for basic competence and work toward intentional competence later. Experience with the microskills model is cumulative, and you will find yourself mastering intentional competencies with greater ease as you gain more practice. For intentional competence leading to skilled counseling, do you:

❏ Understand and manage your own pattern of selective attention.

❏ Change your attending style to meet client individual and cultural differences.

❏ Note topics that clients particularly attend to and topics that they may avoid.

❏ Use attending skills with more challenging clients, while maintaining an empathic style.

❏ Help clients, through empathic attention and inattention, to move from negative, self-defeating conversation to more positive and useful topics. This also includes helping clients who are avoiding issues to talk about them in more depth.

Psychoeducational teaching competence will be reviewed in later chapters.

Personal Reflection on Attending Behavior and Empathy Skills

This chapter has focused on the importance of empathic listening as the foundation of effective counseling practice. When in doubt as to what to do—listen, listen, listen! Individual and cultural differences are central—visual, vocal, verbal, and body language styles vary. Avoid stereotyping any group.

What single idea stood out for you among all those presented in this chapter, in class, or through informal learning? What stands out for you is likely to be important as a guide toward your next steps.

How might you use ideas in this chapter to begin the process of establishing your own style and theory?

What are your thoughts about using attending behavior in psychoeducational practice?

Please turn to your journal and write your thoughts.

CHAPTER 4

Observation Skills

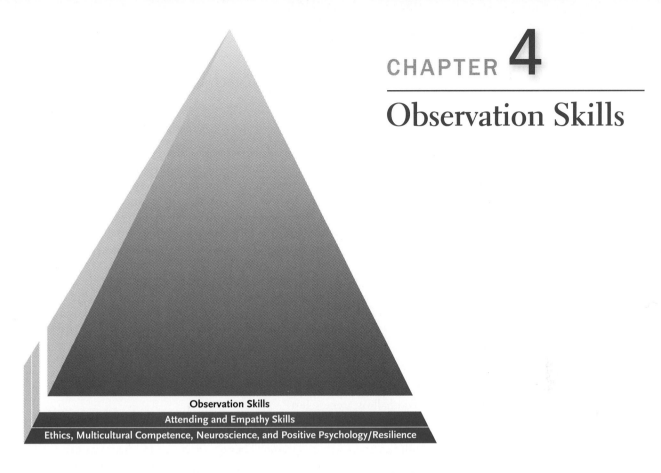

Observation Skills
Attending and Empathy Skills
Ethics, Multicultural Competence, Neuroscience, and Positive Psychology/Resilience

The scientific method may be summed up in the single word "observation."

—E. B. Titchener 1908

Chapter Goals and Competency Objectives

Awareness and Knowledge

▲ Understand nonverbal behavior. How do you and your clients behave nonverbally?

▲ Note verbal behavior. How do you and your clients use language?

▲ Recognize discrepancies and conflict. Much of counseling and psychotherapy is about working through conflict and coping with the inevitable stressful incongruities we all face.

▲ Learn about abstract versus concrete language. Where is the client on the "abstraction ladder"?

▲ Understand varying individual and cultural ways of verbal and nonverbal expression.

Skills and Action

▲ See, hear, and feel the client's world.

▲ Flex intentionally when working with diverse clients, and avoid stereotyping your observations.

▲ Observe your impact on the client: How does what you say change or relate to the client's behavior?

Introduction: Are You a Good Observer?

You can observe a lot by watching.

—Yogi Berra, legendary Yankee player and coach

What can you learn by observing, and why is it important? Through observation you get to know the client and what is conveyed by his or her verbal and nonverbal behavior. Clients' intentions, needs, meanings, and underlying emotions are often conveyed not only with words, but also through nonverbals. In fact, awareness of the close connection between nonverbals and emotions is become clearer. Some authorities say that 85% or more of communication is nonverbal. How something is said can overrule the actual words used by you or your client. A keen observer discovers the many ways clients express their needs, emotions, and motivations.

Observation is the act of watching carefully and intentionally with the purpose of understanding behavior. In spite of what some professionals believe, mastering this skill is not easy. Let's have a little fun as we start. The YouTube examples below show us what Yogi Berra is talking about. They illustrate how much you may be missing as you talk with a client. Sometimes there is more going on than meets the eye. Use the following key words to find the importance of careful observation.

- The Monkey Business Illusion
- The Mentalist: Football Awareness Test
- The Mentalist: Cards Awareness Test

These videos are both fun and teach us a lesson. They are often used in neuroscience demonstrations to show how focused attention can result in missing important things in the background (Chabris & Simon, 2009). Writing in the *Scientific American*, Martinez-Conde and Macknik (2013) note:

> Many experiences in daily life reflect the physical stimuli that send signals to the brain. But the same neural machinery that interprets inputs from our eyes, ears and other sensory organs is also responsible for our dreams, delusions and failings of memory. In other words, the real and the imagined share a physical source in the brain. So take a lesson from Socrates: "All I know is that I know nothing."

In counseling, you may be so focused on a single issue you note in the client that you are missing the underlying meanings. Also, you may miss the cultural issues underlying the client's conversation.

If you use observation skills as defined here, you can anticipate the following.

Observation Skills	Anticipated Client Response
Observe your own and the client's verbal and nonverbal behavior. Anticipate individual and multicultural differences in nonverbal and verbal behavior. Carefully and selectively feed back some here-and-now observations to the client as topics for exploration.	Observations provide specific data validating or invalidating what is happening in the session. Also, they provide guidance for the use of various microskills and strategies. The smoothly flowing session will often demonstrate movement symmetry or complementarity. Movement dyssynchrony provides a clear clue that you are not "in tune" with the client.

Awareness, Knowledge, and Skills: Principles for Observation

> In working with patients (clients), if you miss those nuances—if you misread what they may be trying to communicate, if you misjudge their character, if you don't notice when their emotions, gestures, or tone of voice don't fit what they are saying, if you don't catch the fleeting sadness or anger that lingers on their face for only a few milliseconds as they mention someone or something you might otherwise not know was important—you will lose your patients (clients). Or worse still, you don't.
>
> —Drew Weston

You learn as much about yourself and your counseling skills as you do about your clients during observation. By turning observation inward, you may tune into your own reactions and examine what is happening with you. You have cognitive thoughts, but your body is also reacting emotionally. Out of this process of self-awareness will come growth and change in your skills. Whether the focus is on the client or on you, observation provides a compass to guide you.

An ideal place to start practicing your observation skills is by noting your client's style of attending and how you relate to that client. There are two people in the relationship. What about you? How are you affecting the client verbally and nonverbally? Looking at your way of being can be as important as, or more important than, observing the client. Start by taking a brief inventory of your own nonverbal style. You might begin by thinking back to your natural style of attending (see Chapter 1), but expand those self-observations. Better yet, make a video of you talking with someone about a topic on which you agree. Then make a video in which the two of you have differing opinions. What do you notice, and how might it affect your relationship with others?

Nonverbal Behavior

Nonverbal behavior is often the first clue to what clients are feeling underneath the language they use. Observation skills are basic to the empathic process. As we listen to a client's story, we need to be constantly aware of how he or she reacts to what we say—both verbally and nonverbally. If a client breaks eye contact and/or shifts hands or body, "something" is happening. The client may find the topic uncomfortable, or may not like what you just said, or perhaps just happened to hear a noise outside the window and looked outside. On the other hand, a solid accurate reflection or meaningful interpretation of client behavior can produce the same result. There is also the "recognition response": You will find that many clients look down as they realize that what you just said is true and makes a difference. Your ability to observe will help you anticipate and understand what is happening with your client, but be careful to watch for individual and cultural differences.

"The voice of the therapist, regardless of what he or she says, should be warm, professional—competent, and free from fear" (Grawe, 2007, p. 411). Your vocal tone, perhaps more than anything else you do, conveys the emotional dimensions of your words and body. A "prosodic" vocal tone, both individually and culturally appropriate, is an important key to the relationship. You can learn about nonverbal behavior research in Box 4.1. Periodically observe yourself on video, and carefully note your own nonverbal style. Continue this practice long after you become a practicing professional to avoid falling into bad habits.

| BOX 4.1 | **Research and Related Evidence That You Can Use** |

Observation

In the fields of observation chance favors only the prepared mind.

—Louis Pasteur

Research on nonverbal behavior has a long and distinguished history. Edward Hall's *The Silent Language* (1959) is a classic of anthropological and multicultural research and remains the place to start. Early work in nonverbal communication was completed by Paul Ekman, and his 1999 and 2007 summaries of this work are basic to the field.

A good place to start this chapter and learn about empathy and observing research on emotion is to view Paul Ekman's interview on YouTube. Use the search term "Paul Ekman talks about Empathy with Edwin Rutsch."

Eye contact and forward trunk lean were found to be highly correlated with ratings of empathy (Sharpley & Guidara, 1993; Sharpley & Sagris, 1995). Hill (2014; Hill & O'Brien, 2004) noted fewer head nods when reactions were negative. Using an empathic accuracy paradigm, Hall and Schmid Mast (2007) found that participants shifted attention toward visual nonverbal cues and away from verbal cues when asked to infer feelings; when asked to infer thoughts, they did the reverse.

Miller (2007) studied communication in the workplace and found that "connecting," an important aspect of communicating compassion, included empathy and perspective taking. Haskard et al. (2008) examined physicians' nonverbal communication and found that their tone of voice plays an important role in the relationship with their patient, affecting patient satisfaction and adherence to treatment. Health care providers need to be aware of the power of their tone of voice, which may inadvertently communicate their emotions and affect clients' satisfaction.

Two studies of White and African American doctors reveal the depth of unconscious racial prejudice. These two studies have profound implications for your behavior as a counselor and your potential impact on your clients. Elliott, Alexander, Mescher, Mohan, and Barnato (2016) found that African American patients were more likely than White patients to die in intensive care. They found significantly less positive and encouraging statements for African Americans. Conversely, Irina Stepanikova (2012) found that African American physicians gave White patients fewer positive and supportive statements.

Another study found that when African Americans had a strong sense of their own racial identity, they showed more openness to Whites. But Whites who had a strong identity with their own whiteness responded negatively (Kaiser, Drury, Malahy, & King, 2011). This seems to be so even though verbal and nonverbal behaviors of African Americans have changed considerably (Greene & Stewart, 2011).

Together these studies have profound implications for our practice of counseling and therapy. Racial and other multicultural differences are there in the session. We all need to learn more about one another and the importance of respect and empathy for those different from us in any way.

Also relevant to this chapter is a study that examined women's attitudes toward real and imagined gender harassment. When they were asked to imagine being harassed in an interview, they said that they were angry. But those who were actually harassed felt *fear* (Woodzicka & LaFrance, 2002). From a neuroscience perspective, anger is a secondary emotion to the more basic fear (see Chapter 7). As a partial explanation for what happened in this study, earlier research by LaFrance and Woodzicka (1998) found that women's nonverbal reactions to sexist humor are quite marked and different from those of men.

Facial Expressions

For you, as the counselor, facial expression and smiling are good indicators of your warmth and caring. Your ability to develop a relationship will often carry you through difficult problems and situations. When it comes to observing the client, here are some things to notice: The brow may furrow, lips may tighten or loosen, flushing may occur, or a client may smile at an inappropriate time. Even more careful observation will reveal subtle color changes in the face as blood flow reflects emotional reactions. Breathing may speed up or stop temporarily. The lips may swell, and pupils may dilate or contract. These seemingly small responses are important clues to what a client is experiencing; to notice them takes work and practice. You may want to select one or two kinds of facial expressions and study them for a few days in your regular daily interactions, then move on to others as part of a systematic program to heighten your powers of observation.

Neuroscience offers critical data that we all need to consider. Grawe (2007, p. 78) reviews key literature and points out that the amygdala, critical center of emotional experience, appears to be highly sensitive to "fearful, irritated, and angry faces . . . even when the faces have not been perceived consciously. . . . We can be certain that in psychotherapy the patient's amygdala will respond to even the tiniest sign of anger in the facial expressions of therapists." Self-awareness of your own being is obviously as important as awareness of client behavior.

Smiling is a sign of warmth in most cultures, but in some situations in Japan, smiling may indicate discomfort and embarrassment. In all cultures, there can be a difference between a genuine smile and one that seeks to cover up real feelings. Eye contact may be inappropriate for the traditional Navajo, but highly appropriate and expected for a Navajo official who interacts commonly with European American Arizonans. In Nigeria, direct eye contact can be seen as intrusive, so looking at the shoulder is more appropriate.

Be careful not to assign your own ideas about what is "standard" and appropriate nonverbal communication. It is important for the helping professional to begin a lifetime of study of nonverbal communication patterns and their variations. In terms of counseling sessions, you will find that changes in style may be as important as, or more important than, finding specific meanings in communication style. Box 4.2 presents an interesting example of this.

BOX 4.2 | **National and International Perspectives on Counseling Skills**

Can I Trust What I See?
Weijun Zhang

James Harris, an African American professor of education, and I were invited by a national Native American youth leadership organization to give talks in Oklahoma. I attended Dr. Harris's first lecture with about 60 Native American children. Dynamic and humorous, Dr. Harris touched the heartstrings of everyone. But much to my surprise, when the lecture was over, he was very upset. "A complete failure," he said to me with a long face; "they are not interested." "No," I replied, "It was a great success. Don't you see how people loved your lecture?" "No, not at all." After some deliberation, I came to see why he could have such an erroneous impression.

"There were not many facial expressions," Professor Harris said. I said, "You may be right, using African American standards. However, that is not a sign that your audience was not interested. Native American people, in a way, are programmed to restrain their feelings, whether positive or negative, in public; as a result, their facial expressions would be hard to detect. Native American people have always valued restraint of emotion, considering this a sign of maturity and wisdom, as I know. Actually, in terms of emotional expressiveness, African American culture and Native American culture may represent two extremes on a continuum."

"They did not ask a single question, though I repeatedly asked them to," he said. I replied, "Well, they didn't because they respect you." "Come on, you are kidding me." I told him, "Many Native Americans are not accustomed to asking questions in public, probably for the following reasons. (a) If you ask an intelligent question, you will draw attention away from the teacher and onto yourself. That is not an act of modesty and may be seen as showing off. (b) If your question

is silly, you will be seen as a laughingstock and lose face. (c) Whether your question is good or bad, one thing is certain: You will disturb the instructor's teaching plan, or you may suggest the teacher is unclear. That goes against the Native American tradition of being respectful to the senior. So you can never expect a Native American audience to be as active as African Americans in asking questions. In today's situation, some kids probably did want to ask you questions, for you repeatedly encouraged them to. But, unfortunately, they still couldn't do it." He asked why. I said, "You waited only a few seconds for questions before you went on lecturing, which is far from enough. With Native Americans, you have to adopt a longer time frame. European Americans and African Americans may ask questions as soon as you invite them to; American Indians may wait for about 20 to 30 seconds to start to do so. That period of silence is a necessity for them. You might say that Native Americans are true believers in the saying 'Speech is silver, silence is golden.'" He asked, "Well, why didn't you tell me that on the spot, then?" "If it is respectful for those Native American kids not to ask you questions, Dr. Harris," I said, "how could you expect this humble Chinese to be so disrespectful as to come to the stage to correct you?"

The director of the Native American organization, who overheard my conversation with James Harris, approached me with the question "How do you possibly know all this about us Native Americans?" "Well, I don't think it is news for you that Native Americans migrated from Asia some thousands of years ago. You don't mind that your Asian cousins still share your ethnic traits, do you?" The three of us all laughed.

Body Language

Particularly important are discrepancies in nonverbal behavior. When a client is talking casually about a friend, for example, one hand may be tightly clenched in a fist and the other relaxed and open, possibly indicating mixed feelings toward the friend or something related to the friend.

Hand and arm gestures may give you an indication of how you and the client are organizing things. Random, discrepant gestures may indicate confusion, whereas a person seeking to control or organize things may move hands and arms in straight lines and point with fingers authoritatively. Smooth, flowing gestures, particularly those in harmony with the gestures of others, such as family members, friends, or even you as counselor, may suggest openness.

Often people who are communicating well mirror each other's body language. Mirror neurons in the brain enable counselors to become empathic with their clients. When empathy is at a height, client and counselor may unconsciously sit in identical positions and make complex hand movements together as if in a ballet. This is termed *movement synchrony*. *Movement complementarity* is paired movements that may not be identical, but still harmonious. For instance, one person talks and the other nods in agreement. You may observe a hand movement at the end of one person's statement that is answered by a related hand movement as the other takes the conversational "ball" and starts talking.

Brain scans have shown that when two people are in a close working relationship, similar brain structures are simultaneously activated. We can now extend mirroring beyond our external observations and realize that internal observations of the brain are also parallel. Unconscious body movement synchrony has been discovered in neural correlates and connectivity among and within specific brain regions (Yun, Watanabe, & Shimojo, 2012).

Some expert counselors and therapists deliberately "mirror" their clients. Experience shows that matching body language, breathing rates, and key words of the client can heighten counselor understanding of how the client perceives and experiences the world. You may find it possible to produce this synchronicity in the interview, but it is far more likely to occur spontaneously in moments of true empathy.

Couples who disagree typically are not expected to have synchronous nonverbal or brain behavior. Being able to observe discrepant nonverbals between couples will provide you with valuable information, which may lead to new insights in the counseling relationship.

Equipment for recording interviews was new when microcounseling research began. In Box 4.3 are frame-by-frame pictures of a grainy black-and-white video that illustrates mirroring in an early microcounseling training session. You will find that slow motion and even frame-by-frame analysis can teach a lot about session behavior of both client and counselors. What you see in Box 4.3 takes about a second.

Movement synchrony occurs when client and counselor demonstrate body movements that mirror one another, as illustrated in Box 4.1, and suggests the presence of an empathic, authentic relationship. Look for examples of movement synchrony in your own sessions. Slow frame-by-frame viewing will be needed to catch the details. Also watch for examples of movement dyssynchrony where nonverbals appear to be in opposition.

But be careful with deliberate mirroring. A practicum student reported difficulty with a client, noting that the client's nonverbal behavior seemed especially unusual. Near the end of the session, the client reported, "I know you guys; you try to mirror my nonverbal behavior. So I keep moving to make it difficult for you." You can expect that some clients will know as much about observation skills and nonverbal behavior as you do. What should you do in such situations? Use the skills and concepts in this book with *honesty* and *authenticity*. Talk with your clients about their observations of you without being defensive. Openness works!

| BOX 4.3 | Mirroring in the Session, Frame by Frame |

Courtesy of Allen Ivey

Another example of movement synchrony comes from studies of the average number of times friends of different cultural groups touch each other in an hour while talking in a coffee shop. The results showed that English friends did not touch each other at all, French friends touched 110 times, and Puerto Rican friends touched 180 times (cited in Asbell & Wynn, 1991). Gallace and Spence (2010) cite a parallel study in which students at the University of Florida, Gainesville, touched twice, compared with 180 instances of touching in San Juan, Puerto Rico.

Acculturation Issues in Nonverbal Behavior: Avoid Stereotyping

We have stressed that there are many differences in individual and cultural styles. Acculturation is a fundamental concept of anthropology with significant relevance for the session.

Acculturation is the degree to which an individual has adopted the norms or standard way of behaving in a given culture. Because of the unique family, community, economic status, and part of the country in which a person is raised (and many other factors), no two people will be acculturated to general standards in the same way. In effect, "normative behavior" does not exist in any single individual. Thus, stereotyping individuals or groups needs to be avoided at all costs.

An African American client raised in a small town in upstate New York in a two-parent family has different acculturation experiences from those of an otherwise similar person raised in Los Angeles or East St. Louis. If one were from a single-parent family, the acculturation experiences would change further. If we alter only the ethnic/racial background of this client to Italian American, Jewish American, or Arab American, the acculturation experience changes again. Many other factors, of course, influence acculturation—religion, economic bracket, and even being the first or second child in a family. Awareness of diversity in life experience is critical if we are to recognize uniqueness and specialness in each individual. If you define yourself as White American, Canadian, or Australian and you think of others as the only people who are multicultural, you need to rethink your awareness. All of us are multicultural beings with varying and singular acculturation experiences.

Finally, consider biculturality and multiculturality. Many of your clients will have more than one significant community cultural experience. A Puerto Rican, Mexican, or Cuban American client is likely to be acculturated in both Hispanic and U.S. culture. A Polish Canadian client in Quebec, a Ukrainian Canadian client in Alberta, and an Aboriginal client in Sydney, Australia, may also be expected to represent biculturality. And all Native Americans and Hawai'ians in the United States, Dene and Inuit in Canada, Maori in New Zealand, and Aboriginals in Australia exist in at least two cultures. There is a culture among people who have experienced cancer, AIDS, war, abuse, and alcoholism. All of these issues and many others deeply affect acculturation.

In short, stereotyping any one individual is not only discriminatory; it is also naïve!

Verbal Behavior

Counseling and psychotherapy theory and practice have an almost infinite array of verbal frameworks within which to examine the session. Three useful concepts for session analysis are presented here: key words, concreteness versus abstractions, and "I" statements versus "other" statements. We will also discuss some key multicultural issues connected with verbal behavior.

Key Words

If you listen carefully to clients, you will find that certain words appear again and again in their descriptions of situations. Noting their key words and helping them explore the facts, feelings, and meanings underlying those words may be useful. Key descriptive words are often the constructs by which a client organizes the world; these words may reveal underlying meanings. Verbal underlining through vocal emphasis is another helpful clue in determining what is most important to a client. Through intonation and volume, clients tend to stress the single words or phrases that are most closely related to central issues for discussion.

Joining clients by using their key words facilitates your understanding and communication with them. If their words are negative and self-demeaning, reflect those perceptions early in the session but later help them use more positive descriptions of the same situations or events. Help the client change from "I can't" to "I can."

Many clients will demonstrate problems of verbal tracking and selective attention. They may either stay on a single topic to the exclusion of other issues or change the topic, either subtly or abruptly, when they want to avoid talking about a difficult issue. Perhaps the most difficult task for the beginning counselor or psychotherapist is to help the client stay on the topic without being overly controlling. Observing the changes in client's topic is essential. At times it may be helpful to comment—for instance, "A few minutes ago we were talking about X." Another possibility is to follow up that observation by asking how the client might explain the shift in topic.

Concreteness Versus Abstraction

Where is the client on the "abstraction ladder"? Two major styles of communication presented in Box 4.4 provide information on how you can best talk to the client. Observe, and be prepared for client talk or when they move up the abstraction ladder (Ivey, Ivey, Myers, & Sweeney, 2005). Clients who talk with a concrete/situational style are skilled at providing specifics and examples of their concerns and problems. The language of these clients forms the foundation or "bottom" of the abstraction ladder. These clients may have difficulty reflecting on themselves and their situations and seeing patterns in their lives.

Clients who are more abstract and formal operational, on the other hand, have strengths in self-analysis and are often skilled at reflecting on their issues. They are at the "top" of the ladder, but you will find that getting specific concrete details from them as to what is actually going on may be difficult.

Thus, neither concrete nor abstract is "best." Both are necessary for full communication. Of course, most adult and many adolescent clients will talk at both levels. Children, however, can be expected to be primarily concrete in their talk—and so are many adolescents and adults.

Clients with a concrete/situational style will provide you with plenty of specifics. The strength and value of these details is that you know relatively precisely what happened, at least from their point of view. However, they will often have difficulties in seeing the point of view of others. Some with a concrete style may tell you, for example, what happened to them when they went to the hospital from start to finish, with every detail of the operation

---

BOX 4.4   **The Abstraction Ladder**

Abstract/Formal Operational

Clients high on the abstraction ladder tend to talk in a more reflective fashion, analyzing their thoughts and behaviors. They are often good at self-analysis. These clients may not easily provide concrete examples of their issues. They may prefer to analyze, rather than to act. Self-oriented, abstract theories, such as person-centered or psychodynamic, are often useful with this style.

Concrete/Situational

Clients who talk in a more concrete/situational style tend to provide specific examples and stories, often with

considerable detail. You'll hear what they see, hear, and feel. Helping some clients to reflect on their situations and issues may be difficult. In general, they will look to the counselor for specific actions that they can follow. Concrete behavioral theories may be preferred.

More specific examples and extensions of these concepts can be found in *Developmental Counseling and Therapy: Promoting Wellness Over the Lifespan* (Ivey, Ivey, Myers, & Sweeney, 2005).

and how the hospital functions. Or ask a 10-year-old to tell you about a movie, and you will practically get a complete script. Concretely oriented clients with a difficult interpersonal relationship may discuss the situation through a series of endless stories full of specific facts: "He said . . . and then I said. . . ." If asked to reflect on the meaning of their story or what they have said, they may appear puzzled.

Here are some examples of concrete/situational statements:

Child, age 5:   Jonnie hit me in the arm—right here!

Child, age 10:   He hit me when we were playing soccer. I had just scored a goal and it made him mad. He snuck up behind me, grabbed my leg, and then punched me when I fell down! Do you know what else he did? Well, he . . .

Man, age 45:   I was down to Myrtle Beach—we drove there on 95 and was the traffic ever terrible. Well, we drove into town and the first thing we did was to find a motel to stay in, you know. But we found one for only $60 and it had a swimming pool. Well, we signed in and then . . .

Woman, age 27:   You asked for an example of how my ex-husband interferes with my life? Well, a friend and I were sitting quietly in the cafeteria, just drinking coffee. Suddenly, he came up behind me, he grabbed my arm (but didn't hurt me this time), then he smiled and walked out. I was scared to death. If he had said something, it might not have been so frightening.

The details are important, but clients who use a primarily concrete style in their conversation and thinking may have real difficulty in reflecting on themselves and seeing patterns in situations.

Abstract/formal operational clients are good at making sense of the world and reflecting on themselves and their situations. But some clients will talk in such broad abstract generalities that it is hard to understand what they are really saying. They may be able to see patterns in their lives and be good at discussing and analyzing themselves, but you may have difficulty finding out specifically what is going on in their lives. They may prefer to reflect rather than to act on their issues.

Examples of abstract/formal operational statements include these:

Child, age 12:   He does it to me all the time. It never stops. It's what he does to everyone all the time.

Man, age 20:   As I think about myself, I see a person who responds to others and cares deeply, but somehow I feel that they don't respond to me.

Woman, age 68:   As I reflect back on my life, I see a pattern of selfishness that makes me uncomfortable. I think a lot about myself.

Many counselors and therapists tend to be more abstract/formal operational themselves and may be drawn to the analytical and self-reflective style. They may conduct entire sessions focusing totally on analysis, and an observer might wonder what the client and counselor are talking about.

In each of the conversational styles, the strength is also possibly a weakness. You will want to help abstract/formal clients to become more concrete ("Could you give me an example?"). If you persist, most of these clients will be able to provide the needed specifics.

You will also want to help concrete clients become more abstract and pattern oriented. This is best effected by a conscious effort to listen to their sometimes lengthy stories very carefully. Paraphrasing and summarizing what they have said can be helpful (see Chapter 6). Just asking them to reflect on their story may not work ("Could you tell me what the story means to you?" "Can you reflect on that story and what it says about you as a person?"). More direct questions may be needed to help concrete clients step back and reflect on their stories. A series of questions such as these might help: "What one thing do you remember most about this story?" "What did you like best about what happened?" "What least?" "What could you have done differently to change the ending of the story?" Questions like these that narrow the focus can help children and clients with a concrete orientation move from self-report to self-examination.

Essentially, we all need to match our own style and language to the uniqueness of the client. If you have a concrete style, abstract clients may challenge or even puzzle you. If you are more abstract, you may not be able to understand and reach those with a concrete orientation (and they are the majority of our clients). Abstract counselors often are bored and impatient with concrete clients. If the client tends to be concrete, listen to the specifics and enter that client's world as he or she presents it. If the client is abstract, listen and join that client where he or she is. Consider the possibility of helping the client look at the concern from the other perspective.

"I" Statements and "Other" Statements

Clients' ownership and responsibility for issues will often be shown in their "I" and "other" statements. Consider the following:

"I'm working hard to get along with my partner. I've tried to change and meet her/him halfway."

versus

"It's her/his fault. No change is happening."

"I'm not studying enough. I should work harder."

versus

"The racist insults we get on this campus make it nearly impossible to study."

"I feel terrible. If only I could do more to help. I try so hard."

versus

"Dad's an alcoholic. Everyone suffers."

"I'm at fault. I shouldn't have worn that dress. It may have been too sexy."

versus

"No, women should be free to wear whatever they wish."

"I believe in a personal God. God is central to my life."

versus

"Our church provides a lot of support and helps us understand spirituality more deeply."

Review these five pairs of statements. Some of them represent positive "I" and "other" statements; some are negative. Some clients attribute their difficulties solely to themselves; others see the outside world as the issue. A woman may be sexually harassed and see clearly that others and the environment are at fault; another woman will feel that somehow she provoked the incident. Counselors need to help individuals look at their issues but also help them consider how these concerns relate to others and the surrounding environment. There is a need to balance internal and external responsibility for issues.

The alcohol-related statements above may serve as an example. Some children of alcoholics see themselves as somehow responsible for a family member's drinking. Their "I" statements may be unrealistic and ultimately "enable" the alcoholic to drink even more. In such cases, the counselor needs to help the client learn to attribute family difficulties to alcohol and the alcoholic. In work with alcoholics themselves who may deny the problem, one goal is often to help move them to that critical "I" statement, "I am an alcoholic." Part of recovery from alcoholism, of course, is recognizing others and showing esteem for others. Thus, a balance of "I" and "other" statements is a useful goal.

You can also observe "I" statements as a person progresses through treatment. For example, the client at the beginning of counseling may use many negative self-statements: "It's my fault." "I did a bad thing, I'm a bad person." "I don't respect myself." "I don't like myself." If your sessions are effective, expect such statements to change to "I'm still responsible, but I now know that it wasn't all my fault." "Calling myself 'bad' is self-defeating. I now realize that I did my best." "I can respect myself more." "I'm beginning to like myself."

Multicultural differences in the use of the word "I" need to be considered. We should remember that English is one of the very few languages that capitalizes "I." A Vietnamese immigrant comments:

> There is no such thing as "I" in Vietnamese. . . . We define ourselves in rela-
> tionships. . . . If I talk to my mother, the "I" for me is "daughter," the "you" is
> for "mother." Our language speaks to relationships rather than to individuals.

Observe: Is This Interview About Studying or Racial Harassment?

Kyle Yellowhorse is a second-semester junior business major at a large university in the Northwest. He was raised in a relatively traditional Lakota family on the Rosebud Reservation in South Dakota. Native American Indians are unlikely to come to counseling unless they are referred by others. Derek, the counselor, has established himself previously as a person who can be trusted; he has often been seen on campus and has attended many powwows.*

Kyle did well in his first two years at the university, but during the fall term, his "B" average dropped to barely passing. His professor of marketing has referred him to the campus counseling center.

In this case, you will see a slow start, due at least partially to multicultural differences. See how the counselor uses observation skills to open Kyle to discussing his issues.

---

*Many schools, elementary through university, have a small population of Native American Indians. For example, the Chicago public schools have slightly over 1,000 Native Americans mixed in schools throughout a large system. Most often, the Native American Indian population is invisible, and you may never know this cultural group exists unless you indicate through your behavior and actions in the community that you are a person who can be trusted and who wishes to know the community.

Counselor and Client Conversation	Process Comments
1. *Derek:* Kyle, come on in; I'm glad to see you.	Derek walks to the door; he smiles, shakes hands, and uses direct eye contact.
2. *Kyle:* Thanks. (Pause)	Kyle gives the counselor eye contact for only a brief moment. He sits down quietly.
3. *Derek:* You're from the Rosebud Reservation, I see.	Derek knows that contextual and family issues are often important to Native American Indian clients. Rather than focusing on the individual and seeking "I" statements, he realizes that more time may be needed to develop a relationship. Derek's office decorations include artwork from the Native American Indian, African American, and Mexican traditions, as well as symbols of his own Irish American heritage.
4. *Kyle:* Yeah. (Pause)	While his response is minimal, Kyle notes that Derek relaxes slightly in the chair.
5. *Derek:* There've been some hard times here on campus lately. (Pause, but there is still no active response from Kyle.) I'm wondering what I might do to help. But first, I know that coming to this office is not always easy. I know Professor Harris asked you to come in because of your grades dropping this last term.	During the fall term, the Native American Indian association on campus had organized several protests against the school mascot—"the fighting Sioux." As a result, the campus has been in turmoil with recurring events of racial insults and several fights. Noting Kyle's lack of eye contact, Derek has reduced the amount of direct gaze and he also looks down. Among traditional people, the lack of eye contact generally indicates respect. At the same time, many clients who are depressed use little eye contact. (Interchangeable empathy)
6. *Kyle:* Yeah, my grades aren't so good. It's hard to study.	Kyle continues to look down and talks slowly and carefully.
7. *Derek:* I feel honored that you are willing to come in and talk, given all that has happened here. Kyle, I've been upset with all the incidents on campus. I can imagine that they have affected you. But first, how do you feel about being here talking with me, a White counselor?	Derek self-discloses his feelings about campus events. He makes an educated guess as to why Kyle's grades have dropped. As he talks about the campus problems, Kyle looks up directly at him for the first time and Derek notes some fire of anger in his eyes. Kyle nods slightly when Derek says, "I feel honored." (Potentially additive)
8. *Kyle:* It's been hard. I simply can't study. (Pause) Professor Harris asked me to come and see you. I wouldn't have come, but I heard from some friends that you were OK. I guess I'm willing to talk a bit and see what happens.	People who are culturally different from you may not come to your office setting easily. This is where your ability to get into the community is important. In this case, college counseling center staff have been active in the campus community, leading discussions and workshops seeking to promote racial understanding. As Kyle talks, Derek notices increased relaxation and senses that the beginning of trust and rapport has occurred. With some clients, reaching this point may take a full session. Kyle is bicultural in that he has had wide experience in White American culture as well as in his more traditional Lakota family. (The self-disclosure was additive.)
9. *Derek:* Thanks, maybe we could get started. What's happening?	We see that Kyle's words and nonverbal actions have changed in the short time that he has been in the session. Derek, for the first time, asks an open question. Questions, if used too early in the session, might have led Kyle to be guarded and say very little.
10. *Kyle:* I'm vice president of the Native American Indian Association—see—and that's been taking a lot of time. Sometimes there are more important things than studying.	Kyle starts slowly and as he gets to the words "more important things," the fire starts to rise in his eyes.

(continued)

Counselor and Client Conversation	Process Comments
11. *Derek:* More important things?	The restatement encourages the client to elaborate on the critical issue that Derek has observed through Kyle's eyes.
12. *Kyle:* Yeah, like last night, we had a march and demonstration against the Indian mascot. It's so disrespectful and demeaning to have this little Indian cartoon with the big teeth. What does that have to do with education? They talk about "liberal education." I think it's far from liberal; it's constricting. But worse, when we got back to the dormitories, the car that belongs to one of our students had all the windows broken out. And inside was a brick with the words "You're next" painted on it.	Kyle is now sitting up and talking more rapidly. Anger and frustration show in his body—his fists are clenched and his face shows strain and tension. Women, gays, or other minorities who experience disrespect or harassment may feel the same way and demonstrate similar verbal and nonverbal behaviors. The car-bashing incident clearly illustrates that the harassing students had a lack of "other" esteem and respect.
13. *Derek:* That's news to me. The situation on campus is getting worse. Your leadership of the association is really important, and now you face even more challenges.	Derek shares his knowledge of the situation and paraphrases what Kyle has just said. He sits forward in his chair and leans toward Kyle, almost mirroring his pose. However, if Kyle were less fully acculturated in White European American culture, the whole tone of the conversation would be quite different. It probably would have taken longer to establish a relationship, Kyle would have spoken more carefully, and his anger and frustration would likely not have been visible. The counselor, in turn, would be likely to spend more time on relationship development, use more personal self-disclosure, use less direct eye contact, and—especially—be comfortable with longer periods of silence. (Interchangeable)
14. *Kyle:* Yeah. (Pause) But we're going to manage it. We won't give up. (Pause) But—I've been so involved in this campus work that my grades are suffering. I can't help the association if I flunk out.	Kyle feels heard and supported by the counselor. Being heard allows him to turn to the reasons he came to the counselor's office. He relaxes a bit more and looks to Derek, as if asking for help.
15. *Derek:* Kyle, what I've heard so far is that you've gotten caught up in the many difficulties on campus. As association vice president, it's taken a lot of time—and you are very angry about what's happened. I also understand that you intend to "hang in" and that you believe you can manage it. But now you'd like to talk about managing your academics as well. Have I heard you correctly?	Derek has used his observation skills so that he knows now that Kyle's major objective is to work on improving his grades. He wisely kept questioning to a minimum and used some personal sharing and listening skills to start the session. (Interchangeable)
16. *Kyle:* Right! I don't like what's going on, but I also know I have a responsibility to my people back home on the Rosebud Reservation, the Lakota people, and to myself to succeed here. I could talk forever on what's going on here on campus, but first, I've got to get my grades straightened out.	Here we see the importance of self-esteem and self-focus if Kyle is to succeed. But his respect for others is an important part of who he is. He does not see himself as just an individual. He also sees himself as an extension of his group. Kyle has considerable energy and perhaps a need to discuss campus issues, but his vocal tone and body language make it clear that the first topic of importance for him today is staying at the university.
17. *Derek:* OK . . . If you'd like, later we might come back to what's going on. You could tell me a bit about what's happening here with the mascot and all the campus troubles. But now the issue is what's going on with your studies. Could you share what's happening?	Derek observes that Kyle's mission right now is to work on staying in school. He asks an open question to change the focus of the session to academic issues, but he keeps open the possibility of discussing campus issues later in the session. (Potentially additive)

This session has now started. Kyle has obviously been observing and deciding if he can trust Derek. Fortunately, Derek has a good reputation on campus. He regularly attended Native American powwows and other multicultural events. You will find if you work in a school or university setting that clients will be very aware of the effectiveness and trustworthiness of counseling staff.

Although focused on Native American Indian issues, this session has many parallels with other native people who have been dispossessed of their land and whose culture has been belittled—Hawai'ians, Aboriginals in Australia, Dene and Inuit in Canada, Maori in New Zealand, and Celtic people in Great Britain. Moreover, the session in many ways illustrates what might happen in any type of cross-cultural counseling. A European American student meeting a counselor who is Latina/o, African American, or another Person of Color (or vice versa if the roles of counselor and client are switched) might also have early difficulty in talking and establishing trust.

You will find that many counseling sessions start slowly, regardless of the cultural background of the client. Your skill at observing nonverbal and verbal behavior in the here and now of the session will enable you to choose appropriate things to say. Some approaches that may help you as you begin sessions include patience, a good sense of humor, and a willingness to disclose, share stories, and talk about neutral subjects such as sports or the weather. You will also find an early exploration of positive assets useful at times: "Before we start, I'd like to get to know a little bit more about you. Could you tell me specifically about something from your past that you feel particularly good about?" "What are some of the things you do well?" "What types of things do you like?" As time permits, consider conducting a full wellness review.

Discrepancies, Mixed Messages, and Conflict

We all live with contradiction, conflict, incongruity, and discrepancies that provide challenges. This is what brings most clients to counseling and therapy—and we devote the whole of Chapter 10 on empathic confrontation to resolution of these issue. At the same time, conflict provides us with opportunities and openings for growth and change. The variety of discrepancies clients may manifest is perhaps best illustrated by the following statements:

"My son is perfect, but he just doesn't respect me."

"I deserve to pass the course." (from a student who has done no homework and just failed the final examination)

"That question doesn't bother me." (said with a flushed face and a closed fist)

"I love my partner, but we just can't get along anymore."

Once the client is relatively comfortable and some beginning steps have been taken toward rapport and understanding, a major task of the counselor or therapist is to identify basic discrepancies, mixed messages, conflicts, or incongruities in the client's behavior and life. A common goal in most sessions is to assist clients in working through discrepancies and conflict, but first these have to be identified clearly.

Examples of Conflict Internal to the Client

Discrepancies in Verbal Statements. In a single sentence, a client may express two completely contradictory ideas ("My son is perfect, but he just doesn't respect me" or "This is a lovely office you have; it's too bad that it's in this neighborhood"). Most of us have mixed feelings toward our loved ones, our work, and other situations. It is helpful to aid others in understanding their ambivalences.

Discrepancies Between Statements and Nonverbal Behavior. Very important are discrepancies between what one says and what one does. A parent may talk of love for a child but be guilty of child abuse. A student may say that he or she deserved a higher grade than the time actually spent studying suggests. A client may verbalize support for multicultural, women's, or ecology causes, but fail to "walk the walk." A client may be talking of a desire to repair a troubled relationship while simultaneously picking at his or her clothes, or make small or large physical movements away from the counselor when confronted with a troubling issue.

Examples of Conflict Between the Client and the External World

Discrepancies and Conflict Between People. "I cannot tolerant my neighbors." Noting interpersonal conflict is a key task of the counselor or psychotherapist.

Discrepancies Between a Client and a Situation. "I want to be admitted to medical school, but I didn't make it." "I can't find a job." In such situations, the client's ideal world is often incongruent with what really is. Many People of Color, gays, women, or people with disabilities find themselves in a contextual situation that makes life difficult for them. Discrimination, heterosexism, sexism, and ableism represent situational discrepancies.

Discrepancies in Goals

Goal setting is an important part of the *empathic relationship—story and strengths—goals—restory—action* model. As part of establishing the purposes of counseling or psychotherapy, you will often find that a client seeks incompatible goals. For example, the client may want the approval of friends and parents, but winning acceptance from peers may mean that academic performance suffers, and pleasing the parents may be considered "selling out."

Discrepancies Between You and the Client

One of the more challenging issues occurs when you and the client are not in synchrony. And this can occur on any of the dimensions above. Your nonverbal communication may be misread by the client. The client may avoid really facing issues. You may be saying one thing, the client another. A conflict of values or goals in counseling may be directly apparent or a quiet, unsaid thing that still impacts the client—and either can destroy your relationship. When you get too close to the truth, clients may wipe their nose. Does the client turn toward you or turn away?

In any of these situations, it is helpful to aid clients in understanding their ambivalences by summarizing the conflict—the client's own thoughts, emotions, and behaviors and/or conflicting issues posed by someone else or life's situation. The summary of the conflict can then be followed by a variation of the basic challenge, such as "I hear you saying one side of the issue is (insert appropriate comment representing part of the conflict or discrepancy). But, I also hear the other side as (insert the opposing side of the conflict)." Then, through further listening and observation, the client may come up with her or his own unique solution.

The issues of conflict, discrepancies, and contradiction will be explored in much more detail in Chapter 10, which discusses confrontation as a constructive challenge.

Action: Key Points and Practice of Observation Skills

Importance of Observation. The self-aware counselor is constantly aware of the client and of the here-and-now interaction in the session. Clients tell us about their world by nonverbal and verbal means. Observation skills are a critical tool in determining how the client interprets the world.

Nonverbal Behavior. Your own and your clients' eye contact patterns, body language, and vocal qualities are, of course, important. Shifts and changes in these may be indicative of client interest or discomfort. A client may lean forward, indicating excitement about an idea, or cross his or her arms to close it off. Facial clues (brow furrowing, lip tightening or loosening, flushing, pulse rate visible at temples) are especially important. Larger-scale body movements may indicate shifts in reactions, thoughts, or the topic.

Verbal Behavior. Noting patterns of verbal tracking for both you and the client is particularly important. At what point does the topic change, and who initiates the change? Where is the client on the abstraction ladder? If the client is concrete, are you matching his or her language? Is the client making "I" statements or "other" statements? Do the client's negative statements become more positive as counseling progresses? Clients tend to use certain key words to describe their behavior and situations; noting these descriptive words and repetitive themes is helpful.

Discrepancies. Conflict, discrepancies, incongruities, mixed messages, and contradictions are manifest in many and perhaps all sessions. The effective counselor is able to identify these discrepancies, to name them appropriately, and sometimes to feed them back to the client. These discrepancies may be between nonverbal behaviors, between two statements, between what clients say and what they do, or between incompatible goals. They may also represent a conflict between people or between a client and a situation. And your own behaviors may be positively or negatively discrepant.

Simple, careful observation of the session is basic. What can you see, hear, and feel of the client's world? Note your impact on the client: How does what you say change or relate to the client's behavior? Use these data to adjust your microskill technique.

Multicultural Issues. Observation skills are essential with all clients. Note individual and cultural differences in verbal and nonverbal behavior. Always remember that some individuals and some cultures may have a different meaning for a movement or use of language from your own personal meaning. Use caution in your interpretation of nonverbal behavior.

Mirroring. When two people are talking together and communicating well, they often exhibit movement synchrony or movement complementarity in that their bodies move in a harmonious fashion. Increased movement synchrony suggests implicit social interaction in observable behavior and in brain scans. Counselors sometimes mirror their clients deliberatively. When people are not communicating clearly, movement dyssynchrony may appear: body shifts, jerks, and pulling away. Brain regions parallel what we observe in client nonverbal behavior.

Concreteness Versus Abstraction. Clients who talk with a concrete/situational style are skilled at providing specifics and examples of their concerns and problems. They may have difficulty reflecting on themselves and their situations and seeing patterns in their lives. Clients who are more abstract and formal operational have strengths in self-analysis and are often skilled at reflecting on their issues. They may experience difficulty reporting the concretes and specifics of what is actually going on.

Neither concrete nor abstract is "best." Both are necessary for full communication.

Practice and Feedback: Individual, Group, and Microsupervision

Additional resources can be found by going to CengageBrain.com and logging into the MindTap course created by your professor. There you will find a variety of study tools and useful resources that include quizzes, videos, interactive counseling and psychotherapy exercises, case studies, the Portfolio of Competencies, and more.

Many concepts have been presented in this chapter; it will take time to master them and make them a useful part of your work. Therefore, the exercises here should be considered introductory. Further, it is suggested that you continue to work on these concepts throughout the time that you read this book. If you keep practicing the concepts in this chapter throughout the book, material that might now seem confusing will gradually be clarified and become part of your natural style.

Many of you now have small cameras that have both video and sound capability, or you can use a smartphone or computer. We urge you to take the time to observe closely what you are doing. Practicing with clear video feedback is the original and most effective route toward mastery and competence.

Individual Practice

Exercise 4.1 Observation of Nonverbal Patterns

Observe 10 minutes of a counseling session, a television interview, or any two people talking. Video record what you observe for repeated viewing.

Visual/eye contact patterns. Do people maintain eye contact more while talking or while listening? Does the "client" break eye contact more often while discussing certain subjects than others? Can you observe changes in pupil dilation as an expression of interest?

Vocal qualities. Note speech rate and changes in intonation or volume. Give special attention to speech "hitches" or hesitations.

Attentive body language. Note gestures, shifts of posture, leaning, patterns of breathing, and use of space. Give special attention to facial expressions, such as changes in skin color, flushing, and lip movements. Note appropriate and inappropriate smiling, furrowing of the brow, and so on.

Movement harmonics. Note places where movement synchrony occurred. Did you observe examples of movement dyssynchrony?

Use your video recording to view the session several times. Be sure to separate behavioral observations from impressions on the Feedback Form: Observation (Box 4.5).

BOX 4.5	Feedback Form: Attending Behavior

_____
(DATE)

_____   _____
(NAME OF COUNSELOR)   (NAME OF PERSON COMPLETING FORM)

Instructions: Observe the client or counselor carefully during the role-play session and immediately afterward complete the nonverbal feedback portion of the form. As you view the video or listen to the audio recording, give special attention to verbal behavior and note discrepancies. If no recording equipment is available, one observer should note nonverbal behavior and the other verbal behavior.

Nonverbal Behavior Checklist

1. _Visuals._ At what points did eye contact breaks occur? Staring? Did the individual maintain eye contact more when talking or when listening? Changes in pupil dilation?

2. _Vocals._ When did speech hesitations occur? Changes in tone and volume? What single words or short phrases were emphasized?

3. _Body language._ General style and changes in position of hands and arms, trunk, legs? Open or closed gestures? Tight fist? Playing with hands or objects? Physical tension: relaxed or tight? Body oriented toward or away from the other? Sudden body shifts? Twitching? Distance? Breathing changes? At what points did changes in facial expression occur? Changes in skin color, flushing, swelling or contracting of lips? Appropriate or inappropriate smiling? Head nods? Brow furrowing?

4. _Movement harmonics._ Examples of movement complementarity, synchrony, or dyssynchrony? At what times did these occur?

(continued)

BOX 4.5 **(continued)**

5. *Nonverbal discrepancies.* Did one part of the body say something different from another? With what topics did this occur?

Verbal Behavior Checklist

1. *Verbal tracking and selective attention.* At what points did the client or counselor fail to stay on the topic? To what topics did each give most attention? List here the most important key words used by the client; these are important for deeper analysis.

2. *Abstract or concrete?* Which word represents the client? How did the counselor work with this dimension? Was the counselor abstract or concrete?

3. *Verbal discrepancies.* Write here observations of verbal discrepancies in either client or counselor.

1. Present the context of observation. Briefly summarize what is happening verbally at the time of the observation. Number each observation.

2. Observe the interview for the following, and describe what you see as precisely and concretely as possible: visual/eye contact patterns, vocal qualities, attentive body language, and movement harmonics. From this, what do you notice about emotional tone?

3. Record your impressions. What interpretations of the observation do you make? How do you make sense of each observation unit? And—most important—are you cautious in drawing conclusions from what you have seen and noticed?

Exercise 4.2 Examining Your Own Verbal and Nonverbal Styles
Video record yourself with another person in a real conversation or session for at least 20 minutes. Do not make this a role-play. Then view your own verbal and nonverbal

behavior and that of the person you are talking with in the same detail as in Exercise 4.1. Spend some time watching in slow motion, and perhaps even frame by frame. What do you learn about yourself?

Exercise 4.3 Classifying Statements as Concrete or Abstract

Following are examples of client statements. Classify each statement as primarily concrete (C) or primarily abstract (A). You will gain considerably more practice and thus have more suggestions for interventions in later chapters. (Answers to this exercise may be found at the conclusion of this chapter.)

C A 1. I cry all day long. I didn't sleep last night. I can't eat.

C A 2. I feel rotten about myself lately.

C A 3. I feel very guilty.

C A 4. Sorry I'm late for the session. Traffic was very heavy.

C A 5. I feel really awkward on dates. I'm a social dud.

C A 6. Last night my date said that I wasn't much fun. Then I started to cry.

C A 7. My father is tall, has red hair, and yells a lot.

C A 8. My father is very hard to get along with. He's difficult.

C A 9. My family is very loving. We have a pattern of sharing.

C A 10. My mom just sent me a box of cookies.

Group Practice and Microsupervision

Exercise 4.4 Practicing Observation and Microsupervision in Groups

Many observation concepts have been discussed in this chapter. It is obviously not possible to observe all of them in a single role-play session. However, practice can serve as a foundation for elaboration at a later time. This exercise has been selected to summarize the central ideas of the chapter.

Step 1: Divide into practice groups. Groups of three or four are most appropriate.

Step 2: Select a group leader.

Step 3: Assign roles for the first practice session.

❑ Client, who responds naturally and is talkative.

❑ Counselor, who will seek to demonstrate a natural, authentic style.

❑ Observer 1, who observes client communication, using the Feedback Form: Observation (Box 4.5).

❑ Observer 2, who observes counselor communication, using the Feedback Form. Ideally you have a video recording available for precise feedback.

Step 4: Plan. State the goals of the session. As the central task is observation, the counselor should give primary attention to attending and open questions. Use other skills as you wish. After the role-play is over, the counselor should report personal observation of the client made during that time and demonstrate basic or active mastery skills. The client will report on observations of the counselor.

The suggested topic for the practice role-play is "Something or someone with whom I have a present conflict or have had a past conflict." Alternative topics include the following:

My positive and negative feelings toward my parents or other significant persons

The mixed blessings of my work, home community, or present living situation

The two observers may use this session as an opportunity both for providing feedback to the counselor and for sharpening their own observation skills.

Step 5: Conduct a 6-minute practice session. As much as possible, both the counselor and the client will behave as naturally as possible discussing a real situation.

Step 6: Review the practice session and provide feedback for 14 minutes. Remember to stop the audio or video recording periodically and listen to or view key items several times for increased clarity. Observers should give special attention to careful completion of the feedback sheet throughout the session, and the client can give important feedback via the Client Feedback Form in Chapter 1.

Step 7: Rotate roles.

Portfolio of Competencies and Personal Reflection

Determining your own style and theory can be best accomplished on a base of competence. Each chapter closes with a reflective exercise asking your thoughts and feelings about what has been discussed. By the time you finish this book, you will have a substantial record of your competencies and a good written record as you move toward determining your own style and theory.

Assessing Your Level of Competence: Awareness, Knowledge, Skills, and Action

Use the following checklists to evaluate your present level of mastery. Check those dimensions that you currently feel able to do. Those that remain unchecked can serve as future goals. Do not expect to attain intentional competence on every dimension as you work through this book. You will find, however, that you will improve your competencies with repetition and practice.

Awareness and Knowledge. Can you define and discuss the following concepts?

❑ Describe the importance of multicultural differences in verbal and nonverbal communication.

❑ Note attending nonverbal behaviors, particularly changes in behavior in visuals/eye contact, vocal tone, and body language.

❑ Note movement harmonics.

❑ Note verbal tracking and selective attention.

- ❏ Note key words used by the client and yourself.
- ❏ Note distinctions between concrete/situational and abstract/formal operational conversation.
- ❏ Note discrepancies in verbal and nonverbal behavior.
- ❏ Note discrepancies in the client.
- ❏ Note discrepancies in yourself.
- ❏ Note discrepancies between yourself and the client.

Basic Competence. Nonverbal and verbal observation skills are things that you can work on and improve over a lifetime. Therefore, use the intentional competence list below for your self-assessments.

Intentional Competence. You will be able to note client verbal and nonverbal behaviors in the session and use these observations at times to facilitate session conversation. You will be able to match your behavior to the client's. When necessary, you will be able to mismatch behaviors to promote client movement. For example, if you first join the negative body language of a depressed client and then take a more positive position, the client may follow and adopt a more assertive posture. You will be able to note your own verbal and nonverbal responses to the client. You will be able to note discrepancies between yourself and the client and work to resolve those discrepancies.

Developing mastery of the following areas will take time. Come back to this list later as you practice other skills in this book. For the first stages of basic and intentional mastery, the following competencies are suggested as most important:

- ❏ Mirror nonverbal patterns of the client. The counselor mirrors body position, eye contact patterns, facial expression, and vocal qualities.
- ❏ Identify client patterns of selective attention and use those patterns either to bring talk back to the original topic or to move knowingly to the new topic provided by the client.
- ❏ Match clients' concrete/situation or abstract/formal operational language and help them to expand their stories in their own style.
- ❏ Identify key client "I" and "other" statements and feed them back to the client accurately, thus enabling the client to describe and define what is meant more fully.
- ❏ Note discrepancies and feed them back to the client accurately. The client in turn will be able to accept and use the feedback for further effective self-exploration.
- ❏ Note discrepancies in yourself, and act to change them appropriately.

Psychoeducational Teaching Competence. You will demonstrate your ability to teach others observation skills. Your achievement of this level can be determined by how well your students can be rated on the basic competencies of this self-assessment form. Certain of your clients in counseling may be quite insensitive to obvious patterns of nonverbal and verbal communication. Teaching them beginning methods of observing others can be most helpful to them. Do not try to introduce more than one or two concepts to a client per session.

- ❏ Teach clients in a helping session the social skills of nonverbal and verbal observation and the ability to note discrepancies.
- ❏ Teach these skills to small groups.

Personal Reflection on Observation Skills

This chapter has focused on the importance of verbal and nonverbal observation skills, and you have experienced a variety of exercises designed to enhance your awareness in this area.

What single idea stood out for you among all those presented in this chapter, in class, or through informal learning? What stands out for you is likely to be important as a guide toward your next steps.

What are your thoughts on multicultural differences?

What other points in this chapter struck you as important?

How might you use ideas in this chapter to begin the process of establishing your own style and theory?

Correct Responses for Exercise 4.3

1 — C	5 — A	8 — A
2 — A	6 — C	9 — A
3 — A	7 — C	10 — C
4 — C		

SECTION II

The Basic Listening Sequence

Organizing a Session to Be More Fully Empathic and to Promote Creative Solutions

The **basic listening sequence (BLS)** will enable you to elicit empathically the major thoughts, feelings, and behaviors of the client. Through the use of questions, encouragers, paraphrases, reflection of feelings, and summaries, you will draw out and understand the way clients see their issues.

The BLS will help you reach a more comprehensive view of the client, and as you gain knowledge of client stories, your clients will also understand themselves more fully. Be prepared to develop new stories and behavioral changes through this process as the BLS facilitates organization and new integrations.

Attending behavior and observation skills are basic to all the communication skills of the microskills hierarchy—they are also *central* in all counseling and psychotherapy. Without individually and culturally appropriate attending behavior, there can be no counseling or psychotherapy. Listening lights up the brain; the basic listening sequence facilitates organization and the potential for a creative new integration.

Chapter 5. Questions: Opening Communication We encounter questions every day. Most theories of counseling now use questions rather extensively. Examples include cognitive behavioral therapy (CBT), solution-focused brief counseling, and motivational interviewing. This chapter explains open and closed questions and their place in communication. Nonetheless, opinions vary as to their use as we want to hear the client's story accurately, not our ideas about what the story means.

Chapter 6. Encouraging, Paraphrasing, and Summarizing: Active Listening and Cognition Here you will explore the clarifying skills of paraphrasing, encouraging, and summarizing, which are foundational to developing a relationship and working alliance with your client. They are central also for drawing out the story as seen and experienced by the client.

Chapter 7. Observing and Reflecting Feelings: The Heart of Empathic Understanding
This skill gets at the heart of the issue and truly personalizes the session. You will learn how to bring out the rich emotional world of your clients. Reflecting feelings is a challenging skill to master fully and requires special attention, but many believe that real and lasting change is founded on emotions and feelings.

Chapter 8. The Five-Stage Interview: Empathically Integrating Skills for Creative Change
Once you become competent in observation skills and the basic listening sequence, you will be prepared to conduct a full, well-formed session. You will be able to conduct this session using only attending and the skills of the basic listening sequence. Furthermore, you will be able to evaluate your skills and those of others for level of empathic understanding.

This section, then, has ambitious goals. By the time you have completed Chapter 8, you will have attained several major objectives, enabling you to move on to the influencing skills of interpersonal change, growth, and development. At an intentional level of competence, you may aim to accomplish the following in this section:

1. Develop competence with the basic listening sequence, enabling the client to tell the story more fully. In addition, draw out key thoughts, feelings, and behaviors related to client issues.

2. Observe clients' reactions to your skill usage and modify your skills and attending behaviors to complement each client's uniqueness.

3. Conduct a complete session using only listening and observing skills.

4. Evaluate your counseling session for its level of empathy and examine your ability to communicate warmth, positive regard, and other more subjective dimensions of counseling and psychotherapy.

5. Encourage your clients to establish meaningful goals that enable and motivate them to resolve their issues for meaningful change.

When you have accomplished these tasks, you may find that your clients have a surprising ability to solve their own difficulties, issues, concerns, or challenges. You may also gain a sense of confidence in your own ability as a counselor or psychotherapist. Knowing what to do enables you to do it.

"When in doubt, listen!" This is the motto of this section and the entire microskills framework.

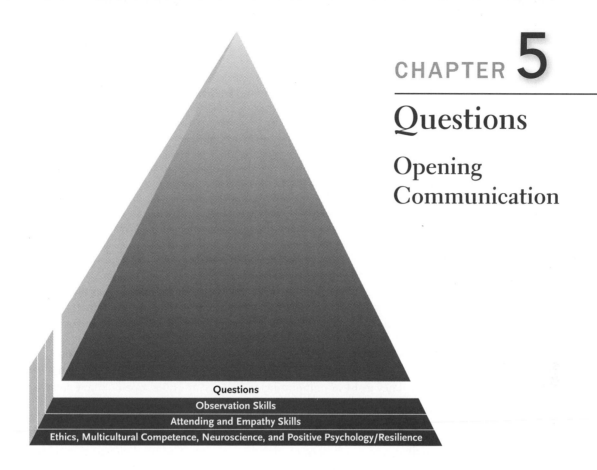

CHAPTER 5

Questions

Opening Communication

Asking questions of your client can be helpful in establishing a basis for effective communication. Effective questions open the door to knowledge and understanding. The art of questioning lies in knowing which questions to ask when. Address your first question to yourself: if you could press a magic button and get every piece of information you want, what would you want to know? The answer will immediately help you compose the right questions.

—Robert Heller and Tim Hindle

Chapter Goals and Competency Objectives

Awareness and Knowledge

▲ Understand the two key styles of questioning: open and closed questions.

▲ Choose the question stem and style that are most likely to achieve a useful anticipated result that clarifies the client's story. For example, *what* questions often lead to talk about facts, *how* questions to feelings or process, and *why* questions to reasons. *Could/would* questions tend to be the most open.

Skills and Action

▲ Draw out and enrich client stories by bringing out a more complete description, including background information and needed details.

▲ Open or close client talk, intentionally, according to the individual needs of the client.

▲ Balance discussion of clients' concerns in a more positive way using the positive asset search, strength emphasis, positive psychology, and wellness.

▲ Use questions in a culturally sensitive and respectful way.

Introduction: Questions

Benjamin is in his junior year of high school, in the middle third of his class. In this school, each student must be interviewed about plans after graduation—work, the armed forces, or college. You are the high school counselor and have called Benjamin in to check on his plans after graduation. His grades are average, he is not particularly verbal and talkative, but he is known as a "nice boy."

Reread the quotation that introduced this chapter. What are some questions that you could use to draw Benjamin out and help him think ahead to the future? What might happen if you ask too many questions? To compare your thoughts with ours, see page 130–131 at the end of this chapter.

Skilled attending behavior is the foundation of the microskills hierarchy; questioning provides a useful framework for focusing the session. Questions help a session begin and move along smoothly. They assist in pinpointing and clarifying issues, open up new areas for discussion, and aid in clients' self-exploration.

Questions are an essential component in many theories and styles of helping, particularly cognitive behavioral therapy (CBT), motivational interviewing, solution-focused brief counseling, and much of career decision making. The employment counselor facilitating a job search, the social worker conducting an assessment interview, and the high school guidance counselor helping a student work on college admissions all need to use questions. Moreover, the diagnostic process, while not counseling, uses many questions.

This chapter focuses on two key styles of questioning: open and closed questions.

Open questions are those that can't be answered in a few words. They tend to facilitate deeper exploration of client issues. They encourage others to talk and provide you with maximum information. Typically, open questions begin with *what, how, why,* or *could/ would*. For example, "Could you tell me what brings you here today?"

Closed questions enable you to obtain specifics and can usually be answered in very few words. They may provide useful information, but the burden of guiding the talk remains on the counselor. Closed questions often begin with *is, are,* or *do*. For example, "Are you living with your family?" Not always, however. "What is your job?" might better be asked as "Could you tell me about your work?"

If you use open questions effectively, the client may talk more freely and openly. Closed questions elicit shorter responses and may provide you with specific information. Following are some of the results you can anticipate when using questions.

Open Questions	Anticipated Client Response
Begin open questions with the often useful *who, what, when, where,* and *why. Could, can,* or *would* questions are considered open but have the additional advantage of giving more power to the client, who can more easily say that he or she doesn't want to respond.	Clients will give more detail and talk more in response to open questions. Could, would, and can questions are often the most open of all, because they give clients the choice to respond briefly ("No, I can't") or, much more likely, explore their issues in an open fashion.

Closed Questions	Anticipated Client Response
Closed questions may start with *do*, *is*, or *are*.	Closed questions may provide specific information but may close off client talk. As such, they need to be asked carefully. But if the relationship is solid and the topic important, the client may talk as much as if given an open question.

Effective questions encourage more focused client conversations with more pertinent detail and less wandering.

A beginning framework for drawing out client issues and stories is the model used in newspaper reporting, a mixture of open and closed questions: who, what, when, where, how, and why.

- Who is the client? What are key personal background factors? Who else is involved?
- What is the issue? What are the specific details of the situation?
- When does the problem occur? What immediately preceded and followed the situation?
- Where does the issue occur? In what environments and situations?
- How does the client react? How does he or she feel about it?
- Why does the issue or concern happen? What are your thoughts as why that happened?

As you listen to answers and have a general understanding of the story or life challenge, this may be followed up by listening further questions using the newspaper framework.

Questioning Questions

Carl Rogers was not fond of questions and worried that they could get in the way of the client's real story. Some other theorists and practitioners are also concerned about the use of questions and believe that they are best studied, learned, and practiced *after* expertise is developed in the reflective listening skills of Chapters 6 and 7. They point out that once questions are presented to those beginning counseling, the listening skills of paraphrasing, reflecting feelings, and summarizing may receive insufficient attention. Certainly, excessive use of questions takes the focus from the client and can give too much power to the counselor.

Reflective Exercise **Why do some people object to questions?**

Take a minute to recall and explore some of your own experiences with questions in the past. Perhaps you had a teacher or a parent who used questions in a manner that resulted in your feeling uncomfortable or even attacked. Write here one of your negative experiences with questions and the feelings and thoughts the questioning process produced in you.

- How was your difficult personal experience with questions?
- How did it make you feel?
- What thoughts did you have?
- What do you think about it now, and how does it make you feel about questions?

People often respond by describing situations in which they were put on the spot, accused, or grilled by someone. They may associate questions with feelings of anger and guilt. Many of us have had negative experiences with questions. Furthermore, questions may be used to direct and control client talk. School discipline and legal disputes typically use questions to control the person being interviewed. If your objective is to enable clients to find their own way, questions may inhibit your reaching that goal, particularly if they are used ineffectively. For these reasons, some helping authorities—particularly those who are humanistically oriented—object to questions.

Additionally, in many non-Western cultures, questions are inappropriate and may be considered offensive or overly intrusive. Nevertheless, questions remain a fact of life in our culture. We encounter them everywhere. The physician or nurse, the salesperson, the government official, and many others find questioning clients basic to their profession. Most counseling theories use questions extensively. The issue, then, is how to use questions wisely and intentionally. Your central task in this chapter is to find an appropriate balance in using questions with clients.

Awareness, Knowledge and Skills: Questions for Results

> Questions make the session work for me. I searched through many questions and found the ones that I thought most helpful in my own practice. I then memorized them and now I always draw on them as needed. Being prepared makes a difference.
>
> —Norma Gluckstern Packard

Questions can be facilitative, or they can be so intrusive that clients want to say nothing. Used carefully, questioning is a valuable skill. The ideas presented here will guide you to question effectively. Furthermore, they will help you define your own questioning techniques and strategies and how questioning fits with your natural counseling style.

In Some Situations Questions Are Essential: "What Else?"

An incident in Allen's life illustrates the importance of questions. His father became blind after open heart surgery. Was that a result of the surgery? No, it was because the physicians failed to ask the basic open question "Is anything else happening physically or emotionally in your life at this time?" If that question had been asked, the physicians would have discovered that Allen's father had developed severe and unusual headaches the week before surgery was scheduled, and they could have diagnosed an eye infection that was easily treatable with medication.

Clients do not always spontaneously provide you with all necessary information, and sometimes the only way to get at missing data is by asking questions. For example, the client may talk about being depressed and unable to act. As a helper, you could listen to the story carefully but still miss underlying issues relating to the client's depression. The open question "What is happening in your life right now or with your family?" might bring out information about an impending separation or divorce, a lost job, or some other issue underlying the concern. What you first interpreted as a classical clinical depression becomes modified by what is occurring in the client's life, and treatment takes a different direction.

In counseling, a client may speak of tension, anxiety, and sleeplessness. You listen carefully and believe the problem can be resolved by helping the client relax and plan changes in her work schedule. However, you ask the client "What else is going on in your life?" Having developed trust in you because of your careful listening and interest, the client finally opens up and shares a story of sexual harassment. At this point, the goals of the session change.

Questions Can Help Begin the Session

With verbal clients and a comfortable relationship, the open question facilitates free discussion and leaves plenty of room to talk. Here are some examples:

"What would you like to talk about today?"

"Could you tell me what prompted you to see me?"

"How have things been since we last talked together?"

"The last time we talked, you planned to talk with your partner about your sexual difficulties. How did it go this week?"

The first three open questions provide room for the client to talk about virtually anything. The last question is open but provides some focus for the session, building on material from the preceding week. These types of questions will work well for a highly verbal client. However, such open questions may be more than a less talkative client can handle. It may be best to start the session with more informal conversation—focusing on the weather, a positive part of last week's session, or a current event of interest to the client. You can turn to the issues for this session as the client becomes more comfortable.

The First Word of Open Questions May Determine Client Response

Question stems often, but not always, result in anticipated outcomes. Use the following guidelines and you'll be surprised how effective these simple questions can be in gathering information.

What questions most often lead to facts.

"What happened?"

"What are you going to do?"

How questions may lead to an exploration of process or feeling and emotion.

"How could that be explained?"

"How do you feel about that?"

Why questions can lead to a discussion of reasons. Use *why* questions with care. While understanding reasons may have value, a discussion of reasons can also lead to sidetracks. In addition, many clients may not respond well because they associate *why* with a past experience of being grilled.

"Why is that meaningful to you?"

"Why do you think that happened?"

Could, can, or *would* questions are considered maximally open and also contain some advantages of closed questions. Clients are free to say "No, I don't want to talk about that." *Could* questions suggest less counselor control.

"Could you tell me more about your situation?"

"Would you give me a specific example?"

"Can you tell me what you'd like to talk about today?"

Give it a try and you'll be surprised to see how effective these simple guidelines can be.

Open Questions Help Clients Elaborate and Enrich Their Story

A beginning counselor often asks one or two questions and then wonders what to do next. Even more experienced therapists can find themselves hard-pressed to know what to do next. To help the session start again and keep it moving, ask an open question on a topic the client presented earlier in the session.

"Could you tell me more about that?"

"How did you feel when that happened?"

"Given what you've said, what would be your ideal solution to the problem?"

"What might we have missed so far?"

"What else comes to your mind?"

Questions Can Reveal Concrete Specifics from the Client's World

To be empathic with a client requires that you understand specifically what the client is saying to you. Concreteness is valuable in empathic understanding. Seek specifics rather than vague generalities. As counselors, we are most often interested in specific feelings, specific thoughts, and specific examples of actions. One of the most useful of all open questions here is "Could you give me a specific example of . . . ?" Concreteness helps the session come alive and clarifies what the client is saying. Likewise, communication from the counselor—the directive, the feedback skill, and interpretation—needs to be concrete and understandable to the client.

Suppose, for example, that a client says, "Ricardo makes me so mad!" Some open questions that aim for concreteness and specifics might be

"Could you give me a specific example of what Ricardo does?"

"What does Ricardo do, specifically, that brings out your anger?"

"What do you mean by 'makes me mad'?"

"Could you specify what you do before and after Ricardo makes you mad?"

Closed questions can bring out specifics as well, but even well-directed closed questions may take the initiative away from the client. However, at the discretion of the counselor, closed questions may prove invaluable.

"Does Ricardo demand a lot from you?"

"Did Ricardo show his anger by striking you?"

"Is he on drugs?"

Questions like these may encourage clients to say out loud what they have only hinted at before. When you suspect abuse or bullying, more direct questions may be needed to protect the client.

Difficulties Versus Stories of Strength to Help Build Client Resilience

Stories presented in the helping interview are often negative and full of serious challenges. Carl Rogers, the founder of client-centered counseling, was always able to find something positive in the client. He considered positive regard and respect for the client essential for future growth. This is illustrated clearly in neuroscience research—energized by the amygdala and distributed by the hippocampus, memories are located in many sections of the brain. Negative emotions and feelings originate primarily in the amygdala, deep in the limbic system brain. Positive emotions (developed later in human evolution) are located in many areas, but the nearby nucleus accumbens sends out signals to the prefrontal cortex, enabling focus on the positive (Ratey, 2008a). See Appendix IV for further discussion.

It is said that it takes at least five positive comments to balance one negative; it often can takes many more if the "neggie" hits deep. If the put-down experience is so much more powerful, think how long it might take to recover from these (some of them we never forget):

Parent:	How could you do that? You are the worst child I have ever seen. I won't speak to you until you apologize.
Peers at school:	You're a sissy, a freak, ugly; you're a whore, fatso, string bean; you act like a girl/boy; you can't play with us; now we are all going to beat you up.
Partner:	I see a muffin top. Why are you so messy/smelly? Why can't you get along with anyone? Making love with you is like _____ (fill in the blank).
A friend:	Your hair looks funny.

Our lives are full of challenges, and the repeated hassle of thoughtless comments and put-downs can result in permanent memories that you will encounter in your counseling practice. This makes positive psychology, wellness, and therapeutic lifestyle changes all the more central to interviewing, counseling, and therapy. Our task is to help clients rewrite these negative experiences.

Thus, we strongly recommend that every session include time spent on strengths and positives. "What's new and good?" is used in some counseling theories to start the session. Likely more helpful is your ability to search out hidden strengths that have helped the person survive difficulties and trauma. Commenting periodically on clients' abilities and drawing out stories of strength are important in counteracting the inevitable negatives we experience daily. Don't forget spiritual and family strengths.

Other Potential Difficulties with Questions

Bombardment/Grilling. Too many questions may give too much control to the interviewer and tend to put clients on the defensive.

Multiple Questions. Another form of bombardment, throwing out too many questions at once may confuse clients. However, it may enable clients to select which question they prefer to answer.

Questions as Statements. Some interviewers may use questions to sell their own points of view. "Don't you think it would be helpful if you studied more?" This question clearly puts the client on the spot. On the other hand, "What occurs to you as you think about improving your grades?" might be helpful to get some clients thinking in new ways. Consider alternative and more direct routes of reaching the client. A useful standard is this: If you are going to make a statement, do not frame it as a question.

Why Questions. *Why* questions can put interviewees on the defensive and cause discomfort. As children, most of us experienced some form of "Why did you do that?" Any question that evokes a sense of being attacked can create client discomfort and defensiveness.

Observe: Questions in the Interview

Virtually all of us have experienced conflict on the job: angry or difficult customers, insensitive supervisors, lazy colleagues, or challenges from those we supervise. In the following set of transcripts, we see an employee assistance counselor, Jamila, meeting with Kelly, a junior manager who has a conflict with Peter.

Closed Question Example

The first session illustrates how closed questions can bring out specific facts but can sometimes end in leading the client, even to the point of putting the counselor's ideas into the client's mind.

Counselor and Client Conversation	Process Comments
1. *Jamila:* Hi, Kelly. What's happening with you today?	Jamila has talked with Kelly once in the past about difficulties she was having in her early experiences supervising others for the first time. She begins with an open question that could also be seen as a standard social greeting.
2. *Kelly:* Well, I'm having problems with Peter again.	Jamila and Kelly have a good relationship. Not all clients are so ready to discuss their issues. More time for developing rapport and trust will be necessary for many clients, even on return visits.
3. *Jamila:* Is he arguing with you?	Jamila appears interested, is listening and demonstrating good attending skills. However, she asks a closed question; she is already defining the issue without hearing the latest story and discovering Kelly's thoughts and feelings. (Subtractive empathy)
4. *Kelly:* (hesitates) Not really, he's so difficult to work with.	Kelly sits back in her chair and waits for the counselor to take the lead.
5. *Jamila:* Is he getting his work in on time?	Jamila tries to diagnose the problem with Peter by asking a series of closed questions. It is much too early in the session for a diagnosis. (Subtractive; note that the counselor is supplying concreteness, not the client.)
6. *Kelly:* No, that's not the issue. He's even early.	
7. *Jamila:* Is his work decent? Does he do a good job?	Jamila is starting to grill Kelly. (Subtractive)
8. *Kelly:* That's one of the problems; his work is excellent, always on time. I can't criticize what he does.	
9. *Jamila:* (hesitates) Is he getting along with others on your team?	Jamila frowns and her body tenses as she thinks of what to ask next. Counselors who rely on closed questions suddenly find themselves having run out of questions to ask. They continue searching for another closed question usually further off the mark.

Counselor and Client Conversation	Process Comments
10. *Kelly:* Well, he likes to go off with Daniel, and they laugh in the corner. It makes me nervous. He ignores the rest of the staff—it isn't just me.	
11. *Jamila:* So, it's you we need to work on. Is that right?	Jamila has been searching for an individual to blame. Jamila relaxes a little as she thinks she is on to something. Kelly sits back in discouragement. (Very subtractive and off the mark)
12. *Kelly:* (hesitates and stammers) . . . Well, I suppose so . . . I . . . I . . . really hope you can help me work it out.	Kelly looks to Jamila as the expert. While she dislikes taking blame for the situation, she is also anxious to please and too readily accepts the counselor's diagnosis.

Closed questions can overwhelm clients and can be used as evidence to force them to agree with the counselor's ideas. While the session above seems extreme, encounters like this are common in daily life and even occur in counseling and psychotherapy sessions. There is a power differential between clients and counselors. It is possible that a counselor who fails to listen can impose inappropriate decisions on a client.

Open Question Example

The session is for the client, not the counselor. Using open questions, Jamila learns Kelly's story rather than the one she imposed with closed questions in the first example. Again, this interview is in the employee assistance office.

Counselor and Client Conversation	Process Comments
1. *Jamila:* Hi, Kelly. What's happening with you today?	Jamila uses the same easy beginning as in the closed question example. She has excellent attending skills and is good at relationship building.
2. *Kelly:* Well, I'm having problems with Peter again.	Kelly responds in the same way as in the first demonstration.
3. *Jamila:* More problems? Could you share more with me about what's been happening lately?	Open questions beginning with *could* provide some control to the client. Potentially a *could* question may be responded to as a closed question and answered with "yes" or "no." But in the United States, Canada, and other English-speaking countries, it usually functions as an open question. (Aiming toward concreteness, this is interchangeable empathy.)
4. *Kelly:* This last week Peter has been going off in the corner with Daniel, and the two of them start laughing. He's ignoring most of our staff, and he's been getting under my skin even more lately. In the middle of all this, his work is fine, on time and near perfect. But he is so impossible to deal with.	We are hearing Kelly's story. The anticipated result from open questions is that Kelly will respond with information. She provides an overview of the situation and shares how it is affecting her.
5. *Jamila:* I hear you. Peter is getting even more difficult and seems to be affecting your team as well. It's really stressing you out, and you look upset. Is that pretty much how you are feeling about things?	When clients provide lots of information, we need to ensure that we hear them accurately. Jamila summarizes what has been said and acknowledges Kelly's emotions. The closed question at the end is a **perception check**, or **checkout**. Periodically checking with your client can help you in two ways: (a) It communicates to clients that you are listening and encourages them to continue; (b) it allows the client to correct any wrong assumptions you may have. (Interchangeable empathy)

(continued)

Counselor and Client Conversation	Process Comments
6. *Kelly:* That's right. I really need to calm down.	Even though discussing a difficult issue, Jamila notices Kelly relaxing a bit.
7. *Jamila:* Let's change the pace a bit. Could you give me a specific example of an exchange you had with Peter last week that didn't work well?	Jamila asks for a concrete example. Specific illustrations of client issues are often helpful in understanding what is really occurring. (Potentially additive empathy)
8. *Kelly:* Last week, I asked him to review a bookkeeping report prepared by Anne. It's pretty important that our team understand what's going on. He looked at me like, "Who are you to tell *me* what to do?" But he sat down and did it that day. Friday, at the staff meeting, I asked him to summarize the report for everyone. In front of the whole group, he said he had to review this report for me and joked about me not understanding numbers. Daniel laughed, but the rest of the staff just sat there. He even put Anne down and presented her report as not very interesting and poorly written. He was obviously trying to get me. I just ignored it. But that's typical of what he does.	Specific and concrete examples can be representative of recurring issues that repeat themselves over and over. The concrete specifics from one or two detailed stories can lead to a better understanding of what is really happening. Now that Jamila has heard the specifics, she is better prepared to be helpful.
9. *Jamila:* Underneath it all, you're furious. Kelly, why do you imagine he is doing that to you?	She begins with a reflection of feeling. Will the *why* question lead to the discovery of reasons? (Potentially subtractive, and note Kelly's response. Later this might be appropriate.)
10. *Kelly:* (hesitates) Really, I don't know why. I've tried to be helpful to him.	The microskill did not result in the expected response. This is, of course, not unusual. Likely this is too soon for Kelly to know why. This illustrates a common problem with *why* questions.
11. *Jamila:* Gender can be an issue; men do put women down at times. Would you be willing to consider that possibility?	Jamila carefully presents her own hunch. But instead of expressing her own ideas as truth, she offers them tentatively with a *would* question and reframes the situation as a "possibility." (Potentially additive)
12. *Kelly:* Jamila, it makes sense. I've halfway thought of it, but I didn't really want to acknowledge the possibility. But it is clear that Peter has taken Daniel away from the team. Until Peter came aboard, we worked together beautifully. (pause) Yes, it makes sense for me. I think he's out to take care of himself. I see Peter going up to my supervisor all the time. He talks to the female staff members in a demeaning way. Somehow, I'd like to keep his great talent on the team, but how when he is so difficult?	With Jamila's help, Kelly is beginning to obtain a broader perspective. She thinks of several situations indicating that Peter's ambition and sexist behavior are issues that need to be addressed. (Here we see Kelly beginning to be able to create new thoughts to help her look at the situation. Jamila's previous comment was additive.)
13. *Jamila:* So, the problem is becoming clearer. You want a working team and you want Peter to be part of it. We can explore the possibility of assertiveness training as a way to deal with Peter. But, before that, what do you bring to this situation that will help you deal with him?	Jamila provides support for Kelly's new frame of reference and ideas for where the session can go next. She suggests that time needs to be spent on finding positive assets and wellness strengths. Kelly can best resolve these issues if she works from a base of resources and capabilities. (Potentially additive; looks for concrete specifics of what Kelly can do rather than what she can't do)

Counselor and Client Conversation	Process Comments
14. *Kelly:* I need to remind myself that I really do know more about our work than Peter. I worked through a similar issue with Jonathan two years ago. He kept hassling me until I had it out with him. He was fine after that. I know my team respects me; they come to me for advice.	Kelly smiles for the first time. She has sufficient support from Jamila to readily come up with her strengths. However, don't expect it always to be that easy. Clients may return to their weaknesses and ignore their assets. Clients can often resolve their own issues if they remember their strengths and abilities.
15. *Jamila:* Could you tell me specifically what happened when you sat down and faced Jon's challenge directly?	This *could* question searches for concrete specifics when Kelly handled a difficult situation effectively. Jamila can identify specific skills that Kelly can later apply to Peter. At this point, the session can move from problem definition to problem solution.

In this excerpt, we see that Kelly has been given more talk time and room to explore what is happening. The questions, focused on specific examples, clarify what is happening. We also see that question stems such as *why, how,* and *could* have some predictability in expected client responses. The positive asset search is a particularly relevant part of successful questioning. Issues are best resolved by emphasizing strengths.

You are very likely to work with clients who have similar interpersonal issues wherever you may practice. The previous case examples focus on the single skill of questioning as a way to bring out client stories. Questioning is an extremely helpful skill, but do not forget the dangers of using too many questions.

Multiple Applications of Questions

Behavioral psychology's **antecedent-behavior-consequence (ABC)** model is particularly useful when a challenging specific situation is unclear. The ABC pattern brings out key facts of the event. We suggest that this sequence is one of several that are well worth memorizing.

1. *Antecedent:* Draw out the linear sequence of the story: "What happened first? What happened next? What was the result?"

2. *Behavior:* Focus on observable concrete actions: "What did the other person say? What did he or she do? What did you say or do?"

3. *Consequence:* Help clients see the result of an event: "What happened afterward? What did you do afterward? What did he or she do afterward?" Sometimes clients are so focused on the event that they don't yet realize it is over.

Albert Ellis's rational emotional behavior therapy (REBT) adds two more often essential issues: T for thoughts and cognitions (the prefrontal cortex TAP executive system) and E for emotions (relating to emotional regulation, involving the limbic system and hormonal impact on the brain and body). Usually it is most helpful to start with the rational thoughts, followed by exploration of emotions.

Thoughts: "What was going on in your mind—what were you thinking?" And later, "What might the other person have been thinking?" Variations of these questions can be used at all three phases when drawing out the ABC behaviors.

Emotions: "How did you feel and what were your emotions just before it happened? During? After?" Later, "How do you think the other person felt?"

Note that each of the ABC questions requires a relatively short concrete answer. Do not expect your less verbal client to give you full answers to these questions. You may need to ask closed questions to fill in the details and obtain specific information. "Did he say anything?" "Where was she?" "Is your family angry?" "Did they say 'yes' or 'no'?" As you move to the thought and feeling/emotional questions, longer answers are to be expected. Many children can share their thoughts and feelings during teasing, bullying, or other troubling events.

In short, in drawing out any story or concern, the ABC-TF framework of antecedents, behaviors, consequences, and the accompanying thoughts and feelings will give you a comprehensive picture of the client's experiential world. Again, use this model to draw out client strengths as well as concerns.

A Cross-Cultural Example of How Questions May Be Inappropriate

If your life background and experience are similar to your client's, you may be able to use questions immediately and freely. If you come from a significantly different cultural background, your questions may be met by distrust and given only grudging answers. Questions place power with the interviewer. A poor client who is clearly in financial jeopardy may not come back for another interview after receiving a barrage of questions from a clearly middle-class interviewer. If you are African American or European American and working with an Asian American or a Latina/o, an extreme questioning style can produce mistrust. If the ethnicities are reversed, the same problem can occur.

Allen was conducting research and teaching in South Australia with Aboriginal social workers. He was seeking to understand their culture and their special needs for training. Allen is naturally inquisitive and sometimes asks many questions. Nonetheless, the relationship between him and the group seemed to be going well. But one day, Matt Rigney, whom Allen felt particularly close to, took him aside and gave some very useful corrective feedback:

> You White fellas! . . . Always asking questions! My culture considers many questions rude. But I know you, and that's what you do. But this is what goes on in my mind when you ask me a question. First, I wonder if I can trust you enough to give you an honest answer. Then, I realize that the question you asked is too complex to be answered in a few words. But I know you want an answer. So I chew on the question in my mind. Then, you know what? Before I can answer the first question, you've moved on to the next question!

Allen was lucky he had developed enough trust and rapport that Matt was willing to share his perceptions. Many People of Color have said that this kind of feedback represents how they feel about interactions with White people, but they do not share their feelings and stay quiet. People with disabilities, gays/lesbians/bisexuals/questioning/transgendered people, spiritually conservative persons, and many others—anyone, in fact—may be distrustful of the interviewer who uses too many questions.

On the other hand, questions can be useful in group discussions to help at-risk youth redefine themselves in a more positive way, as suggested in Box 5.1.

Focus on Hesitant Clients and Children

Generally in the interview, open questions are preferred over closed questions. Yet it must be recognized that open questions require a verbal client, one who is willing to share with you. Here are some suggestions to encourage clients to talk with you more freely.

BOX 5.1 National and International Perspectives on Counseling Skills

Using Questions with Youth at Risk
Courtland Lee, Past President, American Counseling Association, University of Maryland

Malik is a 13-year-old African American male who is in the seventh grade at an urban junior high school. He lives in an apartment complex in a lower-middle-class (working-class) neighborhood with his mother and 7-year-old sister. Malik's parents have been divorced since he was 6, and he sees his father very infrequently. His mother works two jobs to hold the family together, and she is not able to be there when he and his sister come home from school.

Throughout his elementary school years, Malik was an honor roll student. However, since starting junior high school, his grades have dropped dramatically, and he expresses no interest in doing well academically. He spends his days at school in the company of a group of seventh- and eighth-grade boys who are frequently in trouble with school officials.

This case is one that is repeated among many African American early teens. But this problem also occurs among other racial/ethnic groups, particularly those who are struggling economically. And the same pattern occurs frequently even in well-off homes. There are many teens at risk for getting in trouble or using drugs.

While still a boy, Malik has been asked to shoulder a man's responsibilities as he must pick up things his mother can't do. Simultaneously, his peer group discounts the importance of academic success and wants to challenge traditional authority. And Malik is making the difficult transition from childhood to manhood without a positive male model.

I've developed a counseling program designed to empower adolescent Black males that focuses on personal and cultural pride. The full program is outlined in my book *Empowering Black Males* (1992) and focuses on the central question, "What is a strong Black man?" Although this question is designed for group discussion, it is also one for adolescent males in general, who might be engaging

in individual counseling. The idea is to use this question to help the youth redefine in a more positive sense what it means to be strong and powerful. Some related questions that I find helpful are

- What makes a man strong?
- Who are some strong Black men that you know personally? What makes these men strong?
- Do you think that you are strong? Why?
- What makes a strong body?
- Is abuse of your body a sign of strength?
- Who are some African heroes or elders that are important to you? What did they do that made them strong?
- How is education strength?
- What is a strong Black man?
- What does a strong Black man do that makes a difference for his people?
- What can you do to make a difference?

Needless to say, you can't ask an African American adolescent or a youth of any color these questions unless you and he are in a positive and open relationship. Developing sufficient trust so that you can ask these challenging questions may take time. You may have to get out of your office and into the school and community to become a person of trust.

My hope for you as a professional counselor is that you will have a positive attitude when you encounter challenging adolescents. They are seeking models for a successful life, and you may become one of those models yourself. I hope you think about establishing group programs to facilitate development and that you'll use some of these ideas with adolescents to help move them toward a more positive track.

Build Trust at the Client's Pace. A central issue with hesitant clients is trust. Children are generally trusting, but extensive questioning too early can make trust building a slow process with some clients. If the client is required to meet with you or is culturally different from you, he or she may be less willing to talk. Trust building and rapport need to come first, and your own natural openness and social skills are essential. With some clients, trust building may take a full session or more.

Search for Concrete Specifics. Some interviewers and many clients talk in vague generalities. We call this "talking high on the abstraction ladder." This may be contrasted with concrete and specific language, where what is said immediately makes sense. If your client is talking in very general terms and is hard to understand, it often helps to ask questions from the antecedent-behavior-consequence (ABC) pattern above. Add thoughts and feelings as appropriate.

As the examples become clearer, ask even more specific questions: "You said that you are not getting along with your teacher. What specifically did your teacher say (or do)?" Your chances for helping the client talk will be greatly enhanced when you focus on concrete events in a nonjudgmental fashion, avoiding evaluation and opinion.

At the same time, remember that this sequence is also invaluable for drawing stories of success that can be used as a basis for resilience and growth. *Try to balance what the child or client can't do with what he or she can do.* This is the best way to build resilience and strength to solve current issues and future concerns that will inevitably arise.

A *leading* closed question is dangerous, particularly with children. ABC questions need to be phrased carefully and empathically. You have seen that a long series of closed questions may bring out the story, but may provide only the client's limited responses to *your* questions rather than what the client really thinks or feels. Worse, the client may end up adopting your way of thinking or may simply stop coming to see you. Or false memories or distortions can arise.

Clients also ask us questions, and answering many of them is a challenge. Here are some questions that Mary received when she worked as an elementary school counselor. They are not easy to answer and illustrate the importance of listening, intentionality, and being active and ready for anything. Note how family centered the children's issues are. These are predominantly cognitive questions, but imagine the power of emotions underlying each. Children's issues are fully as challenging as those that teens and adults experience—and they don't have the same level of experience that we do.

> What do you do when your little sister bugs you and you get into trouble?
>
> How can I keep from fighting my brother and sister? Why do people always fight?
>
> My brother always throws me down the stairs. I scream and I get into trouble.
>
> My big brother calls me names, but I never call him names. How can I live with him?
>
> I hate my mom's boyfriend. What should I do?
>
> My mom has a boyfriend and we would fight and get my mom mad. But now we are getting along. But he seems far away from me and I think he feels unsure with me. What should I do?
>
> Last year my parents were not getting along at all. What do you say if your parents ask you who you would rather live with if they got a divorce? How do I answer that?
>
> Why do parents get divorced? Is it because of me?
>
> How can my parents get back together?

Mary points out that we always need to listen. For example, with the client who is being bugged by a sibling, empathically draw out the story of what happened. Consider the ABC pattern and possibly the thoughts and feelings as well. But listening is often not enough. You will often have to turn to the influencing skills of the last half of this book—and to your own wisdom and personal experience, including even providing commonsense advice.

It helps children talk if they have something to do with their hands, such as drawing, playing with small toys, or playing with a house that includes child and adult figures. A sandbox can be an important play situation for full involvement. If the problem seems to be turning into bullying, this may be reflected in the classroom or other personal relations. Observe the child in the classroom and on the playground. Parents may sometimes need to be involved.

The more difficult issues occur when a parental separation is involved. Children often think that they are the cause. Or they may think that they are somehow responsible for patching things up. Your first task is to reassure them that it is definitely not their fault. They love their parents, and their parents love them. This is an issue between their parents, and they will be there for support. (If not, be prepared for more complex issues.)

These situations require endless empathy and understanding. Sorting out the conflict and working with the child's varied emotions—which can range from anger and fears to tears—will be essential. At this time, watch for issues on the playground and in the classroom. The child may be depressed or may act out. Informing teachers about the situation and helping them support the child is important. Where possible, talk to the parents and help them understand what is occurring for their child or children. Children may have "abnormal" responses in these situations, but they really are having normal responses to an abnormal situation. Avoid any blaming of the child.

| **BOX 5.2** | **Research and Related Neuroscience Evidence That You Can Use** |

Questions

Research results can be a guide or a confirmation for what types of questions work best in certain situations or with particular theoretical orientations. Not too surprising is confirmation that open questions produce longer client responses than those that are closed (Ivey & Daniels, 2016; Tamase, Torisu, & Ikawa, 1991).

Different theoretical orientations to helping vary widely in their use of questions. Person-centered leaders tend to use very few questions, whereas 40% of the leads of a problem-solving group leader were questions (Ivey, Pedersen, & Ivey, 2001; Sherrard, 1973). You will find that cognitive behavioral therapy, motivational interviewing, and brief counseling use questions extensively (Chapters 13 and 14). Your decision about which theory of helping you use most frequently will in part determine your use of questioning skills.

Understanding emotions is central to the helping process. Clients may most easily talk about feelings when they are asked about them directly (Hill, 2014; Tamase, 1991). If you are to reflect feelings and explore background emotions (Chapter 7), you will often have to ask questions.

Drawing Out Stories

At this point, it may helpful to visit pages 392–394 of Appendix IV. Questions dealing with the facts in the antecedent-behavior-consequence framework, as well as those focusing on thoughts, are related most closely to areas of the prefrontal cortex (PFC). Feeling and emotions are related to the limbic system, particularly the amygdala.

Negative emotions are most associated with the amygdala, positives with the PFC, which is also concerned with emotional regulation. A decision is generally not fully effective unless there is a feeling of satisfaction.

Questions are also a good way to help a client discuss issues from the past residing in long-term memory, lodged primarily in the executive cortex and hippocampus (Kolb & Wishaw, 2009). The goal of questions is to obtain information that will enhance client growth and ultimately generate new, more positive and accurate neural networks.

The Danger of False Memories

However, questions that lead clients too much can result in their constructing stories of things that never happened. In a classic study, Loftus (1997, 2011) found that false memories could be brought out simply by reminding people of things that never happened. In other studies, brain scans have revealed that false memories activate different patterns than those that are true (Abe et al., 2008). Obviously, you as a counselor or therapist don't have that information available, and you may not know whether the memories a client reports are true or false. Be careful of putting your ideas into the client's head by means of probing questions.

If you use too many questions, you may have the best of intentions, but the possibility is that clients will end up taking your point of view rather than developing their own. We want clients to find their own direction. And directive intrusive questions may even lead to new false client memories.

Children whose parents are separating or divorced are not alone. Mary found that bringing such children together in friendship/support groups was often helpful. This often occurred at lunch, adding a friendly atmosphere. Here the child learns that others feel and think as they do. They may discover new ideas for their own lives, but always it is support from others that will be critical.

With challenging child cases, a team approach and meeting may be necessary, including the teacher, principal, school social worker, special education teacher, and others. Depending on the situation, you may meet individually with the parents one at a time, or possibly together. At times, they will respond better to members of the team.

We began this chapter by asking you to think carefully about your personal experience with questions. Clearly, their overuse can damage the relationship with the client. On the other hand, questions facilitate conversation and help ensure that a complete picture is obtained. Questions can help the client bring in missing information. Among such questions are "What else?" "What have we missed so far?" and "Can you think of something important that is occurring in your life right now that you haven't shared with me yet?"

Person-centered theorists and many professionals sincerely argue against the use of any questions at all. They strongly object to the control implications of questions. They point out that careful attending and use of the listening skills can usually bring out major client issues. If you work with someone culturally different from you, a questioning style may develop distrust. In such cases, questions need to be balanced with self-disclosure and listening.

Our position on questions is clear: We believe in questions, but we also fear overuse and the fact that they can reduce equality in the session. We are impressed by the brief solution-focused counselors who seem to use questions more than any other skill but are still able to respect their clients and help them change. On the other hand, we have seen students who have demonstrated excellent attending skills regress to using only questions. Questions can be an easy "fix," but they require listening to the client if they are to be meaningful.

Action: Key Points and Practice of Questions

How are you going to use questions in your own practice? The key points below can serve as beginning guidelines.

Act. Apply what you have learned in your everyday activities. Let the key points guide your actions. Make a contract with yourself to do this. Practice makes perfect!

Value of Questions. Questions are a key component in many theories and styles of helping. Questions help begin the session, open new areas for discussion, assist in pinpointing and clarifying issues, and assist the client in self-exploration.

Open Questions. Questions can be described as open or closed. *Open questions* are those that can't be answered in a few words. They encourage others to talk and provide you with maximum information. Typically, open questions begin with *what*, *how*, *why*, or *could*. One of the most helpful of all open questions is "Could you give a specific example of . . . ?"

Closed Questions. *Closed questions* are those that can be answered in a few words or sentences. They have the advantage of focusing the session and bringing out specifics, but they place the prime responsibility for talk on the counselor. Closed questions often begin with *is*, *are*, or *do*. An example is "Where do you live?"

Note that a question, open or closed, on a topic of deep interest to the client will often result in extensive talk time if interesting or important enough. If a session is flowing well, the distinction between open and closed questions becomes less relevant.

"What Else?" Questions. *What else* is there to add to the story? Have we missed anything? "What else?" questions bring out missing data. These are maximally open and allow the client considerable control.

Promoting Client Elaboration. Open questions can help clients elaborate and enrich their story. Questions can reveal concrete specifics from the client's world.

The Negative Approach. Counseling and psychotherapy are typically seen as focused on life challenges and problems. But this focus needs to be balanced with questions that bring out client strengths, supports in the family or friendship group, and past and present accomplishments. Counseling session training can overemphasize concerns and difficulties. A positive approach is needed for balance.

The Positive Approach. Emphasizing only negative issues results in a downward cycle of depression and discouragement. The positive asset search, strength emphasis, positive psychology, and wellness need to balance discussion of client issues and concerns. What is the client doing right? What are the exceptions to the problem? What are the client's new options? How would these options enrich the client's life?

Multiple Applications of Questions. The antecedent-behavior-consequence (ABC) model helps draw out key facts about events, especially in unclear and challenging situations. By moving to ABC-TF, we bring in thoughts and feelings about the event or personal experience.

Multicultural Issues. All these questions may turn off some clients. **Some** cultural groups find North American rapid-fire questions rude and intrusive, particularly if asked before trust is developed. Yet questions are very much a part of Western culture and provide a way to obtain information that many clients find helpful. Questions help us find the client's personal, family, and cultural/contextual resources. If properly structured and your clients know the real purpose is to help them reach their own goals, questions may be used more easily.

Practice and Feedback: Individual, Group, and Microsupervision

Additional resources can be found by going to CengageBrain.com and logging into the MindTap course created by your professor. There you will find a variety of study tools and useful resources that include quizzes, videos, interactive counseling and psychotherapy exercises, case studies, the Portfolio of Competencies, and more.

Take time to master the many concepts and skills presented in this chapter and make them a useful part of your counseling or psychotherapy. These exercises will help you achieve this goal, but you should continue to work on these concepts throughout the book and beyond. With practice, all these materials will become clearer and, most important, will become a part of your natural style.

Individual Practice

Exercise 5.1 Writing Closed and Open Questions

Select one or more of the following client stories and then write open and closed questions to elicit further information. Can you ask closed questions designed to bring out specifics of the situation? Can you use open questions to facilitate further elaboration of the topic, including the facts, feelings, and possible reasons? What special considerations might be beneficial with each person as you consider age-related multicultural issues?

> *Jordan* (age 15, African American): I was walking down the hall and three guys came up to me and called me "queer" and pushed me against the wall. They started hitting me, but then a teacher came up.
>
> *Alicja* (age 35, Polish American): I've been passed over for a promotion three times now. Each time, it's been a man who has been picked for the next level. I'm getting very angry and suspicious.
>
> *Dominique* (age 78, French Canadian): I feel so badly. No one pays any attention to me in this "home." The food is terrible. Everyone is so rude. Sometimes I feel frightened.

Write open questions for one or more of the above. The questions should be designed to bring out broad information, facts, feelings and emotions, and reasons.

Could . . . ?

What . . . ?

How . . . ?

Why . . . ?

Now generate three closed questions that might bring out useful specifics of the situation.

Do . . . ?

Are . . . ?

Where . . . ?

Finally, write a question designed to obtain concrete examples and details that might make the problem more specific and understandable.

Exercise 5.2 Observation of Questions in Your Daily Interactions

This chapter has talked about the basic question stems *what, how, why,* and *could,* and how clients respond differently to each. During a conversation with a friend or acquaintance, try these five basic question stems sequentially. Note that this is another way to bring out the ABC behaviors plus thoughts and emotions.

Could you tell me generally what happened?

What are the critical facts?

How do you feel about the situation?

Why do you think it happened?

What else is important? What have we missed?

Record your observations. Were the anticipated results or outcomes fulfilled? Did the person provide you, in order, (1) a general picture of the situation, (2) the relevant facts, (3) personal feelings about the situation, and (4) background reasons that might be causing the situation?

Exercise 5.3 Individual Practice with Your Own Video or Audio Feedback

Make a recording of a practice interview. Placing your phone, computer camera, or camera is a bit of a challenge, but "it works." If possible use the recording first for client feedback as you play though the video. How does the client react, and what does he or she notice? Then review the video by yourself, noting how the client responded to each of the questions.

Please write down or photocopy the specific antecedent-behavior-consequence and the thoughts and feelings questions on page 119. With a volunteer client who is willing to share a positive story of strength and resilience, use those questions to have the client elaborate on his or her experience. Some clients may also want to explore current or past issues, but do not do that until you have focused first on strengths.

Group Practice and Microsupervision

Exercise 5.4 Group Practice and Microsupervision

The following exercise is suggested for practice with questions. The objective is to use both open and closed questions. The instructional steps for practice are abbreviated from those described in Chapter 3 on attending behavior. As necessary, refer to those instructions for more detail on the steps for systematic practice.

Step 1: Divide into practice groups.

Step 2: Select a group leader.

Step 3: Assign roles for the first practice session. Client, counselor, and the microsupervision observers who use the Feedback Form. Remember to focus on counselor strengths as well as areas for improvement.

Step 4: Plan. The counselor should plan to use both open and closed questions. Include in your practice session the key *what*, *how*, *why*, and *could* questions. Add *what else* for enrichment.

Discuss a work challenge. The client may share a present or past interpersonal job conflict. The counselor first draws out the conflict, then searches for positive assets and strengths.

Suggested alternative topics might include the following:

❑ A friend or family member in conflict
❑ A positive addiction (such as jogging, health food, biking, team sports)
❑ Strengths from spirituality or ethnic/racial background

Step 5: Conduct a 3- to 6-minute practice session using only questions. The counselor should use mostly questions during this practice. When reviewing the session, note both the nature of the questions and if the client responded as anticipated.

Step 6: Review the practice session and provide feedback for 6 to 7 minutes. Remember to stop the audio or video recording periodically and listen to or view key items several times for increased clarity. Observers should give special attention to careful completion of the Feedback Form in Box 5.3 throughout the session, and the client can give important feedback via the Client Feedback Form in Chapter 1.

Step 7: Rotate roles.

BOX 5.3	Feedback Form: Questions

_____
(DATE)

_____   _____
(NAME OF COUNSELOR)   (NAME OF PERSON COMPLETING FORM)

Instructions: On the lines below, list as completely as possible the questions asked by the counselor. At a minimum, indicate the first key words of the question (*what, why, how, do, are,* and so on). Indicate whether each question was open (O) or closed (C). Use additional paper as needed. Does the session focus on strengths and goal attainment? How well did the counselor listen as well as ask questions?

1. _____

2. _____

3. _____

4. _____

5. _____

6. _____

7. _____

8. _____

9. _____

10. _____

1. Which questions seemed to provide the most useful client information?

2. Did the client respond as anticipated, or did the client go in a different direction?

3. Provide specific feedback on the attending skills of the counselor.

4. Discuss the use of the positive asset search and wellness, as well as the use of questions.

Portfolio of Competencies and Personal Reflection

Each chapter closes with a reflective exercise asking about your skills, thoughts, and feelings about what has been discussed. By the time you finish this book, you will have a substantial record of your competencies and a good written record as you move toward determining your own style and theory.

Assessing Your Level of Competence: Awareness, Knowledge, Skills, and Action

Use the following checklists to evaluate your present level of mastery. Check those dimensions that you currently feel able to do. Those that remain unchecked can serve as future goals. Do not expect to attain intentional competence on every dimension as you work through this book. You will find, however, that you will improve your competencies with repetition and practice.

Awareness and Knowledge. Can you identify and write the following?

❑ Identify and classify open and closed questions.

❑ Discuss, in a preliminary fashion, issues in diversity that occur in relation to questioning.

❑ Write open and closed questions that might anticipate what a client will say next.

Basic Competence. Aim for this level of competence before moving on to the next skill area.

❑ Ask both open and closed questions in a role-played session.

❑ Obtain longer responses to open questions and shorter responses to closed questions.

❑ Describe the strength emphasis.

❑ Identify wellness needs.

Intentional Competence. Can you do the following? Work toward intentional competence throughout this book. All of us can improve our skills, regardless of where we start.

❑ Use closed questions to obtain necessary facts without disturbing the client's natural conversation.

❑ Use open questions to help clients elaborate their stories.

❑ Use *could* questions and, as anticipated, obtain a general client story ("Could you tell me generally what happened?" "Could you tell me more?").

❑ Use *what* questions to facilitate discussion of facts.

❑ Use *how* questions to bring out feelings ("How do you feel about that?") and information about process or sequence ("How did that happen?").

❑ Use *why* questions to bring out client reasons ("Why do think your spouse/lover responds coldly?").

❑ Focus on searching for resilience and strengths.

❑ Bring out concrete information and specifics ("Could you give me a specific example?").

❑ Use the antecedent-behavior-consequence plus thoughts and feelings (ABC—TF) to being out concrete specifics of clients stories, including stories of strengths and resilience.

Psychoeducational Teaching Competence. As stated earlier, do not expect to become skilled in teaching skills to groups or peer counselors at this point. You may find, however, that some clients benefit from direct instruction in open questions focusing on others' thoughts and opinions rather than their own. Those who talk too much about themselves find this skill useful in breaking through their self-absorption. At the same time, please point out the dangers of too many questions, especially that *why* question, which can put others on the spot and make them defensive.

❑ Teach clients in a helping session the social skill of questioning. You may either tell clients about the skill or practice a role-play with them.

❑ Teach small groups the skills of questioning.

Personal Reflection on Questions

This chapter has focused on the pluses and minuses of using questions in the session. While we, as authors, obviously feel that questions are an important part of the counseling process, we have tried to point out that there are those who differ from us. Questions clearly can get in the way of effective relationships in counseling and psychotherapy.

Regardless of what any text on counseling and psychotherapy says, the fact remains that it is *you* who will decide whether to implement the ideas, suggestions, and concepts.

What single idea stands out for you among all those presented in this chapter, in class, or through informal learning? What stands out for you is likely to be important as a guide toward your next steps.

What are your thoughts on multiculturalism and how it relates to your use of questions? How might you use ideas in this chapter to begin the process of establishing your own style and theory?

Complement your self-assessment with direct feedback from those affected by your use of the skill in the real world. What was their reaction, how do they feel, how they describe your interaction? Use their feedback to further assess your skills.

Our Thoughts About Benjamin

We would probably start the session by explaining to Benjamin that we'd like to know what he is thinking about his future after he completes school. We would begin the session with some informal conversation about current school events or something personal we know about him. The first question might be stated something like this: "You'll soon be starting your senior year; what have you been thinking about doing after you graduate?" If this question opens up some tentative ideas, we'd listen to these and ask him for elaboration. If he focuses on indecision between volunteering for the army and entering a local community college or the state university, we'd likely ask him some of the following questions:

"What about each of these appeals to you?"

"Could you tell me about some of your strengths that would help you in the army or college?"

"If you went to college, what might you like to study?"

"How do finances play a role in these decisions?"

"Are there any negatives about any of these possibilities?"

"How do you imagine your ideal life 10 years from now?"

On the other hand, Benjamin just might say to any of these, "I don't know, but I guess I better start thinking about it" and look to you for guidance. You sense a need to review his past likes and dislikes as possible clues to the future.

"What courses have you liked best in high school?"

"What have been some of your activities?"

"Could you tell me about the jobs you've had in the past?"

"Could you tell me about your hobbies and what you do in your spare time?"

"What gets you most excited and involved?"

"What did you do that made you feel most happy in the past year?"

Out of questions such as these, we may see patterns of ability and interest that suggest actions for the future.

If Benjamin is uncomfortable in the counseling office, all of these questions might put him off. He might feel that we are grilling him and perhaps even see us as intruding in his world. Generally speaking, getting this type of important information and organizing it requires the use of questioning. But questions are effective only if you and the client are working together and have a good relationship.

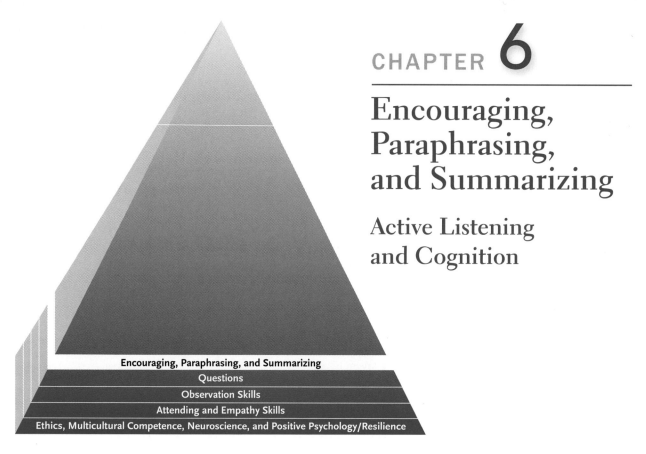

CHAPTER **6**

Encouraging, Paraphrasing, and Summarizing

Active Listening and Cognition

Encouraging, Paraphrasing, and Summarizing
Questions
Observation Skills
Attending and Empathy Skills
Ethics, Multicultural Competence, Neuroscience, and Positive Psychology/Resilience

Emotion is the system that tells us how important something is. Attention focuses us on the important and away from the unimportant things. Cognition tells us what to do about it. Cognitive skills are whatever it takes to do those things.

—Alvaro Fernandez

Chapter Goals and Competency Objectives

Awareness and Knowledge

▲ Value active listening in the communication process.

▲ Identify the role of intentional participation, decision making, and responding to client conversation.

Skills and Action

▲ Help clients talk in more detail about their issues of concern and help prevent the overly talkative client from repeating the same facts. Clarify for the client and you, the interviewer, what is really being said during the session.

▲ Check on the accuracy of what you hear by saying back to clients the essence of their comments and providing periodic summarizations.

▲ Develop cognitive empathy and facilitate client cognitive understanding for clearer decision making and more effective action.

▲ Through cognitive empathy, promote development of the brain's executive functions, central for organizing thoughts, regulating emotions, planning action, and implementing planned actions.

Introduction: Encouraging, Paraphrasing, and Summarizing

> To show that you understand exactly, make a sentence or two which gets exactly at the personal meaning the person wanted to put across. This might be in your own words, usually, but use that person's own words for the touchy main things.
>
> —Eugene Gendlin and Marion Hendricks-Gendlin

Encouraging, paraphrasing, and summarizing are active listening skills that are the cognitive center of the basic listening sequence and are key in building the empathic relationship. When we attend and clients sense that their story is heard, they open up and become more ready for change. This leads to more effective **executive brain functioning**, which in turn improves **cognitive** understanding, organization of issues, and decision making.

Emotional regulation is the second critical aspect of executive brain functioning. The next chapter on reflecting feelings discusses drawing out client emotions and balancing feelings with cognitive reality.

Active listening is a communication process that requires intentional participation, decision making, and responding to client conversation. What we listen to (selective attention) and respond to have a profound influence on how clients talk to us about their concerns. When a client shares with us a lot of information all at once and talks rapidly, we can find ourselves confused and even overwhelmed by the complexity of the story. We need to hear this client accurately and often slow the story down a bit. Our accurate listening, in turn, leads to client understanding and synthesis, providing clients with a clearer picture of their own stories.

Active listening is central in facilitating our brain's executive functioning—the cognitive understanding and making sense of the emotional underpinnings of the story. Without organization of our own stories and client stories, we would continue to live in indecision, confusion, and even chaos. Box 6.1 further explains the neuroscientific basis of empathy. This chapter focuses on the cognitive executive part of active listening, while the following chapter on reflection of feeling examines the role of emotion.

Encouraging, paraphrasing, and summarizing are basic to empathic understanding and enable you to communicate to clients that they have been heard. When using empathic listening skills, be sure not to mix in your own ideas with what the client has been saying. You say back to clients what you have heard, using their key words. You help clients by distilling, shortening, and clarifying what has been said. Accurate empathic listening is not as common, nor as easy, as it may sound, but its impact is often profound.

Following are the responses you can expect from your client when you use these active listening skills. Remember to use the checkout frequently to obtain feedback on the accuracy of your listening skills.

BOX 6.1   The Neuroscience of Empathy: Cognition, Emotion, and Theory of Mind (Mentalizing)

Appendix IV identifies some critical brain structures involved in executive functioning and emotional regulation. Your listening skills can be made more precise with an understanding of how your counseling leads affect regions of the client's brain.

Neuroscience takes us to a deeper level of understanding of the meaning and value of empathy through brain fMRI studies Extensive literature on empathy and the brain concludes that at the most basic level, there are two kinds of empathy—*cognitive empathy* and *affective empathy*—each located in different parts of the brain (Fan et al., 2011). A third type of empathic understanding is *mentalizing*, better understood as the way you think about and integrate in your mind what the client is thinking and feeling holistically.

Cognitive empathy involves *understanding* the other's emotions and activity and is centered in the prefrontal cortex (PFC), most specifically in the midcingulate cortex (MCC) and dorsal medial prefrontal cortex (DMPFC). We see the world through others' eyes and understand how they think— their thought process.

Affective empathy, the central focus of the next chapter, is related to increased activity in the insula and experiencing another person's emotional state. Often this happens at the unconscious level, and one of our tasks to enable the client to become more cognitively aware of emotional life. The insula, also central in attentional processes, appears to be central in experiencing emotions vicariously. A study by Eres, Decety, Louis, and Molenberghs (2015) found that among those with higher scores on a test of empathy, those oriented to affective empathy showed increased fMRI gray matter density in the insula, whereas those leaning more to cognitive empathy showed gray matter density increases in the MCC and DMPFC.

Theory of Mind (ToM), also known as **mentalizing**, involves a holistic cognitive view of clients. Most counselors try to understand how and why their clients are thinking and behaving. We often attribute mental states (intentions, beliefs, and cognitive/emotional perspectives) to others; these are associated with many brain areas. The **temporal parietal junction** integrates many distinct brain areas, including internal and external input. Among several other areas, the medial prefrontal cortex is associated with executive decision making and emotional regulation through its close contact with the amygdala. ToM, from another perspective, could be called "the working of the mind."

Attending behavior and skilled observation are, of course, basic to all three empathic frameworks. But the skill of paraphrasing is most closely related to cognitive empathy, while reflection of feeling refers to affective empathy. When you use the summary, particularly of extensive client comments, you are verging on Theory of Mind. The goal there is to understand the client's cognitive and affective worlds, but also integrate them in a way that requires mentalizing—understanding more fully the client's mental state.

Encouraging	Anticipated Client Response
Encourage with short responses that help the client keep talking. These responses may be verbal (repeating key words and short statements) or nonverbal (head nods and smiling).	Clients elaborate on the topic, particularly when encouragers and restatements are used in a questioning, supportive tone of voice.
Paraphrasing	**Anticipated Client Response**
Shorten or clarify the essence of what has just been said, but be sure to use the client's main words when you paraphrase. Paraphrases are often fed back to the client in a questioning tone of voice.	Clients will feel heard. They tend to give more detail without repeating the exact same story. They also become clearer and more organized in their thinking. If a paraphrase is inaccurate, the client has an opportunity to correct the interviewer. Paraphrasing of client statements is important in cognitive empathy.

Summarizing	Anticipated Client Response
Summarize client comments and integrate thoughts, emotions, and behaviors. Summarizing is similar to paraphrasing but used over a longer time span.	Clients will feel heard and discover how their complex and even fragmented stories are integrated. The summary helps clients make sense of their lives and will facilitate a more centered and focused discussion. Secondarily, the summary also provides a more coherent transition from one topic to the next or a way to begin and end a full session. As a client organizes the story more effectively, we see growth in brain executive functioning and better decision making.
Checkout/Perception Check	**Anticipated Client Response**
Periodically check with your client to discover how your interviewing lead or skill was received. "Is that right?" "Did I hear you correctly?" "What might I have missed?"	Interviewing leads such as these give clients a chance to pause and reflect on what they have said. If you indeed have missed something important or have distorted their story and meaning, they have the opportunity to correct you. Without an occasional checkout, it is possible to lead clients away from what they really want to talk about.

Awareness, Knowledge, and Skills: Encouraging, Paraphrasing, and Summarizing

A client, Jennifer, enters the room and starts talking immediately:

> I really need to talk to you. I don't know where to start. I just got my last exam back and it was a disaster, maybe because I haven't studied much lately. I was up late drinking at a party last night and I almost passed out. I've been sort of going out with a guy for the last month, but that's over as of last night. . . . [pause] But what really bothers me is that my mom and dad called last Monday and they are going to separate. I know that they have fought a lot, but I never thought it would come to this. I'm thinking of going home, but I'm afraid to. . . .

Jennifer continues for another three minutes in much the same manner, repeating herself, and she seems close to tears. Information is coming so fast that it makes it hard to follow her. Finally, she stops and looks at you expectantly.

Imagine you are listening to Jennifer's detailed and emotional story. What are you thinking about her at this moment? Write down what you could say and do to help her feel that you empathize with her and understand her concerns, but perhaps also to help her focus. Compare your ideas with the discussion that follows.

Jennifer is a client who does not have cognitive control of her thoughts and feelings. She needs help in organizing and making sense of her world and then deciding how to act appropriately.

When working with Jennifer, a useful first step is to summarize the essence of her several points and say them back to her. As part of this initial response, use a checkout (e.g., "Have I heard you correctly?") to see how accurate your listening was. The **checkout** (sometimes called the "perception check") offers clients a chance to think about what they said and the accuracy and completeness of your summary. You could follow this by asking her, "You've talked about many things. Where would you like to start today?"

Choosing to focus first on the precipitating crisis is another possible strategy. We could start with Jennifer's parents' separation, as that seems to be the immediate precipitating crisis, and restate and paraphrase some of her key ideas. Doing this is likely to help her focus on one key issue before turning to the others. The other concerns clearly relate to the parental separation and will have to be dealt with as well.

Reflective Exercise Comparing responses

How do our thoughts for counseling Jennifer compare to what you would do? What might you see as the place to respond?

Basic Techniques and Strategies of Encouraging, Paraphrasing, and Summarizing

Encouraging

Encouragers are verbal and nonverbal expressions the counselor or therapist can use to prompt clients to continue talking. Encouragers include minimal verbal utterances ("ummm" and "uh-huh"), head nods, open-handed gestures, and positive facial expressions that encourage the client to keep talking. Silence, accompanied by appropriate nonverbal communication, can be another type of encourager. These encouragers are not meant to direct client talk; rather, they simply encourage clients to keep talking. They help the client elaborate both cognitive and affective thought.

Repetition of key words can encourage a client and has more influence on the direction of client talk. Consider the following client statement:

> "And then it happened again. The grocery store clerk gave me a dirty look and I got angry. It reminded me of my last job, where I had so much trouble getting along. Why are they always after me?"

The counselor could use a variety of short encouragers in a questioning tone of voice ("Angry?" "Last job?" "Trouble getting along?" "After you?" "Tell me more"), and in each case the client would likely talk about a different topic. Note your selection of single-word encouraging responses as they may direct clients more than you think.

A **restatement** is a type of extended encourager in which the counselor or interviewer repeats short statements, two or more words exactly as used by the client. "The clerk gave you a dirty look." "You got angry." "You had trouble getting along in your last job." "You wonder why they are always after you." Restatements can be used with a questioning tone of voice; they then function much like the single-word encourager. Like short encouragers, different types of restatements lead the client in different directions.

Well-timed encouragers maintain flow and continually communicate to the client that you are listening. All types of encouragers facilitate client talk unless they are overused or used badly. Picking out a key word or short phrase to use as an encourager often leads clients to provide you with their underlying thoughts, feelings, or behaviors related to that word or phrase. Just one well-observed word or restatement can open important new avenues in the session. On the other hand, the use of too many encouragers can seem wooden and lacking expression, whereas too few encouragers may suggest to clients that you are not interested.

Always remember smiling is an effective encourager. The warmth and caring you demonstrate may be the most important part of the relationship, even more important than what you say. Facial expression and vocal tone are the nonverbal components of encouraging.

Paraphrasing

Paraphrasing is the most important cognitive empathic listening skill. At first glance, paraphrasing appears to be a simple skill, only slightly more complex than encouraging. In encouraging and restating, exact words and phrases are fed back to the client. Paraphrasing covers more of what the client has just said, usually several sentences. Paraphrasing continues to feed back key words and phrases, but catches and distills the cognitive essence of what the client has said. Paraphrasing clarifies a confusing client story.

When you paraphrase, the tone of your voice and your body language indicate to the client whether you are interested in listening in more depth or would prefer that the client move on to another topic.

If your paraphrase is accurate, the client is likely to reward you with a "That's right" or "Yes . . ." and then go on to explore the issue in more depth. Once clients know they have been heard, they are often able to move on to new topics. The goal of paraphrasing is to facilitate client exploration and clarification of issues.

Accurate paraphrasing will help the client stop repeating a story unnecessarily. Some clients have complex problems that no one has ever bothered to hear accurately, and they literally need to tell their story over and over until someone indicates they have been heard clearly.

How do you paraphrase? Observe clients, hear their important words, and use them in your paraphrase much as the client does. You may use your own words, but the main ideas and concepts must reflect the client's view of the world, not yours!

An accurate paraphrase usually consists of four dimensions:

1. A *sentence stem* sometimes using the client's name. Names help personalize the session. Examples are: "Damaris, I hear you saying . . . ," "Luciano, sounds like . . . ," "Looks like the situation is. . . ."

2. The *key words* used by the client to describe the situation or person. Include main cognitions, ideas, and exact words that come from clients. This aspect of the paraphrase is sometimes confused with the encouraging restatement. A restatement, however, covers a very limited amount of client talk and is almost entirely in the client's own words.

3. The *essence of what the client has said* in briefer and clearer form. Identify, clarify, and feed back the client's sometimes confused or lengthy talk into succinct and meaningful statements. The counselor has the difficult task of staying true to the client's ideas but not repeating them exactly.

4. A *checkout* for accuracy. Here you ask the client for feedback on whether the paraphrase (or other skill) was correct and useful.

The following examples illustrate how a brief client statement can be followed by key word encouragers, restatements, and a paraphrase.

> "I'm really concerned about my wife. She has this feeling that she has to get out of the house, see the world, and get a job. I'm the breadwinner, and I think I have a good income. The children view Yolanda as a perfect mother, and I do too. But last night, we really saw the problem differently and had a terrible argument."

- *Key word encouragers:* "Breadwinner?" "Terrible argument?" "Perfect mother?"
- *Restatement encouragers:* "You're really concerned about your wife." "You see yourself as the breadwinner." "You had a terrible argument."
- *Paraphrase:* "You're concerned about your picture-perfect wife who wants to work even though you have a good income, and you've had a terrible argument. Is that how you see it?"

As always, personalize and make your active listening real. A stem is not always necessary and, if overused, can make your comments seem like parroting. Clients have been known to say in frustration, "That's what I just said; why do you ask?" Again, smiling and warmth make a difference.

Summarizing

Summarizing falls along the same continuum as the key word encourager, restatement, and paraphrase, but often includes feeling and emotions as well. Summarizing encompasses a longer period of conversation than paraphrasing; at times it may cover an entire interview or even issues discussed by the client over several interviews. The summary essentially puts together and organizes client conversation, thus supporting the brain's executive functioning.

Summarizing is key to Theory of Mind (ToM) and your ability to mentalize the world of the client. Our goal is to be fully with the client and see the world as he or she experiences it, but we also stay separate and maintain appropriate boundaries.

The summary is primarily cognitive but includes client emotional and feeling tone. This occurs with many primarily cognitive paraphrases. As emotions are often first reactions and typically occur before cognition regulates emotion, consistently think about (mentalize) the possibly underlying unsaid emotions. A major role of executive functioning is emotional regulation, outlined in the next chapter.

In summarizing, the interviewer attends to verbal and nonverbal comments from the client over a period of time and selectively attends to key concepts and dimensions, restating them for the client as accurately as possible. A checkout at the end for accuracy is a key part of summarizing. Following are some examples. The emotional words are in italics, although you can see that the key issue is interpreting the mind of the client—mentalizing.

> *To begin a session:* "Let's see, last time we talked about your *feelings* toward your mother-in-law, and we discussed the argument you had with her when the new baby arrived. You saw yourself as *guilty* and *anxious.* Since then you haven't gotten along too well. We also discussed a plan of action for the week. How did that go?"

> *Midway in the interview:* "So far, I've seen that you *felt guilty* again when you saw the action plan as manipulative. Yet one idea did work. You were able to talk with your mother-in-law about her garden, and it was the first time you had been able to talk about anything without an argument and you *felt more comfortable.* You visualize the possibility of following up on the plan next week. Is that about it?"

> *At the end of the session:* "In this interview we've reviewed more detail about your *feelings* toward your mother-in-law. Some of the following things seem to stand out: First, our plan didn't work completely, but you were able to talk about one thing without yelling. As we talked, we identified some behaviors on your part that could be changed. They include better eye contact, relaxing more, and changing the topic when you start to see yourself getting *angry.* Does that sum it up?"

Observe: Listening Skills and Children

The observation case presented here is to remind us that the listening skills are used with children in somewhat the same way as with adults. Children too often go through life being told what to do. If we listen to them and their singular constructions of the world, we can reinforce their unique qualities and help them develop a belief in themselves and their own value. In this way, we are increasing their executive functioning, a necessary part of growth and development.

Children generally respond best if you seek to understand the world as they do. Smiling, warmth, and the active listening skills are essential. Frequently paraphrase or restate what they have said, using their important words. Under stress, children may be confused, so be careful that the story you bring out is theirs, not yours.

Talk to children at their eye level whenever possible; avoid looking down at them. This may mean sitting on the floor or in small chairs. Be prepared for more topic jumps with children; use attending skills to bring them back to critical issues. They may need to expend excess energy by doing something with their hands; allow them to draw or play with sand or clay as they talk to you, or engage them in a game like Chutes and Ladders or checkers.

Questions can put off some children, although they remain one of the best ways to obtain information. Seek to get the child's perspective, not yours, being careful not to use leading questions, which may bring out false memories. Children may have difficulty with a general open question such as "Could you tell me what happened?" Use short sentences, simple words, and a concrete language style. Break down abstract questions into concrete and situational language, using a mix of closed and open questions as represented by the ABC-TF framework for questioning: "Where were you when the fight occurred?" "What was going on just before the fight?" (Antecedents) "Then what happened?" (Behavior) What happened next?" "What happened afterward" (Consequence) "What were you thinking through all this?" (Thoughts/cognitions) "How were you feeling at each point?" (Feelings/emotions)

In questioning children on touchy issues, be especially careful of leading questions, which easily can lead to inaccurate understanding of situations. Furthermore, leading questions have been known to encourage development of false memories in both children and adults.

The following example interview is an edited version of a videotaped interview conducted by Mary Bradford Ivey with Damaris, role-playing a problem based on a composite of real cases. Damaris is an 11-year-old sixth grader. The session presents a child's problem, but all of us, regardless of age, have experienced nasty teasing and put-downs, often in our closest relationships. If carried on too long, what you read here could easily become bullying. Harassment in the form of bullying can over time be damaging to the brain and cognitive/emotional development.

The case of Damaris demonstrates effective verbal attending through encouraging, paraphrasing, and summarizing, which help the client explore the issues more effectively. Effective questions are used to bring in new data, organize the discussion, and point out positive strengths. Children and adolescents will be more comfortable if you provide something for them to do with their hands. In summary, avoid towering over small children; sit at their level. Avoid abstractions, use short sentences and simple words, and focus on concrete, observable issues and behaviors. As you will see below, children can provide useful information on their thoughts and feelings.

Interviewer and Client Conversation	Process Comments
1. *Mary:* (smiling) Damaris, how're you doing?	The relationship between Mary and Damaris is already established; they know each other through school activities.
2. *Damaris:* Good.	She smiles and sits down.
3. *Mary:* I'm glad you could come down. You can use these markers if you want to doodle or draw something while we're talking. I know—you sort of indicated that you wanted to talk to me a little bit.	Mary welcomes the child and offers her something to do with her hands. Many children get restless just talking. Damaris starts to draw almost immediately. You may do better with an active male teen by taking him to the basketball court while you discuss issues. It can also help to have things available for adults to do with their hands.
4. *Damaris:* In school, in my class, there's this group of girls that keep making fun of my shoes, just 'cause I don't have Nikes.	Damaris looks down and appears a bit sad. She stops drawing. Children, particularly the "have-nots," are well aware of their economic circumstances. Some children have used sneakers.
5. *Mary:* They "keep making fun of your shoes"?	Encourage in the form of a restatement using Damaris's *exact key words*. (Restatements generally are interchangeable Level 2 empathy.)
6. *Damaris:* Well, they're not the best; I mean—they're not Nikes, like everyone else has.	Damaris has a slight angry tone mixed with her sadness. She starts to draw again.
7. *Mary:* Yeah, they're nice shoes, though. You know?	It is sometimes tempting to comfort clients rather than just listen. We already know that Damaris is not satisfied with them. A simple "uh-huh" could have been more effective. However, positive comments and reassurance used later and more appropriately may be very effective. Too early use turns out to be subtractive empathy.
8. *Damaris:* Yeah. But my family's not that rich, you know. Those girls are rich.	Clients, especially children, hesitate to contradict the counselor. Notice that Damaris uses the word "*But* . . ." When clients say, "Yes, *but* . . .," interviewers are off track and need to change their style. Here we see implied that Damaris is less "OK" that her peers. Her self-cognitions leave her with a sense of inadequacy. (Here we begin to get a picture of Damaris's cognitions.)
9. *Mary:* I see. And the others can afford Nike shoes, and you have nice shoes, but your shoes are just not like the shoes the others have, and they tease you about it?	Mary backs off her reassurances and paraphrases the essence of what Damaris has been saying, using her key cognitive words. (Level 2 interchangeable empathy)
10. *Damaris:* Yeah. . . . Well, sometimes they make fun of me and call me names, and I feel sad. I try to ignore them, but still, the feeling inside me just hurts.	If you paraphrase or summarize accurately, a client will usually respond with *yeah* or *yes* and continue to elaborate the story, but this time she adds her emotions as well.
11. *Mary:* It makes you feel hurt inside that they should tease you about shoes.	Mary reflects Damaris's feelings. The reflection of feeling is close to a paraphrase and is elaborated in the following chapter. (Interchangeable empathy, illustrative of affective empathy)
12. *Damaris:* Mmm-hmm. [pause] It's not fair.	Damaris thinks about Mary's statement and looks up expectantly as if to see what happens next. She thinks back on the basic unfairness of the whole situation.

Interviewer and Client Conversation	Process Comments
13. *Mary:* So far, Damaris, I've heard how the kids tease you about not having Nikes and that it really hurts. It's not fair. You know, I think of you, though, and I think of all the things that you do well. I get . . . you know . . . it makes me sad to hear this part because I think of all the talents you have, and all the things that you like to do and—and the strengths that you have.	Mary's brief summary covers most of what Damaris has said so far. Mary also discloses some of her own feelings. Sparingly used, self-disclosure can be helpful. Mary begins the strength-based positive asset search by reminding Damaris that she has strengths to draw from. These strengths support the building of both cognitive and emotional resilience. (Additive empathy, and Mary is showing awareness of Damaris's internal mental state—mentalizing.)
14. *Damaris:* Right. Yeah.	Damaris smiles slightly and relaxes a bit.
15. *Mary:* What comes to mind when you think about all the positive things you are and have to offer?	Here May is empowering executive functioning. An open question encourages Damaris to think cognitively about her strengths and positives. (Potentially additive empathy)
16. *Damaris:* Well, in school, the teacher says I'm a good writer, and I want to be a journalist when I grow up. The teacher wants me to put the last story I wrote in the school paper.	Damaris talks a bit more rapidly and smiles. (It was additive.)
17. *Mary:* You want to be a journalist, 'cause you can write well? Wow!	Mary enthusiastically paraphrases positive comments using Damaris's own key words. (Interchangeably empathy, but the "wow" and enthusiasm are also additive.)
18. *Damaris:* Mmm-hmm. And I play soccer on our team. I'm one of the people that plays a lot, so I'm like the leader, almost, but . . . [Damaris stops in mid-sentence.]	Damaris has many things to feel good about; she is smiling for the first time in the session. Seeing personal strengths facilitates positive cognition and executive functioning.
19. *Mary:* So, you are a scholar, a leader, and an athlete. Other people look up to you. Is that right? So how does it feel when you're a leader in soccer?	Mary is strengthening executive functioning by adding the names *scholar* and *athlete* for clarification and elaboration of the positive asset search. She knows from observation on the playground that other children do look up to Damaris. Counselors may add related words to expand the meaning. Mary wisely avoids leading Damaris and uses the checkout, "Is that right?" Mary also asks an open question about feelings. And we note that Damaris used that important word "but." Do you think that Mary should have followed up on that, or should she continue with her search for strengths? (Involves both cognitive and affective empathy)
20. *Damaris:* [small giggle, looking down briefly] Yeah. It feels good.	Looking down is not always sadness! The spontaneous movement of looking down briefly is termed the "recognition response." It most often happens when clients learn something new and true about themselves. Damaris has internalized the good feelings.
21. *Mary:* So you're a good student, and you are good at soccer and a leader, and it makes you feel good inside.	Mary summarizes the positive asset search using both facts and feelings. The summary of feeling *good inside* contrasts with the earlier feelings of *hurt inside*. (Interchangeable cognitive/affective empathy)

(continued)

Interviewer and Client Conversation	Process Comments
22. *Damaris:* Yeah, it makes me feel good inside. I do my homework and everything [pause and the sad look returns], but then when I come to school, they just have to spoil it for me.	Again, Damaris agrees with the paraphrase. She feels support from Mary and is now prepared to deal from a stronger position with the teasing. Here we see what lies behind the "but" in 18 above. We believe Mary did the right thing in ignoring the "but" the first time. Now it is obvious that the negative feelings need to be addressed. When Damaris's wellness strengths are clear, Mary can better address those negative feelings.
23. *Mary:* They just spoil it. So you've got these good feelings inside, good that you're strong in academics, good that you're, you know, good at soccer and a leader. Now, I'm just wondering how we can use those good feelings that you feel as a student who's going to be a journalist someday and a soccer player who's a leader. Now the big question is how you can take the good, strong feelings and deal with the kids who are teasing. Let's look at ways to solve your problem now.	Mary restates Damaris's last words and again summarizes the many good things that Damaris does well. Mary changes pace and is ready to move to the problem-solving portion of the interview. This additive empathy is setting the stage for stronger executive functioning, emotional regulation, and resolution of the teasing before it becomes serious. (Here we see Mary seeking to help Damaris "rewire" her brain in positive ways.)

Positive stories and identified strengths when put next to the negative cognitions almost inevitably weaken the negative while simultaneously building executive functioning. Note that cognition here is reinforced by the positive feelings. Thinking is often not enough. This is the power of the positive asset search and positive psychology. Mary had an empathic warm relationship and was able to draw out Damaris's story fairly quickly. She focused on positive assets and wellness strengths to address Damaris's issues and challenges. Clients can solve problems best from their strengths. Be positive, but don't minimize why they came to see you.

Situations such as this one with Damaris may sound simple and basic, but what is happening here occurs in parallel form with teens and adults. Someone hassles or bullies us, and this enters our thoughts and cognitions—and over time can interfere with executive functioning. Adult bullying is beginning to be recognized as a significant personal issue for our clients. There is awareness of harassment, but bullying takes the issue even further. The approaches of positive psychology are a critical part of working with bullying and harassment, but they are not the whole solution. Comprehensive approaches to protect the well-being of children are taking place worldwide, guided by international agreements such as the Convention on the Rights of the Child (UNICEF, 2014; see Box 6.2).

School situations such as the one described here often require the counselor to talk to teachers and make sure that the school bullying policy is followed. Mary often brought in those who were teased for small friendship groups. If the teasers were "mean girls," Mary would set up a supportive group of friends for Damaris. Actually, one teacher commented briefly about this on the playground and the teasing ceased. But it is often not that easy. Be prepared to take supportive action outside the interview.

The active listening skills of encouraging, paraphrasing, and summarizing are key to understanding client cognitions. It is the way that clients think about things and their beliefs that we seek to understand. Their stories are the key to their cognitions. If we are to be successful in interviewing and counseling, we need to understand clients' cognitive styles—and when appropriate, we also need to help them change their cognitions and thoughts so that our clients are more comfortable with themselves and with others.

BOX 6.2   **The Convention on the Rights of the Child**

The Convention on the Rights of the Child (CRC) represents the strongest commitment to the well-being of children in recent decades. Based on the belief that each child is born with the right to survival, food and nutrition, health and shelter, education, equal participation, and protection, children under the age of 18 require special legal protections. The four core principles of the CRC are:

1. Nondiscrimination

2. Devotion to the best interests of the child

3. The right to life, survival, and development

4. Respect for the views and opinions of the child

UNICEF believes that helping children reach their potential will positively impact humanity's progress and reduce poverty. Research is clear that poverty and oppression deeply affect the developing brain. This is especially important because children represent the largest percentage of the world's poor. Accordingly, early investments in children's physical, intellectual, and emotional development, as well as the removal of the barriers affecting their physical and mental health, should be a universal priority. Counseling and psychotherapy and the professionals that use these tools are in a privileged position to help children reach their full potential.

United Nations, 1989; UNICEF, 2009, 2014.

Multiple Applications: Additional Functions of the Skills of Encouraging, Paraphrasing, and Summarizing

When we attend to clients and use the active listening skills, we facilitate executive functioning and the development of new **neural networks** that become part of long-term memory in the hippocampus. Moreover, the very act of listening can lead to restorying—the generation of new means and new ways of thinking. That new story leads to more effective executive functioning in your brain, which can also be seen in your overt behavior.

Executive functioning is also critical for **emotional regulation**. We use our cognitive capacity to regulate impulsive emotions and act appropriately in complex situations, especially when we feel challenged. This could range from eating less sugar when tempted by beautifully decorated cupcakes to not saying or doing something hurtful when suddenly angry with a loved one. It enables us to live in a complex social world. Ineffective executive functioning may lead to negative emotions, including feelings of anxiety or depression.

Cognitions may be defined as language-based thought processes underlying all thinking activities, such as analyzing, imaging, remembering, judging, and problem solving. Cognitive behavioral therapy (CBT), rational emotive behavioral therapy (REBT), and dialectical behavior therapy (DBT) are three examples of cognitive theories of counseling and psychotherapy that focus on changing cognitions to achieve client change.

Our interviewing and counseling skills affect the brain and the mind. We can improve cognitive functioning, emotional regulation, relationships with others, and intentional action. In this process, we access memories in the hippocampus and can potentially facilitate the development of new neural networks in the process of generating new stories.

As an example of the importance of how client concerns and issues arise, see Box 6.3. Here we see how the behavior of others, starting with small hurts, damages and limits one's self-concept and ultimately executive functioning as well.

BOX 6.3	Cumulative Stress and Microaggressions

When Do "Small" Events Become Traumatic?

Active listening skills not only demonstrate empathy, they also facilitate organization of chaotic stories and troubling life experiences. In neuroscience terms, listening leads to more effective brain executive functioning, critical for cognitive understanding and behavioral change.

—Carlos Zalaquett

At one level, being teased about the shoes one wears doesn't sound all that serious—children will be children! However, some poor children are teased and laughed at throughout their lives for the clothes they wear. At a high school reunion, Allen talked with a classmate who recalled painful memories, still immediate, of teasing and bullying during school days. Child and adolescent trauma can affect one's whole life experience.

The microaggressions she experienced in high school became part of this classmate's persona and left her with more limited executive functioning and fewer life possibilities. Small slights become big hurts if repeated again and again. Athletes and "popular" students may talk arrogantly and dismissively about the "nerds," "townies," "hicks," or other outgroups. Teachers, coaches, and even counselors sometimes join in the laughter. Over time, these slights mount inside the child or adolescent. Some people internalize their issues as psychological distress; others may act them out in a dramatic fashion—witness the continued school, church, and workplace shootings throughout the United States.

One of Mary's interns, a young African American woman, spoke of a recent racial insult. At a restaurant, she overheard two White people talking loudly about how they hated to see minorities take away their rights. They talked loudly enough that she easily overheard them, and they seemed to be speaking so that she would hear. She related how common racial insults and microaggressions were in her life, directly or indirectly. She could tell how bad things were racially by how much time she spent on the cell phone with her sister or parents to seek support.

Out of continuing indignities can come feelings of underlying insecurity about one's place in the world (internalized oppression and self-blame) and/or tension and rage about unfairness (externalized awareness of oppression).

Either way, the person who is ignored or insulted feels tension in the body, the pulse and heart rate increase, and—over time—hypertension and high blood pressure may result. The psychological becomes physical, and cumulative stress becomes traumatic.

Soldiers, veterans, police officers, firefighters, women who suffer sexual harassment, those who are short or overweight, the physically disfigured through birth or accident, gays and lesbians, and many others are at risk for having cumulative stress build to real trauma or posttraumatic stress.

Be alert for signs of cumulative stress in your clients. Are they internalizing the stressors by blaming themselves? Or are they externalizing and building a pattern of explosive rage and anger? All these people have important stories to tell, and at first these stories may sound routine. The occurrence of posttraumatic stress responses in later life may be alleviated or prevented by your careful listening and support.

Children, adolescents and, adults may internalize harassment and bullying, somehow thinking it is "their fault." Our task is to help these clients name their very real concerns as externally caused, even to naming it as *oppression*. Next, we want to facilitate development of personal strategies and social support systems so that the individual facing these challenges is no longer weak and alone.

Finally, think of yourself as a potential social action agent in your community. What can you do to help groups of clients (such as those described above) deal with microaggressions and stressors more effectively? Understanding broad social stressors is part of being an effective helper—and taking action or organizing groups toward a healthier lifestyle and working with them to take action for betterment represent a challenge for the future.

Multicultural Issues in Encouraging, Paraphrasing, and Summarizing

Language is one of the important issues related to the listening skills. Box 6.4 discusses the importance of developing skills to assist clients whose primary language is not English.

BOX 6.4   Developing Skills to Help the Bilingual Client

Azara Santiago-Rivera

It wasn't that long ago that counselors considered bilingualism a "disadvantage." We now know that a new perspective is needed. Let's start with two fundamental assumptions: *The person who speaks two languages is able to work and communicate in two cultures and, actually, is advantaged. The monolingual person is the one at a disadvantage!* Research actually shows that bilingual children have more fully developed capacities and a broader intelligence (Power & Lopez, 1985).

If your client was raised in a Spanish-speaking home, for example, he or she is likely to think in Spanish at times, even though having considerable English skills. We tend to experience the world nonverbally before we add words to describe what we see, feel, or hear. For example, Salvadorans who experienced war or other forms of oppression *felt* that situation in their own language.

You are very likely to work with clients in your community who come from one or more language backgrounds. Your first task is to understand some of the history and experience of these immigrant groups. Then we suggest that you learn some key words and phrases in their original language. Why? Experiences that occur in a particular language are typically encoded in memory in that language. So, certain memories containing powerful emotions may not be accessible in a person's second language (English) because they were originally encoded in the first language (for example, Spanish). And if the client is talking about something that was experienced in Spanish, Khmer, or Russian, the *key words* are not in English; they are in the original language.

Here is an example of how you might use these ideas in the session:

Social worker: Could you tell me what happened for you when you lost your job?

Maria (Spanish-speaking client): It was hard; I really don't know what to say.

Social worker: It might help us if you would say what happened in Spanish and then you could translate it for me.

Maria: ¡Es tan injusto! Yo pensé que perdí el trabajo porque no hablo el inglés muy bien. Me da mucho coraje cuando me hacen esto. Me siento herido.

Social worker: Thanks; I can see that it really affected you. Could you tell me what you said now in English?

Maria (more emotionally): I said, "It all seemed so unfair. I thought I lost my job because I couldn't speak English well enough for them. It makes me really angry when they do that to me. It hurts."

Social worker: I understand better now. Thanks for sharing that in your own language. I hear you saying that *injusto* hurts and you are very angry. Let's continue to work on this and, from time to time, let's have you talk about the really important things in Spanish, OK?

This brief example provides a start. The next step is to develop a vocabulary of key words in the language of your client. This cannot happen all at once, but you can gradually increase your skills. Here are some Spanish key words that might be useful with many clients:

Respeto: Was the client treated with respect? For example, the social worker might say, "Your boss did not treat you with *respeto.*"

Familismo: Family is very important to many Spanish-speaking people. You might say, "How are things with your *familia*?"

Emotions (see next chapter) are often experienced in the original language. When reflecting feeling, you could learn and use the following key words.

aguantar: endure	*miedo:* fear	*amor:* love
orgullo: proud	*cariño:* affection	*sentir:* feel
coraje: anger		

We also recommend learning key sayings, metaphors, and proverbs in the language(s) of your clients. *Dichos* are Spanish proverbs, as in the following examples:

Al que mucho se le da, mucho se le demanda.
The more people give you, the more they expect of you.

Más vale tarde que nunca.
Better late than never.

No hay peor sordo que el no quiere oir.
There is no worse deaf person than someone who doesn't want to hear.

La unión hace la fuerza.
Union is strength.

(continued)

BOX 6.4	(continued)

Consider developing a list like this, learn to pronounce them correctly, and you will find them useful in counseling Spanish-speaking clients. Indeed, you are giving them *respeto*. You may wish to learn key words in several languages.

Carlos Zalaquett comments: The Spanish version of *Basic Attending Skills, Las Habilidades Atencionales Básicas: Pilares Fundamentales de la Comunicación Efectiva*

(Zalaquett, Ivey, Gluckstern-Packard, & Ivey, 2008), can help both monolingual and bilingual helpers. The attending skills are illustrated with examples provided by Latina/o professionals from different Latin American countries. Using the information and exercises included in the book, you can sharpen your interviewing, counseling, and psychotherapeutic tools to provide effective services to clients who speak Spanish.

North American and European counseling theory and style generally expect the client to get at the problem immediately and may not allow enough emphasis on relationship building. Some traditional Native American Indians, Dene, Pacific Islanders, Aboriginal Australians, and New Zealand Maori may want to spend a full interview getting to know and trust you before you begin. You may find yourself conducting interviews in homes or in other village settings. *Do not expect this to be true of every client who comes from an indigenous background.* If these clients have experienced the dominant culture, they are likely to be more comfortable with your usual style. Nonetheless, expect trust building and rapport to take more time.

Building trust requires learning about the other person's world. In general, if you are actively working as a counselor, involve yourself in positive community activities that will help you understand your clients better. If you are seen as a person who enjoys yourself in a natural way in the village, the community, and at powwows or other cultural celebrations, this will help build general community trust. The same holds true if you are a person of color. It will be helpful for you to visit a synagogue or an all-White church, understand the political/power structure of a community, and view White people as a distinct cultural group with many variations. Each of these activities may help you to avoid stereotyping those who are culturally different from you. Keep in mind the multiple dimensions of the RESPECTFUL model— virtually all interviewing and counseling are cross-cultural in some fashion.

When you are culturally different from your client, self-disclosure and an explanation of your methods may be helpful. For example, if you use only questioning and listening skills, the client may view you as suspicious and untrustworthy. The client may want directions and suggestions for action.

A general recommendation for working cross-culturally is to discuss differences early in the interview. For example, "I'm a White European American and we may need to discuss whether this is an issue for you. And if I miss something, please let me know." "I know that some gay people may distrust heterosexuals. Please let me know if anything bothers you." "Some White people may have issues talking with an African American counselor. If that's a concern, let's talk about it up front." "You are 57 and I [the counselor] am 26. How comfortable are you working with someone my age?" There are no absolute rules here for what is right. A highly acculturated Jamaican, Native American Indian, or Asian American might be offended by the same statements. Also remember that you can, if necessary, refer the client to another colleague if you get the sense that your client is truly uncomfortable and does not trust you.

Some Asian (Cambodian, Chinese, Japanese, Indian) clients from traditional backgrounds may be seeking direction and advice. They are likely to be willing to share their stories, but you may need to tell them why you want to wait a bit before coming up with answers. To establish credibility, there may be times when you have to commit yourself and provide advice earlier than you wish. If this becomes necessary, be assured and confident; just let them know you want to learn more, and that the advice may change as you get to know them better.

Consider possible differences in gender. Even though there are many exceptions to this "rule," women tend to use more paraphrasing and related listening skills; men tend to use questions more frequently. You may notice in your own classes and workshops that men tend to raise their hands faster at the first question and interrupt more often.

Practice, Practice, and Practice

In this chapter, we have stressed the importance of three major listening skills: encouraging, paraphrasing, and summarizing. These skills are central to effective counseling and psychotherapy, regardless of your theory of choice and your personal integration of these microskills into your own natural style.

Intentional competence in these skills requires practice. Basic competence comes when you use the skills in a session and expect them to be helpful to your clients. Every client needs to be heard; demonstrating that you are listening carefully often makes a real difference. Advanced intentional competence requires deliberate and repeated practice.

The positive asset search, the search for strengths, plus an emphasis on wellness and therapeutic lifestyle changes can do much to enhance healthy cognitions. These, in turn, are the basis for helping that focuses on what clients can do, rather than spending all the time on negative issues. You may want to review the positive examples of questions in Chapter 5.

At this point in this book, we want to share the following story, as it really drives home the importance of continuing to practice the skills. You can pass exams without practice, but if you are serious about helping, learning these skills to full mastery is critical.

Amanda Russo was a student in a counseling course at Western Kentucky University taught by Dr. Neresa Minatrea. Amanda shared with us how she practiced the skills and gave us her permission to pass this on to you. As you read her comments, ask yourself if you are willing to go as far as she did to ensure expertise.

> For my final project I selected a practice exercise from the strength-based questions in Chapter 5. I chose this exercise early on in the book because I do not have much experience with counseling and I wanted to try a fairly simple exercise to start out. I performed the same exercise on five different people to see if I would get the same results.
>
> The exercise consists of asking the client what some of their areas of strength are, getting them to share a story regarding that strength, and then for the counselor to observe the client's gestures and be aware of any changes. The first person I tried this exercise on was Raphael, a dormitory proctor. Some of his strengths were family, friends, working out, and that he had a good inner circle/support group. As he talked about his support group and how they reminded him of the positives, he started to sit in a less tense position. He seemed very relaxed, yet excited about his topic of discussion, and I noticed a lot of hand gestures. In a matter of seconds I saw him change from tense and unsure to relaxed and enthusiastic about what he was saying.

The next person I practiced this exercise on was my roommate Karol. She was a bit nervous when we started and had a difficult time thinking of strengths. Once I asked her to share a story with me, she became very animated. As she spoke, I could see a sparkle in her eyes. Her voice became stronger and her hands were moving every which way. She feels strongly about doing well at work, giving advice, working out, playing music, and finishing the song she is currently writing.

Once she gave me a couple of strengths, they started her wheels turning and she was coming up with more and more. She felt very good about jazz practice earlier that day. She introduced a new song to the group and they really enjoyed it. She also shared a story with me about a huge accomplishment at work that day. Karol was definitely the person whose mood/persona changed the most in this exercise.

Amanda went on to interview three more people to practice her skills and reported in detail on each one. If you seek to reach intentional competence, the best route toward this is systematic practice. For some of us, one practice session may be enough. For most of us, it will take more time. What commitments are you willing to make?

Action: Key Points and Practice of Encouraging, Paraphrasing, and Summarizing

Purpose of Listening Skills. The purpose of the listening skills is to hear the client and feed back what has been said. Clients need to know that their story has been heard. Active and accurate listening communicates that you have indeed heard the client fully. It also communicates your interest and helps clarify relevant issues for both the client and you.

Encouragers. Encouragers are a variety of verbal and nonverbal means the counselor can use to encourage others to continue talking. They include head nods, an open palm, "Uh-huh," and the repetition of key words the client has uttered. Restatements are extended encouragers using the exact words of the client and are less likely to determine what the client might say next.

Paraphrases. Paraphrases are key to cognitive empathy. They feed back to the client the essence of what has just been said by shortening and clarifying client cognitive comments. Paraphrasing is not repetition; it is using some of your own words plus the important main words of the client.

Summarizations. Summaries provides the basis of mentalizing. They are similar to paraphrases except that a longer time and more information are involved. Attention is also given to emotions and feelings as they are expressed by the client. Summarizations may be used to begin an interview, for transition to a new topic, to provide clarity in lengthy and complex client stories, and, of course, to end the session. It is wise to ask clients to summarize the interview and the important points that they observed. Cognitive and affective empathy are typically shown in summaries and represent both large and small aspects of mentalizing, which in turn enable the client to self-reflect and integrate.

Active Listening, Cognition, and Executive Functioning. Executive functions are the cognitive mental processes that regulate human behavior. The brain's frontal lobes provide the biological substrate of these functions. Cognitive skills help clients become aware of their emotions and regulate their emotional reactions. Cognitive-based counseling and therapy focus on changing clients' cognitions in order to achieve change.

Theory of Mind (ToM, or mentalizing) can be called "the working of the mind" and involves a holistic cognitive view of clients. We attribute mental states to others (their intentions, beliefs, and cognitive/emotional perspectives). The temporal parietal junction integrates many distinct brain areas, including internal and external input, to help us understand other people's perspectives.

Diversity and Active Listening. The listening skills are widely used cross-culturally and throughout the counseling and psychotherapy world, but more participation and self-disclosure on your part may be necessary. Trust building occurs when you visit the client's community and learn about cultures different from your own. Best of all is having a varied multicultural group of friends.

A Word of Caution. These skills are useful with virtually any client when used in an effective and nonmechanical way. Otherwise, clients will find repetition tiresome and may ask, "Didn't I just say that?" Use your client observation skills when you use these skills. Seek to maintain a nonjudgmental, accepting attitude. Even the most accurate paraphrasing or summarizing can be negated if you lack supporting nonverbal behaviors.

Practice and Feedback: Individual, Group, and Microsupervision

Additional resources can be found by going to CengageBrain.com and logging into the MindTap course created by your professor. There you will find a variety of study tools and useful resources that include quizzes, videos, interactive counseling and psychotherapy exercises, case studies, the Portfolio of Competencies, and more.

Many concepts have been presented in this chapter; it will take time to master them and make them a useful part of your counseling and therapy. Therefore, the exercises here should be considered introductory.

Individual Practice

Exercise 6.1 Self-Reflection: You as an Active Listener
The active listening skills could be termed specific refinements of attending behavior. Earlier, we asked you to think about yourself and your ability to attend. There we gave some focus to talk time. In a conversation, do you talk most of the time? Or are you mainly a listener? Or perhaps you evenly balance talk time. Think back and start observing yourself even more carefully on these dimensions.

Then, as a second part of self-reflection, ask yourself the following questions: (1) Am I an encouraging person, indicating to the person that I am there through smiling and short minimal encourages (uh-huh, yes, tell me more). (2) Do I paraphrase to make sure that I understand the cognitive frame of my friends and family, particularly when we see thing differently or even have conflict. (3) Less likely, have you ever listened so carefully to someone that you summarize what they have said so that they (and you) truly understand?

Exercise 6.2 Generating Written Encouragers, Paraphrases, and Summarizations

> "Chen and I have separated. I couldn't take his drinking any longer. It was great when he was sober, but it wasn't that often he was. Yet that leaves me alone. I don't know what I'm going to do about money, the kids, or even where to start looking for work."

Write three different types of key word encouragers for this client statement:

_____

Write a restatement/encourager:

_____

Write a paraphrase (include a checkout):

_____

Write a summarization (generate data by imagining previous sessions):

_____

> "And in addition to all that, we worry about having a child. We've been trying for months now, but with no luck. We're thinking about going to a doctor, but we don't have medical insurance."

Write three different types of key word encouragers for this client statement:

_____

Write a restatement:

_____

Write a paraphrase (include a checkout):

_____

Write a summarization (generate data by imagining previous sessions):

_____

Group Practice and Microsupervision

Exercise 6.3 Group or Individual Practice

The seven steps of group practice were defined in Chapter 5 and need not be repeated here. In groups of three or four, make a video or audio recording of an interview focusing on encouraging, restating, paraphrasing, and summarizing. If a group is not available, then find a volunteer and record the session and review it afterwards with the client.

Establish and state clear goals for the practice session. For real mastery, try to use only the three skills in this chapter; use questions only to begin the session or as a last resort. Plan a role-play in which open questions are used to elicit the client's concern. Once this is done, use encouragers to help bring out more details and deeper meanings. Use open and closed questions as appropriate, but give primary attention to the paraphrase and the encourager. End the session with a summary (this is often forgotten). Check the accuracy of your summary with a checkout ("Am I hearing you correctly?").

The suggested topic for this practice session is the story of a past or present stressful experience that may relate to the idea of accumulative stress from microaggressions. Examples include teasing, bullying, or being made the butt of a joke; an incident in which you were seriously misunderstood or misjudged; an unfair experience with a teacher, coach, or counselor; or a time you experienced prejudice or oppression of some type.

The following will make the practice experience more meaningful. Do what Amanda Russo did as described previously. Ask an open question seeking to elicit client strengths, resources, and positive assets. Seek to develop at least one story of success. Use the active listening skills of this chapter to enrich the story, and include a summary at the end.

Debrief your client on how he or she felt about the two experiences.

Observers should use the form in Box 6.5 to analyze the session and provide feedback to the counselor.

BOX 6.5	Feedback Form: Encouraging, Paraphrasing, and Summarizing

_____
(DATE)

_____   _____
(NAME OF COUNSELOR)   (NAME OF PERSON COMPLETING FORM)

Instructions: Write below as much as you can of each counselor statement. Then classify the statement as a question, an encourager, a paraphrase, a summarization, or other. Rate each of the last three skills on a scale of 1 (low) to 5 (high) for its accuracy.

Counselor statement	Open question	Closed question	Encourager	Paraphrase	Summarization	Other	Accuracy rating
1. _____							
2. _____							
3. _____							
4. _____							
5. _____							
6. _____							
7. _____							
8. _____							
9. _____							
10. _____							
11. _____							
12. _____							
13. _____							
14. _____							

1. What were the key discrepancies demonstrated by the client?

2. General session observations. Was responsibility for the concern placed internally or externally, or with some balance between the two?

Portfolio of Competencies and Personal Reflection

Active listening is one of the core competencies of intentional counseling and psychotherapy. Please take a moment to review where you stand and where you plan to go in the future.

Assessing Your Level of Competence: Awareness, Knowledge, Skills, and Action

Use the following checklists to evaluate your present level of competence. Check those dimensions that you currently feel able to do. Those that remain unchecked can serve as future goals. Do not expect to attain intentional competence on every dimension as you work through this book. You will find, however, that you will improve your competencies with repetition and practice.

Awareness and Knowledge. Can you do the following?

❑ Identify and classify encouragers, paraphrases, and summaries.

❑ Discuss issues in diversity that occur in relation to these skills.

❑ Write encouragers, paraphrases, and summaries that might predict what a client will say next.

Basic Competence. Aim for this level of competence before moving on to the next skill area.

❑ Use encouragers, paraphrases, and summaries in a role-played session.

❑ Encourage clients to keep talking through the use of nonverbals and the use of silence, minimal encouragers ("uh-huh"), and the repetition of key words.

❑ Discuss cultural differences with the client early in the session, as appropriate to the individual.

Intentional Competence

❑ Use encouragers, paraphrases, and summaries accurately to facilitate client conversation.

❑ Use encouragers, paraphrases, and summaries accurately to keep clients from repeating their stories unnecessarily.

❑ Use key word encouragers to direct client conversation toward significant topics and central ideas.

❑ Summarize accurately longer periods of client utterances—for example, an entire session or the main themes of several sessions.

❑ Communicate with bilingual clients using some of the key words and phrases in their primary language.

Psychoeducational Teaching Competence. Teaching competence in these skills is best planned for a later time, but a client who has particular difficulty in listening to others may benefit from careful training in paraphrasing. Some individuals often fail to hear accurately and distort what others have said to them.

❑ Teach clients in a helping session the social skills of encouraging, paraphrasing, and summarizing.

❑ Teach these skills to small groups.

Personal Reflection on the Active Listening Skills

This chapter has focused on the active listening skills of encouraging, paraphrasing, and summarizing, which are critical to obtaining a solid understanding of what clients want and need. Active listening is central, and these three skills are key.

What single idea stands out for you among all those presented in this chapter, in class, or through informal learning? What stands out for you is likely to be important as a guide toward your next steps.

What do you think about the use of checkouts? How are you planning to engage in intentional and deliberate practice?

What are your thoughts on race/ethnicity? What other points in this chapter struck you as most useful and interesting? How might you use ideas in this chapter to begin the process of establishing your own style and theory?

If you are keeping a journal, what trends do you see as you progress this far?

Our Thoughts About Jennifer

Active listening requires actions and decisions on our part. What we listen to (selective attention) will have a profound influence on how clients talk about their concerns. When a client comes in full of information and talks rapidly, we often find ourselves confused and, we admit, a bit overwhelmed. It takes a lot of active listening to hear this type of client accurately and fully.

If our work was personal counseling, we would most likely focus on Jennifer's parents' separation and use an encourager by restating some of her key words, thus helping her focus on what may be the most central issue at the moment (e.g., "Your Mom and Dad called earlier this week and are separating"). We'd likely get a more focused story and could learn more about what's happening. As we understand this issue more fully, we could later move to discussing some of the other problems.

Another possibility would be to summarize the main things that Jennifer was saying as succinctly and accurately as possible. We'd do this by catching the essence of her several points and saying them back to her. Most likely, we'd use a checkout to see if we have come reasonably close to what she thinks and feels (e.g., "Have I heard you correctly so far?"). We would then ask her, "You've talked about many things. Where would you like to start today?"

If we were academic counselors, not engaging in personal issues, we'd likely selectively attend to the area of our expertise (study issues) and refer Jennifer to an outside source for personal counseling.

How does this compare to what you would have done?

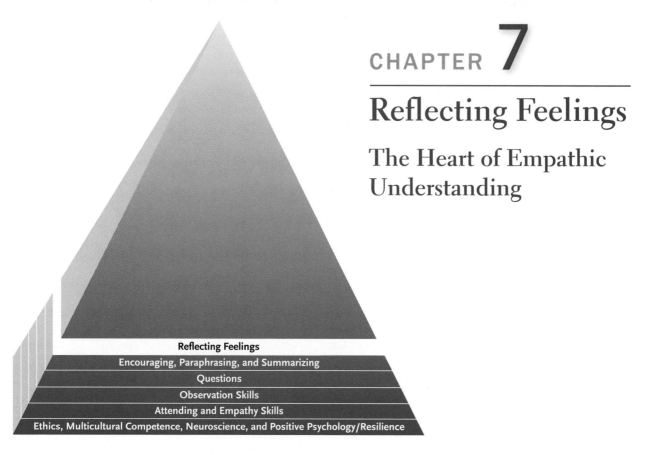

CHAPTER 7

Reflecting Feelings

The Heart of Empathic Understanding

Reflecting Feelings

Encouraging, Paraphrasing, and Summarizing

Questions

Observation Skills

Attending and Empathy Skills

Ethics, Multicultural Competence, Neuroscience, and Positive Psychology/Resilience

No cognition without emotion and no emotion without cognition.

—Jean Piaget

Emotions and feelings underlie our cognitions, thoughts, and behavior. Emotions are our biological engines; without the support of our feelings, change in cognitions and behavior becomes far more difficult. Our words, thoughts, and behaviors are all intertwined with emotions—and the emotions often take the lead in what we say and do. Reflecting feelings is recognized by many as the most significant listening skill after attending behavior and is central to emotional regulation.

At the same time, we need to remember that the brain is holistic and cognitive changes affect the limbic emotional system. It is not an "either-or" situation. Cognitive behavioral therapy by itself can change underlying emotions.

Chapter Goals and Competency Objectives

Awareness and Knowledge

▲ Discover the nature and central importance of reflecting feeling and what to expect when you use this skill.

▲ Understand and appreciate affective empathy and its relationship to cognitive empathy and mentalizing.

Skills and Action

▲ Facilitate client awareness of their emotional world and its effect on their thoughts and behavior.

▲ Help clients sort out and organize their mixed feelings, thoughts, and behaviors toward themselves, significant others, or events.

▲ Clarify emotional strengths and use these to further client resilience.

▲ Center the counselor and client in fundamental emotional experience basic to resolving issues and achieving goals.

▲ Facilitate executive brain functioning through emotional regulation and affective empathy.

Introduction: Reflection of Feeling

The artistic counselor catches the feelings of the client. Our emotional side
often guides our thoughts and action, even without our conscious awareness.

—Allen Ivey

For practical counseling and therapy purposes, you will find this chapter using the words *feelings* and *emotions* interchangeably, as practice does not usually separate the two words.

What is reflecting feeling, and what can the counselor expect when they use this skill? How might the client respond? Not all clients are comfortable with exploring emotions, so be ready for another response to keep the interviewing progressing—be intentional!

Reflection of Feelings	Anticipated Client Response
Identify the key emotions of a client and feed them back to clarify affective experience. With some clients, the brief acknowledgment of feelings may be more appropriate. Affective empathy is often combined with paraphrasing and summarizing. Include a search for positive feelings and strengths.	Clients will experience and understand their emotional states more fully and talk in more depth about feelings. They may correct the counselor's reflection with a more accurate descriptor. In addition, client understanding of underlying feelings leads to emotional regulation with clearer cognitive understanding and behavioral action. Critical to lasting change is a more positive emotional outlook.

Reflective Exercise Compare emotional reflection of feeling with cognitive paraphrasing/summarizing

Reflection of feelings involves observing emotions, naming them, and repeating them back to the client. More cognitively oriented, paraphrasing and summarizing focus primarily on words and concepts. The critical distinction is how much one emphasizes cognitive content (paraphrase) or emotion (reflection of feelings). At the same time, you will often find yourself combining paraphrases with reflections of feeling. In summarizing, this is all the more so. Note both the confused cognitions and emotions expressed in Thomas's story:

My Dad drank a lot when I was growing up, but it didn't bother me so much until now. [Pause] But I was just home and it really hurts to see what Dad's starting to do to my Mum—she's awful quiet, you know. [Looks down with brows furrowed and tense]

(continued)

> Why she takes so much, I don't figure out. [Looks at you with a puzzled expression] But, like I was saying, Mum and I were sitting there one night drinking coffee, and he came in, stumbled over the doorstep, and then he got angry. He started to hit my mother and I stopped him. I almost hit him myself, I was so angry. [Anger flashes in his eyes.] I worry about Mum. [A slight tinge of fear seems to mix with the anger in the eyes, and you notice that his body is tensing.]

Paraphrase Thomas's main ideas, the cognitive content of his conversation; then focus on emotion and reflect his feelings. Use your attention and intuition to note the main feeling expressed in words or nonverbally. Here are two possible sentence stems for your consideration:

- Paraphrase: "Thomas, I hear you saying . . ."
- Reflection of feelings: "Thomas, sounds like you feel . . ."

Thomas's cognitive content includes Dad's history of drinking, Mum's quietness and submission, and the difficult situation when Thomas was last home. The paraphrase focuses on the content, clarifies the essence of what the client said, indicates that you heard what was said, and encourages him to move the discussion further. The key cognitive concepts and words (besides mother and father) are *drinking*, *violence*, and *hit*.

> "Thomas, I hear you saying that your father has been *drinking* a long time, and your Mum takes a lot. But now he's started to be *violent*, and you've been tempted to *hit* him yourself. Have I heard you right?"

Emotion is found both nonverbally and verbally through words and tone of voice. Nonverbals and vocal tone are often the first and clearest sign of emotion. As the story is told, the counselor sees Thomas look down with furrowed brow, followed by a puzzled look, all possibly indicating deep concern for his Mum. There is facial and body tension, and we see anger in his eyes. Underlying the puzzlement and confusion, we find key feeling words include the relatively mild *bother*, but more significant are *hurt*, *fear*, and *anger*. We can assume all of these exist in quiet Mum and in Thomas as well. Here you see the body's emotions described in words.

> "Thomas, Sounds like you *fear* for your Mum, and I see the *hurt*, *confusion*, and I see some *fear* in your eyes. A lot is happening inside you. I sense some real *tension*. Am I close to what you're feeling right now?"

Emotional regulation, a key aspect of prefrontal cortex executive functioning, is about using the cognitive portions of the brain to *control* and interpret the more immediate and reactive emotional body experience. Thus, increased and appropriate emotional regulation is one of the long-term goals of counseling. Reflecting and clarifying feelings is basic to affective empathy.

However, emotional regulation is equally about emotional freedom, the ability to experience feelings in the here and now, rather than just controlling or suppressing them. With Thomas, the counselor would seek to help Thomas use his valid anger in useful ways, rather than letting the anger spill over in an emotional and cognitive disaster. In addition, he likely needs to talk about his sadness. Unless we empathically acknowledge the difficult emotions, clients are less likely to respond to the positive approach.

Coupled with this, supported by positive psychology and therapeutic lifestyle changes, we need to look for strengths and resources from the past and present. These are important to build Thomas's resilience and emotional regulation as he finds new ways of thinking, feeling, and behaving. The positive language of hope and gladness needs encouragement. As we work with client indecision and the many confusing things we face daily, understanding underlying emotions will enable better decisions and personal satisfaction.

Awareness, Knowledge, and Skills: The Emotional Basis of Counseling and Therapy

Reason is the horse we ride after our emotions have decided where to go!

—Carlos Zalaquett

Emotional regulation allows us to manage the horse, its direction, and its speed (most of the time).

—Mary Bradford Ivey

First, let us examine observation of emotions. This will be followed by specifics of the skill.

Observing the Verbal and Nonverbal Language of Emotions

As a first step toward naming and understanding emotions, seek to establish and increase your vocabulary of emotions and your ability to observe and name them accurately. Counseling and therapy traditionally have focused on four primary emotions—*sad, mad, glad, scared*—as a mnemonic for memory, but two more need to be added—*disgust* and *surprise*. These six are the **primary emotions**, and their commonality, in terms of facial expression and language, has been validated throughout the world in many cultures (Ekman, 2007). Make special note that our field historically listed four "negative" emotions and one that is positive. Surprise can be either pleasant or very scary, depending on what is happening at the moment.

Expanding Emotional Vocabulary

Sad/unhappy, mad/angry, scared/fear are also the center of much of the work we do in counseling and therapy. Sadness can lead to depression, which leads to a cycle of inaction. Although anger sometimes motivates people toward positive ends and against oppression, all too often it leads to impulsive behavior that is destructive of self and others. Anger appears in spousal abuse, those who bully, oppositional defiant children and teens, and sociopathic behavior. Fear is related to anxiety, phobias, and an avoidant personality style. Clearly, if we are to work to improve executive functioning and cognitive competence, we must deal effectively with these challenging emotions constantly, but this is best done from a base of positives and strengths. As Nobel Prize winner Daniel Kahneman (2011) notes, quoting a classic paper titled "Bad Is Stronger Than Good" (Baumeister, Bratslavsky, Finkenauer, & Vohs, 2001, p. 323):

> Bad emotions, bad parents, and bad feedback have more impact than good ones, and bad information is processed more thoroughly than good. The self is more motivated to avoid bad self-definitions than to pursue good ones. Bad impressions and bad stereotypes are quicker to form and more resistant to disconfirmation than good ones.

When we experience a significant loss of a loved one, the emotion of sadness needs to be worked through rather than denied. While anger management will be one of your important counseling and therapy skill sets, there are times when anger is appropriate and a motivator for action. Injustice, unfairness, bullying, and harassment are four situations in which client anger is suitable. For some clients, enabling them to recognize underlying emotional anger will be a breakthrough, enabling cognitive change.

The so-called "negative" emotions of sad/mad/fear are primarily located in the limbic system. But calling them negative is not fully accurate. Fear in the face of danger is protective.

For example, you see a car coming toward you, and you don't have time to think. Protective fear enables you to short-circuit the time it takes to think and swerve to save your life. When we duck at a baseball coming toward our head, protective fear is there again. When a woman fears abuse, it can serve as a motivator to find safety. As mentioned earlier, anger often leads us to action. Emotional regulation of anger leads to impulse control and less acting out. Combined with the strengths of the prefrontal cortex and the anterior cingulate cortex, anger is often useful in motivating us to combat unfairness, bullying, and other forms of oppression, but in a planful manner.

Disgust is an interesting addition to the original four primary emotions. It is thought to have evolved as a way to ensure avoidance of unhealthy objects, particularly when we recognize a "disgusting" odor. We say that "it stinks" or a certain person is "rotten" and a "rat." Disgust evokes dimensions of anger, but overlaps with fear as well. If you are counseling a couple thinking of separation or divorce, you likely will see anger. However, if the word "anger" turns to "disgust," saving the relationship is a larger challenge.

Two examples related to surprise are shock and wonder, one often negative, the other usually positive. When you counsel clients and hear them accurately, their surprise at being heard opens the way to change. Confrontation is a basic skill presented in Section IV that can lead to change. The surprise behind confronting clients with the mixed messages and unseen conflicts in their lives can be an important moment. A good reframe/interpretation also produces surprise. An interesting nonverbal measure of the accuracy of your intervention is the **recognition response**. You will see some clients look down briefly, sometimes with an embarrassed look. You likely have reminded them of something they know is right.

Limbic Brain Structures Central in Affective Empathy

It is time to visit and review Appendix IV, which presents a simplified picture of key brain structures. There you will see the limbic system located and discussed. The following serves as a basic outline of what you can view there. Note the key brain structures that you are reaching in reflecting feeling.

1. *Amygdala:* Our emotional (and cognitive) driver, taking information from the senses and passing it on. The thalamus, of course, is a key factor in distributing observations, thoughts, and experiences. The amygdala is closely related to the total limbic system and has close connections with the prefrontal cortex.

2. *Prefrontal cortex (PFC):* With very close connections to the amygdala, the PFC labels emotions as feelings and, when possible, regulates action. In times of emergency, the amygdala emotional foundation will override the PFC, and emotional regulation breaks down. Those with impulse control issues are often ruled by regions associated with the amygdala, and the PFC follows—in effect, actions are now regulated by the emotions, a failure of emotional regulation.

3. *Hippocampus:* Our memory center that holds and distributes information throughout the brain. Both cognitive and emotional decisions are related to short- and long-term memory.

4. *Hypothalamus, pituitary, and adrenal glands (HPA):* Important for understanding the physical role of emotion, they produces the hormones for our brain and body. The hypothalamus controls release of hormones as stimulated by the amygdala, while the pituitary stimulates the adrenals at the top of the kidney. In turn, the adrenals produce cortisol, which in a balanced amount is essential for learning. But with stress and trauma, too much cortisol is delivered with harm to both brain and body.

All these and other structures are activated when you reflect feelings. As the limbic system received information from the senses first, it operates ahead of the structures in the prefrontal cortex, which seeks to regulate emotions. Thus, many of our decisions are made before we are cognitively aware of what we are about to decide.

The Importance of Building Solid Positive Emotions and Expressing Them in Feelings and Action

You can enable clients to bring out stories of positive emotions and thoughts; this is particularly valuable when you work with clients who have a cognitive style of primarily negative thinking. Love and caring are complex positive social emotions. The left prefrontal cortex is the primary location of positive emotional experience (e.g., glad/happy) and is also where our executive decision-making functions lie. Both love and caring represent thoughts and feelings that are expressed in relationship with others, although self-love is real as well. Think of love as a positive emotion, but with a cognitive component relating to a loved one, animal, or even an object related to joyful or caring experiences.

If you use a strength-based approach based on positive psychology and therapeutic lifestyle changes, you can help clients gain mental health and effective problem solving much more quickly. There is evidence that positive emotions also lead to physical well-being, as well as life satisfaction (Carl, Soskin, Kerns, & Barlow, 2013). Historically, the counseling and therapy fields have focused on negative issues ("What's your problem?") with insufficient attention to client resources and personal assets. As stated in Chapter 2, your client can best deal with the challenges of negative emotions and negative cognitions if we constantly build resilience based on their capabilities.

Confusion, Frustration, and Mixed Feelings

Clients often express emotions in unclear ways, demonstrating mixed and conflicting emotions. They may experience many feelings all at once. A client going through a difficult separation or divorce may express feelings of love toward the partner at one moment and extreme anger the next. You can help clients sort through these more complex feelings.

How we deal with emotional experience, of course, depends very much on our learning history. Words such as *puzzled, sympathy, embarrassment, guilt, pride, jealousy, gratitude, admiration, indignation,* and *contempt* are social emotions that come from primary emotions. Basic emotions appear to be universal across all cultures, but the social emotions appear to be learned from one's community, culture, family members, and peers. They are made more complex by the multifaceted and challenging world that surrounds us; their meanings may change from day to day and even in one session.

While not recognized as part of the basic cross-cultural emotions, most of us are likely to recognize these words and the body feelings that often go with them. Stop for just a moment and recall when you might have been confused, frustrated, or undecided as to where to go. If you can take time to note the possible feelings in your stomach, the tension in your shoulders, and maybe even the clenching of your fists, it becomes apparent that these words do carry a heavy load of emotion. Clients will benefit if you help them become aware of what is occurring in the body along with their emotions and cognitions. That knowledge will help strengthen emotional regulation.

When clients are torn and experience confusion and frustration, at another level they typically come with mixed feelings of sadness, anger, or fear. On the other hand, their feeling of being torn may come from an optimistic positive desire for resolution and deciding between good things. Internal emotions may metaphorically "tear apart" the body as well. Relatively few clients will come to you with single, distinct emotions. Even in the most discouraged of clients, you will want to search out pieces and past stories of feelings of success and pride and then use these emotional strengths to develop more positive thoughts and cognitions.

Many clients will say, "I feel confused" or "I feel frustrated" early in the session. This is your opportunity to reflect that feeling, and as you hear the story and concerns around these words, you will discover a fuller understanding of the mixed emotions usually involved. In short, for practical purposes, confusion, frustration, and indecision need to be treated as emotional reactions to a literally confusing and often overwhelming world. As you listen to the cognitive content of the story, you encourage, paraphrase, and summarize, and the underlying basic emotions will appear.

We need to understand the facts, issues, and stories around the emotions we discover. When we untangle cognitions and thoughts, we and the client may be expected to become clearer as to what is going on. With cognitive understanding, emotions also become better defined.

The Skill Dimensions of Reflection of Feeling

The first task, of course, is constant awareness and observation of emotions and feelings as clients show them explicitly through words in their conversation and nonverbally through their patterns of eye contact, facial expression, and body language. Vocal tone, speech hesitations, and variations in loudness are also useful clues to underlying feelings.

At the most basic level, reflection of feelings involves the following set of verbal responses:

1. *Sentence stem.* Choose a sentence stem such as "I hear you are feeling . . .," "Sounds like you feel . . .," "I sense you are feeling . . ." Unfortunately, these sentence stems have been used so often they can sound like comical stereotypes. As you practice, you will want to vary sentence stems and sometimes omit them completely. Using the client's name and the pronoun you help soften and personalize the sentence stem.

2. *Feeling label.* Add an emotional word or feeling label to the stem ("Angelica, you seem to feel sad about . . .," "Looks like you're happy," "Sounds like you're discouraged today; you look like you feel really down"). For mixed feelings, more than one emotional word may be used ("Miguel, you appear both glad and sad . . ."). Note that if you use the term "you feel . . ." too often that you may sound repetitious.

3. *Context or brief paraphrase.* You may add a brief paraphrase to broaden the reflection of feelings. The words *about*, *when*, and *because* are only three of many that add context to a reflection of feelings ("Angelica, you seem to feel angry about all the things that have happened in the past two weeks"; "Miguel, you seem to be excited and glad when you think about moving out and going to college, but also sad as you won't be with your friends and parents."

4. *Tense and immediacy.* Reflections in the present tense ("Right now, you look very angry") tend to be more useful than those in the past ("You felt angry when . . ."). Some clients have difficulty with the present tense and talking in the "here and now." "There and then" review of past feelings can be helpful and may feel safer for the client.

5. *Checkout.* Check to see whether your reflection of feelings is accurate. This is especially helpful if the feeling is unspoken ("You really feel angry and frustrated today—am I hearing you correctly?").

6. *Bring out positive emotional stories and strengths* to counter the negatives and difficulties. If you only focus on fearful, angry, sad emotions, you may find yourself reinforcing negative cognitions as well as negative emotions. If we employ positive psychology and bring out strengths close to the time we discuss serious concerns and issues, you will find that you have helped clients strengthen their abilities, leading to resolution.

Your humanness and ability to be with the client through empathic understanding, attending behavior, and showing interest and compassion are basic when you deal with client emotional experience. Your vocal tone, your nonverbal behavior, and—particularly—your facial expressions together communicate appropriate levels of support and caring.

Reflection of feeling is a skill, but as we note at the beginning of the chapter, it is also an art. Please do not find yourself in a set routine that takes the authenticity out of the relationship. Being your authentic, natural self is equally important to the skill.

Acknowledgment of Feelings

Sometimes a simple, brief recognition of feelings can be as helpful as a full reflection. In acknowledging feelings, you state the feeling briefly ("You seem to be sad about that," "It makes you happy") and move on with the rest of the conversation. With a harried and perhaps even rude busy clerk or restaurant server, try saying in a warm and supportive tone, "Being that rushed must make you tense." Often this is met with a surprised or relaxed look with an implicit thanks of appreciation. Other times the surprise for clients is unsettling. The same structure is used in an acknowledgment as in a full reflection, but with less emphasis and time given to the feeling.

Acknowledging and naming feelings may be especially helpful with children, particularly when they are unaware of what they are feeling. Children often respond well to the classic reflection of feelings, "You feel . . . [sad, mad, glad, scared] because . . ."

The Nonverbal Language of Emotion: Micro and Macro Feelings

Macro nonverbals are those that are relatively easy to see. The client may drop the eyes downward, twist away from you, and talk very quietly, a fairly clear indication of some difficulty in talking about an issue. Some specific dimensions of expected nonverbal presentation of emotion are summarized in Box 7.1.

But micro nonverbals can be equally or more important. Not surprisingly, this is an area to which the Federal Bureau of Investigation has given special attention in its search for clues to deception and subtle behaviors that may be revealing. As we go through the day, these micro nonverbals occur in front of us constantly, but for the most part we don't notice them—and if we do, we don't say anything.

"Microexpressions are fleeting expressions of concealed emotion, sometimes so fast that they happen in the blink of an eye—as fast as one-fifteenth of a second. . . . This results from the individual's attempt to hide them. . . . [They are] a powerful tool for investigators because facial expressions of emotion are the closest thing humans have to a universal language" (Matsumoto, Hwang, Skinner, & Frank, 2011). With practice, we can learn to observe these as they can be as reliable indicators of underlying feelings as macro nonverbals—and they can be even more valuable than more overt, easily observable client behavior.

While observing micro nonverbals is valuable, reflecting them needs to be done carefully, as often the client will not be aware of these underlying feelings. Generally, it is best simply to note them and then watch for a time that these observations may be shared in the session.

BOX 7.1	Nonverbal Examples of Underlying Emotions

Sad   The mouth curves down and the upper eyelids droop. A raise of the inner brows is considered one of the best indications of sadness. The body may slump or the shoulders drop, while vocal tone may be soft and speech rate slow. The arms may be crossed along with the hunching behavior.

Mad   Anger is typically expressed with an upright body position, frowning, and a louder or forced vocal tone; the mouth and jaws may be tense and lips tightened, fists clenched or the palms down. Other nonverbals may be rapid foot tapping, hands on hips, and in situations of danger, possibly moving toward you. There is also the "anger grin," which may indicate a desire to hide underlying feelings.

Glad   Happiness shows itself in smiling, a general picture of relaxation, open body posture, and direct eye contact, typically with pupil dilation. It tends to be a holistic state that usually does not show in micro nonverbals. The smiling face's mouth curves up. The client will often move forward in the chair and use open gestures with the palms up.

Scared   Fear may be indicated by general tension and increased breathing rate, averted eyes or raised eyebrows, furrowed brow, biting the lips, crossed arms, or anxious playing with fingers. The pupils may contract (sometimes this may be the only real clue to fear or the desire to avoid a subject). Vocal tone may waver, with possible stammering or clearing of the throat. There is also the fear grin, which is closely related to the anger grin (i.e., not real).

Disgust   The nose tends to be wrinkled and the upper lip raised as the lips are pursed. Some believe that this evolved through the smell and awareness of rotten food and served as a protective reaction. Cultures vary in what they find disgusting. Disgust shows in interpersonal communication through "disgusting" behavior or ideas, showing up even in politics. If you work with a client who feels disgust for her or his partner, repair of the relationship may be difficult.

Contempt   Though closely related to disgust, the feeling of contempt has slightly different but identifiable facial features, which reflect an attitude of disdain and disrespect toward another person. The chin is raised, which gives the appearance of looking down one's nose at the other. One lip corner may be tightened and slightly raised. A slight smile is often interpreted as a sneer.

Surprise   This emotion typically lasts for only a few seconds, significantly less if the person seeks to hide the surprise. Eyes wide open, the eyebrows raised, and a crinkly forehead are typical. Think of the "jaw dropping" experiences that have surprised you from time to time. Surprise may show as a fleeting micro nonverbal when you have helped the client discover a new insight or following the effective use of the skill of confrontation (Chapter 10).
Surprise can lead to cognitive and emotional change.

Micro nonverbals may be examples of the major underlying issues, or they may be minor parts of a larger story.

Box 7.1 provides some examples of what you might expect nonverbally with the major emotions. Facial expressions are considered the most important, as the clearest indication of underlying feelings. And the fleeting, quick micro nonverbals may be most easily noted in the face.

Diversity and Reflection of Feeling

Many of your clients of diverse backgrounds will come to you having experienced varying types of discrimination and prejudice, leaving them with both conscious and unconscious hurts and emotions of anger, sadness, and fear. As noted in Chapter 2, many minorities have learned to fear White people and their power. Frequent, even daily, experience with microaggressions leads to a feeling of being unsafe. We have seen how microaggressions have injured the lives of Jenny Galbraith at Harvard and Talia Aligo, the high school student in Atlanta. Chapter 2 provides some suggestions for counseling clients who have experienced racism and microaggressions.

BOX 7.2 National and International Perspectives on Counseling Skills

Does He Have Any Feelings?
Weijun Zhang

Videotaped session: A student from China comes in for counseling, referred by his American roommate. According to the roommate, the client quite often calls his wife's name out loud while dreaming, which usually wakes the others in the apartment, and he was seen several times doing nothing but gazing at his wife's picture. Throughout the session the client is quite cooperative in letting the counselor know all the facts concerning his marriage and why his wife is not able to join him. But each time the counselor tries to identify or elicit his feelings toward his wife, the client diverts these efforts by talking about something else. He remains perfectly polite and expressionless until the end of the session.

No sooner had the practicing counselor in my practicum class stopped the videotape than I heard comments such as "inscrutable" and "He has no feelings!" escape from the mouths of my European American classmates. I do not blame them, for the Chinese student did behave strangely, judged from their frame of reference. "How do you feel about this?" "What feelings are you experiencing when you think of this?" How many times have we heard questions such as these? The problem with these questions is that they stem from a European American counseling tradition, which is not always appropriate.

For example, in much of Asia, the cultural rationale is that the social order doesn't need extensive consideration of personal, inner feelings. We make sense of ourselves in terms of our society and the roles we are given within the society. In this light, in China, individual feelings are ordinarily seen as lacking social significance. For thousands of years, our ancestors have stressed how one behaves in public, not how one feels inside. We do not believe that feelings have to be consistent with actions. Against such a cultural background, one might understand why the Chinese student was resistant when the counselor showed interest in his feelings and addressed that issue directly.

I am not suggesting here that Asians are devoid of feelings or strong emotions. We are just not supposed to telegraph them as do people from the West. Indeed, if feelings are seen as an insignificant part of an individual and regarded as irrelevant in terms of social importance, why should one send out emotional messages to casual acquaintances or outsiders (the counselor being one of them)?

What is more, most Asian men still have traditional beliefs that showing affection toward one's wife while others are around, even verbally, is a sign of being a sissy, being unmanly or weak. I can still vividly remember when my child was 4 years old, my wife and I once received some serious lecturing on parental influence and social morality from both our parents and grandparents, simply because our son reported to them that he saw "Dad give Mom a kiss." You can imagine how shocking it must be for most Chinese husbands, who do not dare even touch their wives' hands in public, to see on television that American presidential candidates display such intimacy with their spouses on the stage! But the other side of the coin is that not many Chinese husbands watch television sports programs while their wives are busy with household chores after a full day's work. They show their affection by sharing the housework!

There is a need to respect individual and cultural diversity in the way people express feelings. The student from China discussed in Box 7.2 is an example of cultural emotional control. Emotions are obviously still there, but they are expressed differently. Do not expect all Chinese or other Asians to be emotionally reserved, however. Their style of emotional expression will depend on their individual upbringing, their acculturation, and other factors. Many people, including New England Yankees as well as the English, may be as reserved or even more cautious in emotional expression than the Chinese student described by Weijun Zhang. But again, it would be unwise to stereotype all New Englanders in this fashion.

Openness to emotional roots varies widely from culture to culture and group to group. Generally speaking, citizens of the United States more freely express opinions and emotions than do those from other countries. Thus, Americans may be seen as rude and intrusive, while they may see other groups as cold and indifferent. English and German people are generally seen as reserved, while those of Italian or Latino/a background deal more easily

BOX 7.3   Research and Related Evidence That You Can Use

Emotional processing—the working through of emotions and the ability to examine feelings and body states—has been found to be fundamental in effective experiential counseling and therapy. People with lower emotional awareness are more likely to make errors than those who are in touch with their emotions (Szczygieł, Buczny, & Bazińska, 2012). Gains in treatment of depressed clients were found to be highly related to emotional processing skills (Pos, Greenberg, Goldman, & Korman, 2003). As you work with all your clients, your skill in reflecting feelings can be a basic factor in helping them take more control of their lives.

Dealing with feeling and emotion is not only a central aspect of counseling and psychotherapy; it is also key to high-quality interviewing with medical patients (Bensing, 1999b; Bensing & Verheul, 2009, 2010). Working with emotion requires attention to nonverbal dimensions. Head nodding, eye contact, and especially smiling are facilitative. Clearly, warmth, interest, and caring are communicated nonverbally as much as, or more than, verbally. Moreover, Hill (2009) found that using questions oriented toward affect increased client expression of emotion. However, once

a client has expressed emotion, continued use of questions may be too intrusive and the more reflective approach will be more useful.

"Several studies have shown that between 30% and 60% of patients in general practice present health problems for which no firm diagnosis can be made" (Bensing, 1999a). Be ready to look to emotions in clients who have medical issues. Older persons tend to manifest more mixed feelings than others (Carstensen, Pasupathi, Mayr, & Nesselroade, 2000). Perhaps this is because life experience has taught them that things are more multifaceted than they once thought. Helping younger clients become aware of emotional complexity may also be a goal of some counseling sessions.

Tamase, Otsuka, and Otani (1990), through their work in Japan, have provided clear indication that the reflection of feelings is useful cross-culturally. Reporting on a series of studies in this area, Hill (2009) notes the facilitating impact of reflective responses. She reports that clients are usually not aware when helpers are using good restatements and reflections. Effective listening facilitates exploration.

with their emotional world. In turn, they may see the United States as "uptight" and remote. The nonverbal and emotional communication of different cultures obviously varies widely.

Regardless of multicultural background and experience, all clients have limbic systems and emotions. See Box 7.3. They breathe and experience varying heart rates. Their standards of what is appropriate emotional regulation, however, will vary. In general, expect emotional openness to be maximized with family and friends, but even here cultures vary markedly.

Observe: Reflecting Feelings in Action

The discovery of cancer, AIDS, or other major physical illness brings with it an immense emotional load. Busy physicians and nurses, operating primarily at the executive cognitive level, may fail to deal with patients' emotions or the worries of family members. In fact, the diagnosis of a life-threatening disease increases the risk of depression, general anxiety disorder, and suicide. Lorelei Mucci of Harvard University, best known for her wide-ranging research on suicide and cancer of various types, found, for example, that prostate cancer doubles the risk of suicide in affected patients (Fall et al., 2009).

The following transcript illustrates reflection of feelings in action. This is the second session, and Jennifer has just welcomed the client, Stephanie, into the room. After a brief personal exchange of greetings, it is clear, nonverbally, that the client is ready to start immediately.

Counselor and Client Conversation	Process Comments
1. *Jennifer:* So, Stephanie, how are things going with your mother?	Jennifer knows what the main issue is likely to be, so she introduces it with her first open question.
2. *Stephanie:* Well, the tests came back and the last set looks pretty good. But I'm upset. With cancer, you never can tell. It's hard . . . [pause]	Stephanie speaks quietly and as she talks, her voice becomes even softer. At the word "cancer," she looks down. An issue with virtually all clients facing crisis is maintaining emotional balance and effective executive decision making.
3. *Jennifer:* Right now, you're really worried and upset about your mother.	Jennifer uses the emotional word ("upset") used by the client, but adds the unspoken emotion of worry, a word that we associate with basic fear and avoidance. With "right now," she brings the feelings to here-and-now immediacy. She did not use a checkout. Was that wise? (Interchangeable empathy)
4. *Stephanie:* That's right. Since she had her first bout with cancer . . . [pause], I've been really concerned and worried. She just doesn't look as well as she used to, she needs a lot more rest. Colon cancer is so scary.	Often if you help clients name their unspoken feelings, they will verbally affirm or nod their head. Naming and acknowledging emotions helps clarify the total situation.
5. *Jennifer:* Scary?	Repeating the key emotional words used by clients often helps them elaborate in more depth. (Interchangeable empathy)
6. *Stephanie:* Yes, I'm scared for her and for me. They say it can be genetic. She had Stage 2 cancer and we really have to watch things carefully.	Stephanie confirms the intentional prediction and elaborates on the scary feelings. She looks frightened and physically exhausted. This is a clear example of how cognitions affect emotional experience. Frightening thoughts and experiences have a significant impact on bodily functioning.
7. *Jennifer:* So, we've got two things here. You've just gone through your mother's operation, and that was scary. You said they got the entire tumor, but your mom really had trouble with the anesthesia, and that was frightening. You had to do all the caregiving, and you felt pretty lonely and unsupported. And the possibility of inheriting the genes is pretty terrifying. Putting it all together, you feel overwhelmed. Is that the right word to use, overwhelmed?	Jennifer summarizes what has been said. She repeats key feelings. She uses a new word, "overwhelmed," which comes from her observations of the total situation. Bringing in a new word to describe emotions needs to be done with care. In this example, Jennifer asked the client if that word made sense, and we see in the next exchange that Stephanie could use "overwhelmed" to discuss what was going on. Note how the counselor balances emotion and cognition. (Additive empathy due to the accuracy of the summary)
8. *Stephanie:* [immediately] Yes, I'm overwhelmed. I'm so tired, I'm scared, and I'm furious with myself. [Pause] But I can't be angry; my mother needs me. It makes me feel guilty that I can't do more. [Starts to sob]	Here we see emotion of basic fear tied in with feelings of anger and guilt. Stephanie is now talking about her issues at the here-and-now level. Stephanie has not cried in the interview before, and she likely needs to allow herself to cry and let the emotions out. Caregivers often burn out and need care themselves. Continual listening to sad stories and traumatic events affect the mind/body.
9. *Jennifer:* [sits silently for a moment] Stephanie, you've faced a lot and you've done it alone. Allow yourself to pay attention to you for a moment and experience the hurt. [As Stephanie cries, Jennifer comments.] Let it out . . . that's OK.	Stephanie has held it all in and needs to experience what she is feeling. If you are personally comfortable with emotional experience, this ventilation of feelings can be helpful. There will be a need to return later to a discussion of Stephanie's situation from a less emotional frame of reference. Jennifer hands Stephanie a box of tissues. A glass of water is available. (Additive empathy as Stephanie is there to support.)

(continued)

Counselor and Client Conversation	Process Comments
10. *Stephanie:* [continues to cry, but the sobbing lessens]	Sadness is a valid emotion that needs to be acknowledged and worked through. A later section of this chapter provides ideas for helping clients deal with deeper emotional experience.
11. *Jennifer:* Stephanie, I really sense your hurt and aloneness. I admire your ability to feel—it shows that you care. Could you sit up a little straighter now and take a deep breath? [Pause] How are you doing?	With the supportive positive reflection, the client sits up, the crying almost stops, and she looks cautiously at the counselor. She wipes her nose and takes a deep breath. Jennifer did three things here: (1) She reflected Stephanie's here-and-now emotions; (2) she identified a positive asset and strength; and (3) she suggested that Stephanie take a breath. Conscious breathing often helps clients pull themselves together. (Additive empathy)
12. *Stephanie:* I'm OK. [Pause]	She wipes her eyes and continues to breathe. She seems more relaxed now that she has let out some of her emotions. At this point, she can explore the situation more fully—both emotionally and content-wise—as she moves toward decisions. Now emotional regulation and balance have returned.
13. *Jennifer:* You really feel deeply about what's happened with your mother, and it is so sad. At the same time, I see a lot of strength in you. We've talked earlier about how you supported her both before and after the operation. You did a lot for her. You care, and you also show her how you care by what you do.	Jennifer brings in the positive asset search, bringing the positive of strength and possibility to the session, the beginning of positive rewiring of the brain. In the middle of a difficult time, Stephanie has shown she has the ability to handle the situation, even in the midst of worry and anxiety. This response from Jennifer reflects feelings, but also summarizes strengths that Stephanie can use to continue to deal with her concerns. (Additive empathy)

The major skill Jennifer used in this session was reflection of feelings, with a few questions to draw out emotions. Because human change and development are rooted in emotional experience, reflection of feelings is used in all theories of counseling and therapy. Humanistic counselors often consider eliciting and reflecting feelings the central skill and strategy of counseling and therapy. The cognitive behavioral theories (e.g., CBT) use the skill, but their focus is on changing cognitions, which in turns often leads to emotional change.

You are most likely beginning your work and starting to discover the importance of reflecting feelings. It may take you some time before you are fully comfortable using this skill because it is seldom a part of daily communication. We suggest you start by first simply noting emotions and then acknowledging them through short reflections indicating that the emotions have been observed. As you gain confidence and skill, you will eventually decide the extent and place of emotional exploration in your helping repertoire.

Multiple Applications of Reflecting Feelings

Once you have a basic sense of the centrality of emotions in the helping session and the relationship of cognition and emotions to executive functioning and emotional regulation, the following may be useful elaborations of how to use the skill in multiple settings for maximal impact.

Helping Clients Increase or Decrease Emotional Expressiveness*

You can assist your clients to get more accurately in touch with their emotions. Ultimately, this awareness will help both overall executive cognitive functioning and their own emotional regulation.

Observe Nonverbals. Breathing directly reflects emotional content. Rapid or frozen breath indicates contact with intense emotion, indicative of both limbic and reptilian brain involvement and potential loss of emotional regulation. Anticipate heart rate and blood pressure increase. Also note facial flushing, eye movement, body tension, and changes in vocal tone. Especially, note hesitations. Attend to these clues and follow up at an appropriate point.

At times you may also find an apparent absence of emotion when discussing a difficult issue. This might be a clue that the client is avoiding dealing with feelings or that the expression of emotion is culturally inappropriate for this client. You can pace clients by listening and acknowledging feelings and then gradually lead them to fuller expression and awareness of affect. Many people get right to the edge of a feeling, and then back away with a joke, change of subject, or intellectual analysis.

Pace and Encourage Clients to Express More Emotion.
- Say to the client that she looked as though she was close to something important. "Would you like to go back and try again?"
- Discuss some positive resource that the client has. This base can free the client to face the negative. You as counselor also represent a positive asset.
- Consider asking questions. Used carefully, questions may help some clients explore emotions.
- Use here-and-now strategies, especially in the present tense: "What are you feeling right now—at this moment?" "What's occurring in your body as you talk about this?" Use the word *do* if you find yourself uncomfortable with emotion and move the client slightly away from depth: "What did you feel?" or "What are you thinking about this now?"

When Tears, Rage, Despair, Joy, or Exhilaration Occur. Your comfort level with your own emotional expression will affect how clients face their feelings. If you aren't comfortable with a particular emotion or topic, your clients will likely avoid it, resulting in their handling the issue less effectively. Keep a balance between being very present with your own breathing and showing culturally appropriate and supportive eye contact, but still allowing room for the client to sob, yell, or shake. You can also use phrases such as these:

> I'm here.
>
> I've been there, too.
>
> Let it out . . . that's OK.
>
> These feelings are just right.
>
> I hear you.
>
> I see you.
>
> Breathe with it.

Seek to keep deep emotional expression within a reasonable time frame; 2 minutes is a long time when you are crying. Afterward, help the person reorient to the here-and-now present moment before reflecting and discussing the strong emotions.

---

*Leslie Brain, graduate student at the University of Massachusetts, Amherst provided specific ideas for the recent updating provided in this section. We encourage all of you to share your ideas for enriching this book with us.

Reorient the Session Toward Emotional Regulation.

- Help the client use slowed, rhythmic breathing.
- This may be accompanied by a brief acknowledgment of feeling.
- Discuss the client's positive strengths.
- Discuss direct, empowering, self-protective steps that the client can take in response to the feelings expressed.
- Stand and walk or center the pelvis and torso in a seated position.
- Explore the emotional outburst as appropriate to the situation. Cognitively reframe the emotional experience in a positive way.
- Comment that it helps to tell the story many times.

Caution. As you work with emotion, there is the possibility of reawakening issues in a client who has a history of painful trauma. When you sense this possibility, decide with the client in advance and obtain permission for the desired depth of emotional experiencing. The beginning interviewer, counselor, or therapist needs to seek supervision and/or refer the client to a more experienced professional.

The Place of Positive Emotions in Reflecting Feelings and Resilience

> Contrasting the negative emotions such as sadness with joy can lead to inner peace and equanimity.
>
> —Antonio Damasio

> When we learn how to become resilient, we learn how to embrace the beautifully broad spectrum of the human experience.
>
> —Jaeda DeWalt

We grow best and become resilient with what we "can do" rather than what we "can't do." Positive emotions, whether joyful or merely contented, are likely to color the ways people respond to others and their environments. Research shows that positive emotions broaden the scope of people's visual attention, expand their repertoires for action, and increase their capacities to cope in a crisis. Research also suggests that positive emotions produce patterns of thought that are flexible, creative, integrative, and open to information (Gergen & Gergen, 2005). *Sad, mad, glad, scared* is one way to organize the language of emotion. But perhaps we need more attention to glad words, such as *pleased, happy, love, contented, together, excited, delighted, pleasured,* and the like.

Reflective Exercise Here-and-now contrast of positive and negative emotions

Take a moment now to think of specific situations in which you experienced each of the positive emotions or "glad words" listed immediately above. What changes did you notice in your body and mind? Compare these with the bodily experience of negative feelings and emotions.

   If you smiled, your body tension would very likely be reduced, and even your blood pressure could lower. Perhaps you even found a little bit of inner peace. Positive psychology shows us that attention to strengths lessens the load, takes our mind off our concerns, and empowers us when we recall our strengths and good experiences.

- How was your experience?
- What do you think of the changes you noticed in your body? In your mind?

When you experience emotion, your brain signals bodily changes. With positive emotions, reinforcing neurotransmitters such as dopamine and serotonin increase. When you feel sad or angry, potentially destructive cortisol and hormones are released in the limbic system. These chemical changes will show nonverbally. Emotions change the way our body functions and thus are a foundation for our thinking experience. As you help your clients experience more positive emotions, you are also facilitating wellness and a healthier body. The route toward health, of course, often entails confronting negative emotions.

Research examining the life of the long-living Mankato nuns found that women who had expressed the most positive emotions in early life lived longer than those who expressed a difficult past (Danner, Snowdon, & Friesen, 2001). Research on stress reactions to the 9/11 bombing disaster found that those who had access to the most positive emotions were more resilient and showed fewer signs of depression (Fredrickson, Tugade, Waugh, & Larkin, 2003). A resilient affective lifestyle results in a faster recovery and lower damaging cortisol levels.

These examples of well-being and wellness are located predominantly in the executive prefrontal cortex, with lower levels of activation in the amygdala (Davidson, 2004, p. 1395). Drawing on long-term memory for positive experiences is one route toward well-being and stress reduction. These strengths enable us to deal more effectively with life challenges.

Depression, Emotion, and the Body

Depression is a biological disease.

—Robert Sapolsky

The leading Stanford neuroscientist, Robert Sapolsky, discusses the brain/body interaction in an important and foundational YouTube presentation. (Search "Sapolsky, depression" on your browser.) Figure 7.1 shows the level of brain activity of a depressed and a nondepressed

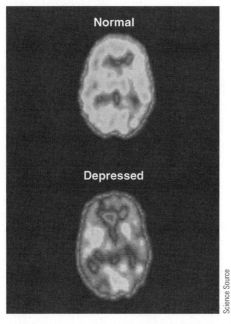

FIGURE 7.1 Brain activation in normal and depressed individuals.

individual. In this picture, we see what happens in the absence of positive feelings and emotions. Executive functioning and emotional regulation have broken down. The lower brain limbic system has taken over.

Working with depression can be challenging, but our goal is to increase positive functioning. If we only focus on negative, problematic stories and reflect these negative feelings, we are reinforcing depression. While it is necessary to hear these stories, ultimately our goal with any client expressing sadness or depression is to find positives. (Needless to say, major depression such as you see in Figure 7.1 will often require medication before counseling and therapy can be fully effective.)

Some Limitations of Reflecting Feelings

Reflection has been described as a basic skill of the counseling process, yet it can be overdone. With friends, family, and fellow employees, a quick acknowledgment of feelings ("If I were you, I'd feel angry about that . . ." or "You must be tired today") followed by continued normal conversational flow may be most helpful in developing better relationships. Similarly, with many clients a brief acknowledgment of feelings may be more useful. However, with complex issues, identifying unspoken feelings may be necessary. Sorting out mixed feelings is key to successful counseling, be it vocational counseling, personal decision making, or in-depth individual counseling and therapy.

Be aware that not all clients will appreciate your comments on their feelings. Exploring the emotional world can be uncomfortable for those who have avoided looking at feelings in the past. An empathic reflection can have a confrontational quality that causes clients to look at themselves from a different perspective; therefore, it may seem intrusive to some. Timing is particularly important with reflection of feelings. Clients tend to disclose feelings only after rapport and trust have been developed. Less verbal clients may find reflection puzzling or may say, "Of course I'm angry; why did you say that?" Some men may believe that expression of feelings is "unmanly." Brief acknowledgment of feelings may be received with appreciation early on and can lead to deeper exploration in later sessions.

Action: Key Points and Practice

Emotions and Feelings. The words are used interchangeably in most counseling and therapy practice. Emotions are the source of many of our thoughts and actions, most deeply based in the body. Feelings are the words we use to cognitively describe variations and mixtures of emotion. Words like *frustration*, *mixed up*, and *confused* represent conflicting feelings and emotions underneath. If we can identify and sort out clients' feelings, we have a foundation for further action.

Identifying Emotions and Naming Feelings. Emotions and feelings can be identified by attending to:

1. Emotional/feeling words used by the client
2. Implicit emotions and feelings not actually spoken
3. Nonverbally expressed emotions discovered through the observation of body movement and facial expressions

4. Micro nonverbal emotional expressions identified by research

5. Mixed verbal and nonverbal emotional cues, which may represent a variety of discrepancies

Expanding the Emotional Vocabulary. *Sad*, *mad*, *glad*, and *scared* are primary emotions used as root words for building a vocabulary of emotion. They appear to be universal across cultures. Social emotions (embarrassment, guilt, pride) are modified and built on primary emotions. They are learned in a family and cultural context. Everyone has complex emotions associated with people and events in their lives. Helping clients sort out these feelings is an important part of interviewing, counseling, and psychotherapy.

Naming. In naming client feelings, note the following:

- Emotional and feeling words used by the client
- Implicit emotional and feelings not actually spoken
- Observation of nonverbally expressed emotions and feelings
- Mixed verbal and nonverbal emotional cues, which often represent conflict

Reflection of Feeling. Emotions may be observed directly, drawn out through questions ("How do you feel about that?" "Do you feel angry?"), and then reflected through the following steps:

1. Begin with a sentence stem such as "You feel . . ." or "Sounds like you feel . . ." or "Could it be you feel . . .?" Use the client's name.

2. Feeling word(s) may be added (sad, happy, glad).

3. The context may be added through a paraphrase or a repetition of key content ("Looks like you feel happy about the excellent rating").

4. In many cases a present-tense reflection is more powerful than one in the past or future tense ("You feel happy right now" rather than "you felt" or "you will feel").

5. Following identification of an unspoken feeling, the checkout may be most useful. ("Am I hearing you correctly?" "Is that close?") This lets the client correct you if you are either incorrect or uncomfortably close to a truth that he or she is not yet ready to admit. Unspoken feelings may be seen in the client's nonverbal expression, may be heard in the client's vocal tone, or may be inferred from the client's language.

Checkouts can help confirm the accuracy of paraphrases and reflections of feelings.

Acknowledgment of Feelings. The acknowledgment of feeling puts less pressure on clients to examine their emotions and may be especially helpful in the early stages of counseling or with clients who are less verbal or clients who are culturally different from you. Later, as trust develops, you can explore emotion and feelings in more depth.

Diversity and Emotions. Respect individual and cultural diversity in the way people express feelings. Social emotions (for example, guilt, compassion, love) are developed in a cultural context and may be blends of basic feelings. Individuals from different cultures may differ in their style of emotional expression, but this will depend on their individual upbringing, their acculturation, and other factors. Be mindful of such possibilities, but avoid stereotyping. Diverse clients may have experienced varying types of discrimination and prejudice.

Emotional Regulation and Affective Empathy. Appropriate emotional regulation as well as the capacity to experience feelings in the here and now are long-term goals of counseling.

Unless we empathically acknowledge the positive or difficult emotions, clients will be less likely to achieve emotional regulation. Affective empathy facilitates this process.

The Limbic System. The limbic system holds structures essential to our emotional life. The amygdala is the prime location of fear, considered the most basic of emotions. It is also sensitive to positive emotions. Some would describe the most basic of emotions as simple approach and avoidance. The hippocampus relates to memory, particularly long-term memory. The hypothalamus links the nervous system to the endocrine system and controls our master gland, the pituitary, which affect the stress response.

Positive Emotions in Reflecting Feelings. Positive emotions affect the way people respond to others and their environments. They activate our brains, broaden our perceptions, let us think more flexibly, increase our capacity to deal with crises, make us happier, and improve our wellness. Seek out these emotions constantly. Help the client activate the positive areas of the brain in the prefrontal cortex.

Reflecting feelings helps us reach both negative and positive emotions and feelings. Recognize and reflect the negative, but search for positive strengths and feelings. It can take five or more positives to counteract a negative (Gottman, 2011). A single emotionally laden damaging comment or negative life experience can last a lifetime and change one's self-view. If we wish to build effective emotional and self-regulation, as well as intentionality, the positive approach becomes essential.

Interview Lessons. In the example interview, Jennifer reflects the main emotional words actually used by the client. She also points out unspoken feelings and checks out with the client whether the identified feeling is accurate. For example, "Is that close to what you feel?" Because human change and development are rooted in emotional experience, reflection of feelings is critical in all theories of counseling and therapy. Reflection of feelings clarifies the client's emotional state, leads clients in new directions, and results in new discoveries. Identify positive qualities and emotions to help clients deal more effectively with negative emotions.

Practice and Feedback: Individual, Group, and Microsupervision

Additional resources can be found by going to CengageBrain.com and logging into the MindTap course created by your professor. There you will find a variety of study tools and useful resources that include quizzes, videos, interactive counseling and psychotherapy exercises, case studies, the Portfolio of Competencies, and more.

We observe feelings in many daily interactions, but we usually ignore them. In counseling and helping situations, however, they can be central to the process of understanding another person. Further, you will find that increased attention to feelings and emotions may enrich your daily life and bring you to a closer understanding of those with whom you live and work. These exercises will help you master reflection of feelings, but you should continue to work on these concepts throughout this book and beyond. With practice, all these materials will become clearer and, most important, will become a part of your natural style.

Individual Practice

Exercise 7.1 Distinguishing a Reflection of Feeling from a Paraphrase

The key feature that distinguishes a reflection of feeling from a paraphrase is the affective word. Many paraphrases contain reflection of feeling; such counselor statements are classified as both. In the example that follows, indicate which of the responses is an encourager (E), which a paraphrase (P), and which a reflection of feeling (RF).

> "I am really discouraged. I can't find anywhere to live. I've looked at so many apartments, but they are all so expensive. I'm tired and I don't know where to turn."

Mark the following counselor responses with an E, P, RF, or a combination if more than one skill is used:

_____ "Where to turn?"

_____ "Tired . . ."

_____ "You feel very tired and discouraged."

_____ "Searching for an apartment simply hasn't been successful; they're all so expensive."

_____ "You look tired and discouraged; you've looked hard but haven't been able to find an apartment you can afford."

For the next example, write an encourager, a paraphrase, a reflection of feeling, and a combination paraphrase/reflection of feeling in response to the client.

> "Right, I do feel tired and frustrated. In fact, I'm really angry. At one place they treated me like dirt!"

Exercise 7.2 Positive Emotions

Take a moment to think of specific situations in which you experienced positive emotions. Very likely when you think of these situations, you will smile, which will help reduce your overall body tension. Encouraging positive memories is a central route toward emotional regulation and executive functioning. It is even likely that your blood pressure will change in a more positive direction. Tension produces damaging cortisol in the brain, and we all need to learn to control our bodies more effectively. Positive imaging is a useful strategy for both you and your clients.

Exercise 7.3 Acknowledgment of Feeling

We have seen that the brief reflection of feeling (or acknowledgment of feeling) may be useful in your interactions with busy and harried people during the day. At least once a day, deliberately tune in to a server/wait staff person, teacher, service station attendant, telephone operator, or friend, and give a brief acknowledgment of feeling ("You seem terribly busy and pushed"). Follow this with a brief self-statement ("Can I help?" "Should I come back?" "I've been pushed today myself, as well") and note what happens in your journal.

Assume you are working with one of the preceding clients who avoids really looking at here-and-now emotional experiencing. How would you help this client increase affect and feeling?

Group Practice and Microsupervision

Exercise 7.4 Practicing Reflection of Feelings and Microsupervision

One of the most challenging skills is reflection of feeling. Mastering this skill, however, is critical to effective counseling and psychotherapy. As always, divide into practice groups, select a leader, and assign roles. See Box 7.4.

BOX 7.4   Feedback Form: Observing and Reflecting Feelings

_____ (DATE)

_____   _____
(NAME OF COUNSELOR)   (NAME OF PERSON COMPLETING FORM)

Instructions: Observer 1 will give special attention to client feelings in notations of verbal and nonverbal behavior below. On a separate sheet, Observer 2 will write down the wording of counselor reflections of feeling as closely as possible and comment on their accuracy and value.

1. Verbal feelings and emotions expressed by the client. List here all related words.

2. Nonverbal indications of feeling states in the client. Facial flush? Body movements? Others? Later check this out with the client. What does the client recall feeling? What did you as counselor feel emotionally through the process of listening and reflecting?

3. Implicit feelings not actually spoken by the client. Check these out with the client later for validity.

4. Reflections of feelings used by the counselor. As closely as possible, use the exact words and record them on a separate sheet of paper.

5. Comments on the reflections of feeling. What were the strengths of the session? Was the counselor's use of the skill accurate and valid? Was the checkout used?

6. How were checkouts used in the session?

If you are working individually with a volunteer client, record the session, look at skill usage and fill out the feedback form with the client.

We suggest examining a past or present story of a stressful experience (bullying, teasing, being seriously misunderstood, an unfair situation in school, with friends, parents, or work). Establish clear goals for the session and draw out the associated emotions. The practice session should end with a summarization of both the feelings and the facts of the situation.

Finally, discuss the concepts of cognitive and emotional empathy as they might relate to this practice session. Give some attention to mentalizing and understanding the situation further as seen by the client.

Portfolio of Competencies and Personal Reflection

Skill in reflection of feeling rests in your ability to observe client verbal and nonverbal emotions. Reflections of feeling can vary from brief acknowledgment to exploration of deeper emotions. You may find this a central skill as you determine your own style and theory.

Assessing Your Level of Competence: Awareness, Knowledge, Skills, and Action

Use the following checklist to evaluate your present level of mastery. As you review the items below, ask yourself, "Can I do this?" Check those dimensions that you currently feel able to do. Those that remain unchecked can serve as future goals. Do not expect to attain intentional competence on every dimension as you work through this book. You will find, however, that you will improve your competencies with repetition and practice.

Awareness and Knowledge. Can you do the following?
❑ Generate an extensive list of affective words.
❑ Distinguish a reflection of feeling from a paraphrase.
❑ Identify and classify reflections of feeling.
❑ Discuss, in a preliminary fashion, issues in diversity that occur in relation to this skill.
❑ Write reflections of feeling that might encourage clients to explore their emotions.

Basic Competence. Aim for this level of competence before moving on to the next skill area.
❑ Acknowledge feelings briefly in daily interactions with people outside of counseling situations (restaurants, grocery stores, with friends, and the like).
❑ Use reflection of feeling in a role-played session.
❑ Use the skill in a real session.

Intentional Competence. The following skills are all related to predictability and evaluation of the effectiveness of your abilities in working with emotion. These are skill levels that may take some time to achieve. Be patient with yourself as you gain mastery and understanding.

❑ Facilitate client exploration of emotions. When you observe clients' emotions and reflect them, do clients increase their exploration of feeling states?

❑ Reflect feelings so that clients feel their emotions are clarified. They may often say, "That's right, and . . ." and then continue to explore their emotions.

❑ Help clients move out of overly emotional states to a period of calm.

❑ Facilitate client exploration of multiple emotions one might have toward a close interpersonal relationship (confused, mixed positive and negative feelings).

Psychoeducational Teaching Competence. A client who has particular difficulty in listening to others may indeed benefit from training in observing emotions. Many individuals fail to see the emotions occurring all around them—for example, one partner may fail to understand how deeply the other feels. Empathic understanding is rooted in awareness of the emotions of others. All of us, including clients, can benefit from bringing this skill area into use in our daily lives. There is clear evidence that people diagnosed with antisocial personality disorder have real difficulty in recognizing and being empathic with the feelings of others. You will also find this problem in some conduct disorder children. Here psychoeducation on empathy and recognizing the other person's feelings can be a critical treatment. A good place to start is to help them observe and name feelings, followed by acknowledgment of feelings.

❑ Teach clients in a helping session how to observe emotions in those around them.

❑ Teach clients how to acknowledge emotions—and, at times, to reflect the feelings of those around them.

❑ Teach small groups the skills of observing and reflecting feelings.

Personal Reflection on Reflection of Feeling

This chapter has focused on emotion and the importance of establishing a foundation between counselor and client. Special attention has been given to identifying six basic feelings, along with many more examples of the social emotions, as well as how you might help clients express more or less emotion as appropriate to their situation.

What single idea stands out for you among all those presented in this chapter, in class, or through informal learning? What stands out to you is likely to be useful as a guide toward your next steps.

What are your thoughts on diversity?

What other points in this chapter strike you as particularly useful in your future practice?

How might you use ideas in this chapter to begin the process of establishing your own style and theory?

Emotional Words

Confused, Frustrated, Mixed Up

General words related to confusion ("I'm undecided"): undecided, uncertain, vague, indistinct, blurred, imperfect, sketchy, unsure, foggy

Suggests negative emotions underlying the confusion ("I feel powerless"): frustrated, torn, lost, drained, empty, numb, exhausted, bewildered, overwhelmed, muddled, addled, befuddled, disoriented, disorientated, unbalanced, unhinged, helpless, stupid, disappointed, dissatisfied

Clients may express confusion and indecision in more positive ways: "I see the move offering a better job, but here I'm pretty comfortable." "I'm undecided between psychology and social work, and both appeal to me." "I want to make my life meaningful, but I'm not sure how." "Should I marry this person?" "Which school is best for me?" "How can I improve my relationship/lifestyle/study habits/nutrition?"

Sad/Unhappy/Depressed. Miserable, down, blue, lonely, pained, devastated, disillusioned, bitter, sorry, hopeless, grief-stricken, guilty, ashamed, dejected, despair, glum, gloomy, dismal, heartbroken, hurting, joyless, cheerless, fragile

Mad/Angry/Explosive. Annoyed, bad, furious, seething, hostile, violent, jealous, vicious, irate, critical, competitive, cross, irritated, indignant, irked, enraged, hostile, pissed off, ticked off, fed up, stormy, volatile, bombed, kill, hit, rub, thoughtless

Glad/Happy/Contented/Love. Comfortable, content, pleased, open, safe, peaceful, aware, relaxed, proud, easy, pleased, supportive, kind, good, up, satisfied, grateful, hope, joy, cheer, sunny, cheer, cheery, strong, overjoyed, gleeful, thrilled, walking on air, fortunate, lucky, warmth, intimacy, attraction, tranquil, untroubled, sooth, soothing, harmony, friendly, nonviolent, sincere, genuine, humorous, empathic, sympathetic, thoughtful

Scared/Fear/Anxious. Nervous, panic, terror, suffocated, panicked, on edge, edgy, horror, alarm, agitation, caged, dread, distress, uptight, surrounded, uneasy, trapped, troubled, tense, uncomfortable, fragile, threatened, vulnerable, unloved, timid, small, insecure, irrelevant, unimportant

Disgust, Revulsion, Distaste. Revolted, revolting, repugnant, aversive, nausea, vomit, sicken, nauseating, loathing, contempt, outrage, shocking, stinks, stinky, tasteless, smells, horrifying, appalling, offending, foul, gross, incredibly bad, terrible, dreadful, grim, monstrous, yucky, "out of it"

CHAPTER 8

How to Conduct a Five-Stage Counseling Session Using Only Listening Skills

How to Conduct a Five-Stage Counseling Session Using Only Listening Skills

Reflecting Feelings

Encouraging, Paraphrasing, and Summarizing

Questions

Observation Skills

Attending and Empathy Skills

Ethics, Multicultural Competence, Neuroscience, and Positive Psychology/Resilience

A leader is best when people barely know that he [or she] exists,
Of a good leader who talks little,
When the work is done,
The aim is fulfilled
They will say, "We did this ourselves."

—Lao Tse

This quote from Lao Tse summarizes a major goal of this book. Our goal as competent interviewers and counselors is to facilitate clients' making their own decisions, finding their own direction, resolving their concerns, and discovering their true selves. Listening is at the heart of interviewing and counseling. If we truly listen, clients will often find their own resolution for many of their concerns and issues, thus finding new possibilities as they meet life challenges.

Chapter Goals and Competency Objectives

Awareness and Knowledge

▲ Understand and become competent in the five stages of the well-formed session: *empathic relationship—story and strengths—goals—restory—action.*

▲ Learn the basics of decision counseling and how it relates to other theories of helping.

▲ Have a good start on Rogerian person-centered counseling, as well as the empathic relationship basics of most other theories of counseling and therapy, through your awareness, knowledge, and skills with the basic listening sequence and the five stages.

Skills and Action

▲ Develop further competence with the basic listening sequence (BLS)—the foundation of effective interviewing, counseling, and psychotherapy,

▲ Conduct a complete decision counseling session using only listening skills and the five stages.

▲ Evaluate your interviewing style and competence.

▲ Become skilled in completing a full five-stage interview using only the listening skills and be able to take these skills to other theories of helping, such as client-centered therapy, cognitive behavioral therapy, and crisis counseling.

Introduction: The Basic Listening Sequence: Foundation for Empathic Listening in Many Settings

To review, the listening skills of the first section of this book contain the building blocks for establishing empathic relationships for effective interviewing, counseling, and psychotherapy. In addition, empathic understanding and careful listening are valuable in all areas of human communication. When you go to see your physician, you want the best diagnostic skills, but you also want the doctor to listen to your story with understanding and empathy. A competent teacher or manager knows the importance of listening to the student or employee.

The microskills represent the specifics of effective interpersonal communication and are used in many situations, ranging from helping a couple communicate more effectively to enabling a severely depressed client to make contact with others. They also are used to train AIDS workers in Africa and in counseling refugees around the world. Furthermore, teaching the social skills of listening has become a standard and common part of individual counseling and psychotherapy.

When you use the BLS, you can anticipate how others are likely to respond.

Basic Listening Sequence (BLS)	Anticipated Client Response
The basic listening sequence (BLS), based on attending and observing, consists of these microskills: using open and closed questions, encouraging, paraphrasing, reflecting feelings, and summarizing.	Clients will discuss their stories, issues, or concerns, including the key facts, thoughts, feelings, and behaviors. Clients will feel that their stories have been heard. In addition, these same skills will help friends, family members, and others to be clearer with you and facilitate better interpersonal relationships.

To review, the basic listening sequence rests on a foundation of ethics, multicultural competence, positive psychology/wellness, therapeutic lifestyle changes, and neuroscience. And none of the BLS skills will be effective or meaningful without skilled attending and observation skills. Attending behavior "lights up the brain," and observation enables you to work with nonverbal and verbal behavior more competently.

- Questioning—open questions followed by closed questions to bring out client stories and concerns.
- Encouraging—used throughout the session to support clients and help them provide specifics around their thoughts, feelings, and behaviors.
- Paraphrasing—catches the cognitive essence of stories and facilitates executive functioning.
- Reflecting feelings—provides a foundation for emotional regulation, enabling examination of the complexity of emotions.
- Summarizing—brings order and makes sense of client conversation, thus facilitating executive functioning and emotional regulation. This integrative skill is particularly critical in mentalizing—your ability to understand and join clients in *their* perspectives and view of the world.

When you add the checkout to the basic listening sequence, you have the opportunity to obtain feedback on the accuracy of your listening. Clients will let you know how accurately you have listened.

The skills of the BLS need not be used in any specific order, but it is wise to ensure that all are used in listening to client stories. Each person needs to adapt these skills to meet the requirements of the client and the situation. The competent counselor uses client observation skills to note client reactions and intentionally flexes to change style, thus providing the support the client needs.

Examples of how the BLS is used in counseling, management, medicine, and general interpersonal communication are shown in Table 8.1.

TABLE 8.1   Four Examples of the Basic Listening Sequence

Skill	Counseling and Psychotherapy	Management	Medicine	Interpersonal Communication (Listening to Others, Friends, or Family)
Using open questions	"Thomas, could you tell me what you'd like to talk to me about . . .?"	"Ajay, tell me what happened when the production line went down."	"Ms. Santiago, could we start with what you think is happening with your headache? Is this OK?"	"Kiara, how did the session with the college loan officer go?"
Using closed questions	"Did you graduate from high school?" "What specific careers have you looked at?"	"Who was involved with the production line problem?" "Did you check the main belt?"	"Is the headache on the left side or on the right?" "How long have you had it?"	"Were you able to get a loan that covers what you need?" "What interest rate are they using?"
Encouraging	Repetition of key words and restatement of longer phrases.			
Paraphrasing	"So you're considering returning to college."	"Sounds like you've consulted with almost everyone."	"I hear you saying that the headache may be worse with red wine or too much chocolate."	"Wow, I can understand what you are saying, Kiara, two loans are a lot."
Reflecting feelings	"You feel confident of your ability but worry about getting in."	"I sense you're upset and troubled by the supervisor's reaction."	"You say that you've been feeling very anxious and tense lately."	"I sense you feel somewhat anxious and worried when you think of paying it back."
Summarizing	In each case, the effective listener summarizes the cognitive/emotional story from the client's or other person's point of view *before* bringing in the listener's own point of view or perhaps an influencing skill.			

Awareness, Knowledge, and Skills: The Five-Stage Model for Structuring the Session

Carl Rogers identified the characteristics that make an ideal relationship and working alliance (Fielder, 1950b; Rogers, 1951).

Most Characteristic

The therapist is able to participate completely in the patient's communication.

Very Characteristic

The therapist's comments are always right in line with what the patient is trying to convey.

The therapist sees the patient as a coworker on a common problem.

That therapist treats the patient as an equal.

The therapist is well able to understand the patient's feelings.

The therapist always follows the patient's line of thought (cognitions).

The therapist's tone of voice conveys a complete ability to share the patient's feelings.

This is also a good summary of the first half of this book, as these characteristics are needed for meaningful use of the influencing skills. Also known as the *working alliance*, they are equally important in any theory of counselor and therapist practice.

The five stages of the well-formed session (*empathic relationship—story and strengths—goals—restory—action*) provide an organizing framework for using the microskills with multiple theories of counseling and psychotherapy. All counselors and therapists need to establish an empathic relationship and draw out the client's story. In different ways, explicit or implicit goals are established, and all seek to develop new ways of thinking, feeling, and behaving. The structure here will also serve as a way to organize your work with theories as varying as client-centered therapy, cognitive behavioral therapy (CBT), psychodynamic, and more.

At this point, please review Table 8.2, which summarizes the five stages of the session. A detailed discussion follows. Note that listening skills are central at each stage. If you are sufficiently skilled in the microskills and the five stages, you are ready to complete a full session using only attending, observation, and the BLS. Many clients can resolve concerns and make decisions without your direct intervention. This was one of Carl Rogers's major goals in his person-centered approach.

The second half of this book focuses on influencing skills, which are used primarily in stages 4 and 5. Thus, brief mention of some of the influencing skills is included in the table.

The five stages can be used as a checklist to ensure that you have covered all the bases in any session. However, following the stages in a specific order is not essential. Many clients will discuss their issues moving from one stage to another and then back again, and you will frequently want to encourage this type of recycling. New information revealed in later counseling stages might result in the need for more data about the basic story, thus redefining client concerns and goals in a new way. Often you will want to draw out more strengths and wellness assets.

TABLE 8.2   The Five Stages of the Microskills Session

Stage	Function and Purpose	Commonly Used Skills	Anticipated Client Response
1. *Empathic relationship.* Initiate the session. Develop rapport and structuring. "Hello, what would you like to talk about?" "What might you like to see as a result of our talking today?"	Build a working alliance and enable the client to feel comfortable with the counseling process. Explain what is likely to happen in the session or series of sessions, including informed consent and ethical issues. Discover client reasons for coming to you.	Attending, observation skills, BLS, information giving to help structure the session. If the client asks you questions, you may use self-disclosure.	The client feels at ease with an understanding of the key ethical issues and the purpose of the session. The client may also know you more completely as a person and a professional—and has a sense that you are interested in his or her concerns.
2. *Story and strengths.* Gather data. Use the BLS to draw out client stories, concerns, problems, or issues. "I'd like to hear your story." "What are your strengths and resources?"	Discover and clarify why the client has come to the session and listen to the client's stories and issues. Identify strengths and resources as part of a strength-based positive psychology approach.	Attending and observation skills, especially the basic listening sequence and the positive asset search.	The client shares thoughts, feelings, and behaviors; tells the story in detail; presents strengths and resources.
3. *Goals.* Set goals mutually. The BLS will help define goals. "What do you want to happen?" "How would you feel emotionally if you achieved this goal?" One possible goal is exploration of possibilities, rather than focusing immediately.	If you don't know where you are going, you may end up somewhere else. In brief counseling (later in this chapter), goal setting is fundamental, and this stage may be part of the first phase of the session. All the same, openness to change and exploration are good places to start.	Attending skills, especially the basic listening sequence; certain influencing skills, especially confrontation (Chapter 10), may be useful.	The client will discuss directions in which he or she might want to go, new ways of thinking, desired feeling states, and behaviors that might be changed. The client might also seek to learn how to live more effectively with stressful situations or events that cannot be changed at this point (rape, death, an accident, an illness). A more ideal story might be defined.
4. *Restory.* Explore alternatives via the BLS. Confront client incongruities and conflict. "What are we going to do about it?" "Can we generate new ways of thinking, feeling, and behaving?"	Generate at least three alternatives that might resolve the client's issues. Creativity is useful here. Seek to find at least three alternatives so that the client has a choice. One choice at times may be to do nothing and accept things as they are. The system of restorying will vary extensively with different theories and approaches.	Summary of major discrepancies with a supportive confrontation. More extensive use of influencing skills, depending on theoretical orientation (e.g., psychoeducation, interpretation, reflection of meaning, feedback). But this is also possible using only listening skills. Use creativity to solve problems.	The client may reexamine individual goals in new ways, solve problems from at least those alternatives, and start the move toward new stories and actions.
5. *Action.* Plan for generalizing session learning to "real life." "Will you use what you decided to do today, tomorrow, or this coming week?"	Generalize new learning and facilitate client changes in thoughts, feelings, and behaviors in daily life. Commit the client to homework and an action plan. As appropriate, plan for termination of sessions.	Influencing skills, such as directives and information/explanation, plus attending and observation skills and the basic listening sequence to check out client understanding.	The client demonstrates changes in behavior, thoughts, and feelings in daily life outside of the interview conversation. Or the client explores new alternatives and reports back discoveries.

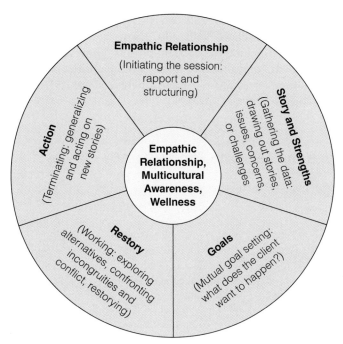

FIGURE 8.1 The circle of counseling stages.

The circle of the five stages of a counseling session in Figure 8.1 reminds us that helping is a mutual endeavor between client and counselor. We need to be flexible in our use of skills and strategies. A circle has no beginning or end; rather, a circle is a symbol of an egalitarian relationship in which counselor and client work together. The hub of the circle is empathy, positive assets, and wellness, a central part of all stages.

Decision Counseling and the Five Stages

The five stages are also a structure for decision making. Eventually, all clients will be making decisions about behavior, thoughts, feelings, and meanings. Each theory gives different attention to these, and they use varying language, names, and techniques. But the client makes the decisions, not us.

How are the five stages specifically related to decisions? Many see Benjamin Franklin as the originator of the systematic decision-making model. He suggested three phases of what he termed problem solving: (1) identify the problem clearly (draw out the story and strengths, along with goal setting); (2) generate alternative answers (restory); and (3) decide what action to take (action). However, the ancient Franklin model misses the importance of empathic relationship (stage 1), the need for clearer goal setting, and ensuring that the client takes action (stage 5) after the session in the real world.

Virtually all interviewing, counseling, and psychotherapy sessions involve decisions, often deciding what actions to take in intrapersonal or interpersonal conflict. Whether you become oriented to person-centered, cognitive behavior, narrative, behavior therapy, neurocounseling, coaching, or others, decisions will always be part of the counseling and therapy process. Neuroscience foundations of decisions are under study (see Box 8.1). Decisions are a central issue in crisis counseling. Decisions are involved in millions of interviews conducted daily throughout the world in counseling, therapy, medicine, business, sports, government, and many other fields.

Problem-solving therapy (PST) also builds on the Franklin framework, as does the decision counseling model (D'Zurilla & Nesu, 2007; Nesu & Nesu, 2013). PST books can be useful supplements for enriching the decision process. However, PST focuses on the problematic word "problem," and thus may ignore a more positive wellness approach, as well as critical multicultural issues. Despite these differences, the PST framework is a useful addition to your practice.

BOX 8.1   Neuroscience Informs the Decision Process

The body registers a decision before the cognitive mind.

—David Eagleton

Imagine that you have a decision to make—say, a choice between two alternatives: eating or not eating ice cream or some other sugary food that you know is not good for you. Information on which a final decision is to be made will be found in memory and neural networks. For example, the ice cream neural network contains memories of chocolate, the feeling of cold taste on the tongue, and the general joy of eating. Another network, more cognitive, is focused on emotional regulation and the need to maintain a healthy and proper weight. It is the one with awareness of long-term consequences of eating the treat. The cognitive network says "no" while the more affective emotional network is insistently saying "yes"—let's enjoy it in the here and now, future consequences be damned. Which will win and make the decision? In truth, the body reacts before the conscious mind.

Think of two (or more) neural networks fighting for control. Both are obviously involved in a process of weighing the alternatives, but the more fatigued and hungry you are, the more likely that here-and-now emotional and physical demands will take over from the prefrontal cortex. One can gain a lot of weight through a weak PFC. Or one

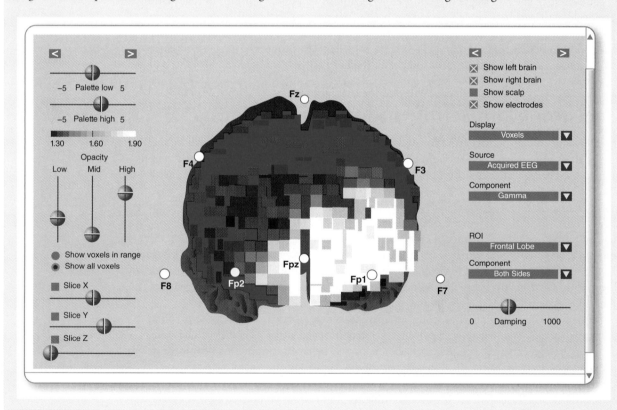

can be so rigidly strict that there is little fun in life. We need to realize that emotions are involved deeply in our decisions. To think that cognition is the sole controlling factor is naïve. Thus, understanding emotions and feelings, along with the ability to draw out and reflect feelings, is considered by many our primary tools in counseling and psychotherapy.

Research now shows that the decision is already made before we become aware and say or take an action. Below you see a photo taken in a research project on how the brain functions when it makes moral choices. The volunteer subject in this case is deciding which ethical choice is more appropriate. As with the ice cream decision, a variety of neural networks, both affective and cognitive, will be involved. Emotional regulation and the current physical status of the subject make a difference. But what is interesting here is that the *decision is made slightly before we become cognitively aware.* The French psychoanalyst, Jacques Lacan, would say, "We do not speak, we are spoken."

The neural networks were developed by both short- and long-term memory, and how they are expressed will be determined by context and current mental and physical levels.

Given this, all counseling, including basic decision counseling, requires us to be aware that clients who deal with decisions—determining what to do in conflict situations or deciding among alternatives—can benefit from clearly defining the issues. In this process, we are strengthening emotional regulation and executive functioning.

Neurons, neural networks, connections among networks, and brain and body parts are all interconnected, and any small piece can ultimately be "the decider."

Collura, T. F., Zalaquett, C. P., Bonnstetter, R. J., & Chatters, S. J. (2014). Toward an operational model of decision making, emotional regulation, and mental health impact. *Advances in Mind-Body Medicine, 28*(4), 18–33.

While neurologically complex, decision counseling is straightforward; you already have a solid basic understanding through the BLS and the five stages. Decision counseling will be elaborated in more detail in Chapters 12 and 13, and in Chapter 14 with an example interview between Mary and Allen. Perhaps the reason you will not find the words *decision counseling* in most books on counseling and therapy is that it is so basic that it is easily overlooked. This is so despite the fact the making decisions is a major part of counseling, regardless of theory, particularly for day-to-day decisions such as work and career, school and college counseling, counseling the elderly around life planning, and many other areas.

As decision counseling and problem-solving therapy are the most widely practiced forms of helping, we urge you to take some time to master the concepts, which will come easily with an understanding of the microskills and the five-stage interview. In addition, you will find that this foundation enables you to become competent more rapidly in the many other theories of counseling and psychotherapy.

The practice of decision counseling is very concrete and specific. Once you accept the idea that most counseling and therapy involve decisions, you will be able to use the structure and methods of decision counseling in many situations, as it can easily accompany other theoretical approaches.

Following is an illustration of an interview using the decision counseling framework.

Observe: Using the Five Stages of Interviewing in Decision Counseling

It requires a verbal, cooperative client to work through a complete session using only listening skills. This interview has been edited to show portions that demonstrate skill usage and levels of empathy. Robert, the client, is 20 and a part-time student who is in conflict with his boss at work. Machiko, the student counselor in a course like this, finds him relatively verbal and willing to work on the problem with her assistance. Completing a session with a client who is less willing to talk can be rather challenging, and there the use of influencing skills will likely be required.

Stage 1: Empathic Relationship

Machiko and Robert know each other from class, so little time is spent on relationship in her transcript, although they did talk while her video camera, microphone, and recorder were being set up and tested for picture and sound. If available, quality video cameras and recorders are ideal, although smartphones and your computer are workable. At times you will want to use Skype, FaceTime, or a similar system so that a supervisor can watch your interviewing style.

Counselor/Client Statement	Process Comments
1. *Machiko:* Thanks for volunteering to enable me to do a practice session for class. After we review this session, we can plan for you to interview me. Robert, do you mind if we tape this interview? It's for a class exercise in counseling skills. I'll be making a transcript of the session, which the professor will read. Okay? We can turn the recorder off at any time. I'll show you the transcript if you are interested. I won't use the material if you decide later you don't want me to use it. Could you sign this consent form? (Robert signs and looks up.)	A small video cam has been set up. Machiko opens with a closed question followed by structuring information. It is critical to obtain client permission and offer client control over the material before recording. As a student you cannot legally control confidentiality, but it is your responsibility to protect your client.
2. *Robert:* Sounds fine; I do have something to talk about. Okay, I'll sign it. [Pause as he signs]	Robert seems at ease and relaxed. As the taping was presented casually, he is not concerned about the use of the recorder. Rapport was easily established.
3. *Machiko:* What would you like to share?	The open question, almost social in nature, is designed to give maximum personal space to the client.
4. *Robert:* My boss at A&B Electronics. He's pretty awful and on me all the time.	Robert indicates clearly through his nonverbal behavior that he is ready to go. At the same time, we see flashes of anger. Machiko observes that he is comfortable and decides to move immediately to gather data (stage 2). With some clients, several sessions may be required to reach this level of rapport.

Stage 2: Story and Strengths

Counselor/Client Statement	Process Comments
5. *Machiko:* Could you tell me about it?	This open question is oriented toward obtaining a general outline of the concern (the boss) the client brings to the session. (Potentially additive empathy)
6. *Robert:* Well, he's impossible . . . (Hesitates and looks at Machiko)	Instead of the expected general outline of the concern, Robert gives a brief answer. The predicted consequence didn't happen, but we see in the next two exchanges that 5 and 7 together are indeed additive.
7. *Machiko:* Impossible? . . . Go on . . .	Encourager with warm, supportive vocal tone. Intentional competence requires you to be ready with another follow-up response. Tone of voice is especially critical here in communicating to the client.

Counselor/Client Statement	Process Comments
8. *Robert:* Well, he's impossible. Yeah, really impossible. It seems that no matter what I do he is on me, always looking over my shoulder. I don't think he trusts me.	We are seeing the story develop. Clients often elaborate on the specific meaning of a concern if you use the encourager. In this case, the prediction holds true. Here we see the need for a decision as to how Robert can handle this difficult situation.
9. *Machiko:* Could you give me a more specific example of what he is doing to indicate he doesn't trust you?	Robert is a bit vague in his description. Machiko asks an open question eliciting concreteness in the story. (A search for concreteness is typically additive.)
10. *Robert:* Well, maybe it isn't trust. Like last week, I had this customer lip off to me. He had a complaint about the TV he bought. I don't like customers yelling at me when it isn't my fault, so I started talking back. No one can do *that* to me! . . . (pause) And of course the boss didn't like it and chewed me out. It wasn't fair.	As events become more concrete through specific examples, we understand more fully what is going on in the client's life and mind. When Robert said, "No one can do *that* to me!" he briefly spoke angrily and clenched his teeth and tightened his fist briefly. After the pause, he spoke more softly and seemed puzzled.
11. *Machiko:* As I hear it, Robert, it sounds as though this guy gave you a bad time and it made you angry, and then the boss came in. I hear some real anger about the customer, but I'm not so clear about the feelings toward your boss.	Machiko's response is relatively similar to what Robert said. Her paraphrase and reflection of feeling represents basic interchangeable empathy. Picking up on the nonverbal mixed message was wise. (Interchangeable at first and then additive in the last sentence as she encourages client to explore emotions.)
12. *Robert:* Exactly! It really made me angry. I have never liked anyone telling me what to do. But when the boss came after me, I almost lost it. I left my last job because the boss was doing the same thing.	Accurate listening often results in the client's saying "exactly" or something similar. Robert loosens up and starts talking about the boss as the real challenge.
13. *Machiko:* Difficult customers are hard to take, but your boss coming in like that is the real issue. He really bugged you. (brief pause) . . . And I hear that your last boss wasn't fair either?	Machiko's vocal tone and body language communicate nonjudgmental warmth and respect. She catches underlying emotions while bringing back Robert's key word *fair* by paraphrasing with a questioning tone of voice, which represents an implied checkout. This is an interchangeable empathic response, again with an additive "tone."

How does this relate to decision counseling? First, one needs to listen and identify client issues and concerns clearly. The interview continues to explore Robert's conflict with customers, his boss, and past supervisors. There appears to be a pattern of conflict with authority figures over the past several years. This is a common pattern among young males in their early careers. After a detailed discussion of the specific conflict situation and several other examples of the pattern, Machiko decides to conduct a positive asset search to discover strengths.

Counselor/Client Statement	Process Comments
14. *Robert:* You got it.	Robert here is speaking to Machiko's understanding of his situation.
15. *Machiko:* Robert, we've been talking for a while about difficulties at work. I'd like to know some things that have gone well for you there. Could you tell me about something you feel good about at work?	Paraphrase, structuring, open question, and beginning positive asset search. (Additive positive empathy)

(continued)

Counselor/Client Statement	Process Comments
16.　*Robert:* Yeah; I work hard. They always say I'm a good worker. I feel good about that. I am one of the top salespeople there.	Robert's increasingly tense body language starts to relax with the introduction of the positive asset search. He talks more slowly.
17.　*Machiko:* Sounds like it makes you feel good about yourself to work hard.	Reflection of feeling, emphasis on positive regard (Interchangeable, with positive additive dimensions)
18.　*Robert:* Yeah. For example, . . .	

The positive strengths can be used in helping Robert make decisions for what he wants to do. Robert continues to talk about his accomplishments. In this way, Machiko learns some of the positives Robert has in his past and not just his difficulties. She has used the basic listening sequence to help Robert feel better about himself. She has also emphasized strengths and positive emotions. Machiko learns that Robert has positive assets, such as determination and willingness to work hard, to help him resolve his own problems. He takes pride in his ability to work hard.

Stage 3: Goals

Counselor/Client Statement	Process Comments
19.　*Machiko:* Robert, given all the things you've talked about, could you describe an ideal solution? How would you like things to be?	Open question. The addition of a new possibility for the client represents additive empathy. It enables Robert to think of something new. Here we start exploring possibilities for a satisfactory decision.
20.　*Robert:* Gee, I guess I'd like things to be smoother, easier, with less conflict. I come home so tired and angry.	
21.　*Machiko:* I hear that. It's taking a lot out of you. Tell me more specifically how things might be better.	Paraphrase, open question oriented toward concreteness. Restating "taking a lot out of you" keeps awareness of emotions present. (Interchangeable; the last sentence is potentially additive.)
22.　*Robert:* I'd just like less hassle. I know what I'm doing, but somehow that isn't helping. I'd just like to be able to resolve these conflicts without always having to give in.	Robert is not as concrete and specific as anticipated. But he brings in a new aspect of the conflict—giving in.
23.　*Machiko:* Give in?	Encourager. (Another potentially additive encourager)
24.　*Robert:* Yeah. And, you know what . . .	

Machiko learns another dimension of Robert's conflict with others. Subsequent use of the basic listening sequence brings out this pattern with several customers and employees. As new data emerge in the goal-setting process, you may find it necessary to change the definition of the concern and perhaps even return to stage 2 for more data gathering.

Counselor/Client Statement	Process Comments
25. *Machiko:* So, Robert, I hear two things in terms of goals. One, that you'd like less hassle, but another, equally important, is that you don't like to give in. But all this makes you tired, irritable, and discouraged. Have I heard you correctly?	Machiko uses a summary of both cognitions and emotions to help Robert clarify his problem, even though no resolution is yet in sight. From a neuroscience frame of reference, she is seeking to help Robert's pattern of emotional regulation lead to better executive decision functioning. She checks out the accuracy of her hearing. (Additive empathy)
26. *Robert:* You're right on, but what am I going to do about it?	At this point, it is clear that Robert feels heard and listened to. He leans forward with some anticipation of working toward resolution of his concerns.

Stage 4: Restory

Counselor/Client Statement	Process Comments
27. *Machiko:* So, Robert, on the one hand, I heard you have a long-term pattern of conflict with supervisors and customers who give you a bad time. On the other hand, I also heard just as loud and clear your desire to have less hassle and not give in to others. We also know that you are a good worker and like to do a good job. Given all this, what decisions might you want to make to better things?	Machiko remains nonjudgmental and appears to be very congruent with the client in terms of both words and body language. In this clear, positive, additive summary, she distills and clarifies what the client has said. (Clear summaries typically include both interchangeable and additive empathy, hopefully including positive dimensions.)
28. *Robert:* Well, I'm a good worker, but I've been fighting too much. I let the boss and the customers control me too much. I think the next time a customer complains, I'll keep quiet and fill out the refund certificate. Why should I take on the world?	Here we see the beginning of a new decision style. Robert talks more rapidly. He, too, leans forward. However, his brow is furrowed, indicating some tension. He is "working hard." At the same time, Robert starts with a positive self-statement that has earlier been reinforced by Machiko.
29. *Machiko:* So, one thing you can do is keep quiet. You could maintain control in your own way, and you would not be giving in.	Paraphrase, interchangeable empathy. Machiko is using Robert's key words and feelings from earlier in the session to reinforce his present thinking. But she waits for Robert's response.
30. *Robert:* Yeah, that's what I'll do, keep quiet.	Whoops! He has identified one way to realize his goals, but clearly is not happy with that one decision. He sits back, his arms folded. This suggests that the "good" response above was in some way actually subtractive. There is more work to do. This is definitely not what we look for in decision counseling.
31. *Machiko:* Sounds like a good beginning, but I'm sure you can think of other things as well, especially when you simply can't be quiet. Can you brainstorm more ideas?	Machiko gives Robert brief feedback. Her open question is an additive response. She is aware that his closed nonverbals suggest more is needed.

Clients are often too willing to seize the first idea as a way to agree and avoid looking fully at issues. It is helpful to use a variety of questions and listening skills to further draw out the client. Later in the session, Robert was able to generate two other suggestions that he decided might be useful: (1) to talk frankly with his boss and seek his advice and (2) to plan an exercise program to blow off steam and energy. In addition, Robert began to realize that his problem with his boss was only one example of a continuing problem with anger.

With these two ideas, we are seeing the beginning of the development of new neural networks, more positive and useful. We have talked about changing and rewiring the client's brain, and this is one example. At the same time, Machiko's neural networks are developing as she practices empathic understanding and learns how to cope with clients who have underlying anger.

Robert and Machiko discussed the possibility of continuing their discussions so that he could seek to keep his temper under control. Robert decided he'd like to talk with her a bit more. A contract was made: If the situation did not improve within 2 weeks, Robert could participate in an anger management program and seek someone with more experience at the campus counseling center.

Stage 5: Action—Generalization and Transfer of Learning

Counselor/Client Statement	Process Comments
32. *Machiko:* So you've decided that the most useful step is to talk with your boss. But the big question is "Will you do it?" "What specifically could you try next week?"	Paraphrase, open question. (Potentially additive) Note that many questions are potentially additive, but questions can also be subtractive—we need to wait and see how useful they are as the client responds.
33. *Robert:* Sure, I'll do it. The first time the boss seems relaxed and has time to talk.	He appears more confident with the decision and sits up.
34. *Machiko:* As you've described him, Robert, that may be a long wait. Could you set up a specific plan so we can talk about it the next time we meet?	Paraphrase, open question. To generalize from the counseling conversation, encourage specific and concrete action in your client so that something actually does happen. Too many of us are far too willing to accept that early acknowledgment of willingness to act. It typically is not enough.
35. *Robert:* I suppose you're right. Okay, occasionally he and I drink coffee in the late afternoon at Rooster's. I'll bring it up with him tomorrow. I know that overall he likes me, but it seems to be his nature to find fault. I'll make a big effort to listen to him and see what happens.	A decision for a specific plan to take action on the new story is developing.
36. *Machiko:* I hear you ready to act. You've got the confidence and ability. What, specifically, are you going to say?	Paraphrase, open question, again eliciting concreteness. The statement on confidence is an interpretation of his overall verbal and nonverbal style. (Interchangeable followed by a potentially additive question)
37. *Robert:* I could tell him that I like working there, but I'm concerned about how to handle difficult customers. I'll ask his advice and how he does it. In some ways, it worries me a little; I don't want to give in to the boss . . . but maybe he will have a useful idea.	Robert is able to plan something that might work. With other clients, you may role-play, give advice, actually assign homework with an action plan. You will also note that Robert is still concerned about "giving in."
38. *Machiko:* Would you like to talk more about this the next time we meet? Maybe through your talk with your boss we can figure out how to deal with this in a way that makes you feel more comfortable. Sounds like a good contract. Robert, you'll talk with your boss, and we'll meet later this week or next week.	Open question, structuring. If Robert does talk to his boss and listens to his advice—and actually changes his behavior—then this interview could be rated holistically at Level 3, additive empathy. If not, then a lower rating is obvious.

Theoretically and philosophically, this decision counseling session using only listening skills is closely related to Carl Rogers's person-centered counseling (Rogers, 1957). Rogers developed empathic guidelines for the "necessary and sufficient conditions of therapeutic personality change." In the Machiko-Robert transcript, you saw a decision counseling model combined with a modified person-centered approach.

The homework/action plan is not included in the Machiko interview, but it will be explored in detail in Chapter 12. Many would argue that what comes after a counseling and therapy session, during the next day and week, is where learning and possible change will occur. In former days, counseling and therapy gave little attention to taking the interview home to the real world. Now this is a central issue and is a key part of decision counseling. You will find more information on decision counseling in Chapters 12 and 13.

Rogers originally was opposed to the use of questions but in later life modified his position so that in some interviews a very few questions might be asked. These would be quite open and as "nondirective" as possible. "What is your goal?" and "What meaning does that have for you?" are two examples of very open questions that tend not to box the client into the counselor's perspective or theory.

Person-centered counseling often requires a client with good verbal skills, and it often takes many sessions for success. Listening is clearly the best way to understand a client, but the action skills of influencing and theoretical approaches such as cognitive behavioral therapy are critical additions so that you can reach many types of clients.

It was Carl Rogers who truly brought the ideas of empathic understanding to the counseling and psychotherapy process. Many would say that he humanized counseling and psychotherapy by stressing the importance of relationship, respect, authenticity, and positive regard, also called the working alliance. *Working alliance* is a useful term as it stresses the way we need to *work with*, rather than *work on*, the client. A good relationship and working alliance may be in itself sufficient to produce positive change.

If you practice listening and seeking to experience the client's world, you will gain a better understanding of what Carl Rogers is seeking—to enable the client to discover the self more fully and reach the power of self-actualization. At the same time, we need to recall that facilitating change often requires a more active stance on your part.

Multiple Applications: Integrating Microskills with Stress Management and Social Justice

> Poverty in early childhood poisons the brain. . . . neuroscientists have found that many children growing up in very poor families with low social status experience unhealthy levels of stress hormones, which impair their neural development. The effect is to impair language development and memory—and hence the ability to escape poverty—for the rest of the child's life.
>
> —Paul Krugman

This chapter brings together the listening skills and presents them as basic to virtually all counseling and therapy. Here we continue this integration with a reminder of the need to remember our foundation of ethics, multicultural competence, and wellness.

Counseling and therapy are conducted on a consistent ethical base. And a fully aware ethical base suggests the need for awareness of social justice and stress management.

Stress is an issue in virtually all client issues and problems. It may not be the presenting issue, but be prepared to assess stress levels and provide education and treatment as needed, using some of the positive psychology/wellness strategies of Chapter 2. You will find that Chapters 12 gives this further attention. Stress will show in body tension and nonverbal behavior, as well as in client verbal conversation. Cognitive/emotional stress is demonstrated in vocal hesitations, emotional difficulties, and the conflicts/discrepancies clients face in their lives.

A moderate amount of stress, if not prolonged, is required for development and for physical health. For example, repeated stressing of a muscle through weight lifting breaks down muscle fibers, but after rest the rebuilding muscle gains extra strength. A similar pattern occurs for running and other physical exercise. If there is no stress, neither physical development nor learning will occur. Stress should also be seen as a motivator for focus and as a condition for change. The amygdala needs to be energized. For example, the skill of confrontation can result in stress for the client. But used as a supportive challenge, this stressor becomes the basis for cognitive and emotional change.

In all of the above, note the word *moderate* and the need for rest between stressors. Ratey (2008b) explains:

> *Cortisol* is the long-acting stress hormone that helps to mobilize fuel, cue attention and memory, and prepare the body and brain to battle challenges to equilibrium. Cortisol oversees the stockpiling of fuel, in the form of fat, for future stresses. Its action is critical for our survival. At high or unrelenting concentrations such as post-traumatic stress, cortisol has a toxic effect on neurons, eroding their connections between them and breaking down muscles and nerve cells to provide an immediate fuel source. (p. 277)

Toxic and long-term stress is damaging and can shrink the brain, enlarge the amygdala, and shrink the hippocampus (Hanson et al., 2015). Making the same point a decade earlier, an article titled "Excessive Stress Disrupts the Architecture of the Developing Brain" (National Scientific Council on the Developing Child, 2005) includes the following useful points:

1. In the uterus, the unborn child responds to stress in the mother, while alcohol, drugs, and other stimulants can be extremely damaging.

2. For the developing child, neural circuits are especially plastic and amenable to growth and change, but again excessive stress results in lesser brain development. In adulthood, that child is more likely to have depression, an anxiety disorder, alcoholism, cardiovascular problems, and diabetes.

3. Positive experiences in pregnancy seem to facilitate child development.

4. Caregivers are critical to the development of the healthy child.

5. Children of poverty or who have been neglected tend to have elevated cortisol levels that can slow or destroy cognitive/emotional development over a lifetime.

Just as counseling can change the brain, oppression changes the brain in negative ways. Incidents of oppression, such as racism, disrespect, and bullying of all types, place the brain on hypervigilance, thus producing significant stress, with accompanying hyperfunctioning of the amygdala and interference with memory and other areas of the brain. We need to be aware that many environmental issues, ranging from poverty to toxic environments to a dangerous community, all work against neurogenesis and the development of full potential. And let us expand this to include trauma.

Recall that the infant, child, and adolescent brain can only pay attention to what is happening in the immediate environment. Again, think of the varying positive and negative environments that your clients come from. One of the purposes of the community is to help you and the client understand how we as individuals relate and are related to individuals, family, groups, and institutions around us. The church that welcomes you helps produce positive development; the bank that refuses your parents a loan or peers that tease and harass you harm development.

Clients need to be informed about how social systems affect personal growth and individual development. Our work here is to help clients understand that the problem does not lie in them but in a social system or life experience that treated them unfairly and did not allow an opportunity for growth.

Finally, there is social action. What are you doing in your community and society to work against social forces that bring about poverty, war, and other types of oppression? Are you teaching your clients how they can work toward social justice themselves? A social justice approach includes helping clients find outlets to prevent oppression and working with schools, community action groups, and others for change.

Neuroscience research provides a biological foundation for understanding the impact of our work. Brain research is not in opposition to the cognitive, emotional, behavioral, and meaning emphasis of counseling and therapy. Rather, it will help us pinpoint types of interventions that are most helpful to the client. In fact, one of the clearest findings is that the brain needs environmental stimulation to grow and develop. You can offer a healthy atmosphere for client growth and development. We advocate the integration of counseling, psychotherapy, neuroscience, molecular biology, and neuroimaging and the infusion of knowledge from such integrated fields of study to practice, training, and research.

Taking Notes in the Session

Beginning helpers typically question whether they should take notes during the session, and it's not hard to find opinions for and against note taking. We are going to share our opinions based on our experience, recognizing that individual views vary on this issue. It is essential to follow the policies of your agency.

Intentionality in counseling and psychotherapy requires accurate information. Therefore, we recommend that you listen intentionally and take notes. This is our opinion, but some authorities will disagree. You and your client can usually work out a suitable arrangement. If you personally are relaxed about note taking, it will seldom become an issue. If you are worried about taking notes, it likely will be a concern. When working with a new client, obtain permission early about taking notes.

Audio or video recording the session follows the same guidelines. If you are relaxed and provide a rationale to your client, making this type of record of the session generally goes smoothly. If you do not take notes during a session, remember that records typically need to be kept, and we recommend writing session summaries shortly after the session finishes. In these days of performance accountability for your actions, a clear record can be helpful to you, the client, and the agency with which you work.

Some expert counselors object to note taking, arguing that when the counselor is closely in tune and has great listening skills, there is no need for notes. Indeed, too much attention to note taking detracts from the central issues of rapport building and active listening. The most famous listener of all, Carl Rogers, found that when one records an interview, one often finds that what actually happened is different from what is found in notes. Thus, a balance is clearly needed, particularly in a time of increased scrutiny and possible legal issues.

Presented in more detail in Appendix II, HIPAA (Health Insurance Portability and Accountability Act) legal requirements regarding note taking are not always clear. Some rules give certain aspects of counseling more detailed protection than general medical records, but these rules are sometimes written vaguely. The agency you work with in practicum or internship can guide you in this area. You will also find Zur's (2015) revised edition of *The HIPAA Compliance Kit* helpful.

Action: Key Points and Practice

The Basic Listening Sequence. The basic listening sequence (BLS) is built on attending and observing the client, but the key skills are using open and closed questions, encouraging, paraphrasing, reflecting feelings, and summarizing. When we listen to clients using the BLS, we want to obtain the overall background of the client's story and learn about the facts, thoughts, feelings, and behaviors that go with that story.

The Five Stages Model. The five stages of the session provide an organizing framework for using the microskills with multiple theories of counseling and psychotherapy.

Stage 1: Empathic Relationship includes initiating the session, establishing rapport, building trust, structuring the session, and establishing early goals.

Stage 2: Story and Strengths focuses on gathering data. Draw out stories, concerns, and strengths.

Stage 3: Goals establishes goals in a collaborative way.

Stage 4: Restory includes working with the client to explore alternatives, confronting client incongruities and conflict, and rewriting the client's narrative.

Stage 5: Action involves collaborating with the client to take steps toward achieving desired outcomes and achieving change.

Use the five stages as a checklist to be covered in each session.

Decision Counseling. Decision counseling follows the five-stage structure of the interview. Along with the microskills, it provides a foundation that you can use to become competent more easily in other theories of helping. Regardless of varying counseling and therapy theories, most sessions involve making some sort of decision, including defining the key issues, defining the goal, and selecting from alternatives.

Social Justice and Stress Management. A fully aware ethical base suggests the need for awareness of social justice and stress management. Unjust social systems, poverty, and toxic and long-term stress are damaging to children and adults and can negatively affect the brain.

Note Taking. Accurate records are essential in counseling and therapy. If you personally are relaxed about note taking, it will seldom become an issue. If you are worried about taking notes, it likely will be a challenging session.

Practice and Feedback: Individual, Group, and Microsupervision

Additional resources can be found by going to CengageBrain.com and logging into the MindTap course created by your professor. There you will find a variety of study tools and useful resources that include quizzes, videos, interactive counseling and psychotherapy exercises, case studies, the Portfolio of Competencies, and more.

This integrative chapter offers the opportunity to organize and think through the listening skills in a context of ethics, multicultural competence, plus positive psychology.

Exercise 8.1 Five Stages and Decision Counseling

This should come fairly easily, as the basic skills of brief counseling are similar to those you have learned through earlier chapters. Audio or video recorded role-playing practice and feedback are essential.

❏ Work with a partner, switching the roles of client and counselor. Plan for a minimum session of 15 minutes.

❏ Select a concern for the role-play. This time the issues need to be very specific—for example, dealing with a specific conflict on the job, in the family, or with a partner. Academic stress is obviously a good topic for practice. Aim for concreteness and clarity throughout the storytelling.

❏ Record the session on audio or video using a computer, cell phone, or personal camera to provide some instant feedback.

❏ Use the Client Feedback Form (Chapter 1) to receive feedback from your partner.

❏ Individually, transcribe the interview and compare your current work with your first interview to note progress.

Portfolio of Competencies and Personal Reflection

A lifetime can be spent increasing one's understanding and competence in the ideas and skills from this chapter. You are asked here to learn and even master the basic ideas of predictability from skill usage in the session, several empathic concepts, and the five stages of the well-formed interview. We have learned that student mastery of these concepts is possible, but most of us (including the authors) find that reaching beginning competence levels makes us aware that we face a lifetime of practice and learning.

You should feel good if you can conduct an interview using only listening skills. Focus on that accomplishment and use it as a building block toward the future. As you do, you are even better prepared for developing your own style and theory.

Assessing Your Level of Competence: Awareness, Knowledge, Skills, and Action

Use the following as a checklist to evaluate your present level of mastery. As you review the items below, ask yourself, "Can I do this?" Check those dimensions that you currently feel able to do. Those that remain unchecked can serve as future goals. Do not expect to attain intentional competence on every dimension as you work through this book. You will improve your competencies with repetition and practice.

Awareness and Knowledge. Can you do the following?

❏ Classify the microskills of listening.

❏ Define empathy and its accompanying dimensions.

❏ Identify the five stages of the structure of the interview.

❏ Discuss, in a preliminary fashion, issues in diversity that occur in relation to these ideas.

Basic Competence. Aim for this level of competence before moving on to the next skill area.

❏ Use the microskills of listening in a real or role-played interview.

❏ Demonstrate the empathic dimensions in a real or role-played interview.

❏ Demonstrate five dimensions of a well-formed interview in a real or role-played session.

Intentional Competence. Ask yourself the following questions, all related to predictability and evaluation of your effectiveness in working with listening skills and the five-stage model. These are skill levels that may take some time to achieve. Be patient with yourself as you gain mastery and understanding.

❏ Anticipate predicted results in clients when using the listening microskills.

❏ Facilitate client comfort, ease, and emotional expression by being empathic.

❏ Enable clients to reach the objectives of the five-stage interview process:

1. *Relationship*: Develop rapport and feel that the interview is structured.

2. *Story and Strengths*: Share data about the concern and also positive strengths to facilitate problem resolution.

3. *Goals*: Identify and even change the goals of the interview.

4. *Restory*: Work toward problem resolution.

5. *Action*: Generalize ideas from the interview to their daily lives.

Psychoeducational Teaching Competence. Teaching competence in these skills is best planned for a later time, but those who run meetings or do systematic planning can profit from learning the five-stage interview process. It serves as a checklist to ensure that all important points are covered in a meeting or planning session.

❏ Teach clients the five stages of the interview.

❏ Teach small groups this skill.

Personal Reflection on Conducting a Five-Stage Counseling Session Using Only Listening Skills

You are now at the stage to initiate construction of your own interviewing process. You certainly cannot be expected to agree with everything we say. You likely have found that some skills work better for you than others, and your values and history deeply affect the way you conduct an interview. Some skills you'd like to keep, and some you might like to change.

We encourage you to look back on these first eight chapters as you consider the following basic question leading toward your own style and theory.

What single idea stood out for you among all those presented in this text, in class, or through informal learning? Allow yourself time to really think through the one key idea or concept—it may be something you discovered yourself. What stands out for you is likely important as a guide toward your next steps.

Continue your development of your own style and theory through writing.

SECTION III

Transitioning from Attending and Listening to Influencing Skills

Focusing and Empathic Confrontation

> Our interaction with clients changes their brain (and ours). In a not too distant future, counseling and psychotherapy will finally be regarded as ideal ways for nurturing nature.
>
> —Oscar Gonçalves
> Harvard Medical School, University of Minho, Portugal

The qualities of listening and relationship form the foundation of helping. Carl Rogers (1957) was correct when he stated that they provide the "necessary and sufficient conditions for psychotherapeutic change." However, not all clients are sufficiently verbal, the issues may be too complex, and the counselor may have other valuable skills and knowledge that the client needs. Thus, the majority of clients will benefit from the action skills of influencing. Sections III, IV, and V are where we explore these skills in detail.

We have frequently noted that counseling changes or rewires the brain. Through neurogenesis, we develop new neurons and neural connections throughout life and manage to keep the number of neurons in the brain at approximately 100 billion. Although there is also evidence regarding areas such as the prefrontal cortex, the olfactory area, and the *nucleus accumbens* (pleasure center), research indicates that the hippocampus, our memory center, is the area that develops most new neurons (Gould, Beylin, Tanapat, Reeves, & Shors, 1999; Seki, Sawamoto, Parent, & Alvarez-Buylla, 2011).

The two chapters of Section III provide transcripts of two live sessions from a DVD that illustrates how counseling skills, using neuroscience concepts, can indeed change memory (Ivey, Ivey, Gluckstern-Packard, Butler, & Zalaquett, 2012). You will see that memories are explored through the listening skills and then changed via dialogue and influencing skills.

Chapter 9. Focusing the Counseling Session: Contextualizing and Broadening the Story Here you will see the client, Nelida, present a troubling and oppressive classroom encounter that, through surprise, was immediately imprinted in her long-term memory. Through the use of listening skills and focusing, the counselor facilitates review of the client's story from multiple perspectives and ensures a comprehensive examination of concerns and issues. Skilled focusing is based on listening; it enables clients to view their stories in new ways without your supplying the answers for them.

Chapter 10. Empathic Confrontation and the Creative New: Identifying and Challenging Client Conflict A second interview with Nelida shows how clarification and empathic confrontation can lead to permanent change in memory. Many theorists and practitioners consider empathic confrontation the key stimulus enabling client change and development. Empathic confrontation builds on your ability to listen empathically and observe client conflict. Skilled supportive confrontation enables resolution of conflict and incongruity, leading toward new behaviors, thoughts, feelings, and meanings.

How Memory Changes Are Enacted in the Session

The following two chapters, on focusing and confrontation, illustrate the drawing out of a negative story and the resulting reframing and change in meaning, enabling the client to feel better about herself and think and act more effectively. Nelida Zamora, a superstar graduate student at the University of South Florida, has kindly given permission for us to share two sessions she videotaped with Allen. In reviewing these sessions, you will see how listening and influencing skills can be combined.

Let us look at this counselor and client from a brain-based skills approach. The client, Nelida, shares a story; Allen listens, reflects, and seeks to help restory the memory and its meaning. Two brains are active in the session, and each person's brain, including short- and long-term memory, may change during the interaction. Two sets of memories in the hippocampus meet in the here and now of conscious conversation. Working memory brings life and the possibility of change to the session. Ultimately, counseling change is an interactive process of influencing client working memory in positive ways. We use working memory as our access to long-term memory, and significant change in long-term memory leads to changes in thoughts, feelings/emotions, and behaviors. Counseling is not a one-way process. We counselors learn and change as we work with clients.

Working memory is the integrated centerpiece of action in counseling conversation. Working memory can be defined as the area where we store high-speed data from here-and-now consciousness, as well as information from short- and long-term memory. Nelida and Allen each likely can store at most 18 items in working memory. However, the amount of information in working memory can change at any moment. For example, a highly emotional experience may leave the client having as few as one or two items in working memory.

In considering the change process that occurs through the interaction of client and counselor, we are also dealing with the relationship of the executive CEO prefrontal cortex to the amygdala and the limbic HPA hormonal axis. Attending behavior (attentional processes heavily controlled by the thalamus and the prefrontal areas) remains foundational in determining whether or not long-term memories are solidified in the hippocampus.

Interaction of Memory Systems of Client and Counselor

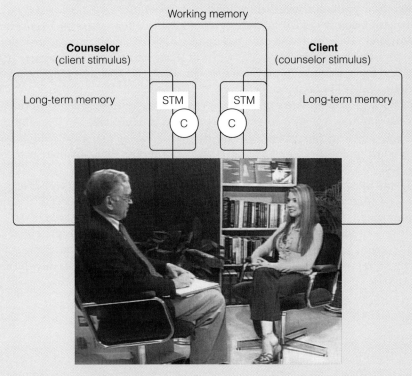

Counselor (Allen) and client (Nelida) in conversation, illustrating nonverbal mirroring (note the almost exact body mirroring) and depicting the interaction of short-term and long-term memories.

© Cengage Learning

But emotional involvement through the energizing amygdala is necessary for working memory to function and to make this happen.

Authorities vary in how many items one can keep in working memory at one time. The original and often used definition of working memory was that we can hold seven plus or minus two (7±2) items (Miller, 1956). Seven is typically the maximum number of digits one can hold in working memory, but with practice, some people can hold 20 or more. Talking about difficult experiences (more so with trauma), as often happens in counseling, may reduce the span of working memory. At times, strong emotions and memories from the long-term memory unconscious may blot out what is occurring in the external here and now.

The figure presents a time-based information-processing approach. Consciousness (C) represents the psychological present, which ranges in length from 100 to 750 milliseconds and has access to short- and long-term memory (Ivey, 2000). A person who meditates, someone experiencing a real "runner's high," or a ballerina, tennis star, or serious painter is very close to living the here and now of consciousness. Of course, few reach these goals without practice, which involves executive functions, limbic HPA hormones, the amygdala's stimulation, and memory in the hippocampus, all operating in the holistic brain. The long-term memory has been automated in procedural memory, thus allowing the person to be fully in the here and now. Interestingly, this is also a goal of counseling and therapy—to help clients learn new ways of being that eventually become so much a part of them that they seldom have to think about their actions.

Short-term memory (STM) holds impressions immediately accessible to consciousness for approximately 10 seconds and can hold up to 100 items. If learning occurs, information moves into long-term memory (LTM). LTM is our storehouse of declarative (episodic, semantic) and nondeclarative (procedural, perceptual, classical conditioning, emotional) information. Deeper in long-term memory is the unconscious, life experience, which is less accessible, but with appropriate stimulation it can be brought to short-term working memory and consciousness. We have potential access to unconscious material if an appropriate event happens or a counselor provides a key stimulus. Working memory could be called the "action" foundation for psychotherapeutic change, integrating the immediate here-and-now consciousness with STM and LTM.

A key task of counseling is to help the client restory past experience and develop new memories and connections (behaviors, thoughts, feelings, meanings). Successful counseling and psychotherapy change the client and LTM in significant ways and even build new neural networks in the brain (brain plasticity). The attending microskills presented earlier provide the cognitive/emotional "charge" to promote understanding and change. The influencing skills introduced in this section of the book both start and solidify the change process.

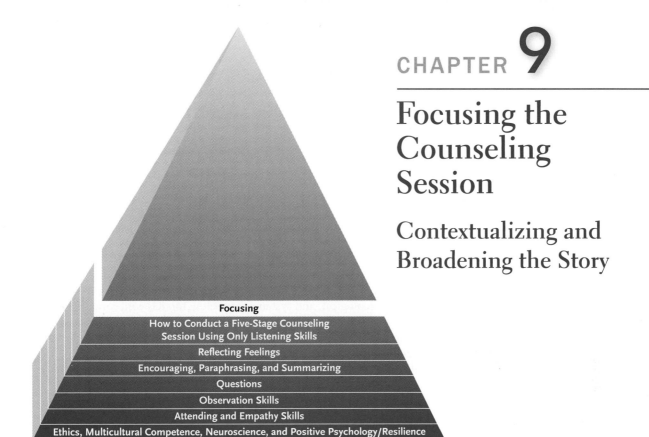

Focusing

How to Conduct a Five-Stage Counseling
Session Using Only Listening Skills

Reflecting Feelings

Encouraging, Paraphrasing, and Summarizing

Questions

Observation Skills

Attending and Empathy Skills

Ethics, Multicultural Competence, Neuroscience, and Positive Psychology/Resilience

CHAPTER **9**

Focusing the Counseling Session

Contextualizing and Broadening the Story

"Yo soy yo y mi circunstancia, y si no la salvo a ella no me salvo yo." "I am I and my circumstances. If I do not save my circumstances, I will not save myself."

—José Ortega y Gasset, Meditations on Quixote

Chapter Goals and Competency Objectives

Awareness and Knowledge

▲ Conceptualize clients as persons-in-relation and persons-in-community.

▲ Identify contextual factors affecting clients' current situation or concern.

Skills and Action

▲ Help clients tell their stories and describe their issues from multiple frames of reference, a valuable method for creative change.

▲ Increase clients' cognitive and emotional complexity, thus expanding their possibilities for restorying and resolving issues.

▲ Enable clients to see themselves as selves-in-relation and persons-in-community through community and family genograms.

203

▲ Facilitate and clarify client cognitive/emotional processes so that they can take action to address their concerns, issues, and challenges.

▲ Include advocacy, community awareness, and social change as part of your counseling or psychotherapy practice.

Introduction: Focusing Essentials

The Spanish philosopher José Ortega y Gasset reminds us that we are immersed in society, culture, family, relations, and historical moment. We are born into a bioecology, and we develop within that context. Through our interconnections, we influence and are influenced by others and by the systems within that context. "I am me and my cultural/environmental/social context."

Couples, family, group, ecological, and systems counseling models conceptualize issues through focusing on personal and contextual factors. Focusing can help counselors and therapists better understand their clients and the multiple systems affecting their success. This is illustrated by elementary or secondary students who may be failing. Is the family distressed by illness, divorce, or job loss caused by the closing of a plant? Is the child being bullied, involved in too much screen time, not getting enough to eat? With this type of understanding, counselors can build a comprehensive and collaborative model to promote student success (McMahon, Mason, Daluga-Guenther, & Ruiz, 2014). Just thinking about the single issue of academic failure is clearly not enough.

Furthermore, death comes five to seven years earlier to those who face multiple issues and concerns, particularly low income. This is called multimorbidity. Studies on best approaches to help clients whose quality of life has been harmed significantly by long-term conditions show that integrative interventions are better. A proposed chronic care model integrates collaborative goal setting and action planning, self-management of physical symptoms and emotional health, social or spiritual support, informed clinical team, and responsive and flexible organizational process (Coventry, Small, Panagioti, Adeyemi, & Bee, 2015).

The following example illustrates the importance of focusing in practice.

Nelida Zamora, a former graduate student at the University of South Florida, has given us permission to use her name and her community genogram in this chapter.* She has a real concern: being made uncomfortable in an introductory counseling class. This is not an unusual situation; many students who come from non-European backgrounds do not always believe that they "fit in." Beyond that, as you know, many "majority" students also may not feel totally welcomed.

Nelida: Here I am, a grad student in counseling. I did well in college in Miami, and thought it was no big deal because I was only four and a half hours away. But my first day of class I raised my hand, made a comment that very first class, and a classmate asked me if I was from America (nervous laugh) or a native (nervous laugh). Yeah, and I said well I'm . . ., I was just four and a half hours away, and

---

*A transcript of the real interview held between Allen Ivey and Nelida Zamora is available on DVD: Ivey, A., Ivey, M., Gluckstern-Packard, N., Butler, K., & Zalaquett, C. (2012). *Basic Influencing Skills* [DVD]. Alexandria, VA: Microtraining/Alexander Street Press. By permission of Microtraining/Alexander Street Press, www.alexanderstreet.com.

he just found it very hard to believe. So, after that comment was made, it kind of made me a little bit more hesitant to participate in discussions. It made me more self-conscious. (Here we see that a single comment affects the amygdala and pre-frontal cortex so that cognitive and emotional memories are stored immediately and permanently in the hippocampus.)

Allen: It made you self-conscious. Could we explore that a little bit more? Ah, first of all, in English, what were the feelings that went with that? (As you will see later in the example transcript, those feelings are soon explored in Spanish.)

Nelida: Well, I was surprised because being from Miami a lot of my family members have recently come from Cuba, so there they look at me as the American girl and they make fun.

Allen: . . . and that embarrasses you. (Encourager focused on feelings. Brings in the emotion underlying the cognitive words, also draws from observation of nonverbals.)

Nelida: Exactly, so when I'm in Miami, my family and friends tease saying that I'm the American who can't speak Spanish a hundred percent correctly 'cause I've forgotten a lot of it because of the English. Then, now, I move here to Tampa, I'm the Cuban girl who can't speak English, so it seems like I'm torn. You know, I don't know where I belong sometimes.

Reflective Exercise How what happens outside affects the inside

What are the issues that Nelida faces? What internal and external factors affect how she thinks, feels, and acts (behaves)? Where should you focus your comment? List as many possibilities as you can before moving on.

_____

_____

_____

We should first note that Nelida faces the stress of not belonging and feeling different, plus being caught between the culture of Cuban Miami and a university in Tampa. These stressors affect her emotionally and imprint her long-term memory with a negative picture of herself. She is in a high state of incongruence. Focusing will help clearly identify the major areas of conflict and discrepancy and then help determine which ones will be approached first. Listening and using supportive challenges will help Nelida clarify her situation and move more readily to problem solution.

Over this and the next chapter on empathic confrontation, we will present an example of how counseling changes memory. The first session introduces the community genogram, a systematic way to review old positive memories and help clients see themselves in social context. The community genogram provides a visual picture that helps us understand the client's personal and cultural background.

A central current issue for Nelida is cultural oppression, which she has internalized; she has come to "blame" herself for being different. Rather than focusing just on Nelida as an individual, if you help her see other perspectives, such as being able to name the oppression of the classroom, she is better prepared to reframe and change the negative memory. In addition, focusing on family and cultural background will facilitate her pride in her Cuban family and culture and provide positive assets, strengths, and resources to deal more effectively with the cutting comments she has experienced.

If you use focusing skills as defined below, you can *anticipate* how clients may respond.

Focusing	Anticipated Result
Intentionally focus the counseling session on the client, theme/concern/issue, significant others (partner/ spouse, family, friends), a mutual "we" focus, the coun- selor, or the cultural/environmental context as necessary to gain a broader understanding of client and issue. You may also focus on what is going on in the here and now of the session.	As the counselor brings in new focuses, the story is elab- orated from multiple perspectives. If you selectively attend only to the individual, the broader dimensions of the social context are likely to be missed, and counseling and therapy may fail in the long run.

Awareness, Knowledge, and Skills of Focusing

Counseling is, first and foremost, for the individual. Thus, the first focus dimension is on the unique client before you. Focus on individual issues, so clients can talk about them- selves from their personal frame of reference. Using the client's name and the word "you" helps personalize the counseling. While it is essential that you draw out the client's story, don't become so fascinated with the details of that story that you forget about the person talking to you.

"Nelida, I hear that class experience made you feel self-conscious. Could explore your thoughts and feelings about that?"

"I see you have many strengths helping you, and they are . . . ."

(Later) "Nelida, you feel much stronger when you recognize the strengths of your family and your Cuban culture."

Attending to the *theme*, or central topic(s), of the session is a second area of focusing. Draw out client stories, issues, or concerns. If a client has gone through a breakup of a sig- nificant relationship, has study difficulties, has cancer or another serious illness, we need to hear the details, and we need to hear a lengthy story. Just telling the story is relieving. We feel better when someone seriously listens and understands. But also focus on the strengths and capabilities clients bring with them. Too many beginners and even professionals be- come transient voyeurs, so interested in the problematic story that they fail to focus on the unique client before them and their personal strengths to facilitate resolving issues. Note that attending to the theme inevitably involves the client as well, and as the session pro- gresses, the presenting issue enlarges.

"Nelida, you said that you hesitated to speak up in classes during the rest or the term. Could you give me some examples of some specific times you wanted to speak up, but didn't."

"I hear that your high school guidance counselor inspired you to go to college."

"You said you really enjoyed studying counseling."

Focus on contextual dimensions. Nelida lives in a broad context of multiple systems. The concept of *self-in-relation* may be helpful. The idea of *person-in-community* was devel- oped from an Afrocentric frame by Ogbonnaya (1994), who pointed out that our family and community history and experiences live within each of us. Since that time, the idea that we are persons-in-community has taken hold, and we often hear "It takes a village to raise a child." Clients bring to you many community voices that influence their view of self and the world. The debriefing of Nelida's community genogram in this chapter shows how this strategy is a useful way to understand your client's history and a good place to identify strengths and resources.

Significant Others. These might include partner or spouse, friends, and family.

"Nelida, tell me a bit more about your relationship with your family in Miami."

"How are your friends helpful to you?"

"Your grandmother was very helpful to you in the past. What would she say to you about all this?"

"Nonetheless, she calls you the "English girl" and thinks your Spanish is slipping. You sometimes feel that you don't fit in either Miami or here."

Mutual Focus. Use "we" statements involving the client, therapist, or group.

(Early in the session) "Nelida, you have something that's been bothering you for over a year, but *we* will work through this. What has been most helpful as we have talked so far. What have I missed?"

Immediacy, Here-and-Now Focus. Talk about what is going on in that moment in the session.

"Nelida, right now I can sense you are hurting still from the comment the first day of class."

Counselor Focus. Share your own experiences and reactions.

"It really bothers me to hear what happened in that first class."

(Later) "I feel good to hear that you are taking charge of your Latina identity and have become aware that it was a form of racism and oppression that you experienced in that class."

Cultural/Environmental/Contextual (CEC) Focus. This includes broader issues, such as the impact of one's culture, life history, and even recent national and world events.

(Near the end of the session)"Nelida, you feel much stronger when you recognize the strengths of your family and Cuban culture."

CEC Counselor Statements Leading to a Positive Conclusion

"The Castro government appeared at your door one day, telling your family they had to leave the next day. Fortunately, you all were able to escape to New York."

"Let's look at your community genogram in Miami so that I can learn more about what makes Nelida, Nelida."

"What are some strengths that you gain from your family, church and community?"

Focus on Physical Health and Therapeutic Lifestyle Issues. Given that we now know that physical and mental health are intimately entwined, focusing reminds us that we need to consider additional issues with each client. This focus area was explored with Nelida in the two interviews in this chapter and the next. Here are some additional issues that are part of collaborative work, as well as matters that need to be considered, as appropriate, in the individual counseling session. Many of them focus on self-management and a positive lifestyle.

- *Medications.* What are they, and what is their purpose? Is the client compliant in their use and able to afford them?

- *Drugs.* As many as 50% of our clients have issues with drugs and potential abuse. More than 700 designer drugs are presently available, thus making legal enforcement nearly impossible.

- *Therapeutic lifestyle changes (TLCs), and the time and money to engage in them.* Checking out exercise, sleep patterns, and other TLCs is critical. The morbidity research discussed above found that self-management (as represented by TLCs), spirituality, and a meaning for living (see Chapter 11) are particularly important issues in maintaining health and building resilience.

As a counselor, be aware of how you focus a counseling session and how you can broaden the session so that clients are aware of themselves more fully in relation to others and social systems: persons-in-relation, persons-in-community. In a sense, you are like an orchestra conductor, selecting which instruments (ideas) to focus on, enabling a better understanding of the whole. Some of us focus exclusively on the client and the issues that the client faces, neglecting to recognize the total context of client concerns. We need to be aware that we are not the only ones who can help clients. Box 9.1 outlines the research illustrating the value of broadened focus on client issues.

Focusing is a skill that enriches our understanding of our clients and their background, plus reminding us of the complexity each of us faces in making decisions in a challenging world.

BOX 9.1   Research and Related Evidence That You Can Use

Focusing

The evidence base for collaborative care is very solid. Collaborative care is more effective than usual care for managing depression and anxiety over the short, medium, and long term.

—Coventry et al., 2015

The recent multimorbidity research cited in some detail at the beginning of this chapter provides solid evidence that a failure to think more broadly and consider multiple dimensions in helping is, frankly, wrong. There is a real need to expand our thinking and practice of helping (Coventry et al., 2015).

Collaborative care, or integrated care, is a team approach involving counselors, physician, social workers, financial advisers, school/community/governmental officials, and others, as appropriate to client and family needs. Typically, it involves both physical and mental health. Depression is now recognized as a biological disease that could be caused by either external psychological challenges or internal imbalances in the body or illness. Focusing enables us to think beyond just a single individual concern and place the situation in broader context, remembering that many concerns at the same time result in multimorbidity and early death.

In a classic review of the contextual focus, Moos (2001) has reviewed much of the literature and points out that the way we appraise a situation can be self-centered or oriented to the cultural/environmental/context. Clients often come to the counseling session with a focus that may work against their own best interest. Too much of an "I"

focus may result in self-blame and lack of awareness of context. On the other hand, too much of a "they" focus may mean that clients are avoiding responsibility or their part in the conflict. As you know, there are two or more stories as people look at the same event. Moos noted that teaching clients the context of their issues helps them understand themselves in new ways and "makes possible a transformative experience."

Training students to focus on cultural/environmental/contextual issues resulted in greater awareness and willingness to discuss racial and gender differences early in the session and to make these issues a consistent part of the counseling or therapy (Zalaquett, Foley, Tillotson, Hof, & Dinsmore, 2008).

Educating and training students in multicultural counseling provides an excellent model for the future (Sue & Sue, 2015). A study that tested a set of multicultural skill training videos found that working with these videotapes of culture-specific counseling increased students' multicultural effectiveness and understanding (Torres-Rivera, Pyhan, Maddux, Wilbur, & Garrett, 2001).

The Community Genogram: Bringing Cultural/Environmental/ Context into the Session

Clients bring us many stories. Most often we tend to work with only one individual story. But stories and issues of many others (e.g., friends, family, unique factors of diversity) deeply affect the client's narrative. There are many other factors we can focus on as well if we are to help the client deal with complexity in living and personal decisions.

A good way to develop an understanding of the value of focusing and enriching client stories is the community genogram, which can give us a good picture of a client's cultural background and history, thus enabling us to view the client in social context. By working with you on their community genograms, your clients will gain a richer understanding of themselves as persons in relation to others.

The community genogram is a "free-form" activity in which clients are encouraged to present their community of origin or their current community, using their own unique style. Some visual examples of community genograms are shown in Box 9.2. Through the community genogram, we can better grasp the developmental history of our clients and identify client strengths for later problem solving. Clients may construct a genogram by themselves or be assisted by you through questioning and listening to the things that they include.

Developing Your Own Community Genogram

Let's start the examination of focusing by having you complete your own community genogram, following the steps outlined below. If you take time to develop the genogram, you will be better prepared to help your clients consider themselves as persons-in-community and see themselves in social context. The community genogram provides a snapshot of the culture from which you and your clients come.

- Select the community in which you were primarily raised. The community of origin is where you tend to learn the most about culture, but any other community, past or present, may be used.

- Use a large poster board or flipchart paper. Representing yourself or the client with a significant symbol, place yourself or the client either at the center or at another appropriate place. Encourage clients to be innovative and represent their communities in a format that appeals to them. Possibilities include maps, constructions, or star diagrams (see Box 9.2).

- Place family or families, nuclear or extended, on the paper, represented by the symbol that is most relevant for you or the client. Different cultural groups define family in varying ways. This may provide sufficient family information, or you may want to add the family genogram shown in Appendix III.

  Place the most influential groups on the community genogram, representing them with distinctive visual symbols. School, family, neighborhood, and spiritual groups are most often selected. For teens, the peer group is often particularly significant. For adults, work groups and other special groups tend to become more central.

  You may wish to suggest relevant aspects of the RESPECTFUL model discussed in Chapter 1. In this way, diversity issues can be included in the genogram. Nelida's Latina background is central to her self-concept. All of your clients are deeply affected by their

| BOX 9.2 | The Community Genogram: Three Visual Examples |

We encourage clients to generate their own visual representations of their "community of origin" and/or their current community support network. The examples presented here are only three of many possibilities.

1. *Nelida Zamora's community genogram.* Nelida gave considerable thought to this genogram, which she then shared with Allen. She used computer-generated images to describe her community of origin. Note that she presents only a few key dimensions, and one gets the sense of the fairly small Latino/a community in which she was raised. Each of these images contains valuable stories that give us a better understanding of Nelida as a holistic person in her home community.

2. *The map.* The client draws a literal or metaphoric map of the community, in this case a rural setting. Note how this view of the client's background reveals a close extended family and a relatively small experiential world. The absence of friends in the map is interesting. Church is the only outside factor noted.

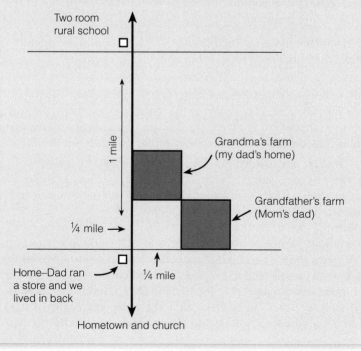

3. *The star.* Janet's world during elementary school tells us a good bit about a difficult time in her life. Nonetheless, pay equal attention to support systems and positive memories.

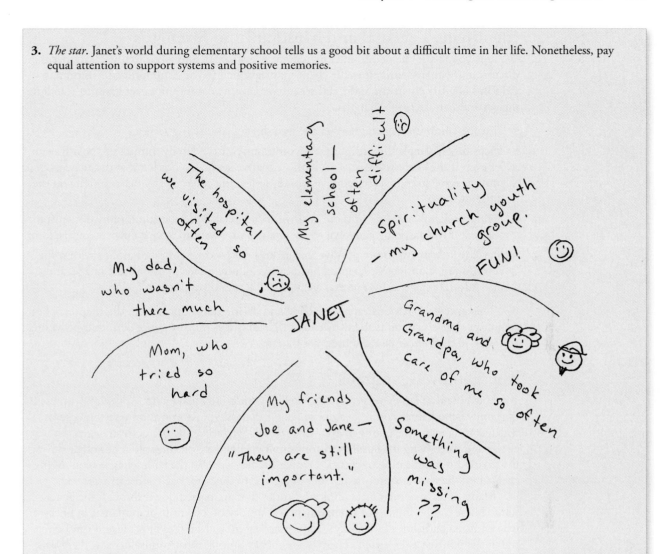

race, ethnicity, social class, and other factors, but they are often unaware of how these factors affect who they are. The community genogram makes it possible to understand where the individual came from.

- Watch for interacting physical and mental health issues and whether or not the client has a healthy lifestyle. This will not always show on the community genogram, but could come out with questioning. Sometimes clients will include a hospital, thus opening the way for this conversation.

We can bring broader understanding and multiple perspectives to the session by what we focus on in the client's life and social context. Part of what leads us to focus on certain issues is our own social context. Your developmental past and present issues can affect the counseling. You may consciously or unconsciously avoid talking about certain subjects that make you uncomfortable. You may do the same thing with clients. Becoming aware of your possible biases will free you to understand the uniqueness of each individual more fully.

Identifying Personal and Multicultural Strengths

We urge you to use the community genogram as a strength and a positive asset. Rather than discussing the many difficulties the client may have had in the community, focus on positives and identify client strengths and resources. Use the community genogram to search for images and narratives of strengths:

- Post the community genogram on the wall during counseling sessions.
- Focus on one single dimension of the community or the family. Emphasize positive stories even if the client wants to start with a negative story. Do not work with the negatives until positive strengths are solidly in mind, unless the client clearly needs to tell you the difficult story.
- Help the client share one or more positive stories relating to the community dimension selected. If you are doing your own genogram, you may want to write it down in journal form.
- Develop at least two more positive images and stories from different groups within the community. Consider developing one positive family image, one spiritual image, and one cultural image so that several areas of wellness and support are included.

This process is demonstrated in Nelida and Allen's transcript later in the chapter. The transcript analysis contains the debriefing of Nelida's community genogram, thus providing you with ideas on how you can use this strategy.

The Family Genogram

You may note that family was central in Nelida's community genogram. Appendix III presents the family genogram, a common strategy taught in most counseling and therapy programs, which can elaborate the family in even more detail. We frequently use both strategies with clients and often hang the family and community genograms on the wall in our office during the session, thus indicating to clients that they are not alone in the counseling session. Many clients find themselves comforted by our awareness of their strengths and social context.

Many of us have memories of family stories that are passed down through the generations. These can be sources of strength (such as a story of a favorite grandparent or ancestor who endured hardship successfully). Family stories are real sources of pride and can be central in the positive asset search. There is a tendency among most counselors and therapists to look for problems in the family history, and of course this is appropriate. But be sure to search for positive family stories as well as problems.

Children often enjoy the family genogram, and a simple adaptation called the "family tree" makes it work for them. The children are encouraged to draw a tree and put their family members on the branches, wherever they wish.

Debriefing a Community Genogram

Debriefing the community genogram is your chance to learn about the developmental history and cultural background of your client. It will provide you with considerable data so that you can spotlight and focus on key issues. Start by asking clients to describe the community and things that they consider most significant in their past development. Obtain an overview of the client's community.

Follow this by asking for a story about each element of the genogram. Seek to obtain positive stories of fun and support, strength, courage, and survival. Bring out the facts, feelings, and thoughts within the client's story. Many of us have been raised in communities that have been challenging and sometimes even oppressive, so a positive orientation can focus on the positives and strengths that have helped the client. That platform of positives makes it possible to explore problematic issues with a greater sense of hope. (For more ideas on the community genogram, see Rigazio-DiGilio, Ivey, Grady, & Kunkler-Peck, 2005.)

Observe: Focusing in Action

Following is a portion of the actual debriefing of Nelida's genogram. Here we see some stories from the genogram and how different issues of focus were brought up. How you work with the community genogram is most important. Clients can learn that their issues were developed in a context. The debriefing that emphasizes strengths helps clients learn that they are able and that you respect them. Armed with these positives, Nelida is better able to face some of the challenges in Tampa, and perhaps other difficulties as well.

Client and Counselor Conversation*	Process Comments
1. *Allen:* Just before the spring break we talked about you doing a community genogram. Were you able to put it together?	Focus on Nelida and the theme and concerns. The theme in this session is the community genogram plus Nelida's individual perceptions of her background. The session began with a brief greeting and check-in as to what was happening with Nelida. She had gone home to Miami. The session starts with an open question focusing on the theme of the session—debriefing the genogram.
2. *Nelida:* I was. Let me show you what I did.	Nelida shows some enthusiasm as she brings her community genogram forward (see Box 9.2).
3. *Allen:* You used the computer to make a gorgeous thing for us. So could you tell me a little bit about what's here?	Focus on individual and theme. Feedback, open question on theme. In using the community genogram, we recommend a focus on positives and strengths, rather than concerns. These positives can be used to ground clients in wellness assets as they move to resolving their challenges.
4. *Nelida:* Sure. I chose to include important parts of my community as I was growing, which shaped me into the person I am today. The solid lines symbolize positive connections, the black lines symbolize both positive and negative connections, and the jagged line is a more negative connection.	There are many ways to construct a community genogram; encourage each person to define it in her or his own way. Nelida used a computer and then enlarged it on poster board. Drawing from family therapy genograms, she added the solid and jagged lines to indicate types of relationships.
5. *Allen:* I see the positive there: the university, the church, the local park, and your grandparents. Your mother's home, which is up and down. Well, we've talked in the past about some of the issues with your parents, but today we really wanted to focus on strengths and positives from your background. I'd like to hear a very brief positive story about each one of these, and then we'll select one or two to look at in a little more depth. So let's take the difficult one first, the high school with that real broken line.	Focus on theme and significant others. Paraphrase of what Allen views on the genogram; again, the focus is on the positives of theme with a minor focus on her mother. He first structures the debriefing by mentioning the plan of a positive focus, but he starts the debriefing with a suggestion that we start with one of the broken lines. (Up to this point, there is no specific empathic response, but the relationship seems solid, so likely we could rate the session as interchangeable empathy so far.)

(continued)

*Slightly edited for clarity, this transcript of a real interview held between Allen Ivey and Nelida Zamora is also available on DVD: Ivey, A., Ivey, M., Gluckstern-Packard, N., Butler, K., & Zalaquett, C. (2012). *Basic Influencing Skills,* 4th ed. [DVD]. Alexandria, VA: Microtraining/Alexander Street Press. By permission of Microtraining/Alexander Street Press. (http://alexanderstreet.com/products/microtraining)

Client and Counselor Conversation	Process Comments
6. *Nelida:* Well, when I was in high school, I wasn't sure if I was going to pursue a college education, just because my family is very traditional. They're Latino, so it's more accepted for a woman to stay at home and be a wife and not pursue an education. So, I really didn't have my parents' and my grandparents' support to pursue that at first. So that's why my senior year was a little more tumultuous.	Here we see Nelida's past conflict with her family before she went to college. Multicultural and gender issues are quite clear here.
7. *Allen:* A little more tumultuous and the Latino tradition was not supportive of women's education. But were there any positive things that happened? Something good must have happened or you wouldn't even be here now.	Allen briefly focuses on the cultural/environmental context, but then moves to asking for an emphasis on positive memories, so critical in the community genogram process. We all have a history of some difficulties in our home communities, and these can be explored later, if appropriate. The goal here is a positive asset search that will bring out strengths that Nelida can use in the future. In later chapters, you will see the cultural/environmental/contextual focus taking central importance.
8. *Nelida:* Well, I was lucky enough to have a very good counselor in high school, who pretty much guided me in the direction that I thought I needed to go. She just put things into perspective for me, so that was very helpful.	The importance of a supportive counselor is mentioned. This is a good illustration of how other people in our historical community have helped us reach where we are. This counselor was a positive resource and still might be helpful now.
9. *Allen:* It's good you found a counselor who helped. Small wonder you end up in the counseling field. So even though you had that jagged line for the high school, there are some real strengths that are there in your community that helped you keep going.	Focus on Nelida and the theme. Allen supports the positive by focusing on a key person in Nelida's past. He adds the word *strengths* as part of his paraphrase. (This is slightly additive empathy.)
10. *Nelida:* Luckily my relationship with my mom right now is much better, so it's always good to visit her whenever I get a chance.	Nelida is talking more positively as she considers memories that support her in the past and present. Too many helpers might search for or underline problematic memories, thus giving a negative tone to the session. We can best work through our issues with the strengths of what we can do and our positive memories.
11. *Allen:* And one thing I hear as you talk about your mom is that even though you put that as sort of semi-conflictual, I see your eyes almost dancing.	Focus on Nelida. Feedback of nonverbal communication. (Additive empathy)
12. *Nelida:* Do you? It honestly feels good for me too because at high school we had a very difficult relationship, but when I decided to pursue college and later graduate school, my mother was one of my main support systems.	Despite the jagged relationship in the genogram, Nelida has reframed her view of her mother as more positive. If Allen had focused on the negative, likely we would have had a series of sad, perhaps even depressing, stories that likely would be of little help to the client. At this point, you can see that the negative story from the classroom is being compared to positive life experiences, a useful way to restory or rewrite negative memories in the brain.
13. *Allen:* So, now things are much better with your mother. Okay, and ah, your grandparents, they were something very special in growing up.	Focus on significant others. Paraphrase, emphasis on positives. (Interchangeable empathy)

Client and Counselor Conversation	Process Comments
14. *Nelida:* Um-hum. Yeah. You know they raised me, so even though they are very traditional and conservative, not too open-minded about certain things such as a woman pursuing a higher level of education. But, regardless, they've always supported me in my decisions and been great strength and support.	Nelida again restates the support she now gets from her family. It is good for clients to repeat positive strengths and assets, thus reinforcing resource development. What is occurring here is building emotional awareness of strengths, thus increasing the possibility of change from a base of strength.
15. *Allen:* They've really been important to you and provided critical support, even though they were so conservative. I imagine that word "conservative" also means you can count on them when you're facing difficulty. Is that right?	Focus on significant others and Nelida. A brief paraphrase followed by a reframing of the word "conservative" in a more positive light. This is an example of respect not only for the client, but also for her family, and for Latino culture. (Additive empathy)
16. *Nelida:* Yes, I can really count on them.	
17. *Allen:* The next thing I see is the local park and church.	Focus on theme—the community genogram. Allen topic-jumps and notes some more positive connections in the community genogram.
18. *Nelida:* These bring back good memories of my childhood. You know, it was right across the street from my grandparents' house. It's not something that you really think about too much, but as I was putting this together, I always kept going back to the park, good memories that I had growing up. I remember when my grandfather would take me to the park and we'd go on bike rides together and things like that. Um, so that definitely made me feel good, thinking about it. I decided to include it.	Respect can be shown by drawing out positive stories, and memories of relationships and good experiences give the client a chance to show the good things in their lives. Our clients are not just a long litany of problems.
19. *Allen:* The park is important with your grandfather. Do you have any particular visual image of the park when you think about it?	Focus on significant others and Nelida. Restatement/paraphrase, open question. Here we see Allen moving to the use of imagery in connection with the community genogram. (See Appendix III for more specifics on this influencing strategy.) Visual memories often encapsulate life's events.
20. *Nelida:* Just how tranquil it was.	Images are not always visual. In this example, Nelida speaks to a feeling that brings her peace.
21. *Allen:* Tranquil.	Focus on Nelida. Encourager. *Tranquil* is obviously what we call a key word, which is often representative of positive here-and-now memories from the past. In our discussion of reflection of feeling, we noted that the technical definitions of emotion and feeling differ. Emotions are partially cognitive constructions, while feelings are more associated with the body. A goal of the imagery exercise is to put people more in touch with their feelings. (Interchangeable empathy)
22. *Nelida:* Um-hum. There was always discord between my grandparents and my mom, which of course consequently affected me, but the park was like a getaway, you know, so it was a just calm, tranquil atmosphere. My grandfather would take me on bike rides, and it was just a chance to kind of leave the issues at home, you know, and just kind of go on a mini-vacation and get away someplace.	We hear the cognitions that go with the feelings.

(continued)

Client and Counselor Conversation	Process Comments
23. *Allen:* A place where you could really feel tranquil and at peace. I'd like to stay with that feeling of tranquility. Can you kind of get a visual image of a time when you're in that park and you have a specific time with your grandfather that you really felt peaceful and tranquil?	Focus on Nelida. Reflection of feeling/paraphrase, followed by a directive associated with imagery. This is additive, as it encourages Nelida to go into her experience in more concrete depth. Seeking concreteness is often associated with additive helping. Note that Allen ignores the discord issue, which has already been explored. (Interchangeable empathy; the request for the visual image is potentially additive.)
24. *Nelida:* It must have been, you know, when I would just take my bike down there and just go on the hills and ride back and forth. Um, I just felt very at peace and free to be able to do that safely with my grandfather there.	Nelida seems to be almost totally "into" the recollection, and it clearly is in her working memory. (As Nelida was able to use the visual image, Allen's lead at 23 was additive.)
25. *Allen:* Could you say those words "at peace and free"?	Focus on Nelida. Encourager in the form of a question seeking more immediacy and in-the-moment experiencing.
26. *Nelida:* I just felt at peace and free.	All the next exchanges between Nelida and Allen are very brief, indicating that the session is in the here-and-now moment.
27. *Allen:* How does it feel when you say those words?	Focus on Nelida. Open question directed toward basic feelings.
28. *Nelida:* Soothing.	This is a clear example of a more basic feeling as compared to emotion.
29. *Allen:* Soothing. Where do you feel that soothing physically in your body?	Focus on Nelida. Encourager. Question directs Nelida to the here-and-now feelings in her body. (Another possibly additive comment)
30. *Nelida:* Here on my chest.	Nelida also shows a real feeling of peace and tranquility, likely very similar to the memory of feelings she had with her grandfather in the park. Feeling and emotions are more than cognitive; they are also felt physically at some level.
31. *Allen:* You feel that tranquility in your chest. One of the purposes of the community genogram is to find strengths that we get from past events or present events and then try to locate them in our body. When we are stressed, we can draw on past stories of support and strength, which can help us deal with difficult issues as they come up. Does that make sense?	Focus on Nelida and her concerns. Reflection of feeling followed by explanation of the value of positive events located in the body. It might have been wise for Allen to have encouraged more time with those positive feelings to anchor them more fully. (Interchangeable empathy)
32. *Nelida:* It does. Um-hum, that feeling in my body just helped me put it into perspective because I didn't realize how much I thought about it till I actually did this, so yeah, I do find myself going back to those memories and those visuals and that calming and soothing feeling that I had when I was there.	This is not a new memory for Nelida, but the exercise has brought its importance to her more fully.
33. *Allen:* Yeah. A couple things as we move on. First of all, when you find yourself feeling stress and tension, you always have that tranquil feeling in your chest of you and your grandfather. If you take a breath (breathing), let it go, and visualize that park, that's what we call our resource.	Focus on Nelida, theme, and concerns. Information giving and suggestion. The emphasis here is to help Nelida generalize this experience for action in the real world. If our clients have several positive physical feeling resources in their bodies, they can draw on them to help them move through times of stress. (Additive empathy)
34. *Nelida:* It feels right. Thanks.	

An Action Plan for Nelida

The vital necessity of taking new thoughts, feelings, and behavior home was first emphasized by Albert Ellis in what he called "homework." *Action plan* may be more acceptable to your clients, but use the term you prefer. Work collaboratively *with* clients and make sure that their take-home plans are what they might want to do. Ellis encouraged clients to engage in jointly decided homework plans for 30 days. Long-term planning like this will ensure change not only in behavior, but also in neural networks that are more protective for the client

The action plan for Nelida was based on the work done with her genogram. She agreed to use the anchored feelings of tranquility when she feels stress and tension. The action plan is best reinforced with at least one clear and specific homework assignment, as this kind of activity encourages clients to take home and act on what was learned in the session (Ellis & Ellis, 2011). Nelida's action plan included attending to tensional situations on a daily basis, identifying the feelings and emotions experienced in those situations, and using her newly acquired capacity to relax in those situations. A daily report sheet can be provided to keep a record of such situations, associated feelings and behaviors, and actions taken.

Nelida's Action Plan		
Past Thoughts, Feelings, and Behavior (Story)	**Goal**	**Action Plan/Homework**
I did not speak up when the student asked where I was from. I just clammed up, and this has caused me to be too quiet during a whole semester. Now, this is contrasted with the feelings of pride as I think of what my grandparents went through.	Speak up and change behavior to get positive results.	Become more aware of how people are reacting to me. I think I am received with respect, and I ought to pay more attention to that.
Internal thought/cognitions focused on myself and my past failures Somehow I saw myself as responsible.	Focus cognitions on my strengths and resources. Focus more on what I can do rather than "can't do." Stop inner self-blame.	Think about specific times that I am proud of my family and journal my successes during the week, likely resulting in less worry.
Feelings in such situations are fear, but underlying anger at the way I am treated. My body even feels tense and awkward.	Pay more attention to pride in my family and Cuban heritage . Move from negative thinking and nervous reactions to being more relaxed, being more sure of myself, and in control of my body and enjoying it.	(#1 Action focused on emotion). In addition to the above, visualize positive scenes and experiences from the past. Note the positive feelings in my mind and body as I think about my family. Focus more on my body—slow and relax, breathe normally.

Multiple Applications of Focusing

Multicultural Issues and Focusing

From social sciences to biology, research continues to confirm the critical role of context. Box 9.3 discusses the issue of where to focus and brings our attention to the multicultural applications of focusing. Bullying and cyberbullying, as well as other forms of violence among students, have short- and long-lasting emotional and physical consequences for those targeted

---

BOX 9.3   National and International Perspectives on Counseling Skills

Where to Focus: Individual, Family, or Culture?
Weijun Zhang

Case study: Carlos Reyes, a Latino student majoring in computer science, was referred to counseling by his adviser because of his recent academic difficulties and psychosomatic symptoms. The counselor was able to discern that Carlos's major concern was his increasing dislike of computer science and growing interest in literature. While he was intrigued about changing his major, he felt overwhelmed by the potential consequences for his family, in which he is the oldest of four siblings. He is also the first person in his family to attend college. Carlos has received some limited financial support from his parents and one of his younger siblings, and the family income is barely above the poverty line. The counseling was at an impasse, for Carlos was reluctant to take any action and instead kept saying, "I don't know how to tell this to my folks. I'm sure they'll be mad at me."

During class discussion of this case, almost everyone argued that Carlos's problem is that he does not give priority to his personal career interests, that he should learn to think about what is good for his own mental health, and that he needs assertiveness training. I did not quite agree with my fellow students, who are all European Americans. I thought they were failing to see a decisive factor in the case: Carlos is Latino!

In traditional Hispanic culture, the extended family, rather than the individual, is the psychosocial unit of cooperation. The family is valued over the individual, and subordination of individual wants to family needs is assumed. Also, traditional Hispanic families are hierarchical in form; parents are authority figures and children are supposed to be obedient. Given this cultural background, to encourage Carlos to make a major career decision totally by himself was impossible. Any counseling effort that does not focus on the whole family is doomed to fail.

Because financial support from the family made his college education possible, Carlos may be expected to contribute to the family when he graduates. This reciprocal relationship is a lifelong expectation in Hispanic culture, and the oldest son is especially responsible in this regard.

Changing his major in his junior year not only means postponing the date when he will be able to help his family financially, but it also means he may not be able to do so at all, for we all understand how hard it is to find a job that pays well in the field of literature. When interdependence is the norm among Hispanic Americans, how can we expect Carlos to focus entirely on his personal interests without giving more weight to his family's pressing economic needs?

If I were Carlos's counselor, rather than focusing immediately on his needs, I would first support him with his family loyalty and then help him understand that there are not just two solutions: either/or. Together, we might brainstorm to generate some alternatives, such as having literature as his minor now and as his pastime after he graduates, changing his career when his younger siblings are off on their own, or exploring possibilities that might combine the two. He could, for example, design computer programs to help schoolchildren learn literature. Each of these takes into account family needs as well as his own.

The professor praised me highly for my "different and sensitive perspective," but I shrugged it off; this is just common sense to most Third World people and probably many Italian and Jewish Americans as well. (I remember years ago, when I was trying to make major career decisions with my parents, at least ten of my relatives were involved. And these days, I am still obligated to help anyone in my extended family who is in financial need.)

If the meaning of family in Hispanic culture is confusing to many counselors, the traditional extended family clan system of Native American Indians, Canadian Dene, or New Zealand Maori can be even more difficult for them to grasp. This family extension at times can include several households and even a whole village. Unless majority group counselors are aware of these differences in family structure, they may cause serious harm through their own ignorance.

(Zalaquett & Chatters, 2014). Similarly, ostracism or social exclusion makes people depressed, helpless, and likely to engage in suicidal ideation or behavior. Different from bullying, ostracism, which negatively affects basic needs for acknowledgment and meaning, is difficult to monitor and intervene (Williams & Nida, 2014). Even telomeres, which are the protective end complexes at the termini of chromosomes, are affected by both genetic (internal) and nongenetic (contextual) factors through our lifespan. Negative effects on telomeres of context via stress, trauma, and toxicity interfere with cell division and tissue replacement and give way to mortality and age-related diseases (Blackburn, Epel, & Lin, 2015). These are some of the factors that lead to multimorbidity and the need for collaborative, integrated care.

Advocacy and Social Justice

What is the role of the counselor or psychotherapist in advocacy and social justice as we face clients with multiple issues, often reaching beyond just personal concerns?

You are going to face situations in which your best counseling efforts are insufficient to help your clients resolve their issues and move on with their lives. The social context of homelessness, poverty, racism, sexism, and other contextual issues may leave clients in an impossible situation. The problem may be bullying on the playground, an unfair teacher, or an employer who refuses to follow fair employment practices. Helping clients resolve issues is much more challenging when we examine the societal stressors that they may face.

Advocacy is speaking out for your clients; working in the school, community, or larger setting to help clients; and also working for social change. What are you going to do on a daily basis to help improve the systems within which your clients live? Following are some examples showing that simply talking with clients about their issues may not be enough.

- As an elementary school counselor, you counsel a child who is being bullied on the playground.
- You are a high school counselor working with a 10th grader who is teased and harassed about being gay while the classroom teacher quietly watches and says nothing.
- As a personnel officer, you discover systematic bias against promotion for women and minorities.
- Working in a community agency, you are counseling a client who speaks of abuse in the home but fears leaving because she sees no future financial support.
- You are working with an African American client who has dangerous hypertension. You know that there is solid evidence that racism influences blood pressure.

The elementary school counselor can work with school officials to set up policies concerning bullying and harassment, actively changing the environment that allows bullying to occur. The high school counselor faces an especially challenging issue as session confidentiality may preclude immediate classroom action. If this is not possible, then the counselor can initiate school policies and awareness programs against oppression in the classroom. The passive teacher may be made more aware through training you offer to all the teachers. You can help the African American client understand that hypertension is not just "his problem," but rather that his blood pressure is partially related to racism in his environment, and you can work to eliminate oppression in your community.

"Whistle-blowers" who name problems that others like to avoid can face real difficulty. The company or agency may not want to have their systematic bias exposed. On the other hand, through careful consultation and data gathering, the human relations staff may be able to help managers develop a more fair, honest, and equitable style. Again, the issue of policy becomes important. Counselors can advocate policy changes in work settings and equal pay for equal work. You can help the client who suffers racial, gender, and sexual orientation harassment. You can speak to employers about how they can employ more people with disabilities.

The counselor in the community agency knows that advocacy is the only possibility when a client is being abused. For clients in such situations, advocacy in terms of support in getting out of the home, finding new housing, and learning how to obtain a restraining order against the abusing person may be far more important than self-examination and understanding.

Counselors who care about their clients also act as advocates for them when necessary. They are willing to move out of the counseling office and seek social change. You may work with others on a specific cause or issue to facilitate general human development and wellness (e.g., pregnancy care, child care, fair housing, shelter for the homeless, athletic fields for low-income areas). These efforts require you to speak out, to develop skills with the media, and to learn about legal issues. *Ethical witnessing* moves beyond working with victims of injustice to the deepest level of advocacy (Ishiyama, 2006). Counseling, social work, and human relations are inherently social justice professions. Speaking out for social concerns needs our time and attention.

Counseling Clients Who Have Internalized Oppression

A step-by-step model for working with internalized oppression is provided by an adaption of the South American theorist/practitioner Paulo Freire in his *Pedagogy of the Oppressed* (1970). Freire met over the campfire with poor campesinos (peasants) in rural Brazil, who worked for astoundingly low wages on huge plantations. In a similar fashion to our description of microskills, he developed a trusting relationship and drew out their stories of pain and poverty. He then focused on the plantation owner, and they told stories of his large house, fancy cars, and trips to Europe, as well as troubling stories about not receiving full pay and their not being allowed to say anything for fear of losing their job.

The next chapter in this book—confrontation of incongruities and discrepancies—describes what can be done next. As Freire summarized and pointed out the serious differences, he encouraged *naming* what was occurring, and some form of the word *oppression* invariably appeared. The compesinos began to reframe the situation and their lives in what is termed *conscientization*—the development of a new consciousness, a new way of making meaning in the world, which can lead to new actions concerning self and others.

Adapting Freire's psychology of liberation offers specific steps to work with internalized oppression in the form of internalized racism, denying ones' worth as a woman, believing that one is at fault when bullied, and so on. Our goal is to change consciousness, seeing the person-in-context and learning to reframe and think about oneself more as a person of resilience. The steps will sound familiar:

1. Develop a relationship. Listen to the concern. Listen!

2. Stop and build individual, family, and cultural strengths, along with stories and body anchoring. The community genogram is one way to encourage new ways of thinking.

3. Body anchoring of positives can be useful in this process. In body anchoring, ask the client to find a positive visual or auditory image within the story. It might be a grandfather speaking supportively at a critical time, it may be personal success, or it can be a hero such as Martin Luther King, Harvey Milk, or Gloria Steinem. Ask clients to notice where they get feelings in their body that they associate with the image. Remind clients that the memory, image, and feeling will always be there for them. Often several images are useful, such as those focused on being cared for and loved, a personal triumph, and the power of a grandmother or the hero.

4. Hear the story again, making sure that you have the concrete details, thoughts, and feelings. Work with the negative feelings gently, but do not use body anchoring.

5. Encourage *naming* of the negative story—Nelida names it as *oppression* in her second interview in Chapter 11. Others may use *racism, sexism, ableism*, or another relevant term.

6. Return to strengths, and anchor them once again.

7. Plan for generalization and taking the new knowledge home. This is the action phase of liberation.

8. Follow up to see if changes have occurred in behavior, thought, and emotion.

Action: Key Points and Practice

The Skill of Focusing. Focusing is a form of attending that enables multiple views of client stories. This skill will help you and clients think of creative new possibilities for restorying and action. It emphasizes the importance of both the individual/issue and the social/cultural context. Focusing enables both the client and the counselor to explore the context of past memories more fully.

The Importance of the Individualistic "I" Focus. Recall that counseling is for the client. Though expanding awareness of context and self-in-relation and understanding alternative stories of a situation can be very useful, ultimately the unique client before you will be making decisions and acting. The bottom line is to assist that client in writing his or her own new story and plan of action.

Selective Attention. The way you listen can and does influence clients' choice of topics and responses. Listening exclusively to "I" statements affects the way clients talk about their issues. Listening to culture, gender, and context also affects the way they respond.

Draw Out Stories with Multiple Focusing. Client stories and issues have many dimensions. It is tempting to accept problems as presented and to oversimplify the complexity of life. Focusing helps counselor and client to develop an awareness of the many factors related to an issue as well as to organize thinking. Focusing can help a confused client zero in on important dimensions. Thus, focusing can be used to either open or tighten discussion. Use selective attention to focus the session on the client, issue/concern, significant others (partner/spouse, family, friends), a mutual "we" focus, the counselor, or the cultural/environmental/contextual issues. You may also focus on what is going on in the here and now of the session.

Seven Focus Dimensions. There are seven types of focuses. The one you select determines what the client is likely to talk about next, but each offers considerable room for further examination of client issues.

- Focus on client: "Tari, you were saying last time that you are concerned about your future."
- Focus on the main theme or problem: "Tell me more about your getting fired. What happened specifically?"
- Focus on others: "So you didn't get along with the sales manager. I'd like to know a little more about him." "How supportive has your family been?"
- Focus on mutual issues: "We will work on this. How can you and I work together most effectively?"
- Focus on counselor: "My experience with difficult supervisors was . . . ."
- Focus on cultural/environmental/contextual issues: "It's a time of high unemployment. Given that, what issues will be important to you as a woman seeking a job?"
- Focus on the here and now (immediacy): "You seem disappointed right now. Can you share with me what came to your mind right now?"

Community and Family Genograms. Genograms are visual maps to help clients gain new perspectives on themselves and their relationships to their families and their communities. They can bring to life the "internalized voices" affecting the client. A community genogram will help you and your clients understand their relation to their environment and show both stressors and assets in their lives. A family genogram will help in understanding a client's family history and current relationships. Both represent useful ways to understand the client's history and identify strengths and resources.

Apply Focusing to Examine Your Own Beliefs. As a counselor, explore your own beliefs and compare these with the views of others. Use the focus dimensions to explore other people's views. What do your family, your friends, and others close to you think? Awareness of your and others' views will help your work with your clients.

Focusing and Other Skills. Focusing can be consciously added to the basic microskills of attending, questioning, paraphrasing, and so on. Careful observation of clients will lead to the most appropriate focus. In assessment and problem definition, consciously and deliberately assist the client to explore issues by focusing on all dimensions, one at a time. Advocacy and social action may be necessary when you discover that the client's issues cannot be resolved through the session alone. Counseling could be described as a social justice profession.

The Action Plan. Planning for action helps clients organize their behavior, act according to agree-upon plans, and achieve desired goals. Work with the client to jointly decide the best ways to move forward, address and remove potential barriers, and ensure all this is in line with what the client wants to do. Long-term action plans promote changes in behavior, as well as in neural networks that will provide change sustainability.

Multicultural Issues. Focusing will be useful with all clients. With most clients the goal is often to help them focus on themselves (client focus), but for many people, particularly those of a Southern European or African American background, the family and community focuses may sometimes be more appropriate. The goal of much North American counseling and therapy is individual self-actualization, whereas among other cultures it may be the development of harmony with others—self-in-relation. Deliberate focusing is especially helpful in problem definition and assessment, where the full complexity of the problem is brought to light. Moving from focus to focus can help increase your clients' cognitive complexity and their awareness of the many interconnecting issues in making decisions. With some clients who may be scattered in their thinking, a single focus may be wise.

Social Justice and Advocacy. Sometimes working only with the client may not be enough. Helping clients navigate an unfair situation, working with the school to provide needed accommodations, and working for social change may be appropriate to help improve the systems within which your clients live.

Practice and Feedback: Individual, Group, and Microsupervision

Additional resources can be found by going to CengageBrain.com and logging into the MindTap course created by your professor. There you will find a variety of study tools and useful resources that include quizzes, videos, interactive counseling and psychotherapy exercises, case studies, the Portfolio of Competencies, and more.

Awareness, knowledge, and skills are central, but action is essential. Mastering the skills of counseling and psychotherapy is achieved through intentional practice and experience. Reading and understanding are at best a beginning. Some find the ideas here relatively easy and think that they can perform the skills, but what makes one competent in basic skills is practice, practice, practice.

Individual Practice

Exercise 9.1 Writing Alternative Focus Statements

A 35-year-old client comes to you to talk about an impending divorce hearing. He says the following:

> I'm really lost right now. I can't get along with Elle, and I miss the kids terribly. My lawyer is demanding an arm and a leg for his fee, and I don't feel I can trust him. I resent what has happened over the years, and my work with a men's group at the church has helped, but only a bit. How can I get through the next 2 weeks?

Write several alternative focus statements. Be sure to brainstorm a number of cultural/environmental/contextual possibilities.

Main issue as presented _____

Client focus _____

Theme, concern, story focus _____

Others focus _____

Family focus _____

Mutual, group, "we" focus (include immediacy focus) _____

Counselor focus _____

Cultural/environmental/contextual focus _____

Focus on physical health and therapeutic lifestyle issues _____

Exercise 9.2 Developing a Community Genogram

This chapter presented specific step-by-step instructions for developing a community genogram. Most of your classmates will have completed a genogram by now. With one of them, debrief the genogram. Also consider the Nelida/Allen session; you may want to try the imagery exercise at the end of that session. Present the completed genogram, and briefly summarize what you learned.

Group Practice and Microsupervision

Exercise 9.3 Practicing Focusing Skills

If you are practicing with a family member, friend, or classroom colleague, follow the procedures here and debrief their community genogram.

Step 1: Divide into groups.

Step 2: Select a group leader.

Step 3: Assign roles for the first practice session.

❏ Client, who has completed community genogram.

❏ Counselor, who will use focusing to bring out past memories using a positive wellness orientation.

❏ Observer 1, who will give special attention to focus of the client, using the Feedback Form in Box 9.4. The key microsupervision issue is to help the counselor continue a central focus on the client while simultaneously developing a comprehensive picture of the client's contextual world.

❏ Observer 2, who will give special attention to focus of the counselor, using the Feedback Form in Box 9.4.

Step 4: Plan. Establish clear goals for the session. The task of the counselor in this case is to go through all seven types of focus, systematically outlining the client's issue. If the task is completed successfully, a broader outline of memories related to the client's concern should be available.

A useful topic for this role-play is a story from your family or community. Your goal here is to help the client see the issues in broader perspective.

Observers should take this time to examine the feedback form and plan their own sessions. The client may fill out the Client Feedback Form from Chapter 1.

Step 5: Conduct a 5-minute practice session using the focusing skill.

Step 6: Review the practice session and provide feedback for 10 minutes. Give special attention to the counselor's achievement of goals and determine the mastery competencies demonstrated.

Step 7: Rotate roles.

General reminders: Be sure to cover all types of focus; many practice sessions explore only the first three. In some practice sessions, three members of the group all talk with the same client, and each counselor uses a different focus.

BOX 9.4	**Feedback Form: Focus**

You can download this form from MindTap at CengageBrain.com.

_____ (DATE)

_____   _____

(NAME OF COUNSELOR)   (NAME OF PERSON COMPLETING FORM)

Instructions: Observer 1 will give special attention to the client and Observer 2 to the counselor. Note the correspondence between counselor and client statements. In the space provided, record the main words used. Classify each statement by checking a box.

Main words	Client							Counselor						
	Client (self)	Concern/problem	Significant others	Family	Mutual "we"	Counselor	Cultural/environmental/contextual	Client	Concern/problem	Significant others	Family	Mutual "we"	Counselor (self)	Cultural/environmental/contextual
1.														
2.														
3.														
4.														
5.														
6.														
7.														
8.														
9.														
10.														
11.														
12.														
13.														
14.														

Observations about client verbal and nonverbal behavior:

_____

_____

Observations about counselor verbal and nonverbal behavior:

_____

_____

Portfolio of Competencies and Personal Reflection

The history of counseling and therapy has provided the field with a primary "I" focus in which the client is considered and treated within a totally individualistic framework. The microskill of focusing is key to the future of culturally competent counseling and psychotherapy, as it broadens the way both counselors and clients think about the world and review memories. This does not deny the importance of the "I" focus. Rather, the multiple narratives made possible by the use of microskills actually strengthen the individual, for we all live as selves-in-relation. We are not alone. The collective strengthens the individual.

Some might disagree with the emphasis of this chapter and argue that only the individual and problem focuses are appropriate. What do you think? As you work through this list of competencies, think ahead to how you would include or adapt these ideas in your own Portfolio of Competence.

Assessing Your Level of Competence: Awareness, Knowledge, Skills, and Action

Use the following checklists to evaluate your present level of mastery of the competencies presented here. As you review the items below, ask yourself, "Can I do this?" Check those dimensions that you currently feel able to do. Those that remain unchecked can serve as future goals. Do not expect to attain intentional competence on every dimension as you work through this book. You will find, however, that you will improve your competencies with repetition and practice.

Awareness and Knowledge. Are you able to identify seven types of focus as counselors and clients demonstrate them? Can you note their impact on the conversational flow of the session.

❑ Identify focus statements of the counselor.

❑ Note the impact of focus statements in terms of client conversational flow.

❑ Write alternative focus responses to a single client statement.

Basic Competence. Are you able to use the seven focus types in a role-play session and in your daily life?

❑ Demonstrate use of focus types in a role-play session and draw out multiple stories.

❑ Use focusing in daily life situations.

Intentional Competence. Can you use the seven types of focus in the session, and will clients change the direction of their conversation as you change focus? Maintain the same focus as your client if you choose (that is, do not jump from topic to topic)? Combine this skill with earlier skills (such as reflection of feeling and questioning) and use each skill with alternative focuses? Check those skills you have mastered, and provide evidence via actual session documentation (transcripts, recordings).

❑ My clients tell multiple stories about their issues.

❑ I maintain the same focus as my clients.

❑ During the session, I observe focus changes in the client's conversation and change the focus back to the original one if it is beneficial to the client.

❑ I combine this skill with skills learned earlier. Particularly, I can use focusing together with summarizations, questions, and genograms to expand client development.

❑ I use multiple focus strategies for complex issues facing a client.

Pychoeducational Teaching Competence. Teaching clients how to explore their stories from multiple focuses will help them expand their understanding, recognize strengths, and seek resources and new solutions. Can you do these?

❑ Teach clients in a helping session how to expand stories by using multiple lenses.

❑ Teach clients how to use the different focus dimensions.

❑ Teach small groups the skills of focusing.

Personal Reflection on Focusing

What single idea stands out for you among all those presented in this chapter, in class, or through informal learning? What stands out for you is likely to be a guide toward your next steps.

What do you think of the concept of selective attention and its role in focusing?

Focusing places attention on individual memories as well as their relations, situation, and context. What are your thoughts and feelings on this approach?

What are your thoughts on multicultural issues and the use of the focusing skill?

What are your thoughts and experiences with regard to the community and family genograms?

How might you use ideas in this chapter in the process of establishing your own style and theory?

What other points in this chapter struck you as most memorable?

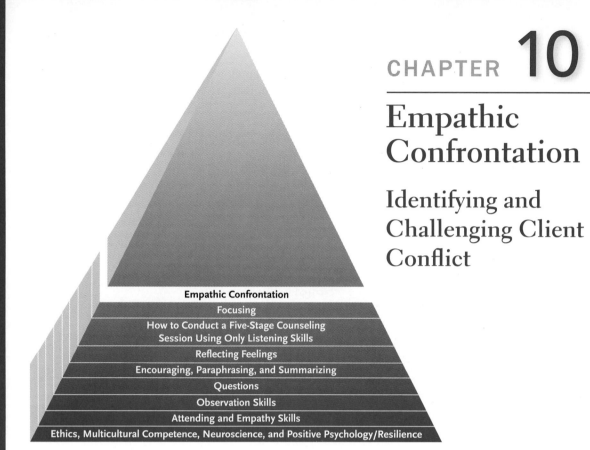

Empathic Confrontation

Focusing

How to Conduct a Five-Stage Counseling
Session Using Only Listening Skills

Reflecting Feelings

Encouraging, Paraphrasing, and Summarizing

Questions

Observation Skills

Attending and Empathy Skills

Ethics, Multicultural Competence, Neuroscience, and Positive Psychology/Resilience

CHAPTER **10**

Empathic Confrontation

Identifying and Challenging Client Conflict

Personal liberation comes from understanding and managing conflict, followed by action to build change in thoughts, feelings, and behavior.

The counselor's task is to listen and clarify, encouraging clients to synthesize and create the New, and then melding their liberation into action.

—Allen Ivey

Chapter Goals and Competency Objectives

Awareness and Knowledge

▲ Identify conflict, incongruity, discrepancies, ambivalence, and mixed messages in behavior, thought, and feelings/emotions.

▲ Consider multicultural and individual differences when using confrontation.

▲ Observe an interview on microaggressions illustrating one approach to positively change cognitions and emotions, and discuss action steps the client might take.

Skills and Action

▲ Encourage and facilitate exploration and creative resolution of conflict and discrepancies.

228

▲ Evaluate client creative change processes occurring during the session and throughout treatment sessions, using the Client Change Scale.

▲ Consider multicultural and individual differences when using confrontation.

Introduction: Empathic Confrontation, Creating the *New*

Many clients come to counseling "stuck"—having limited alternatives for resolving their issues and conflicts. **Internal conflicts** are those that reside primarily within the client's thoughts and feelings. **External conflicts** are those between the client and the surrounding world. Our task is to assist in freeing the client from stuckness and facilitate the development of creative thinking and expansion of choices. *Stuckness* is a term coined by Gestalt theorist Fritz Perls to describe the opposite of intentionality, or a lack of creativity. Other words and phrases that represent stuckness include *immobility and ambivalence, blocks, repetition compulsion, inability to achieve goals, lack of understanding, limited behavioral repertoire, limited life script, impasse,* and *lack of motivation.* Stuckness may also be defined as an inability to resolve conflict, reconcile discrepancies, and deal with incongruity. In short, clients often come to counseling because they are stuck for a variety of reasons and seek the ability to move, expand alternatives for action, and become motivated to do something to rewrite their life stories.

Confrontation is defined in our dictionaries as a hostile or argumentative behavior between opposing persons. This suggests that counselors assertively force the client to look at themselves in an "accurate" or "honest" way. Confrontation requires the ability to engage in self-reflection, something that may not be part the client's world because of various forms of egocentricity—for example, the narcissistic and antisocial personality types. Harsh confrontational practices were used in the past, especially in addiction treatments, but were generally ineffective, potentially harmful, and frequently led to client resistance. In police situations, it has been found that such treatment leads to inaccurate data and even false confessions, but this practice continues in many settings. Research has shown that this type of confrontation does not work in counseling and therapy, leads to dropping out, and leaves lingering negative memories of counseling (Norcross, 2011). The microskill of confrontation is used in a very different way in counseling.

Empathic confrontation is a gentle skill that involves first listening to client stories carefully and respectfully and then encouraging the client to examine self and/or situation more fully. Empathic confrontation is not a direct, harsh challenge, or "going against" the client; rather, it represents "going with" the client, seeking clarification and the possibility of a creative *New*, which enables resolution of difficulties. However, with some clients, you will find that rather direct and assertive behavior will be required before they can hear you.

The theologian Paul Tillich describes the *New being* as a person who has moved from the past to a new present and future. The creation of the *New* is the desired result of empathic conversations between client and counselor/therapist. It is a creative collaboration.

Empathic confrontation is based on listening to conflictual stories carefully and effectively responding using the listening skills. Paraphrasing is particularly useful when the conflict or incongruity involves a cognitive decision and the pluses and minuses of the decision need to be outlined. Reflection of feeling is important for emotional issues, particularly when clients have ambivalent or mixed feelings ("on one hand, you feel . . ., but on the other, you also feel . . ."). A summary is a good choice for bringing together many conflicting strands of thoughts, feelings, and behaviors. At times your observations of verbal and nonverbal behavior and personal awareness of what might be happening with the client may lead you to *carefully* add your thoughts to the confrontation. Your own words can be additive and enrich the client's world—or, if not in tune, subtractive, as conflict is frequently a strong emotional issue.

When you use confrontation with intentionality and effectiveness, you can anticipate the following response.

Confrontation	Anticipated Result
Supportively challenge the client to address observed discrepancies and conflicts. 1. Listen, observe, and note client conflict, mixed messages, and discrepancies in verbal and nonverbal behavior. Give attention to both cognitive and emotional dimensions. 2. Paraphrase and reflect feelings, to clarify internal and external discrepancies. As the issues become clarified, empathically summarize what has been said—for example, "on one hand you feel _____, but on the other hand you feel _____." Bring both cognition and emotions into most summaries. 3. Evaluate how the client responds and whether the confrontation leads to client movement or change. If the client does not change, flex intentionally; try another skill and approach the conflict from another direction.	Clients will respond to effective confrontation of discrepancies and conflict by creating new ideas, thoughts, feelings, and behaviors, and these will be measurable on the five-point Client Change Scale. Again, if no change occurs, *listen.* Then try an alternative style of confrontation.

Awareness, Knowledge, and Skills: Empathic Confrontation for Results

A creative, active, sensitive, accurate, empathic, nonjudgmental listening is for me terribly important in the relationship. It is important for me to provide it; it has been extremely important, especially at certain time in my life, to receive it.

—Carl Rogers

If you don't maintain the relationship, you likely will lose the client. As you draw out client stories and strengths, you will be looking for verbal and nonverbal conflict and discrepancies. An essential part of confrontation is paraphrasing the conflict, reflecting the feelings of mixed messages, and providing an accurate summary of the situation

Even with empathic listening, clients who are challenged or confronted may feel put on the spot, perhaps even that you are attacking them. This is where Carl Rogers's **nonjudgmental** empathy will be most helpful. Closely related to positive regard and respect, a nonjudgmental attitude requires that you suspend your own opinions and attitudes and assume value neutrality in relation to your client. Many clients have attitudes toward their issues and concerns that may be counter to your own cherished beliefs and values. But people who are working through serious difficulties do not need to be judged or evaluated, and your neutrality is necessary if you want to maintain the relationship.

Stop and think for a moment of a person whose behavior troubles or angers you personally. It may be someone whom you regard as dishonest, one who perpetrates violence, or one who clearly demonstrates sexism or racism. These are challenging moments for the nonjudgmental attitude and for empathic understanding. You do not have to give up your personal beliefs to maintain a nonjudgmental attitude; rather, you need to suspend your private thoughts and feelings. You do not have to agree with or approve of the thoughts and behaviors of the client, but if you are to help this person change and become more intentional, presenting yourself as nonjudgmental is critical.

At times, however, judgment may indeed be called for. For example, the Nelida session in the preceding chapter is basically nonjudgmental and supportive, but Allen clearly is judging those who have not respected her Latina heritage. This type of judgmental feedback may be appropriate here, as it tends to provide some safety to continue. But joining clients too soon and agreeing with their views could distort the facts of issues and concerns. It also could be a violation of counseling boundaries.

As you draw out clients' stories through listening skills, observe their reactions and the emotion underlying the conflict. What do you see both nonverbally and verbally? Feed back both cognitions and emotions through paraphrasing and reflecting feelings. Go into the story in necessary depth so that both you and the client have a clear understanding of the conflict. What is going on? What are both sides of a decision? Which feelings are strongest? Are there underlying feelings and thoughts that at first were hidden?

The summary will be used frequently in the process of clarification. Questions often enrich understanding. As the client's conflict becomes clear, an overall summary can provide both you and the client a full picture of the situation. This full picture includes both cognitive and emotional understanding and leads to mentalizing—seeing the client's issue holistically.

Here are some examples of empathic confrontations:

Relationship issue: "Aretha, I hear you saying that you still have deep feelings toward your partner and some of those are good memories of the past and you still have deep feelings of caring, but also sadness. But, on the other hand, (he or she) has hurt you in many ways by what has been said. You feel angry and even afraid. Given these mixed feelings and thoughts, is that accurate?"

Decision about choosing a college: "Jonathon, on one hand, the state university is what you really want and feel is right for you. But, on the other side, the expense and being burdened by loans is almost terrifying."

Observation of nonverbal behavior: "I see you smile and your eyes light up when you talk about the good times you have had, but I also see you sometimes tighten up your body and even look a bit fearful when you talk about that last argument."

At times, psychoeducation and information giving (Chapter 12) may be wise when a client expresses oppressive racist, sexist, or other discriminatory comments. Such clients are typically not aware of the discrepancy between their general kindness to others and what they are saying now. Needless to say, this must be done in a nonjudgmental and supportive fashion, or the client will not return.

Some research speaking to the *how* of empathic confrontation is reported in Box 10.1.

BOX 10.1 Research and Related Evidence That You Can Use

Confront, but Also Support

Relatively little of value in research on confrontation existed until 2011, when Mikecz's qualitative research addressed how to work in difficult counseling situations. He identified the following issues as central:

- Don't confront unless you have trust and relationship.

- Pay attention to and understand the client's point of view, way of thinking, and feeling about the issue. Summarize the client's interpretation of the situation.

- Share with the client only if he or she can listen to and hear you.

- The client needs to be in charge of what happens and how things are interpreted.

- Knowledge of the client's cultural background and general personal style is essential. If you know even a few words of the client's first language (if different from yours), this will help.

- Attending skills such as eye contact are critical in the relationship.

- Maintain neutrality; avoid judgments.

- Follow up, both in the session to examine how the conflict was resolved or not resolved and after the session to see if new knowledge has generalized into action in the real world.

Attending and listening skills are used frequently in the session, but you'll find that confrontation strategies are used only occasionally. Confrontations accounted for only 1% to 5% of counselor statements (Hill & O'Brien, 1999). The reviewers noted that confrontations are useful, but they also often make clients uncomfortable. Empathic listening is required; otherwise, clients frequently become defensive and may not deal fully with issues following a confrontation.

Counselor eye contact affects client perception of rapport in the session. Specifically, less direct eye contact early in the session when discussing sensitive matters is helpful; at this point, clients appreciate a nonconfrontational approach. As the session progresses, more eye contact and more confrontations are acceptable (Sharpley & Sagris, 1995).

The Skills of Empathic Confrontation: An Integrated Three-Step Process

Empathic confrontation is best described as an integrative skill involving both listening and influencing. You can identify the confrontation skill most easily when the counselor paraphrases or summarizes observed ambivalence or conflicts in some form of the classic "On one hand . . ., but on the other hand . . .; how do you put that together?" In this form, the conflict, discrepancy, or mixed message is said back to the client clearly. In addition, some counselors find using their right and left hands along with the summary involves the client more fully.

The story is brought out, and conflict in that story is identified, along with thoughts and feelings. Learning the behaviors of the client and the other person through looking at antecedents, behaviors, and consequences is helpful and will lead to a better understanding of thoughts and feelings. Also, drawing out positive stories is very helpful in the confrontation process. For example, "On one hand, you speak about your inability to defend yourself from your partner, but on the other hand, I also heard you talking earlier about how you were able to handle bullying in high school so successfully. What does that high school story say to the present situation?"

Virtually all counseling and psychotherapy have the goal of enabling clients to explore their ambivalence and conflict, rather than just complaining. Moreover, they also seek to facilitate clients' finding their own resolution—the creative *New*. At times you may have to use reframing or self-disclosure to help the client "put it together," but ultimately resolving discrepancies and conflict almost always is the client's issue.

However, when you face situations of abuse or danger to the client, or the client faces a severe crisis and cannot act, or the client is oppressed by racism, sexism, classism, and the like, then it may be necessary for you to take action and work both inside and outside the session in the community to help find a satisfactory resolution. For example, a school bullying situation may require you to intervene in the school and community.

Observe: Empathic Confrontation in the Interview

Another Allen interview with Nelida Zamora follows; it illustrates the first two steps in empathic confrontation.

The first step is to listen empathically and nonjudgmentally. Here you will see the general story of Nelida's concern brought out primarily through listening skills. The interview on microaggressions illustrates one approach to building increased client resilience. The purpose is for clients to become less distressed about future occurrences and develop their own plans to cope with the next one that occurs.

Nelida's story focuses on a microaggression that affected her total experience of her graduate program. The microaggression goes into long-term memory immediately as it occurs. Why? Emotion often drives cognition. The microaggression "hits" Nelida like a brick.

Many reading about the microaggression that Nelida experienced will consider it minor, which it is compared to other single, more injurious and hostile statements. However, please return to Harvard's Jenny Galbraith's story in Chapter 2 (page 40). Jenny has received many microaggressions, and they "pile up" until they result in feeling unsafe, alienated, and angry. And anger cannot be easily expressed at the aggressor. Thus, a counseling goal is to listen and facilitate awareness of emotional and cognitive experience. The next step is counseling appropriately on emotional regulation so that later the client will cope more effectively with the expected next microaggression.

The second step of the session will emphasize more specifics and clarify internal and external conflicts. In both portions, look for multicultural issues and their impact on Nelida.

Step 1: Listen

Identify conflict by observing incongruities, discrepancies, ambivalence, and mixed messages.

The previous chapter on focusing presented Nelida, who gave permission for us to use her session with Allen, enabling us to present counseling skills as they occur immediately in a real-life session. The first session, based on the community genogram, helps us see Nelida as a person-in-relationship to family and community. You also saw an emphasis on strengths, so important for building resilience as we work with clients on their issues.

In the interview that follows, the emphasis turns to Nelida's internal and external conflicts around a painful microaggression. Again, the transcript here is an edited and markedly shortened version of the original.*

What types of conflict and challenges do you see Nelida facing? Note as many as you can as you read this transcript.

---

*Slightly edited for clarity, this transcript of a real interview held between Allen Ivey and Nelida Zamora is also available on DVD: Ivey, A., Ivey, M., Gluckstern-Packard, N., Butler, K., & Zalaquett, C. (2012). *Basic Influencing Skills*, 4th ed. [DVD]. Alexandria, VA: Microtraining/Alexander Street Press. By permission of Microtraining/Alexander Street Press.

Client and Counselor Conversation	Process Comments
1. *Nelida:* Here I am, a grad student in counseling. I did well in college in Miami, and thought it was no big deal because I was only four and a half hours away. But my first day of class I raised my hand, made a comment . . ., and a classmate asked me if I was from America (nervous laugh) or a native (nervous laugh). Yeah, and I said well I'm . . ., I was just four and a half hours away, and he just found it very hard to believe. So, after that comment was made, it kind of made me a little bit more hesitant to participate in discussions. It made me more self-conscious.	There are several dimensions of conflict in Nelida's words. How many can you identify? Note how powerful one negative microaggression can be. The emotional impact of this comment immediately activated negative emotions in the amygdala and brought them to immediate permanent memory in the hippocampus and prefrontal cortex (PFC). The message in the executive PFC was to keep quiet and not talk in class. Keeping quiet is a form of emotional regulation, although Nelida is paying a personal price. There is already an implicit goal—to facilitate Nelida's being proud of her cultural heritage and her skill in English, to help her build self-esteem and self-confidence and speak up for herself.
2. *Allen:* It made you self-conscious. Could we explore that a little bit more? Ah, first of all, in English, what were the feelings that went with that?	Encouraging in the form of restatement around emotions and open questions. As you will see later, those feelings will soon be explored in Spanish. (Interchangeable empathy with an attempt to add emotional dimensions)
3. *Nelida:* Well, I was surprised because being from Miami a lot of my family members have recently come from Cuba, so there they look at me as the American girl and they make fun.	Can you identify cultural/environmental/contextual issues that add to the conflict?
4. *Allen:* . . . and that embarrasses you.	Drawing on nonverbals, Allen supplies an emotion word that acknowledges feeling, but was that the right word? We shall see in Nelida's next statement. (Interchangeable)
5. *Nelida:* Exactly, so when I'm in Miami, my family and friends tease me saying that I'm the American who can't speak Spanish a hundred percent correctly 'cause I've forgotten a lot of it because of the English. Then, now, I move here to Tampa, I'm the Cuban girl who can't speak English, so it seems like I'm torn. You know, I don't know where I belong sometimes.	What is the cultural conflict here?

While you listen, silently search in your mind for what is "going on now" with the client. Listen and *think* before you help clients clarify their issues. This is where metacognition comes in, as you wish to be thinking about what is happening in the here and now of the session, as well as how to respond to the client. All this is empathic mentalizing on your part as you seek to integrate the client's world in your own mind.

Use the following questions to practice your metacognitive mentalizing skills as you read the interview:

What conflicts came to your mind while reading Nelida's transcript? What did you think about them?

What emotions do you see her experiencing in this process?

Did Nelida's conversation remind you of something from your own life?

How did what Nelida said relate to your own thoughts, feelings, and behaviors?

Perhaps, with a little concentration, you will be able to note feelings in your own body that accompany the emotions. Client emotional experience typically affects your body as you listen.

Step 2: Summarize and Clarify Issues of Internal and External Conflict and Work Toward Resolution Through Further Observation and Listening Skills

Focusing can be very helpful in identifying and working with conflict. While our central focus always is on the client, Nelida's conflicts and internal incongruity relate to cultural/environmental/contextual issues (Cuban American culture and "American" Tampa classroom culture) and her family. Thus, part of the session needs to give these focus areas central attention.

We now return to Nelida's early session comments, this time attending to issues of conflict and discrepancy, which are starting to move to empathic confrontation. This continues the part of the session presented earlier in this chapter. (The conversational exchange in items 4 and 5 are repeated here to provide context for this segment of the session.) Recall that Nelida's community genogram, presented in the previous chapter, served as a background for the session and provided several positive stories on which change could be built. The strengths and resources identified in the community genogram were reviewed again in the following session, but these have been edited out to save space and focus on confrontation skills.

Note that Allen seeks throughout this interchange to draw out further aspects of the conflict surrounding the microaggression, but typically focuses on individual and cultural strengths that Nelida brings for creative resolution—specifically the therapeutic lifestyle change of *cultural health*. Nelida early on responds internally with embarrassment about her Cuban background. There is a need to reframe the encounter so that she is fully aware that she is not "the problem," but rather the external forces of the aggressor and people like him are "the problem." Out of such "problems" come life concerns, issues, and challenges—and the opportunity for change.

We build resilience and solve our difficulties best from our strengths, resources, and positive assets.

Client and Counselor Conversation	Process Comments
4. *Allen:* . . . and that embarrassed you.	Repeat of Allen's comment from prior transcript.
5. *Nelida:* Exactly, so when I'm in Miami, my family and friends tease me saying that I'm the American who can't speak Spanish a hundred percent correctly 'cause I've forgotten a lot of it because of the English. Then, now, I move here to Tampa, I'm the Cuban girl who can't speak English, so it seems like I'm torn. You know, I don't know where I belong sometimes.	Nelida identifies her central internal conflict clearly, but she still needs to tell her story in more detail. If you were to view the video, you would see Nelida moving her hands back and forth as she describes the situation. In addition, tornness—dealing with conflict—often produces internal body tension, exhibited in clenched fists or uncomfortable feelings in the stomach.
6. *Allen:* So, on one hand, you feel challenged about your English here and, on the other hand, when you go home, you get challenged on your Spanish. You feel torn. . . . How would you describe that feeling of tornness in Spanish?	Confrontation in the form of paraphrase and reflection of feeling emphasizes that key word *torn*. Conflict inevitably has an emotional dimension. Working with just the cognitive decision-making issues will be less effective. The open question asks Nelida to describe her feelings in Spanish. One's own natural language is best for fully experiencing emotions. The translation clarified the power of the microaggression. Encouraging clients to use their home language is a sign of empathic respect and shows an authentic openness on the part of the counselor.

(continued)

Client and Counselor Conversation	Process Comments
7. *Nelida: Muy conflictiva.* Just very conflicting.	Nelida shows more tension and frustration in her body language. The confrontation did not enable her to resolve these issues at all, but the tornness and the conflict are now clearer. Here we see clearly that feelings and emotions are physical, as well as cognitive.
8. *Allen:* Did you notice any difference between English and Spanish when you said it?	Closed question to check out the importance of Spanish with this client. Not all clients will feel comfortable with talking about their issues in this way, but most will. This in itself is a confrontation that points out the conflict between the two languages and seeks to facilitate Nelida's learning the power of her own language. She had felt strange and that Spanish was somehow a disability in graduate school, when in truth bilingualism is a strength.
9. *Nelida:* Again, it almost felt more real when I said it in Spanish . . . like truer. I'm comfortable with both languages, but like I said, my primary language is Spanish, so I guess to an extent it does feel more real, truly to myself, when I do say it in Spanish.	A new story is being created here as Nelida creates a new meaning for Spanish. Whereas Spanish was described earlier as kind of a handicap, we see Nelida moving to realizing that her English is good (despite the student comments), and also that her native Spanish is respected and valuable. Bilingualism is a strength and actually builds wider neural networks in the brain than monolingualism.
10. *Allen:* It feels more real when you say it in Spanish. (pause) I'd like to hear more about your story, what it meant for you to come from Miami to Tampa and how it went for you.	Restatement of feelings followed by a statement that really is an open question about *meaning* of the situation. Specifically, how does Nelida frame or interpret what happened? Chapter 11 will discuss issues of reframing and interpreting meaning. (Interchangeable empathy)
11. *Nelida:* Well, I was surprised because being from Miami a lot of my family members have recently come from Cuba, so they look at me as the American girl and they make fun. When I'm over there, I can't speak Spanish as good as they do. When I'm here I can't speak English, so it seems like I'm torn. You know, I don't know where I belong sometimes.	Nelida now is clearer about her understanding of her internal conflict.
12. *Allen:* You seem to face conflict with your family at home, with your classmates here in Tampa—and then the two conflicts actually seem to conflict against each other as well and really add to the tornness.	Summary of the confrontation issues—on one hand, at home; on the other hand, here. Allen almost always moves his own hands in tune with the words. (Interchangeable)
13. *Nelida:* Absolutely. But then I have my grandparents tell me that I'm forgetting my Spanish, you know, also. Kind of confusing and I feel torn. Um-hum.	Expressions like "absolutely," "exactly," or "yes" confirm accuracy of the summary.
14. *Allen:* Um . . . hummm.	Allen leans forward with a minimal encourager anticipating that Nelida will continue processing the issue.
15. *Nelida:* It was pretty bad my first semester in graduate school, maybe even throughout the first year. I think the accent has kind of gone away a little, living on campus here a while, but I still get it every now and then. Not as often, though. And, then, my family in Tampa has gradually become more supportive as they've seen my successes.	Here we see Nelida starting to synthesize discrepancies and resolve part of the issue around language—an example of the creative *New*. She demonstrates that clients can find their own way to resolve contradiction.

Client and Counselor Conversation	Process Comments
16. *Allen:* From your community genogram and the way you talked about the last visit home, it sounds like your grandparents and your mother have become even more supportive, even though they may tease you occasionally about your "American accent."	Summary from previous session. There is a mild confrontation here as we see both sides of Nelida's relationship with her grandparents. Allen follows up with information from the last meeting and the community genogram. (Additive because of linking a past session with the present)
17. *Nelida:* Yes, perhaps I should not feel so torn. Things really are getting better, but I don't know how to deal with those comments about accent. They don't come as often, but . . .	Having worked through a contradiction with her family and remembering the supports and resources in her family genogram, Nelida is prepared to explore the more immediate issues.

As you read, what internal conflicts does Nelida face? "I'm torn" represents a central issue that needs to be addressed. There are mixed and conflicting thoughts of embarrassment, being different, self-consciousness, and not being fully capable. Out of these internal conflicts has come a decision not to speak up in class—another internal conflict, as she would rather say what she thinks. If you have a solid relationship, you may consider asking the client, "Feelings inside our bodies often provide clues to how deeply we are reacting to challenging experiences and issues. Can you notice any part of your body reacting as you say that?"

The key external conflicts include the class members who have made her feel either singled out or excluded because of her accent, and her family and friends in Miami who call her "the American girl." Implicit in this, and explored later, are issues in the cultural/environmental context: What does it mean to be Cuban American? How does my background relate to me where I am now? How do I relate to others in this new context? And, internally, "How do I keep it all together and still feel OK about myself?" because the external conflicts almost always becomes internal as well.

Observe: The Client Change Scale (CCS)*

The third step in empathic confrontation is to determine if what was said affects how clients think and feel about their situation. We can evaluate this process using the **Client Change Scale (CCS)**. We will first present the CCS and then consider how it applies as we observe a continuation of the Nelida and Allen interview.

The effectiveness of a confrontation is measured by how the client responds. If you pay attention in the here and now of the session, you can rate how effective your interventions have been. You will discover if your attempt at confrontation is subtractive, interchangeable, or additive. With a facilitative empathic confrontation, you will see the client change (or not change) language and behavior in the session. When you don't see the change you anticipate or think is needed, it is time for creative intentionality, flexing, and having another response, skill, or strategy available.

Imagine that you have provided an empathic confrontation by summarizing a client conflict ("On one hand, you feel _____ , but on the other hand, you think_____ . How do

---

*A paper-and-pencil measure of the Client Change Scale was developed by Heesacker and Pritchard and later replicated by Rigazio-Digilio (cited in Ivey, Ivey, Myers, & Sweeney, 2005). A factor analytic study of more than 500 students and a second study of 1,200 revealed that the five CCS levels are identifiable and measurable.

you put that together?"). The CCS gives you a framework for evaluating how the client responds to your confrontation. Does the client deny that a conflict, discrepancy, or mixed messages exist; show minor movement toward synthesis; or actually use the confrontation in a way that leads to significant change in thoughts and feelings, so that later these new discoveries can lead to behavioral change?

A summary of the Client Change Scale follows. Figure 10.1 illustrates how clients can move through the various levels of change. For example, Nelida starts the discussion of her issues between denial (she is engaging in self-blame) and partial examination (she is somewhat aware that something is wrong about what happened to her).

Client Change Scale (CCS)	Anticipated Result
The CCS helps you evaluate where the client is in the change process. Level 1. Denial Level 2. Partial examination Level 3. Acceptance and recognition, but no change Level 4. Creation of a new solution Level 5. Transcendence	The CCS can help you determine the impact of your use of skills. This assessment may suggest other skills and strategies that you can use to clarify and support the change process. You will find it invaluable to have a system that enables you to (1) assess the value and impact of what you just said; (2) observe whether the client is changing in response to a single intervention; or (3) examine behavior change over a series of sessions.

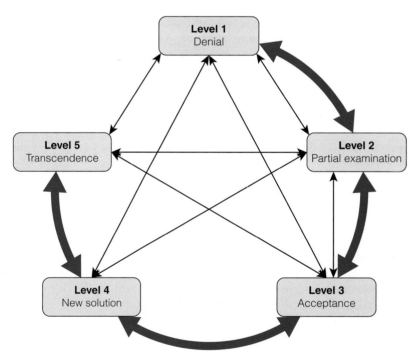

FIGURE 10.1 The Client Change Scale (CCS). The five stages of creative change may occur in order. However, as the arrows indicate, there can be movement back and forth among the stages. In fact, the entire process can go back to the beginning as clients discover new thoughts, feelings, behaviors, and meanings.

Although the progression from denial through acceptance to significant change can be linear and step by step, this is not always the case, as suggested in Figure 10.1. Think of a client working through the expected death of a loved one or a contested divorce. One possibility is for the client to move through the CCS stages one at a time. But often a client who seems to be moving forward will suddenly drop back a level or two. At one session the client may seem acceptant of what is to come, but in the next session move back to partial examination or even denial. Then we might see a temporary jump to transcendence, followed by a return to acceptance.

As you view portions of the following section of the Nelida and Allen interview, please pay special attention to how Nelida moves from lack of awareness to understanding and action. Observe the various steps of movement on the Client Change Scale and how the emphasis first on strengths and positives provides a basis of resilience for change. Change shows when Nelida moves on from internal cognitions and feelings to externalize the situation by saying "It's not right."

Counselor and Client Conversation	Process Comments
1. *Allen:* The grandparents who raised you came from Cuba—could you tell me about their story?	Open question, focused both on family and on the cultural/environmental context.
2. *Nelida:* They were 27 years old and had good jobs when Castro came into power. They weren't in agreement with Communism, so that's why they decided to move to New York. My grandmother was a seamstress. My grandfather was cleaning and mopping floors because that's the only job he was able to get when he moved here. He didn't really have time to learn English properly, so even still today they struggle and they don't speak it fluently at all. And I guess that's something that goes back to my reason for wanting to communicate with them and speak Spanish well . . . (pause)	Client Change Scale (CCS) Level 3, recognition of what is really going on. Client stories may originally start with Level 1 denial of the problem, as Nelida did when she accepted the negative microaggressions such as "Where are you from?" But with a review of strengths provided by the family genogram, Nelida is able to challenge her old thoughts, and her story line now recognizes things "as they are."
3. *Allen:* Talking to them in good Spanish is important.	Restatement. (Interchangeable empathy)
4. *Nelida:* Um-hum, I don't want them to think that I'm forgetting or not valuing the culture as much as I used to, because I still do.	CCS Level 3, recognition. The restatement reinforced Nelida's awareness that she needed to be aware of and value her culture. Because of this, the restatement above has some additive aspects. Here we see that her attitudes and emotions have changed in her long-term memory.
5. *Allen:* Okay. I'd like to go back to strengths for a minute. What do you see are some of the strengths in the Cuban culture, as you've lived with them and it?	This is a longer-term confrontation as Allen is asking Nelida to contrast her former negative beliefs with Cuban strengths. Topic jump, open question, focus on cultural/environmental/context. (Potentially additive as it emphasizes respect, but we need to see how Nelida responds. How additive a statement is depends more on the client's reaction than on what the counselor says.)
6. *Nelida:* They're very persistent. I know my grandparents have always been extremely persistent in terms of moving ahead, you know. They lived in New York and then they moved to Miami, and they're much better off now than what they used to be, but it took them a lot of hard work to get to where they are today.	CCS Level 3. Here we see a very concrete example of the strength of her resources. Allen is also learning more about the cultural context and family in Nelida's background.

(continued)

Counselor and Client Conversation	Process Comments
7. *Allen:* Okay. They are persistent. Who else is persistent?	A positive confrontation in which Nelida is compared to her grandparents. Encourager picking up the key word "persistent" followed by an open question. This is a step toward encouraging Nelida to internalize positive cognitions about herself. (Potentially additive; also indicates a real sense of respect for Nelida and her culture.)
8. *Nelida:* According to the genogram, I am persistent too (laughs). When I get up here in Tampa, I find the culture being denigrated. And when I think of their story, it makes me proud.	CCS Level 4, creation of new thoughts and feelings. Nelida is reframing her thoughts about the relevance of her cultural and family background. She is now able to reframe (see the following chapter) the negative experience by externalizing "the problem" and placing responsibility on outside forces.
9. *Allen:* Proud and persistent. Pretty impressive. You've got a lot to be proud of, given what they went through, and now you seem able to do the same.	Encourager, self-disclosure, feedback with additive empathy. Nelida had previously seen her Miami Cuban background as a problem, but now she is confronted with the idea that she has much to be proud of. The major conflict could be summarized as "On one hand, Nelida, you've been put down for your Cuban American background, but on the other hand, you now are aware that you have much to be proud of and that other people may be the problem, not you." Allen will not lay out the confrontation that clearly, but encourage Nelida to make these discoveries herself.
10. *Nelida:* (pause) Now that I think about it, perhaps I have done better than I thought. They made me feel devalued and, over time, I continued to lose confidence in myself. I did OK in classes, but never felt really good about myself.	CCS Level 3, recognition that will lead to further change of internal positive thoughts and emotions. Nelida herself is stating the confrontation that Allen above hoped for. She is naming the "on one hand . . . , but on the other . . ." herself. Facilitate clients' confronting their own conflict in a positive, strength-based fashion. Allen's comment at 9 is now apparent as additive, and respect is a helpful part of the process.
11. *Allen:* Let's explore that a little bit more. You lost confidence, but now you are seeing yourself a little differently. I'll ask you a question. How do you feel about people that treat others unfairly, particularly if they are talking about one's ethnicity, religion, or sexual orientation?	Confrontation. Allen summarizes the old and new views that Nelida has of herself. Building on that positive self-view, he confronts her moving toward externalization with how she feels about unfairness. (Additive)
12. *Nelida:* It's not right.	CCS Level 4, a new view of the situation. Nelida is succinct and clear, and she is resolving the conflict in her own mind.
13. *Allen:* It's not right. How about their treatment of you?	Encourager and further confrontation through the restatement and question.
14. *Nelida:* It's not right.	CCS Level 4. Affirmation of the change by repetition. Once a new idea is accepted and reinforced, it becomes part of one's self-concept, but needs further reinforcement by both her and Allen.

The conversation between 15 and 21 has been deleted to save space. In those exchanges, Nelida reaffirms her idea that what happened to her was not right, and through Allen's reflecting feeling, she becomes aware of her anger over what she now sees as harassment. She has moved from an internal contradiction ("I'm the one who is inadequate") to awareness that the conflict is external and begins to discover how this situation is related to unconscious (or possibly conscious) classmate verbal oppression. Nelida also becomes aware that she is paying a price for trying to be empathic, rather than facing up to the harassment.

At this point, what are your thoughts about Nelida and this session? Do you think she is overreacting? How might you handle this session differently from Allen? If you are a woman or aware of women's issues, we suspect that you might add some further items to this session. For example, would the students have treated a male Latino the same way?

We now return to Nelida and the process of change. Here we see the beginnings of resolution and a change in meaning. This change in thoughts and feelings will later lead to action where Nelida feels better about herself and her cultural heritage and recognizes valid feelings of anger and frustration toward what happened.

Counselor and Client Conversation	Process Comments
22. *Nelida:* Maybe how to deal with that feeling instead of always being so empathic, you know, what else can I do? (pause) I should probably address it. I don't want to be rude. You know, because like I said, it might be unintentional, but I should address it.	CCS Level 3, recognition and willingness to explore further. Her resilience is beginning to show.
23. *Allen:* Okay. You should address it. So one possibility, then, is to address it when it happens. How would you name it when you get somebody who talks to you denigrating your culture? Ah, what name could we give it?	Encourager/restatement/paraphrase. Allen then supplies a directive oriented to brainstorming plus an open question. (Most likely this is additive empathy, plus acceptance of Nelida "where she is" as well as respect for her and her cultural background.)
24. *Nelida:* Well, it's victimizing . . . and I've allowed myself to be victimized. Reminds me of my multicultural class—I've internalized a negative view of myself. And that actually is a form of racism that I've taken inside myself without thinking.	CCS Level 5, the development of a new and larger view of herself and her situation. Note the use of the transformational word "victimizing." Internal self-blame has become externalized. Note that she used the word "internalized" herself on her own.
25. *Allen:* Victimizing. Okay. So when you're victimized, and historically we find that you've allowed yourself to be victimized, how do you feel about that?	Confrontation in the form of a summary, followed by a checkout/perception check. (Interchangeable empathy)
26. *Nelida:* That's exactly right. I've gone about it the wrong way. (laugh)	CCS Level 4, further integration of the new constructs Nelida created for herself.
27. *Allen:* So you are allowing insensitivity, victimization, and racism oppression to sit inside you and make you feel bad about yourself.	This set of exchanges is grouped together. Allen first summarizes the essence of the conflict with a confrontation. Nelida responds with her awareness of the new way of being. Allen then, at 29, paraphrases what she just said. Most meaningful, we hear the word "Definitely" said strongly for the first time. This consolidates Level 5 thinking and feeling (but is still not behavioral change).
28. *Nelida:* I never thought of it that way.	
29. *Allen:* Okay. So seems to me like your thoughts are changing.	
30. *Nelida:* Yes. Definitely.	

(continued)

Counselor and Client Conversation	Process Comments
31. *Allen:* Now, what's the Spanish word for strength and force and the persistence that you and your grandparents have? What's the Spanish word for that?	CCS Levels 4 and 5. These exchanges represent a further consolidation of Nelida's resources that will enable her to deal behaviorally with the world around her. The strength within the word *fuerza* is becoming a central part of Nelida's being. Allen used the Gestalt directive of repetition to help Nelida reinforce the creation of the *New*, which in turn leads to a more confident self-concept. Again, saying key words, particularly emotional words, in one's native language is highly recommended. Allen's efforts here are additive. This experience was moving for him as a counselor.
32. *Nelida:* Fuerza.	
33. *Allen:* Say it again.	
34. *Nelida:* Fuerza.	
35. *Allen:* Say it loud.	
36. *Nelida:* Fuerza!	
37. *Allen:* Can you think of that word and how it represents you and your grandparents' pride and force the next time you are put down, like you have been?	
38. *Nelida:* I definitely will.	CCS Levels 4 and 5.
39. *Allen:* Fuerza. Okay, *fuerza* is going to protect you from inside. And now how are you going to deal with those who have harassed you? Can we take what you have discovered and use it to make things better for you—and perhaps for others as well?	Summary and a move to go to the action stage of the session, followed by an open question. There is clearly an implied confrontation here: "On one hand, you are feeling better about yourself and your culture, but on the other hand, what are you going to do about it?" All these issues need to be explored in a follow-up third meeting.

Looking back at the first session with Nelida, we have seen her move from CCS Levels 1 and 2 to a significant change in her view of herself and a willingness to take action. We see that Nelida has become much more aware of her personal strengths and resources in her external family and friends. She has a heightened respect for and pride in her cultural heritage and cultural identity. Clearly, she has made considerable progress in her internal feelings about herself and has a better understanding of external factors that have troubled her.

At the end of the session (not shown), Nelida moves to Level 5 on the Client Change Scale. She addresses the need to speak up when others demean her Cuban heritage and educate them when possible. She is aware that bilingualism is an advantage. Finally, she discusses the possibility of bringing other Spanish-speaking students together to provide support for one another. Nelida's developing awareness and knowledge about the microaggression led her to the creative New, increased resilience, and cultural health (see Box 10.2).

BOX 10.2 Confrontation, Creativity, and Neuroscience

Researching creativity is like nailing jelly to a wall.

—Oshin Vartainian

Defining creativity in terms of neural science and the fMRI has turned out to be challenging. In a review of 72 studies, Dietrich and Kanso (2010) leave us with the general impression that "diffuse activity" in the frontal cortex is essential. There is also some limited evidence that inhibiting the left hemisphere enables the right to push through with creative ideas. This idea, while not fully substantiated, is attractive, as confrontation is a way to unbalance thinking and "mix things up," thus resulting in diffuse activity. By challenging preexisting thoughts (existing in the hippocampus), confrontations may enable the prefrontal cortex to "loosen up" and create the *New*. The "right/left brain" terminology is becoming obsolete, however, as there is only one brain (see http://blogs .scientificamerican.com/beautiful-minds/the-real -neuroscience-of-creativity).

Counseling and psychotherapy are very much concerned with helping clients create the *New* and discover pathways to growth, as Nelida did when her developing awareness and knowledge about the microaggression led her to increased resilience and cultural health. As part of this process, fresh neural networks are created. To facilitate significant change, seek an appropriate balance of stress while supporting the client. Too much stress is damaging, but too little stress likely won't lead to change. Maintain awareness that "released adrenaline (resulting from stress) influences almost all regions of the brain—the entire cortex, the hypothalamus, the hind brain, and the brain stem" (Grawe, 2007, p. 220).

Too much stress can flood the brain with damaging cortisol and fix negative memories in the mind (posttraumatic stress). There are a few unethical and charismatic "therapists" who encourage clients to reach strong emotions. They use this here-and-now base to reach back to so-called "long forgotten and repressed" memories of trauma. Unfortunately, this can result in permanently imprinting false memories of things that never actually happened (Loftus, 2003). This type of "therapy" introduces new damaging neural networks in the brain.

The action plan continued some issues from the first session, but moved toward specifics that she could do. Among immediate planned actions were (1) to reflect and even meditate on her personal strengths and that of her Cuban background—*cultural health*; (2) to join the Latina/o group on campus for additional support; (3) during the next home visit, to ask her grandparents for more stories of their move from Cuba to the United States and give them respect, thanks, and love for what they have done for their own lives and for her.

The Client Change Scale is also useful as a broad measure of success and outcome in several types of counseling and psychotherapy, as it provides a useful framework for accountability and measuring client growth. The clearest example is that of clients with substance abuse issues. Such clients may come voluntarily or be referred by the court for cocaine or alcohol abuse (often both). They may also be depressed and use the drugs to alleviate pain. Often these clients start by denying that they really have a problem (Level 1 on the Client Change Scale). If we are successful in challenging, supporting, and confronting, we will see the client move to Level 2, admitting that there may be a problem. Some call this a "bargaining" stage, in which the client moves back and forth between denial and recognizing that something needs to be done.

Acceptance and recognition of the issue occur at Level 3, where the alcoholic admits that he or she is indeed an alcoholic and the cocaine abuser acknowledges addiction. But acceptance of the problem and recognition are not resolution. The client may be less depressed but still continue to drink and use drugs.

Change occurs when the client reaches Level 4 and actually stops the substance use, after which the depression usually lifts. Although substance abuse and depression are often co-occurring mental health issues, there is some risk that depression will remain and continued treatment will be necessary. Even so, this is real success in counseling and therapy; it is not easy to achieve, but clearly observable and measurable. Level 5, transcendence and the development of new ways of being and thinking, may occur, but not all clients will achieve this level.

Level 5 is represented by the user who becomes fully active in support groups, helps others move away from addiction, and continues to work on feelings, behaviors, actions, and relationships that led to the alcohol abuse and addiction. This person achieves changes in life's meaning and a much more positive view of self and the world—far more than just "getting by."

When you confront clients, ask them a key question, or provide any intervention, they may have a variety of responses. Ideally, they will actively generate new ideas and move forward, but much more likely they will move back and forth with varying levels of response. The idea is to note how clients respond (at what level they answer) and then intentionally provide another lead or comment that may help them grow. *Clients do not work through the five levels in a linear, straightforward pattern; they will jump from place to place and often change topic on you.* Be empathic and patient, and keep on task to help them move forward to creating change.

Depending on the issue, change may be slow. For some clients, movement to partial acceptance (Level 2) or acceptance but no change (Level 3) is a real triumph. For a variety of issues, acceptance represents highly successful counseling and therapy. For example, the client may be in a situation in which change is impossible or really difficult. Thus, accepting the situation "as it is" is a good result. A client facing death is perhaps the best example and does not always have to reach new solutions and transcendence. Simply accepting the present situation may be enough. There are some things that cannot be changed and need to be lived with. "Easy does it." "Life is not fair." "There is a need to accept the inevitable." For a client whose partner or parent is an alcoholic, a major step is to realize that sometimes this situation cannot be changed by the client; acceptance is the major breakthrough that will lead later to new solutions. The newly found solutions would facilitate the mediation process and help resolve the conflict.

Cultural Identity Development and the Confrontation Process

We have seen Nelida move from Level 1 to Levels 4 and 5 on the Client Change Scale. Key to this process was her becoming more aware of her identity as a cultural person—a Spanish-speaking Latina, a minority person in a predominantly White environment. Clearly, cultural background is a major part of personal identity, even though she, and many of your own clients, may not be aware of this.

Cultural identity developmental theory has useful parallels to the Client Change Scale. Five levels of identity were first identified by William Cross (1971, 1991), who outlined specific and measurable stages of Black identity development. Since that time, several other theorists have explored the Cross five-stage model and applied it not only to racial/ethnic issues, but also to gender awareness, gay/lesbian identity, people with disabilities, and many other groups. Included in this is White awareness, which focuses on Whiteness and the White experience as a culture.

The five stages of cultural identity theory are:

Conformity stage: The client may be unaware of racial identity and conforms to what he or she sees as societal expectations. There may be lack of awareness or denial of the impact of culture on who we are. This corresponds to Level 1 of the Client Change Scale.

Dissonance stage: This compares to the partial bargaining of the CCS. The client is aware that something "doesn't fit" and moves positions as new discoveries are made. For those in many groups, this leads to conflict between self-appreciation and self-doubt. Awareness of the impact of culture has started.

Resistance and emersion stage: Two things can happen here. Minority clients often become angry at what they see around them. As they grow and mature, they may immerse themselves fully, for example, in African American culture. Nelida for the first time recognized her anger and began to take more pride in her Cuban culture. White clients may also find themselves angry and try to understand other cultural groups' values more fully. Or they may move to active resistance, another form of conformity and denial.

Introspection stage: Clients increasingly think for themselves, whereas before they were embedded in a group view of the world. They focus on themselves and understanding themselves and their own cultural group. Nelida saw that her grandparents represented strength, and she became prouder of her Cuban heritage.

Integrative awareness stage: A fuller sense of caring for oneself and one's cultural heritage appears. Along with this often comes a fuller understanding of other cultures. This may lead to appreciation, or it may lead to a movement to resistance and action, but this time based on pride and awareness. It is a transcendent stage that has many variations.

Moving from one cultural identity stage to another requires confrontation of the discrepancies within life at that stage. For example, the conformity stage is illustrated by the African American who denies racial issues, the woman who accepts male values as "the truth," or the gay male who hides in the closet and denies his sexual orientation. All of these are constantly confronted with the contradictions and discrepancies they see daily as they interact with others. When enough data and emotional impact have come from these encounters, energy to confront the discrepancy mounts and the individual can move to another stage of cultural identity development.

Maintain awareness that each client you meet, whether a Person of Color or White, has some level of cultural identity. Many White people deny that they have a culture or a cultural identity, and this may be an issue for counseling itself where creating the *New* may be challenging. A good place to start identity development with White individuals is ethnicity or region of the nation. Clients are often willing to explore Irish, Polish, or German backgrounds as identity, but often have more difficulty with that word "Whiteness." Reviewing the history of ethnic and religious prejudice is one way to facilitate awareness of societal oppression and lack of tolerance.

Cultural Identity Development, Cultural Health, and the Nelida/Allen Counseling Sessions

Nelida and Allen's two sessions highlight the importance of helping clients address internal and external conflicts related to racial/cultural identity. Clients may report low self-worth and self-esteem, and blame themselves. These feelings and thoughts may be products of oppression and racism. Nelida reported conflictive negative feelings about herself—she "never felt really good about myself" and was unable to stick up for herself. She made excuses for the other student's question because it must have been "unintentional."

With the help of Allen, she begins to appreciate her culture and feels "proud" of her grandparents' persistence and hard work when they came to the United States. She begins to appreciate herself, seeing these positive attributes in herself as well. Nelida gains awareness of her resulting negative feelings and sense of being "devalued" as an issue that should have been addressed during class because it was "victimizing" and self-denigrating. With further progress in her identity development, she will be able to eliminate internalized negative views of self and replace them with positive views and a commitment to eliminate racism and oppression.

To fully achieve complete awareness at the introspection and integrative awareness stages, Nelida will need to take action in the real world. As the session continued from 39 above, Nelida first made it clear that she no longer wanted to continue passive acceptance of things as they were and she wanted to speak up for her grandparents and her culture. Her first action thought was to seek to educate those who might unintentionally* put down her accent and implicitly her culture as well. She commented, "Maybe I should

---

*Showing respect for cultural difference is critical for an empathic relationship. Nelida's fellow students "unintentionally" showed a lack of respect for who she was and where she was from. Those who find themselves in a "minority" culture or group often experience unintentional racism when basically "good" people say things that hurt, which really represent forms of racism or other oppression. For example: "You speak English well" (to a third-generation Asian American, whose English is better than the commenter's). "African Americans have a lot of talent, especially in music and athletics." "I bet you have great Mexican food at home."

educate that person a little bit about where I come from and what my culture is. And hopefully that person will realize what an ignorant question that is to ask somebody. Maybe tell them a little bit about my culture. I see now that I'm speaking not only for me, but also for my grandparents and for my culture. I need to stand up."

Possibilities for educating those around her were explored, and it was clear that this was her major goal. She was not interested in expressing anger. At times, the effort to educate all those who express an oppressive or racist comment can be very tiring, so Nelida realized that sometimes it would be wise for her just to ignore it—and, while ignoring it, think of the good feelings in her body that represent positive family and cultural experiences.

Beyond educating and ignoring, Nelida realized that she could talk about the incidents with her friends and family. Her grandparents had experienced similar racist incidents, particularly in New York. She had not really considered this relating to her own life until she reviewed the community genogram. Further thought led her to join a Latina/o action group on her campus. This group met regularly for support and discussion of issues such as discrimination and microaggressions that they encountered. The group, with the advice of a concerned professor, worked to educate the campus on these issues.

The CCS as a System for Assessing Change Over Several Sessions

We have seen Nelida move, through the community genogram and the session, from CCS Levels 1 and 2 to Levels 3 and 4 with clear beginnings at Level 5. Now let's review the CCS as it might appear in a counseling session with virtually any topic. If the client is in the denial stage, the story may be distorted, others blamed unfairly, and the client's part in the story denied. In effect, the client in *denial* (Level 1) does not deal with reality. When the client is confronted effectively, the story becomes a discussion of inconsistencies and incongruity, and we see Level 2 *bargaining and partial acceptance*—the story is changing. At *acceptance* (Level 3), the reality of the story is recognized and acknowledged, and thus storytelling is more accurate and complete. Moreover, it is possible to create *new solutions* and *transcendence* (Levels 4 and 5). When changes in thoughts, feelings, and behaviors are integrated into a new story, we see the client move into major new ways of thinking accompanied by action after the session is completed.

Virtually any issue or concern a client presents may be assessed at one of the five levels. If your client starts with you at *denial* or *partial acceptance* (Level 1 or 2) and then moves with your help to *acceptance* and *generating new solutions* (Level 3 or 4), you have clear evidence of the effectiveness of your therapy process. The five levels may be seen as a general way to view the change process in counseling and therapy. Each confrontation or other counseling intervention in the here and now may lead to identifiable changes in client awareness.

Small changes in the session will result in larger client change over a session or series of sessions. Not only can you measure these changes over time, but you can also contract with the client in a partnership that seeks to resolve conflict, integrate discrepancies, and work through issues and problems. Specifying concrete goals often helps the client deal more effectively with confrontation.

The CCS provides you with a systematic way to evaluate the effectiveness of each intervention and to track how clients change in the here and now of the session. If you practice assessing client responses with the CCS model, eventually you will be able to make decisions automatically "on the spot" as you see how the client is responding to you. For example, if the client appears to be in denial of an issue despite your confrontation, you can intentionally shift to another microskill or approach that may be more successful.

Action: Key Points and Practice of Applying Empathic Confrontation in the Real World

We all face conflict and issues. Allen's father once said, "If you don't have a problem, you are dead." Life is wonderful, but we need to face the fact that conflict is inevitable. We have the internal conflicts of making personal decisions—where to go to school, what to major in, what job to take in an unpredictable economy, and with whom we wish to develop close relationships. All of these and many more are issues that bring clients to you for counseling and therapy. Underneath, these conflicts typically result in feelings of being torn, confused, and "mixed up."

Then, almost daily, we have the possibility of conflicts with others. These can be decisions on the job, relationship and family differences, or how to deal with a bullying supervisor.

We suggest the following as possible ways to take the concepts of empathic confrontation into the real world:

1. Observe yourself for one day, thinking about the discrepancies, incongruities, and conflicts within you. Pay attention to internal feelings. Do you have mixed feelings about decisions? Are their signs of internal conflict? What are the places where you conflict in some way with others? How are those discrepancies and conflicts worked through?

2. Another day, observe what you see around you. Give special attention to nonverbal behavior that indicates unsaid conflict or discrepancies in others. You will find people who give mixed verbal messages or indicate confusion or indecision. Does this increase your awareness that we live constantly in a world where we have to deal with incongruity and conflict in one form or another?

3. If possible and a friend is interested, sit down and listen to another person's story of conflict or indecision. Using the skills of empathic confrontation, can you help this person understand better what is going on? Perhaps you might help him or her move to a decision or think about a change in behavior or thinking.

Of course, use these observations to make the points and ideas of this chapter more meaningful and useful. Then apply them to counseling and therapy sessions.

Empathic Confrontation. Clients come to us stuck and immobilized in their developmental processes. They experience internal and external conflicts. Through the use of microskills such as empathic confrontation, we facilitate change, movement, and transformation—restorying and action.

Empathic confrontation has been defined as a supportive challenge in which you note incongruities and discrepancies and then feed back or paraphrase those discrepancies to the client, giving appropriate attention to underlying emotional issues. The task is then to work through the resolution of the discrepancy.

Empathic Confrontation and Change Strategies. An explicit empathic confrontation can be recognized by the model sentence, "On one hand . . ., but on the other hand . . . . How do you put those two together?" In addition, many counseling statements contain implicit confrontations that can be helpful in promoting client growth and developmental movement. For example, you may summarize client conversation, pointing out discrepancies, or use an influencing skill such as the interpretation/reframe or feedback to produce change.

The Client Change Scale. The Client Change Scale is a tool to examine the effect that microskills and empathic confrontation have on client verbalizations immediately in the session.

At the lowest level, clients may deny their incongruities; at middle levels, they may acknowledge them; at higher levels, they may transform or integrate incongruity into new stories and action.

Multicultural and Individual Issues. Empathic confrontation is believed to be relevant to all clients, but it must be worded to meet individual and cultural needs. Do not expect individuals in any cultural group always to follow one pattern; avoid stereotyping; and adapt confrontation to individual and cultural differences.

Cultural Identity Theory. Developed by William Cross, the five stages of cultural identify development are conformity, dissonance, resistance and emersion, introspection, and integrative awareness. The Cross model helps the counselor understand where the client is in his or her cultural identity development and determine the role of cultural/contextual factors. Based on this knowledge, counselors can offer interventions to empower clients to embrace their cultural identity, promote life improvements, and assess progress.

Helping Clients Cope With Microaggressions and Related Concerns. (1) Listen empathically and search for internal and external conflict and contradictions in client stories. (2) Validate their cognitions and emotions around the incident(s). (3) Build resilience by focusing on internal strengths as well as external resources that provide strength and support to cope with challenges. (4) Explore and reframe the contradictions in the situation as appropriate. (5) Facilitate an action plan for the next steps toward cultural health.

Practice and Feedback: Individual, Group, and Microsupervision

Additional resources can be found by going to CengageBrain.com and logging into the MindTap course created by your professor. There you will find a variety of study tools and useful resources that include quizzes, videos, interactive counseling and psychotherapy exercises, case studies, the Portfolio of Competencies, and more.

If you master the cognitive concepts of the reading material and engage in deliberate practice of the exercises that follow, you will be able to promote client change and assess the effectiveness of your interventions. Again, this is an area that takes practice and experience. Practice, practice, practice, and apply the ideas here throughout the rest of your work with this book.

Individual Practice

Exercise 10.1 Identifying Discrepancies, Incongruity, and Mixed Messages, Along with Strengths Leading Toward Resolution

Please review your first session (Chapter 1) and other practice exercises completed so far. Viewing video recordings of sessions, especially your own, is an effective way to learn about conflict, discrepancies, contradictions, and confrontation. Unless you can identify incongruity in yourself, seeing it in others may be difficult or even inappropriate. The following exercise will advance your learning of this microskill.

Discrepancies internal to the self. Can you identify specific times when your nonverbal behavior contradicted your verbal statements and gave you away? Are there times when you say two things at once and your verbal statements are incongruous? Have you done one thing while saying another?

Discrepancies between you and the external world. Part of life is living with contradictions. Many of these are unresolvable, but they can give considerable pain. What are some of the discrepancies between you and other individuals? What are some of the mixed messages, contradictions, and incongruities you face in your world of schooling or work?

Discrepancies between you and the client. You may have already experienced this and can easily summarize times when you felt out of tune with and discrepant from a client. Or if you have not engaged in counseling extensively, it may be helpful to think of situations in which you had major differences with someone else. Often we have typical situations that "push our buttons" and move us toward actions that are too quick. Self-awareness in this area can be most helpful.

Specific strengths. Resolution of conflict and discrepancy is often made from a positive frame of reference. Can you identify personal strengths and wellness assets that can help you resolve internal and external differences? What strengths do you admire in others that you might like to add to your repertoire?

Exercise 10.2 Practicing Confrontation of Incongruity and Conflict

Write confrontation statements for the following situations. The model sentence "On the one hand . . ., but on the other hand . . ." provides a useful standard format for the actual confrontation, but you may also use variations such as "You say . . . but you do . . . ." Remember to follow up the confrontation with a checkout.

> A client breaks eye contact, speaks slowly, and slumps in the chair while saying, "Yes, I really like the idea of getting to the library and getting the career information you suggest. Ah . . . I know it would be helpful for me."
>
> "Yes, my family is really important to me. I like to spend a lot of time with them. When I get this big project done, I'll stop working so much and start doing what I should. Not to worry."
>
> "My partner is good to me most of the time—this is only the second time he's hit me. I don't think we should make a big thing out of it."
>
> "My daughter and I don't get along well. I feel that I am really trying, but she doesn't respond. Only last week I bought her a present, but she just ignored it."

Exercise 10.3 Practicing with the Client Change Scale

Here are some statements made by clients. Identify which of the five levels each client statement represents.

1. Denial
2. Partial examination
3. Acceptance and recognition
4. Creation of a new solution
5. Transcendence

Health issues. Look for movement from denial to new ways of taking care of one's body.

———— I can't have a heart attack. It will never happen to me. I need to eat real food.

———— Oh, I suppose I am overweight, but if I cut down a bit on butter and perhaps no more milk shakes, I'll be okay.

———— I guess I can see that I need to balance my diet, but the busy life I lead won't really allow that to happen.

———— I'm now able to cut out fats. At least that's taken care of.

———— I've completely changed my way of doing things. I eat right—no fat at all—I exercise, and I'm even getting to like relaxation and stress management.

Career planning. Look for a movement from inaction or randomness to action.

———— Okay, I guess I see your point. I've been released from two work-study programs because I didn't show up on time. But those were the bosses' fault. They should have made what they wanted clearer.

———— The teacher referred me to you. Everyone has to have a job plan, but I see no need to worry about it so much. I'll be OK.

———— Yes, I need a job plan. I can see now that is necessary. I'll write one and bring it to you tomorrow.

———— I've got a job! The plan worked and I interviewed well and now I'm on my way.

———— The plan has been helpful. I think I see now how to interview more effectively and present myself better.

Awareness of racism, sexism, heterosexism. Look for movement from denial that these issues exist to awareness and action.

———— I feel committed. I've started action at home and at work, and I'm really going to concentrate on a more active approach to work against discrimination.

———— Well, some people do discriminate, but I think that many people are just exaggerating.

———— I don't really believe there is such a thing as racism or sexism. It's just people complaining.

———— I've started working with my family and children on being more tolerant, fair, and understanding of people different from us.

———— There is a fair amount of prejudice, racism, and sexism everywhere.

Exercise 10.4 Thinking About Your Own Cultural Identity Development Stage
Answer the following questions.

❏ Thinking about cultural identity, in what stage do you place yourself?

❏ What does this mean to you? In which ways might advancing your cultural development awareness help you become a better counselor or psychotherapist?

❏ What are some actions you might take to advance your cultural awareness or move higher in the cultural development model?

Group Practice and Microsupervision

Exercise 10.5 Evaluating Confrontation Leads and Client Responses Using the CCS
As always, divide into groups with client, counselor, and observer(s). The counselor's task is to use the basic listening sequence to draw out a conflict in the client and then to confront this conflict or incongruity. The counselor will observe and note discrepancies on the spot during the session and feed them back to the client (see Box 10.3).

Internal conflict often arises around a difficult decision, past or present. External conflict most often appears when one has difficulty in dealing with a family member, a friend, or someone at work. Usually you will find both internal and external conflict in the client. We suggest the following topics as possibilities.

❏ A real microaggression that the client has experienced or a similar role-play. Most of us have had difficult experiences with teasing, bullying, or being made to feel inadequate by friends or family. This practice can be emotionally arousing, so special attention to an egalitarian and ethical approach is essential.

❏ Taking out a loan versus working part-time.

❏ Choosing between two equally attractive majors in college.

❏ A career decision involving a choice between a larger income and work that would be more satisfying.

❏ A conflict with a partner that could lead to a break in the relationship.

BOX 10.3 | **Feedback Form: Confrontation Using the Client Change Scale**

You can download this form from Counseling CourseMate at CengageBrain.com.

_____   _____
(DATE)

_____   _____
(NAME OF COUNSELOR)   (NAME OF PERSON COMPLETING FORM)

Instructions: Video and/or audio recording will be necessary for the best type of feedback. Otherwise, it will be best for the observer(s) to stop the session shortly after a confrontation has occurred and then discuss what was observed. In this practice, we are seeking a review of all leads used by the counselor, but looking for confrontations. Rate how the client responded to the confrontation on the five-point Client Change Scale (1 = denial, 2 = partial examination, 3 = acceptance and recognition, 4 = creation of a new solution, 5 = transcendence and the creation of the *New*).

Counselor Statement (Write key words to help recollection and discussion.)	Client Comment (Write key words to help recollection and discussion.)	CCS Rating
1.		
2.		
3.		
4.		
5.		
6.		
7.		
8.		
9.		
10.		

Portfolio of Competencies and Personal Reflection

Skill in confrontation depends on your ability first to listen and then to take an active role in the helping process. This needs to be done in a nonjudgmental fashion, with respect for differences. As you work through this list of competencies, think ahead to how you would include confrontation skills in your own Portfolio of Competence.

Assessing Your Level of Competence: Awareness, Knowledge, Skills, and Action

Use the following checklist to evaluate your present level of mastery. Check those dimensions that you currently feel able to do. Those that remain unchecked can serve as future goals. Do not expect to attain intentional competence on every dimension as you work through this book. You will find, however, that you will improve your competencies with repetition and practice.

Awareness and Knowledge

❑ Identify discrepancies and incongruities manifested by a client in the session.

❑ Identify client stage of cultural identity development during the session.

❑ Classify and write counselor statements indicating the presence or absence of elements of confrontation.

❑ Identify client change processes on the Client Change Scale through observation.

Basic Competence

❑ Demonstrate confrontation skills in a real or role-played session.

❑ Observe and identify, in the here and now of the session, client responses on the five levels of the Client Change Scale.

❑ Utilize wellness and the positive asset search to help clients find strengths that might help them move forward toward positive change when confronted.

Intentional Competence. You will be able to use confrontational skills in such a manner that clients improve their thinking and behaving as reflected on the CCS.

❑ Help clients change their manner of talking about a problem as a result of confrontation. This may be measured formally by the CCS or by others' observations.

❑ Move clients from initial discussion of issues at the lower levels of the CCS to discussion at higher developmental levels at the end of the session, or when the topic has been fully explored.

❑ Identify client responses inferred from the CCS on the spot in the session, and change counseling interventions to meet those responses.

Psychoeducational Teaching Competence. Are you able to teach change and confrontation concepts to clients and to others?

The basic dimensions of confrontation are really designed more for counselors and psychotherapists than for clients. But there are some very specific ways that psychoeducation will be an important part of your practice. First, those going through the stages of grief associated with death may find it helpful to have the change stages identified for them, thus enabling them to understand their feelings and thoughts more fully. These stages of change will also be helpful in understanding reactions to serious illness, accidents, alcoholism, and traumatic incidents. Second, you can set up change goals with your clients and work with them to discover how far they have progressed in meeting goals and making life changes.

Personal Reflection on Empathic Confrontation

Confrontation is based primarily on listening skills, but it does require you to move more actively in the session by highlighting discrepancies and conflict. The Client Change Scale (CCS) was presented to show that you can assess the influence of your interventions in the here and now of the session. The creative *New* provides a more philosophical dimension to confrontation and change.

What single idea stands out for you among all those presented in this chapter, in class, or through informal learning? What stands out that is likely to be a guide toward your next steps.

How might confrontation relate to diversity issues?

How would you use mediation as a psychoeducational treatment program?

What other points in this chapter strike you as important?

How might you use ideas in this chapter to begin the process of establishing your own style and theory?

Interpersonal Influencing Skills for Creative Change

The key to successful leadership is influence, not authority.

—Kenneth Blanchard

All *Intentional Interviewing and Counseling* skills and strategies are based on attending, observation, and listening. The influencing skills presented here are most useful as supplements to the first half of this text. Couple these two chapters with the preceding two on focusing and confrontation, and you have a rich array of ways to work creatively with your clients, increasing their intentionality and ability to take action on their *New* discoveries.

Through attending and listening, we are influencing the client indirectly. Influencing skills must be based on listening, but they take a more direct approach. Do not assume, however, that you, the counselor, are in charge. Clients are the "deciders"; our task is to help provide options, plus supporting and encouraging change. The influencing skills need to be used judiciously and sparingly.

Given that caveat, you will find that an egalitarian, empathic approach to influencing skills will be welcomed by most of your clients. At issue is seeing that their intentionality and creativity are fostered, rather than yours. Interestingly, to accomplish this effectively will require intentionality and creativity on your part as well.

Look for the following skills and strategies in the two chapters of this section.

Chapter 11. Reflection of Meaning and Interpretation/Reframe: Helping Clients Restory Their Lives
For many of your clients, this may be the most helpful influencing skill for finding meaning and vision in life, as it provides goals that can support

them through many difficulties. Here you will examine the relationship between and among behaviors, thoughts, feelings, and their underlying meaning structure. These skills help you gain a deeper understanding of each client's issues and history. Clients will gain valuable new perspectives on their problems and stories.

Chapter 12. Action Skills for Building Resilience and Managing Stress: Self-Disclosure, Feedback, Logical Consequences, Directives/Instruction, and Psychoeducation The action influencing skills are explored, with specific suggestions for facilitating client restorying and action. Here you will see a wide array of alternatives that actively involve the client in thinking and acting differently. Special attention is paid to stress management strategies and therapeutic lifestyle changes. With all of these strategies, empathic, egalitarian relationships with the client are essential.

As you develop competence in influencing skills, you may expect to develop the ability to

1. Help clients move to deeper levels of self-exploration and self-understanding using the skills of reflection of meaning and interpretation/reframe.

2. Facilitate client restorying by using self-disclosure and feedback, in the process creating a more open egalitarian relationship.

3. Use influencing skills and strategies to assist client developmental progress, particularly when the more reflective listening skills fail to produce change and understanding.

Competence in the influencing skills will further advance your intentional competence. The effective interviewer is always in process—growing and changing in response to new learning.

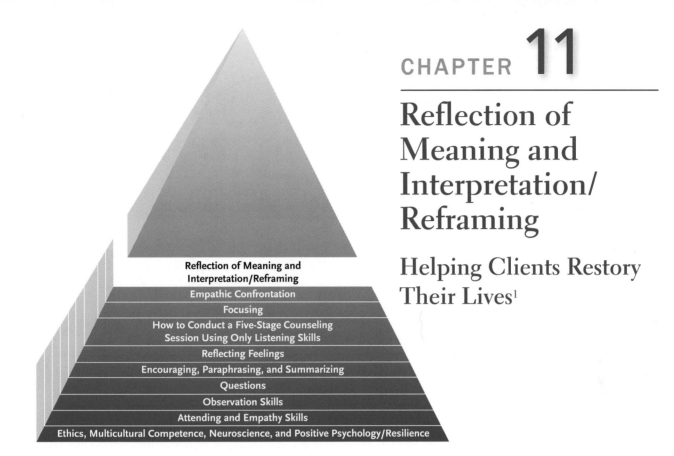

Reflection of Meaning and Interpretation/ Reframing

Helping Clients Restory Their Lives[1]

Reflection of Meaning and
Interpretation/Reframing

Empathic Confrontation

Focusing

How to Conduct a Five-Stage Counseling
Session Using Only Listening Skills

Reflecting Feelings

Encouraging, Paraphrasing, and Summarizing

Questions

Observation Skills

Attending and Empathy Skills

Ethics, Multicultural Competence, Neuroscience, and Positive Psychology/Resilience

Ever more people today have the means to live, but no meaning to live for.

Challenging the meaning of life is the truest expression of the state of being human.

We who lived in concentration camps can remember the men who walked through the huts comforting others, giving away their last piece of bread. They may have been few in number, but they offer sufficient proof that everything can be taken from a man but one thing: the last of the human freedoms—to choose one's attitude in any given set of circumstances, to choose one's own way.

—Viktor Frankl (who helped many Jews survive and find meaning
while imprisoned at Auschwitz during the Holocaust)

Many believe that meaning is the most important influencing skill of all. You and your clients share a strong need for finding meaning in your life experiences—for making sense of their experience and establish goals, values, and a sense of life vision. Furthermore, many of our clients aim to address existential questions, such as Who am I? Why am I here? What

---

[1]This chapter is dedicated to the memory of Viktor Frankl. The initial stimulus for the skill of reflection of meaning came from a 2-hour meeting with him in Vienna shortly after we had visited the German concentration camp at Auschwitz, where he had been imprisoned in World War II. He impressed on us the central value of meaning in counseling and therapy—a topic to which most theories give insufficient attention. It was his unusual ability to find positive meaning in the face of impossible trauma that impressed us most. His thoughts also affected our wellness and positive strengths orientation. His theoretical and practical approach to counseling and therapy deserves far more attention than it receives. We often recommend his gripping short book, *Man's Search for Meaning* (1959), to our clients who face serious life crises. This book remains fully alive today.

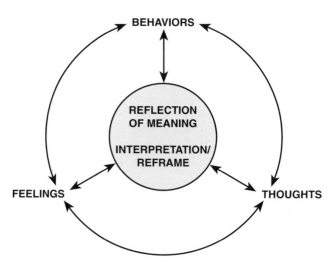

FIGURE 11.1 Meaning as the core of human experience and its relationship to feelings, thoughts, and behaviors. *Meaning can drive action.*

---

is the purpose or significance of it all? What sense does anything make? How much control do I have over my life? What happens when life ends? Earlier philosophers, such as Plato, Socrates, and Aristotle, and current therapists have studied how people understand their life, experiences, and relationships and how this understanding affects their lives. Thus, the pursuit of meaning influences most of what we do.

Two microskills that help clients move to more depth are reflection of meaning and the interpretation/reframe. Clients may come to us making meaning in a negative fashion, creating unnecessary failure in their behavior, emotional distress, and unhappy thoughts and cognitions. Once clients have reinterpreted/reframed their issues and developed a more functional, positive way of making meaning, they see things differently and more clearly than before. This leads to more positive, hopeful, and optimistic views of life, which in turn leads to behavioral, cognitive, and emotional changes. This is illustrated in Figure 11.1.

Reflection of meaning is concerned with helping clients find deeper understanding of significant basic issues, such as life vision and purpose, underlying their thoughts, feelings, and behavior. In turn, finding a deeper meaning leads to new interpretations of life. **Interpretation/reframing** seeks to provide a new way of understanding these thoughts, feelings, and behaviors, and often this also results in perspectives on making meaning. Interpretation often comes from a specific theoretical perspective, such as decisional, psychodynamic, or multicultural. Clients generate their own meanings, whereas reframes usually come from the interviewer.

Chapter Goals and Competency Objectives

Awareness and Knowledge

▲ Understand the skills of reflection of meaning and interpretation/reframing.

▲ Recognize their similarities and differences.

▲ Explore the rationale and some philosophical background of the centrality of these two skills.

▲ Become aware of the critical work of Viktor Frankl, philosopher and practical psychiatrist, who developed logotherapy to change the direction of clients' lives.

▲ Explore discernment, a practical system to aid clients in deciding life direction.

Skills and Action

▲ Assist clients, through reflection of meaning, to explore their deeper meanings and values and to discern their goals and life purpose or mission.

▲ Help clients, through interpretation/reframing, to find alternative ways of thinking that facilitate personal development.

▲ Evaluate clients' progress and change as you use these skills.

▲ Employ the action philosophy and practice of logotherapy.

Introduction: The Skills of Reflecting Meaning and Interpretation/Reframing

Reflection of meaning and interpretation/reframing explore in-depth issues below the surface of client conversation. Reflecting meaning encourages clients to find new and/or clearer visions for understanding themselves and others, as well as clarifying their purpose in life. Interpretation/reframing is the art of supplying the client with new perspectives and ideas to create new ways of thinking, feeling, and behaving. If you use reflection of meaning and interpretation/reframing skills as defined here, you can anticipate how clients will respond.

Reflection of Meaning	Anticipated Client Response
Meanings are close to core experiencing. Encourage clients to explore their own meanings and values in more depth from their own perspective, but also the perspectives of others. Questions eliciting meaning are often a vital first step. A reflection of meaning looks very much like a paraphrase, but focuses beyond what the client says. Appearing often are the words *meaning*, *values*, *vision*, and *goals*.	The client discusses stories, issues, and concerns in more depth, with a special emphasis on deeper meanings, values, and understandings. Clients may be enabled to discern their life goals and vision for the future.
Interpretation/Reframe	**Anticipated Client Response**
Provide the client with a new perspective, frame of reference, or way of thinking about issues. Interpretations/reframes may come from your observations; they may be based on varying theoretical orientations to the helping field; or they may link critical ideas together.	The client may find another perspective or way of thinking about a story, issue, or problem. This new perspective could have been generated by a theory used by the interviewer, from linking ideas or information, or by simply looking at the situation afresh.

He [or she] who has a *why* to live can bear almost any *how*.

—Friedrich Nietzsche

Eliciting and reflecting meaning involve both a skill and a strategy. As a skill, eliciting meaning is fairly straightforward. To elicit meaning, ask the client some variation of the basic question "What does _____ mean to you? At the same time, effective exploration of meaning becomes a major strategy in which you bring out client stories from the past, in the present, or in anticipation of the future. You use all the listening, focusing, and confrontation skills to facilitate this self-examination, yet the focus remains on meaning and finding purpose. Review of life goals and purpose leading toward a more meaningful life can result from this exploration. A decision may be made to spend time helping others and/ or in social justice advocacy.

The case of Charlis will serve as a way to illustrate similarities and differences between reflection of meaning and interpretation.

Charlis, a workaholic 45-year-old middle manager, has a heart attack. After several days of intensive care, she is moved to the floor where you, as the hospital social worker, work with the heart attack aftercare team. Charlis is motivated; she is following physician directives and progressing as rapidly as possible. She listens carefully to diet and exercise suggestions and seems the ideal patient with an excellent prognosis. However, she wants to return to her high-pressure job and continue moving up through the company; you observe some fear and puzzlement about what's happened to her and where she might want to go next.

Reflection of Meaning and Charlis

Charlis has had a heart attack, and you first need to hear her story, with special attention to the many sad, worried, and anxious feelings that she experiences. There is a need to review what happened before the heart attack, including the stressed lifestyle that likely contributed to hospitalization. She will need help and support in her rehabilitation program. Understanding the meaning of the whole event will be critical. You may prescribe lifestyle changes, support her through an exercise program, review nutrition plans, and counsel her on relationships at work.

You recognize that Charlis is reevaluating the meaning of her life. She asks questions that are hard to answer: "Why me?" "What is the meaning of my life?" "What is God saying to me?" "Am I on the wrong track?" "What should I *really* be doing?" You sense that she feels that something is missing in her life, and she wants to reevaluate where she is going and what she is doing.

To elicit meaning, we may ask Charlis some variation of a basic meaning question: "What does the heart attack mean to you, your past, and—perhaps most of all—your future life?" We may also ask Charlis if she would like to examine the meaning of her life through the process of *discernment*, a more systematic approach to meaning and purpose defined in some detail later in this chapter.

We'd share some of the specific questions listed below as a beginning; sometimes that is enough to help the client move to new life directions. We'd ask her to think of questions and issues that are particularly important to her as we work to help her discern the meaning of her life, her work, her goals, and her mission. These questions often bring out emotions, and they certainly bring out meaning in the client's thoughts and cognitions. When clients explore meaning issues, the interview becomes less precise as the client struggles with defining the almost indefinable. As appropriate to the situation, questions such as the following can address the general issue of meaning in more detail:

- What has given you most satisfaction in your job?
- What's *been missing* for you in your present life?
- What do you *value* in your life?
- What *sense* do you make of this heart attack and the future?
- What things in the future will be most *meaningful* to you?
- What is the *purpose* of your working so hard?

- You've said that you wonder what God is saying to you with this trial. Could you share some of your thoughts?
- What gift would you like to leave the world?

Eliciting meaning often precedes reflection. Reflection of meaning as a skill looks very much like a reflection of feeling or paraphrase, but the key words *meaning, sense, deeper understanding, purpose, vision,* or some related concept will be present explicitly or implicitly. "Charlis, I sense that the heart attack has led you to question some basic understandings in your life. Is that close? If so, tell me more." Eliciting and reflecting meaning provide an opening for the client to explore issues in a way that leads not to a final answer but rather to a deeper awareness of the possibilities of life. Both reflecting meaning and interpretation/reframing are designed to help clients look deeper, first by careful listening and then by helping clients examine themselves from a new perspective.

Reflecting meaning involves *client* direction; the interpretation/reframe implies *interviewer* direction. In reflection of meaning, the client provides the new and more comprehensive perspective, whereas an interpretation/reframe offers a new way of being as suggested by the interviewer or counselor.

Comparing Reflection of Meaning and Interpretation/Reframing

Here are some brief examples of how reflection of meaning and interpretation may work for Charlis as she attempts to understand some underlying issues around her heart attack.

> *Charlis:* My job has been so challenging and I really feel that pressure all the time, but I just ignored it. I'm wondering why I didn't figure out what was going on until I got this heart attack. But I just kept going on, no matter what.

Eliciting and Reflecting of Meaning

> *Counselor:* I hear you—you just kept going. Could you share what it feels like to *keep going on* and what it *means* to you? (Encourager focusing on the key words *keep going on*; open question oriented to meaning)
>
> *Charlis:* I was raised to keep going. My mother always prided herself on doing a good job, even in the worst of times. Grandma did the same thing.
>
> *Counselor:* Charlis, I hear that keeping going and persistence have been a key family value that remains very important to you. (Reflection of meaning) "Hanging in" is what you are good at. (Positive asset leading to wellness is mentioned) Could we focus now on how that value around persistence and *keeping going on* relates to your rehab? (Open question that seeks to use the wellness dimensions to help her plan for the future)

Interpretation

> *Counselor:* You could say that you *keep going* until you drop. How does that sound to you? (Mild reframe/interpretation followed by checkout)
>
> *Charlis:* I was raised to keep going. My mother always prided herself on doing a good job, even in the worst of times. Grandma did the same thing.
>
> *Counselor:* Many of us become who we are because of family history. It sounds as if several generations have taught you keep on *no matter what*. You're full of pride as you see the strength in your family that keeps you keeping on. (Interpretation/ reframe, includes reflection of feeling)

Both reflection of feeling and interpretation ended up in nearly the same place, but Charlis is more in control of the process with reflection of meaning. Whichever approach is used, we are closer to helping Charlis work on the difficult questions of the meaning and direction of her future life. If the client does not respond to reflective strategies, move to the more active interpretation. We need to give clients power and control of the session whenever possible. They can often generate new interpretations/reframes and new ways of thinking about their issues.

Interpretations and reframes vary with theoretical orientation, and the joint term *interpretation/reframe* is used because they both focus on providing a new way of thinking or a new frame of reference for the client, but the word *reframe* is a gentler construct. Keep in mind when you use influencing skills that interpretive statements are more directive than reflections of meaning. When we use interpretation/reframing, we are working primarily from the interviewer's frame of reference. This is neither good nor bad; rather, it is something we need to be aware of when we use influencing skills.

Awareness, Knowledge, and Skills of Reflection of Meaning and Interpretation/Reframe

Eliciting Client Meaning

Understanding the client is the essential first step. Consider storytelling as a useful way to discover the background of a client's meaning making. If a major life event is critical, illustrative stories can form the basis for exploration of meaning. Clients do not often volunteer meaning issues, even though these may be central to the clients' concerns. Critical life events such as illness, loss of a parent or loved one, accident, or divorce often force people to encounter deeper meaning issues. Blonna, Loschiavo, and Watter (2011) effectively applied this model to health counseling. If spiritual issues come to the fore, draw out one or two concrete stories of the client's religious heritage. Through the basic listening sequence and careful attending, you may observe the behaviors, thoughts, and feelings that express client meaning.

Fukuyama (1990, p. 9) outlined some useful questions for eliciting stories and client meaning systems. Adapted for this chapter, they include the following:

- When in your life did you have existential or meaning questions? How have you resolved these issues thus far?

- What significant life events have shaped your beliefs about life?

- What are your earliest childhood memories as you first identified your ethnic/cultural background? Your spirituality?

- What are your earliest memories of church, synagogue, mosque, a higher power, of discovering your parents' vital life values?

- Where are you now in your life journey? Your spiritual journey?

Reflecting Client Meaning

Say back to clients their exact key meaning and value words. Reflect their own unique meaning system, not yours. Implicit meanings will become clear through your careful listening and questions designed to elicit meaning issues from the client. Using the client's key words is preferable, but occasionally you may supply the needed meaning word yourself. When you do so, carefully check that the word(s) you use feel right to the client. Simply change "You feel . . ." to "You mean . . . ." A reflection of meaning is structured similarly to a paraphrase or reflection of feeling. "You value . . .," "You care . . .," "Your reasons are . . .," or "Your intention was . . . ." Distinguishing among a reflection of meaning, a paraphrase, and a reflection of feeling can be difficult. Often the skilled counselor will blend the three skills together. For practice, however, it is useful to separate out meaning responses and develop an understanding of their import and power in the interview. Noting the key words that relate to meaning (*meaning, value, reasons, intent, cause,* and the like) will help distinguish reflection of meaning from other skills.

Reflection of meaning becomes more complicated when meanings or values conflict. Here concepts of confrontation (Chapter 10) may be useful. Conflicting values, either explicit or implicit, often underlie mixed and confused feelings expressed by the client. For instance, a client may feel forced to choose between loyalty to family and loyalty to spouse. Underlying love for both may be complicated by a value of dependence fostered by the family and the independence represented by the spouse. When clients make important decisions, sorting out key meaning issues may be crucial.

For example, a young person may be experiencing a value conflict over career choice. Spiritual meanings may conflict with the work setting. The facts may be paraphrased accurately and the feelings about each choice duly noted, yet the underlying *meaning* of the choice may be most important. The counselor can ask, "What does each choice *mean* for you? What sense do you make of each?" The client's answers provide the opportunity for the counselor to reflect back the meaning, eventually leading to a decision that involves not only facts and feelings but also values and meaning. As in confrontation, you can evaluate client change in meaning systems using the Client Change Scale (Chapter 10). Meaning generates a flow of ideas. See Box 11.1 for examples.

Interpretation/Reframe

> You can't connect the dots looking forward; you can only connect them looking backward. So you have to trust that the dots will connect in your future. You have to trust in something—your gut, destiny, life, karma, whatever. This approach will never let you down and it has made all the difference in my life.
> —Steve Jobs

When you use the microskill of interpretation/reframing, you are helping clients to connect the dots of the story and enabling them to look at their issues from a new, more useful perspective. A new way of thinking is central to the restorying and action process. In the microskills hierarchy, the words *interpretation* and *reframe* are used interchangeably. Interpretation reveals new perspectives and new ways of thinking beneath what a client says or does. The reframe provides another frame of reference for considering problems or issues. And eventually the client's story may be reconsidered and rewritten as well.

The basic skill of interpretation/reframing may be described as follows:

- The counselor listens to the client's story, issue, or problem and learns how the client makes sense of, thinks about, or interprets the story or issue.

- The counselor may draw from personal experience and/or observation of the client (reframe) or may use a theoretical perspective, thus providing an alternative meaning or interpretation of the narrative. This may include *linking* together information or ideas discussed earlier that relate to each other. Linking is particularly important as it

BOX 11.1   Research and Related Evidence That You Can Use

Reflection of Meaning and Reframing

Ratey (2008a, p. 41), a leader in neuroscience applications and research, has commented:

> You have to find the right mission, you have to find something that's organic, that's growing, that keeps you focused on and continues to provide meaning and growth and development for yourself.
>
> I see meaning as a big part of neuroscience. We start with neuroscience and now we're talking about transcendence. Spirituality even lights up key centers in the brain. Meaning drives the lower centers and is connected to emotions and motivation areas. It's a huge, huge, human construct that means so much to our race and our species. Obviously it involves memory and learning and remembering the good stuff, remembering what your goals are, remembering what you want to do, and so you need all those things working well to keep you on the right meaning path. If you can get people into a situation where they have the meaning direction provided by their mission or their job or their goal, they don't need medicine.

Carl Rogers brought meaning issues to center stage as part of his work on reflecting feelings. Viktor Frankl provided both philosophical and practical applications of meaning in counseling. A solid relationship with your client helps give meaning to your encounter.

Classic research by Fiedler (1950a) and Barrett-Lennard (1962) set the stage for the present when they found that relationship variables (closely related to the listening skills) were vital to the success of all forms of all counseling and therapy, regardless of theory. Now the relationship issues are termed "common factors," and the idea of relationship as central has become almost universally accepted. Those therapists and researchers working in the Heart and Soul of Change Project cite data suggesting that 30% or more of successful therapy is based on relationship (Miller, Duncan, & Hubble, 2005).

Research has demonstrated that "families that seek support and try to accept what happened after a traumatic injury may experience less injury-related stress and family dysfunction over time" (Wade et al., 2001, p. 412). Turning to religion was the second most used strategy among parents whose children suffered traumatic injury. Connectedness with others and the comfort of spirituality can be a most important positive asset and wellness strength for many clients. A classic and often cited study by Probst (1996) found that religiously oriented clients do better in cognitive behavioral therapy when their spirituality becomes part of the process. Recovery from heart surgery has been found to be more rapid among those with religious involvement, particularly among women (Contrada et al., 2004, p. 227).

Lucas (2007) examined experiences of 19 caregivers and teachers working with traumatized children. She found that learning coping strategies of reframing and realistic goal setting helped them reduce emotional exhaustion and increased their personal sense of accomplishment. Li and Lambert (2008) found positive reframing to be one of the best predictors of job satisfaction among 102 intensive care nurses from the People's Republic of China.

Neuroscience and Meaning

On the surface, the broad idea of meaning would appear to be beyond measurement in a physical sense. Our sense of meaning brings our thoughts, feelings, and behavior into a whole, enabling us to make sense of our experience. A useful exploration of the brain and its relation to meaning is provided by Carter (1999, p. 197):

> Meaningfulness is inextricably bound up with emotion. Depression is marked by wide-ranging symptoms, but the cardinal feature of it is the draining of meaning from life . . . . By contrast, those in a state of mania see life as a gloriously ordered, integrated whole. Everything seems to be connected and the smallest events are bathed in meaning.

Creation of new ideas also means that new neural networks are formed in the brain and long-term memory.

Ratey (2008a) indicates that there is a key moral and spiritual dimension in the brain that we are close to identifying. Stimulation of a portion of the brain appears to bring out spiritual images in many people. Morality may be partially hardwired. *The Political Brain* (Westen, 2007) follows this logic. Westen speaks of how candidates directly affect the mirror neurons of the public, creating empathy and changing neural connections. Morality as described by neuroscientists is awareness of the Other. An interesting challenge in brain science is explaining the individualist mind versus the collectivist mind. Gene expression is clearly part of this, but gene expressions often require environmental events before they are triggered. Some genes may lie dormant throughout a lifetime.

integrates ideas and feelings for clients and frees them to develop new approaches to their issues.

- (Positive reframe from personal experience) "You feel that coming out as gay led you to lose your job, and you blame yourself for not keeping quiet. Maybe you just really needed to become who you are. You seem more confident and sure of yourself. It will take time, but I see you growing through this difficult situation." Here self-blame has been reinterpreted or reframed as a positive step in the long run.

- (Psychoanalytic interpretation with multicultural awareness) "It sounds like the guy who fired you is insecure about anyone who is different from him. He sounds as if he is projecting his own unconscious insecurities on you, rather than looking at his own heterosexism or homophobia."

The value of an interpretation or reframe depends on the client's reaction to it and how he or she changes thoughts, feelings, or behaviors. Think of the Client Change Scale (CCS)—how does the client react to each interpretation? If the client denies or ignores the interpretation, you obviously are working with denial (Level 1 on the CCS). If the client explores the interpretation/reframe and makes some gain, you have moved that client to bargaining and partial understanding (Level 2 on the CCS). Interchangeable responses and acceptance of the interpretation (Level 3) will often be an important part of the gradual growth toward a new understanding of self and situation. If the client develops useful new ways of thinking and behaving (Level 4 on CCS), movement is clearly occurring. Transcendence, perhaps the ultimate creation of the *New* (Level 5), will appear only with major breakthroughs that change the direction of counseling and psychotherapy. But let us recall that movement from denial (Level 1) to partial consideration of issues (Level 2) may be a major breakthrough, beginning client improvement.

Linking is an important part of interpretation, although it often appears in an effective reflection of meaning as well. In linking, two or more ideas are brought together, providing the client with a new insight. The insight comes primarily from the client in reflection of meaning, but almost all from the interviewer in interpretation/reframing. Consider the following four examples:

Interpretation/reframe 1: Charlis, we are all reflections of our family, and it is clear that family history emphasizing success and hard work has deeply affected you, perhaps even to the point of having a heart attack. (Links family history to the heart attack. A family counselor might use this approach.)

Interpretation/reframe 2: Charlis, you seem to have a pattern of thinking that goes back a long way—we could call it an "automatic thought." You seem to have a bit of perfectionism there and you keep saying to yourself (self-talk), "Keep going no matter what." (Links the past to the present perfectionism from a cognitive behavioral perspective.)

Interpretation/reframe 3: It sounds as if you are using hard work as a way to avoid looking at yourself. The avoidance is similar to the way you avoid dealing with what you think you need to change in the future to keep yourself healthier. (Combines confrontation with linking with what is occurring in the interview series. This is close to a person-centered approach.)

Interpretation/reframe 4: The heart attack almost sounds like unconscious self-punishment, as if you wanted it to happen to give you time off from the job and a chance to reassess your life. (Linking interpretation from a psychodynamic perspective.)

Observe: The Skills of Reflection of Meaning and Interpretation/Reframing in Action

In the following session, Travis is reflecting on his recent divorce. When relationships end, the thoughts, feelings, and underlying meaning of the other person and the time together often remain unresolved. Moreover, some clients are likely to repeat the same mistakes in their relationships when they meet a new person. Both the interpretation/reframe and reflection of meaning can help clients draw meaning from their ordeals and gain new perspectives on themselves and their world.

Andreas, the interviewer, seeks to help Travis think about the word *relationship* and its meaning. Note that Travis stresses the importance of connectedness with intimacy and caring. The issue of self-in-relation to others will play itself out very differently among individuals in varying cultural contexts. Many clients will focus on their need for independence.

Interviewer and Client Conversation	Process Comments
1. *Andreas:* So, Travis, you're thinking about the divorce again . . .	Encourager/restatement.
2. *Travis:* Yeah, that divorce has really thrown me for a loop. I really cared a lot about Ashley and . . . ah . . . we got along well together. But there was something missing.	Travis is at Level 2 (partial examination) on the Client Change Scale.
3. *Andreas:* Uh-huh . . . something missing?	Encouragers appear to be closely related to meaning. Clients often supply the meaning of their key words if you repeat them back exactly. (Interchangeable empathy)
4. *Travis:* Uh-huh, we just never really shared something very basic. The relationship didn't have enough depth to go anywhere. We liked each other, we amused one another, but beyond that . . . I don't know . . .	Travis elaborates on the meaning of a closer, more significant relationship than he had with Ashley. (CCS Level 2)
5. *Andreas:* You amused each other, but you wanted more depth. What sense do you make of it?	Paraphrase using Travis's key words followed by a question to elicit meaning. (The paraphrase is interchangeable, the question potentially additive.)
6. *Travis:* Well, in a way, it seems like the relationship was shallow. When we got married, there just wasn't enough depth for a meaningful relationship. The sex was good, but after a while, I even got bored with that. We just didn't talk much. I needed more . . .	Note that Travis's personal constructs for discussing his past relationship center on the word *shallow* and the contrast *meaningful*. This polarity is probably one of Travis's significant meanings around which he organizes much of his experience. (CCS Level 2)
7. *Andreas:* Mm-hmmm . . . you seem to be talking in terms of shallow versus meaningful relationships. What does a meaningful relationship feel like to you?	Reflection of meaning followed by a question designed to elicit further exploration of meaning. (Interchangeable, potentially additive. Please note again that questions have potential, but we don't know whether or not they are additive until we see what the client says next.)
8. *Travis:* Well, I guess . . . ah . . . that's a good question. I guess for me, there has to be some real, you know, some real caring beyond just on a daily basis. It has to be something that goes right to the soul. You know, you're really connected to your partner in a very powerful way.	Connection appears to be a central dimension of meaning. We often believe that connectedness is a female construct, but many men also see it as central. (It was additive, CCS Level 2 but moving toward Level 3.)
9. *Andreas:* So, connections, soul, deeper aspects strike you as really important.	Reflection of meaning. Note that this reflection is also very close to a paraphrase, and Andreas uses Travis's main words. The distinction centers on issues of meaning. A reflection of meaning could be described as a special type of paraphrase. (Interchangeable)

Interviewer and Client Conversation	Process Comments
10. *Travis:* That's right. There has to be some reason for me to really want to stay married, and I think with her . . . ah . . . those connections and that depth were missing. We liked each other, you know, but when one of us was gone, it just didn't seem to matter whether we were here or there.	CCS Level 2–3, as he is building better understanding.
11. *Andreas:* So the relationship did not feel meaningful to either of you. And I hear that closeness and a meaningful relationship are what you missed and what you value.	Reflection of meaning plus some reflection of feeling. Note that Andreas has added the word *value* to the discussion. In reflection of meaning, it is likely that the counselor or interviewer will add words such as *meaning, understanding, sense,* and *value.* Such words lead the client to make sense of experience from the client's own frame of reference. (Interchangeable)
12. *Travis:* Uh-huh. That is really so. (quiet pause, looking puzzled)	
13. *Andreas:* Ah . . . could you fantasize how you might play out those thoughts, feelings, and meanings in another relationship?	Open question oriented to meaning. (Potentially additive)
14. *Travis:* Well, I guess it's important for me to have some independence from a person, but when we were apart, we'd still be thinking of one another. Depth and a soul mate is what I want.	Travis's meaning and desire for a relationship are now being more fully explored. (CCS Level 3+)
15. *Andreas:* Um-hum.	
16. *Travis:* In other words, I don't want a relationship where we always tag along together. The opposite of that is where you don't care enough whether you are together or not. That isn't intimate enough. I really want intimacy in a marriage. My fantasy is to have a very independent partner I care about and who cares about me. We can both be individuals but still have bonding and connectedness.	Connectedness is an important meaning issue for Travis. With other clients, independence and autonomy may be the issue. With still others, the meaning in a relationship may be a balance of the two. (CCS Level 3, moving to 4)
17. *Andreas:* Let's see if I can put together what you're saying. The key words seem to be independence with intimacy and caring. It's these concepts that can produce bonding and connectedness, as you say, whether you are together or not.	This reflection of meaning becomes almost a summarization of meaning. Note that the key words and constructs have come from the client in response to questions about meaning and value. (Additive)

Further counseling would aim to bring behavior or action into accord with thoughts. Other past or current relationships could be explored further to see how well the client's behaviors or actions illustrate or do not illustrate expressed meaning.

Multiple Applications of Reflection of Meaning and Interpretation/Reframing

Theories of Counseling and Interpretation/Reframing

Theoretical interpretations can be extremely valuable, as they provide the interviewer with a tested conceptual framework for thinking about the client. Each theory is itself a story—a story told about what is happening in interviewing, counseling, and therapy and what the story means. Integrative theories find that each theoretical story has some value. As you generate your own natural style, you will most likely develop your own integrative theory, drawing from those approaches that make most sense to you.

Imagine that you have worked with Charlis for some time and she has worked through many of her fears, but she still faces some real challenges. She comments:

> Yesterday my manager gave me a new assignment. I sensed that something was wrong, that he wanted something from me but wouldn't say it. It made me feel very anxious, as now I'm not sure what to do. Where do I go next?

First, keep in mind that the interviewer could reflect back the anxiety concerns and turn the issue back to Charlis, or even ask the meaning of the situation. In interpretation, the counselor supplies an alternative perspective. Below are several examples of how counselors with different theoretical orientations might interpret the same information. Before the actual interpretation, you will see a brief theoretical paragraph that provides some background for the theory-oriented interpretation that follows. Each of the following is slightly exaggerated for clarity.

Decisional Theory. A major issue in interviewing for all clients is making appropriate decisions and understanding alternatives for action. Decisions need to be made with awareness of cultural/environmental context. Interpretation/reframing helps clients find new ways of thinking about their decisions. Linking ideas together is particularly important.

> *Counselor:* Charlis, it sounds as if the manager is again giving you a double message and that causes real anxiety. It's a repetition of some of the things that led to your heart attack. We've spent some time on dealing with your tension. This seems a good place to go over breathing and relaxation again.

Person-Centered. Clients are ultimately self-actualizing, and our goal is to help them find the story that builds on their strengths and helps them find deeper meanings and purpose. Reflection of meaning helps clients find alternative ways of viewing the situation; interpretation/reframing are not used. Linking can occur through effective summarization.

> *Counselor:* Charlis, you are really feeling anxious again, and the manager seems to be giving you a difficult time again. You're wondering what all this mean. (Reflection of feeling and eliciting meaning. The conversation has returned to the client.)

Brief Counseling. Brief methods seek to help clients find quick ways to reach their central goals. The interview itself is conceived first as a goal-setting process, and then methods are found to reach goals through time-efficient methods. Interpretation/reframing will be rare except for linking of key ideas.

> *Counselor:* You're facing some of the old familiar challenges since the heart attack. Let's look at this and think back on what you've done in the past that works in this situation. What comes to mind? (Linking with a mild interpretation and a focus on finding what was effective for Charlis in the past, thus reminding her of her strengths.)

Cognitive Behavioral Theory. The emphasis is on sequences of behavior and thinking and what happens to the client, internally and externally, as a result. Often interpretation/reframing is useful in understanding what is going on in the client's mind and/or linking the client to how the environment affects cognition and behavior.

> *Counselor:* OK, you came in to talk and suddenly the manager seemed almost to be attacking you (antecedent) and then you become anxious (emotional consequence), and this has left you hanging and wondering what to do (behavioral consequence). Now let's analyze the situation; let's look at what he is doing and how he seems to make you feel. Then we can develop a new alternative and more effective way to deal with this in the future.

Psychodynamic/Interpersonal Theory. Individuals are dependent on unconscious forces. Interpretation/reframing is used to help link ideas and enable the client to understand how the unconscious past and long-term, deep-seated thoughts, feelings, and behaviors frame the here and now of daily client experiencing. Freudian, Adlerian, Jungian, and several other psychodynamic theories each tell different stories.

> *Counselor:* Charlis, this seems to go back to the discussions we've had in the past about your issues with authority, particularly with your father. We've even noticed that sometimes you treat me as an authority figure. What sense do you make of this possible pattern in your life? We see the here and now with me, the situation with your boss, and the long-term issues with your father.

Multicultural Counseling and Therapy (MCT). Everyone is always situated in a cultural/environmental context, and we need to help clients interpret and reframe their issues, concerns, and problems in relation to their multicultural background (see the RESPECTFUL model). MCT is an integrative theory and uses all of the methods above, as appropriate, to help clients understand themselves and how the cultural/environmental context affects them personally. The following is from a feminist therapy frame of reference.

> *Counselor:* Sounds to me like simple sexism once again. Charlis, we've got to work on how you can deal with a work environment that seems continually to be hassling you. Perhaps a complaint is in order. But at least we have to engage in more stress management to help you as a woman deal with this productively, so that it does not destroy your health again.

All of the above provide the client with a new, alternative way to reframe the situation. In short, interpretation renames or redefines "reality" from a new point of view. Sometimes just a new way of looking at an issue is enough to produce change. Which is the correct interpretation? Depending on the situation and context, any of these interpretations could be helpful or harmful. The first response deals with here-and-now reality, whereas psychodynamic interpretation deals with the past. The feminist interpretation links the heart attack with sexual harassment on the job.

Multicultural Issues and Reflection of Meaning

For practical multicultural interviewing and counseling, recall the concept of focus (Chapter 9). When helping clients make meaning, focus exploration of meaning not just on the individual but also on the broader life context. In much of Western society, we tend to assume that the individual is the person who makes meaning. But in many other cultures—for example, the traditional Muslim world—the individual will make meaning in accord with the extended family, the neighborhood, and religion. Individuals do not make meaning by themselves; *they make meaning in a multicultural context*. In truth, Western society also draws meaning from family and culture. However, individualism rather than collectivism is generally the focus.

Cultural, ethnic, religious, and gender groups all have systems of meaning that give an individual a sense of coherence and connection with others. Muslims draw on the teachings of the Qur'an. Similarly, Jewish, Buddhist, Christian, and other religious groups will draw on their writings, scriptures, and traditions. African Americans may draw on the meaning strengths of Malcolm X, Martin Luther King Jr., or on support they receive from Black churches as they deal with difficult situations. Women may make meaning out of relationships, whereas men may focus more on issues of personal autonomy and tasks. Witness the conversations in a mixed social group. Often we find women on one side of the room talking about relationships. Men will generally be talking about sports, politics, and their accomplishments.

Viktor Frankl, the Nazi concentration camp survivor, could not change his life situation, but he was able to draw on important strengths of his Jewish tradition to change the meaning he made of it. The Jewish tradition of serving others facilitated his survival and enabled him to help fellow sufferers. When times were particularly bad, when prisoners had been whipped and were not being given food, Frankl (1959, pp. 131–133) counseled his entire barracks, helping them reframe their terrors and difficulties, pointing out that they were developing strengths for the future.

> I quoted from Nietzsche, "That which does not kill me, makes me stronger." I spoke to the future. I said that . . . the future must seem hopeless. I agreed that each of us could guess . . . how small were chances for survival. . . . I estimated my chances at about one in twenty. But I also told them that, in spite of this, I had no intention of losing hope and giving up. . . . I also mentioned the past; all its joys and how its light shone even in the present darkness. . . . Then I spoke of the many opportunities of giving life a meaning. I told my comrades . . . that human life, under any circumstances has meaning. . . . I said that someone looks down on each of us in difficult hours—a friend, a wife, somebody alive or dead, or a God—and He would not expect us to disappoint him . . . . I saw the miserable figures of my friends limping toward me to thank me with tears in their eyes.

You may counsel clients who have experienced some form of religious bias or persecution. As religion plays such an important part in many people's lives, members of dominant religions in a region or a nation may have different experiences from those who follow minority religions. For example, Schlosser (2003) talks of Christian privilege in North America, where people of Jewish and other faiths may feel uncomfortable, even unwelcome, during Christian holidays. Anti-Semitism, anti-Islamism, anti–liberal Christianity, and anti–evangelical Christianity are all possible results when clients experience spiritual and/ or religious intolerance. We also recall that when Christians and other religious groups find themselves in countries where they are a minority, they can suffer similar serious religious persecution, to the point of death.

Frankl, as he sought to survive the German concentration camp at Auschwitz, could not change his life situation, but he was able draw on important strengths of his Jewish tradition to change the meaning he made of it. Shortly after his liberation, Frankl wrote his famous book, *Man's Search for Meaning* (1959), within a 3-week period. This short, emotionally powerful book has remained a constant best seller since that time. Frankl believed that finding positive meanings in the depth of despair was vital to keeping him alive. During the darkest moments, he would focus his attention on his wife and the good things they enjoyed together; or in the middle of extreme hunger, he would meditate on a beautiful sunset.

Mary and Allen spent 2 hours with Dr. Frankl after their lecture tour to Poland, which included a visit to Auschwitz. Frankl shared again the importance of positive meaning for survival. He quoted the German philosopher Nietzsche: "He who has a *why* will find a *how*." If your clients can find a meaningful vision and life direction (the *why*), they often will bear many difficult things as they seek ways to resolve their issues and continue life. Also memorable is Frankl's comment "The best of us did not survive." It was an incredible experience to be in the presence of the man who was the real forerunner of the cognitive behavioral movement (Mahoney & Freeman, 1985). Frankl was fully aware that meaning in itself is not enough—we also must *act* on our meaning and value system.

Logotherapists search for positive meanings that underlie behavior, thought, and action. Dereflection and modification of underlying attitudes are specific techniques that logotherapy uses to uncover meaning and facilitate new actions. Many clients "hyperreflect" (think about something too much) on the negative meaning of events in their lives and may overeat, drink to excess, or wallow in depression. They are constantly attributing a negative meaning to life. When clients focus solely on negatives, dereflection helps to uncover deeper meanings and enables clients to become more positive in outlook.

The direct reflection of meaning may encourage such clients to continue their negative thoughts and behavior patterns. Dereflection, by contrast, seeks to help them discover the values that lie deeper in themselves. This strategy is similar to positive reframing/interpretation, but the client, rather than the counselor, does much of the positive thinking. The goal is to enable clients to think of things other than the negative issue and to find alternative positive meanings in the same event. The questions listed in Box 11.2, later in this chapter, represent first steps in helping clients dereflect and change their attitudes. The following abbreviated example illustrates this approach.

Client: I really feel at a loss. Nothing in my life makes sense right now.

Counselor: I understand that—we've talked about the issues with your partner and how sad you are. Let's shift just a bit. Could you tell me about what has been meaningful and important to you in the past? (The client shares some key supportive religious experiences from the past. The counselor draws out the stories and listens carefully.)

Counselor: (reflecting meaning) So, you found considerable meaning and value in worship and time spent quietly. You also found worth in service in the church. You drifted away because of your partner's lack of interest. And now you feel you betrayed some of your basic values. Where does this lead you in terms of a meaningful way to handle some of your present concerns?

As you may note, the process of dereflection is a special form of the positive asset search. But rather than focusing just on the concretes (spirituality, service to others, walking in the outdoors, enjoying one's friends), the counselor explores the positive meaning of these specifics. "What does spirituality mean to you?" "What sense do you make of a person who finds such joy in walking outdoors and enjoying sunsets?" "What values do you find in service to others?" Out of the exploration of meaning may come data for restorying one's problems and even life-transforming actions.

But Frankl was interested in more than just meaning. He would also discuss specific actions that the client could take in the here and now of daily life. Meaning without implementation and action is not enough. His emphasis on action beyond thinking new thoughts was pathbreaking and innovative.

Resilience, Purpose, and Meaning

Resilience is the ability to recover from a wide variety of difficulties. Two examples of resilience are the child who is teased but bounces back cheerfully the next day, and the adolescent who undergoes being "dumped" by a boyfriend or girlfriend but soon gets over it and moves on. What we see here is the ability to not let bad experiences get one down.

Resilience occurs at a deeper level when the individual suffers a serious or life-threatening trauma. A child is born in poverty, experiences abuse, and somehow manages to put together a successful life. A businessman goes bankrupt, but within two years has put together a successful business. Two women lose their jobs and after both spend six months searching for employment, one goes into a deep depression while the other continues on and eventually finds work. Two soldiers on patrol experience an ambush in which one of their comrades is killed. One soldier leaves the warfront and ends up being treated for posttraumatic stress over a period of months while the other has a short period of mourning, grief, and recovery but soon is back on the front lines.

Meaning and purpose are key to resilience. Viktor Frankl was a survivor of the Nazi death camps. Many around him gave up and died. Frankl kept his focus on the positive and looked for moments of meaning. He enjoyed a beautiful sunset even though hungry, he thought of his beloved wife, and he wrote a book in his mind while doing painful work. Frankl is the theorist who truly brought the importance of meaning to our field. His personal example and his writings remain critical for us today.

Teaching resilience results in less childhood depression (Smith et al., 2008). Seligman (2009), the founder of positive psychology, has shown that teaching purpose and meaning can make a difference among children. One of his major goals was to encourage children to find purpose and positive expectations:

> One exercise involved the students' writing down three good things that happened each day for a week. Examples were: "I answered a really hard question in Spanish class," "I helped my mom shop for groceries" or, "The guy I've liked for months asked me out." Next to each positive event, the students answered the following questions: "What does this mean to you?" and "How can you increase the likelihood of having more of this good thing in the future?"

"What does this mean to you?" is, of course, the basic question to elicit meaning. Meaning is then reflected and synthesized and becomes an important part of the child's cognitive/emotional processing.

African Americans who have suffered trauma but have a sense of purpose and meaning have better mental and physical health (Alim et al., 2008). Similarly, older people who have a clear sense of meaning and purpose have better mental and physical health. Those without a sense of meaning and purpose are two and a half times more likely to suffer the ravages of Alzheimer's disease (Boyle, Buchman, Barnes, & Bennett, 2010). Meaning is not found just in thoughts and feelings; it also affects the body.

The following discussion of discernment is another way to facilitate your clients finding deeper meanings and visions for their lives.

Discernment: Identifying Life Mission and Goals

> Listen. Listen, with intention, with love, with the "ear of the heart." Listen not only cerebrally with the intellect, but with the whole of feelings, our emotions, imaginations, and ourselves.
>
> —Esther de Wall

Discernment is "sifting through our interior and exterior experiences to determine their origin" (Farnham, Gill, McLean, & Ward, 1991). The word *discernment* comes from the Latin *discernere*, which means "to separate," "to determine," "to sort out." In a spiritual or religious sense, discernment means identifying when the spirit is at work in a situation—the spirit of God or some other spirit. The discernment process is important for all clients, regardless of their spiritual or religious orientation or lack thereof. Discernment has broad applications to interviewing and counseling; it describes what we do when we work with clients at deeper levels of meaning. Discernment is also a process whereby clients can focus on envisioning their future as a journey into meaning (see Box 11.2).

BOX 11.2   Questions Leading Toward Discernment of Life's Purpose and Meaning

You may find it helpful to share this list with the client before you begin the discernment process and identify together the most helpful questions to explore. Add topics and questions that occur to you and the client. Discernment is a very personal exploration of meaning. The more the client participates, the more useful it is likely to be. Questions that focus on the *here and now* and intuition may facilitate deeper discovery.

Following is a systematic approach to discernment. First, you or your client may wish to begin by thinking quietly about what might give life purpose, meaning, and vision. Here-and-now body experience and imaging can serve as a physical foundation for intuition and discernment.

- Relax, explore your body, find a positive feeling of strength to serve as an anchor for your search. Build on that feeling and see where it goes.
- Sit quietly and allow an image (visual, auditory, kinesthetic) to build.
- What is your gut feeling? What are your instincts? Get in touch with your body.
- Discerning one's mission cannot be found solely through the intellect. What feelings and thoughts occur to you at this moment?
- Can you recall feelings and thoughts from your childhood that might lead to a sense of direction now?
- What is your felt body sense of spirituality, mission, and life goals?

Concrete questions leading to telling stories can be helpful.

- Tell me a story about that image above. Or a story about any of the here-and-now experiences listed there.
- Can you tell me a story that relates to your goals/vision/mission?
- Can you name the feelings you have in relation to your desires?
- What have you done in the past or what are you doing presently that feels especially satisfying and close to your mission?
- What are some blocks and impediments to your mission? What holds you back?

- Can you tell about spiritual stories that have influenced you?

For self-reflective exploration, the following are often useful.

- Let's go back to that original image and/or the story that goes with it. As you reflect on that experience or story, what occurs for you?
- Looking back on your life, what have been some of the major satisfactions? Dissatisfactions?
- What have you done right?
- What have been the peak moments and experiences of your life?
- What might you change if you were to face that situation again?
- Do you have a sense of obligation that impels you toward this vision?
- Most of us have multiple emotions as we face major challenges such as this. What are some of these feelings, and what impact are they having on you?
- Are you motivated by love/zeal/a sense of morality?
- What are your life goals?
- What do you see as your mission in life?
- What does spirituality mean to you?

The following questions place the client in larger systems and relationships—the self-in-relation. They may also bring multicultural issues into the discussion of meaning.

- Place your previously presented experiences and images in broader context. How have various systems (family, friends, community, culture, spirituality, and significant others) related to these experiences? Think of yourself as a self-in-relation, a person-in-community.
- *Family.* What do you learn from your parents, grandparents, and siblings that might be helpful in your discernment process? Are they models for you that you might want to follow, or even oppose? If you now have your own family, what do you learn from them, and what is the implication of your discernment for them?

(continued)

BOX 11.2   **(continued)**

- *Friends.* What do you learn from friends? How important are relationships to you? Recall important developmental experiences you have had with peer groups. What do you learn from them?

- *Community.* What people have influenced you and perhaps serve as role models? What group activities in your community may have influenced you? What would you like to do to improve your community? What important school experiences do you recall?

- *Cultural groupings.* What is the place of your ethnicity/race in discernment? Gender? Sexual orientation? Physical ability? Language? Socioeconomic background? Age? Life experience with trauma?

- *Significant other(s).* Who is your significant other? What does he or she mean to you? How does this person relate to the discernment process? What occur to you as the gifts of relationship? The challenges?

- *Spiritual.* How might you want to serve? How committed are you? What is your relationship to spirituality and religion? What does your holy book say to you about this process?

Discernment questions from Ivey, A., Ivey, M., Myers, J., & Sweeney, T. (2005). *Developmental counseling and therapy: Promoting wellness over the lifespan.* Boston: Lahaska/Houghton Mifflin. Reprinted by permission of Allen Ivey.

Neuroscience and Ethical Decision Making

To change your life, master your brain.

—Dotti Dixon Schmeling

Many think this chapter on meaning and interpretation is the most important, as thoughts, feelings, and behaviors often stem from central meaning issues. We think that Viktor Frankl is correct when he points out that the person who has a *why for living* will discover *how to live.*

Ethics are the moral principles that define our lives. No matter what we do, our ethics and morals determine our decisions and our behavior. This can be for good or ill. Many would say that ethics, values, and meaning determine who we really are. We may vocalize one thing, but do something else. Do we "walk the walk" or just "talk the talk"? Do we do what is comfortable, or even opposite to what we say?

Neuroscience demonstrates that moral and ethical decision making emerges from a complex interaction among multiple neural systems distributed across brain regions (Greene, 2009; Roskies, 2012). As you might anticipate, brain scans (fMRI) reveal that decision making in the area of values and ethics depends on the executive prefrontal cortex (PFC) in interaction with limbic emotional systems.

The energizing amygdala and the emotional system are fast, automatic, and influenced by the here and now. The executive system is deliberative and both enhanced and limited by memory of past experience (Xu et al., 2013). Value decisions that lead to action rest on the simplistic idea of "good" and "bad." The "good" of the here and now may well conflict with what is considered valuable in the longer term in which meaning and personal values are more deeply considered.

Serious exploration of the meaning of client stories, as well as the discernment exercise presented here, will strengthen the executive meaning-making system and prepare it for emotional challenges in the future. But the executive cannot and must not be separated from the emotional. Unless meaning and discernment have an emotional base, the cognitive meaning is unlikely to hold and lead to action. This is exemplified when we recall that the amygdala has been activated in relation to personal moral dilemmas (Greene, Nystrom, Engell, Darley, & Cohen, 2004). This emotional component is thought to function as an alarm that directly influences behavior or to add motivational energy to different alternatives considered during the reasoning process.

While reflection of meaning, interpretation, and discernment are often central in ethical, value, and life vision decisions, drawing out clients' stories from multiple perspectives (focusing) and confronting decisional conflict are essential. Exploring emotions is also key to cognitive meaning decisions,

Action: Key Points and Practice of Applying Reflection of Meaning and Interpretation/ Reframing Skills in the Real World

Meaning is not observable behavior, although it could be described as a special form of cognition that reaches the core of our being. Helping clients discern the meaning and purpose of their lives can serve as a motivator for change and provide a compass as to the direction of that change. Meaning organizes life experience and can serve as a metaphor from which clients generate thoughts, feelings, and behaviors. Reflections of meaning are often used with more verbal clients. Clients who learn to interpret and think about their lives in new ways gain a new sense of meaning. Reframing one's life in a positive direction enables the development of a new vision. A person with a sense of meaning and a vision for the future can often work through and live with the most difficult issues and problems.

Following are practical key points that may enable you to take the concepts of this chapter into your own life and daily practice.

Eliciting Meaning. "What does _____ mean to you?" Insert the key important words of the client that will lead to meanings and important thoughts underlying key words. "What sense do you make of it?" "What values underlie your actions?" "Why is that important to you?" "Why?" (by itself, used carefully)

Reflecting Meaning. Essentially, this looks like a reflection of feeling except that the words *meaning*, *values*, or *intentions* substitute for feeling words. For example, "You mean . . .," "Could it mean that you . . .?" "Sounds like you value . . .," or "One of the underlying reasons/intentions of your actions was . . . ." Then use the client's own words to describe his or her meaning system. You may add a paraphrase of the context and close with a checkout.

Interpretation/Reframe. The counselor helps clients gain new perspectives, new frames of reference, and sometimes new meanings, all of which can facilitate clients' changing their view and way of thinking about their issues. This skill comes primarily from the counselor's observations and occasionally from the client.

Theoretical Interpretations. These come from a specific counseling theory such as psychodynamic, interpersonal, family therapy, or even Frankl's logotherapy. Clients tell their stories or speak about their problems and issues. The counselor then makes sense of what they are saying from a particular theoretical perspective. "That dream suggests that you have an unconscious wish to run away from your husband." "Sounds like an issue of what we call boundaries—your husband/wife is not respecting your space." "I hear you saying that you don't know where you are going; it sounds like you lack meaning in your life."

Reframes. These tend to come from here-and-now experience in the session, or they might be larger reframes of major client stories. The reframes are based on your experience in providing the client with another interpretation of what has happened or how the story is viewed. Effective reframes can change the meaning of key narratives in clients' lives. The positive reframe is particularly important. "Charlis, what stands out to me at this moment is how able you are, and we can use your 'smarts' and ability to understand situations to find new, more comfortable directions." Positive reframes in the here and now are often the most useful.

Discernment. This is a form of listening that goes beyond our usual descriptions and could be termed "listening with the heart." Both you and the client seriously search for deeper life goals and direction. Specific discernment questions are listed in Box 11.2.

Multicultural and Family Issues and Stories. These may be key in helping clients discover personal meaning. Eliciting meaning and focusing reflection on contextual issues beyond the individual will enhance and broaden one's understanding of life's deeper concerns.

Focusing is the most certain ways to bring multicultural issues into the interview. A woman, a gay or lesbian, or a Person of Color may be depressed over what is considered a personal failure. By helping the client see the cultural/environmental context of the issue, a new perspective will appear, providing a totally new and more workable meaning.

Practice and Feedback: Individual, Group, and Microsupervision

Additional resources can be found by going to CengageBrain.com and logging into the MindTap course created by your professor. There you will find a variety of study tools and useful resources that include quizzes, videos, interactive counseling and psychotherapy exercises, case studies, the Portfolio of Competencies, and more.

The concepts of this chapter build on previous work. If you have solid attending and client observation skills, can use questions effectively, and can demonstrate effective use of the encourager, paraphrase, and reflection of feeling, you are well prepared for the exercises that follow.

Individual Practice

Exercise 11.1 Identification of Skills
Read the following client statement. Which of the following counselor responses are paraphrases (P), reflections of feeling (RF), reflections of meaning (RM), or interpretations/reframes (I/R)?

> I feel very sad and lonely. I thought Jose was the one for me. He's gone now. After our breakup, I saw a lot of people but no one special. Jose seemed to care for me and make it easy for me. Before that I had fun, particularly with Carlos. But it seemed at the end to be just sex. It appears Jose was it; we seemed so close.

——— "You're really hurting and feeling sad right now."

——— "Since the breakup you've seen a lot of people, but Jose provided the most of what you wanted."

———— "Sounds like you are searching for someone to act as the father you never had and Jose was part of that."

———— "Another way to look at it is that you unconsciously don't really want to get close; and when you get really close, the relationship ends."

———— "Looks like the sense of peace, caring, ease, and closeness meant an awful lot to you."

———— "You felt really close to Jose and now are sad and lonely."

———— "Peace, caring, and having someone special mean a lot to you. Jose represented that to you. Carlos seemed to mean mainly fun, and you found no real meaning with him. Is that close?"

List possible single-word encouragers for the same client statement. You will find that the use of single-word encouragers, perhaps more than any other skill, leads your client to talk more deeply about the unique meanings underlying behavior and thought. A good general rule is to search carefully for key words, repeat them, and then reflect meaning.

Exercise 11.2   Discernment: Examining One's Purpose and Mission

Using the suggestions in Box 11.2, work through each of the four sets of questions. You may do this by yourself, using a meditative approach and journaling, or you may want to do it with a classmate or close acquaintance. Allow yourself time to think carefully about each area. Add questions and topics that occur to you—make this exercise fully personal.

What do you learn from this exercise about your own life and wishes?

Exercise 11.3   Individual Practice in Interpretation/Reframing

Interpretations provide alternative frames of reference or perspectives for events in a client's life. In the following examples, provide an attending response (question, reflection of feeling, or the like) and then write an interpretation. Include a checkout in your interpretation.

> "I was passed over for promotion for the third time. Our company is under fire for sex discrimination, and each time a woman gets the job over me. I know it's not my fault at all, but somehow I feel inadequate."

Listening response _____

_____

_____

Interpretation/reframe from a psychodynamic frame of reference (i.e., an interpretation that relates present behavior to something from the past): _____

_____

_____

Interpretation/reframe from a gender frame of reference _____

_____

_____

Interpretation/reframe from your own frame of reference in ways that are appropriate for varying clients _____

_____

_____

> "I'm thinking of trying some pot. Yeah, I'm only 13, but I've been around a lot. My parents really object to it. I can't see why they do. My friends are all into it and seem to be doing fine."

Listening response _____

_____

_____

Reframe from a conservative frame of reference (one that opposes the use of drugs) _____

_____

_____

Reframe from an occasional user's frame of reference _____

_____

_____

Interpretation from your own frame of reference on this issue _____

_____

The preceding examples of interpretations and reframes are representations of meaning and value issues that you will encounter in counseling and therapy. What are the value issues involved in these examples, and what is your personal position on these issues? Finally, how do you reconcile the importance of a client's responsibility for her or his own behavior with your position? What would you actually do in these situations?

Group Practice and Microsupervision

Exercise 11.4 Practice with Reflection of Meaning and Interpretation/Reframe
Set up the video practice session as in the past.

For practice with these skills, it will be most helpful if the session starts with the client's completing one of the following model sentences. The session will then follow along, exploring the attitudes, values, and meanings to the client underlying the sentence.

"My thoughts about spirituality are . . ."

"My thoughts about moving from this area to another are . . ."

"The most important event of my life was . . ."

"I would like to leave to my family . . ."

"The center of my life is . . ."

"My thoughts about divorce/abortion/gay marriage are . . ."

A few alternative topics are "My closest friend," "Someone who made me feel very angry (or happy)," and "A place where I feel very comfortable and happy." Again, a decision conflict or a conflict with another person may be a good topic.

A useful sequence of microskills for eliciting meaning from the model sentence is as follows:

1. An open question, such as "Could you tell me more about that?" "What does that mean to you?" or "How do you make sense of that?"

2. Encouragers and paraphrases focusing on key words to help the client continue

3. Reflections of feeling to ensure that you are in touch with the client's emotions

4. Questions that relate specifically to meaning (see Box 11.2)

5. Reflecting the meaning of the event back to the client, using the framework outlined in this chapter

It is quite acceptable to have key questions and this sequence in your lap to refer to during the practice session. Box 11.3 provides feedback forms that are helpful in practice sessions.

Some general reminders: When we use interpretation/reframing, we are working primarily from the counselor's frame of reference. Your goal is to help the client to restory or look at the frustrating problem or concern from a new perspective. To accomplish this goal, you need to listen before you provide your interpretation or reframe. Respect clients' frame of reference before interpreting or reframing their words and frustrating situations in new ways. Provide clients with a new perspective or way of thinking about issues.

BOX 11.3   Feedback Form: Reflecting Meaning and Interpretation/Reframe

_____   (DATE)

_____
(NAME OF COUNSELOR)   (NAME OF PERSON COMPLETING FORM)

Instructions: Observer 1 will complete the Client Change Scale in Chapter 10. Observer 2 will complete the items below.

1. Did the counselor use the basic listening sequence to draw out and clarify the client's story or concern? How effectively?

2a. (For reflection of meaning) How effectively did the counselor elicit client meaning issues, and were they further explored?

2b. (For interpretation/reframe) How effectively did the counselor use these skills, and were they further explored?

3. How did the client react to the use of these two skills?

4. Did the counselor check out the client's reaction to the intervention? Did the client move on the Client Change Scale?

5. Provide nonjudgmental, factual, and specific feedback for the counselor on the use of reflection of meaning and interpretation/reframe skills.

Portfolio of Competencies and Personal Reflection

As you work through this list of competencies, think about how you would include the ideas related to reflection of meaning in your own Portfolio of Competence.

Use the following checklist to evaluate your present level of mastery. As you review the items below, ask yourself, "Can I do this?" Check those dimensions that you currently feel able to do. Those that remain unchecked can serve as future goals. Do not expect to attain intentional competence on every dimension as you work through this book. You will find, however, that you will improve your competencies with repetition and practice.

Awareness and Knowledge. You will be able to differentiate reflection of meaning and interpretation/reframing from the related skills of paraphrasing and reflection of feeling. You will be able to identify questioning sequences that facilitate client talk about meaning. You will be able to provide new ways for clients to think about their issues through interpretation/reframing.

❑ Identify and classify the skills.

❑ Identify and write questions that elicit meaning from clients.

❑ Note and record key client words indicative of meaning.

Basic Competence. You will be able to demonstrate the skills of eliciting and reflecting meaning and interpretation/reframing. You will be able to demonstrate an elementary skill in dereflection.

❑ Elicit and reflect meaning in a role-play session.

❑ Examine yourself and discern more fully your life direction.

❑ Use dereflection and attitude change in a role-play interview.

❑ Use interpretation/reframing.

Intentional Competence. You will be able to use questioning skill sequences and encouragers to bring out meaning issues and then reflect meaning accurately. You will be able to use the client's main words and constructs to define meaning rather than reframing in your own words (interpretation). You will not interpret but rather will facilitate the client's interpretation of experience.

With interpretation/reframing, you will be able to provide clients with new and fresh perspectives on their issues.

❑ Use questions and encouragers to bring out meaning issues.

❑ When you reflect meaning, use the client's main words and constructs rather than your own.

❑ Reflect meaning in such a fashion that the client starts exploring meaning and value issues in more depth.

❑ In the session, switch the focus of the conversation as necessary from meaning to feeling (via reflection of feeling or questions oriented toward feeling) or to content (via paraphrase or questions oriented toward content).

❑ Help others discern their purpose and mission in life.

❑ When a person is hyperreflecting on the negative meaning of an event or person, find something positive in that person or event and enable the client to dereflect by focusing on the positive.

❑ Provide clients with appropriate new ways to think about their issues, helping them generate new perspectives on their behavior, thoughts, and feelings.

❑ Provide a new perspective via interpretation/reframing, using your own knowledge, and help your clients use these ideas to enlarge their thinking on their issues.

❑ Use various theoretical perspectives to organize your reframing.

Psychoeducational Teaching Competence

❑ Teach clients how to examine their own meaning systems.

❑ Facilitate others' understanding and use of discernment questioning strategies.

❑ Teach reflection of meaning to others.

❑ Teach clients how to interpret their own experience from new frames of reference and to think about their experiences from multiple perspectives.

❑ Teach interpretation/reframing to others.

Personal Reflection on Reflection of Meaning and Interpretation/Reframing

Meaning has been presented as a central issue in counseling and psychotherapy. Interpretation has been presented as an alternative method for achieving much the same objective but with more counselor involvement.

What single idea stands out for you among all those presented in this chapter, in class, or through informal learning? What stands out for you is likely to be important as a guide toward your next steps.

What are your thoughts on multicultural issues and the use of this skill? Are you able to find new meanings and reinterpret/reframe your own life experience? In particular, what have you learned about discernment and its relation to your own life?

What other points in this chapter strike you as important?

How might you use ideas in this chapter to begin the process of establishing your own style and theory?

Our Thoughts About Charlis

Eliciting and reflecting meaning are both skills and strategies. As skills, they are fairly straightforward. To elicit meaning, we'd want to ask Charlis some variation of the basic question, "What does the heart attack mean to you, your past and future life?" As appropriate to the situation, questions such as the following can address the general issue of meaning in more detail:

"What has given you the most satisfaction in your job?"

"What's been missing for you in your present life?"

"What do you find of value in your life?"

"What sense do you make of this heart attack and the future?"

"What things in the future will be most meaningful to you?"

"What is the purpose of your working so hard?"

"You've said that you have been wondering what God is saying to you with this trial. Could you share some of your thoughts here?"

"What would you like to leave the world as a gift?"

Questions such as these do not usually lead to concrete behavioral descriptions. They may often bring out emotions, and they certainly bring out certain types of thoughts and cognitions. Typically, these thoughts are deeper in that they search for meanings and understandings. When clients explore meaning issues, the therapy session, almost by necessity, becomes less precise. Perhaps this is because we are struggling with defining the almost indefinable.

As part of our work with Charlis, we'd ask if she wants to examine the meaning of her life in more detail through the process of discernment. This is a more systematic approach to meaning and purpose defined in some detail in this chapter. If she wishes, we'd share the specific questions of discernment presented here and ask her which ones she'd like to explore. In addition, we'd ask her to think of questions and issues that are particularly important to her, and we would give these special attention as we work to help her discern the meaning of her life, her work, her goals, and her mission.

Reflection of meaning as a skill looks very much like reflection of feeling or paraphrasing, but the key words *meaning, sense, deeper understanding, purpose, vision,* or some related concept will be present explicitly or implicitly. "Charlis, I sense that the heart attack has led you to question some basic understandings in your life. Is that close? If so, tell me more."

It can be seen that we regard eliciting and reflecting meaning as an opening for the client to explore issues where there often is not a final answer but rather a deeper awareness of the possibilities of life. At the same time, effective exploration of meaning becomes a major strategy in which you bring out client stories, past, present, and future. You will use all the listening, focusing, and confrontation skills to facilitate this self-examination. Yet the focus remains on the client's finding meaning and purpose in his or her life.

Self-Disclosure, Feedback, Logical
Consequences, Directives/Instruction,
and Psychoeducation

Reflection of Meaning and
Interpretation/Reframing

Empathic Confrontation

Focusing

How to Conduct a Five-Stage Counseling
Session Using Only Listening Skills

Reflecting Feelings

Encouraging, Paraphrasing, and Summarizing

Questions

Observation Skills

Attending and Empathy Skills

Ethics, Multicultural Competence, Neuroscience, and Positive Psychology/Resilience

CHAPTER 12

Action Skills for Building Resilience and Managing Stress

Self-Disclosure, Feedback, Logical Consequences, Directives/Instruction, and Psychoeducation

Blessed is the influence of one true, loving person on another.

—George Eliot

Do you want to know who you are? Don't ask. Act! Action will delineate and define you.

—Thomas Jefferson

Chapter Goals and Competency Objectives

Awareness and Knowledge

▲ Understand stress management and how the action influencing skills can be central in building resilience.

▲ Explore the nature of interpersonal influence, its specific skills, and our responsibility to work with a client on an egalitarian basis with an emphasis on listening before influencing.

▲ Further understand decision counseling and its relevance to influencing skills and action with varying clients and theories of counseling and therapy.

Skills and Action

▲ Facilitate client self-understanding and empowerment through self-disclosure and feedback.

▲ Enable the client to look at the possible positive and negative results of alternative actions (logical consequences).

▲ Present new information and ideas to clients in a timely and appropriate fashion—for example, career information, teaching about sexuality, and results of test scores (directives, instruction).

▲ Empower clients with specifics for action leading to physical and mental health through stress management. Help them restory and take concrete action in their issues (psychoeducation).

▲ Develop action plans collaboratively with clients to facilitate taking home learning and new skills from the session to the "real world."

Introduction: Action Skills for Resilience and Stress Management

There is an old Zen fable that goes something like this, updated for today.

> A woman is hiking along a California Sierra trail along the edge of a 15-foot drop. As she rounds a bend, she sees a bear, who starts to charge. Surprised but still able, she grabs a wild vine and swings over the edge. As she thankfully hangs and looks for a safe place to jump, she sees another bear below! There are summer strawberries growing on the vine so she decides to hold on with one hand and reaches for a few berries with the other. How sweet they taste!

Your clients face bears of stressful decisions. One bear promises one thing, while the other choice may bring something else. We can help clients taste the sweetness of strawberries and the importance of the moment before they jump.

This is the place for decisions to be taken into action. While all microskills emphasize skills and action, this influencing skill chapter concentrates on how you can encourage, support, and supply answers to help clients deal with stress on their journey to resilience. Stages 4 (restory) and 5 (action) are where you will use these skills most often. Of course, influencing skills will be most useful, and even more powerful, if used on a base of listening and empathic understanding. Throughout these two final stages of the interview, maintaining a solid egalitarian relationship with a focus on client goals is essential.

Clients come to us stressed, with pieces of their lives sometimes literally "all over the place." Understanding and managing the impact of stress is necessary for building client resilience, thus enabling effective, rational, and emotionally satisfying decisions. The "magic" of creative resilience comes from the spontaneous generation of something new out of what already exists. Restorying and decision making are creative practices. The action influencing skills are a significant part of enabling clients to take creative new thoughts, feelings, and meanings into real-world behavior.

This chapter begins with a review of stress and its impact on the client. The concept of allostasis is introduced as a key element of resilience. Specific theories and strategies for managing stress are reviewed. Then, building on this foundation, we present the action influencing skills of self-disclosure and feedback, logical consequences, directives, instruction, and psychoeducation.

Awareness, Knowledge, and Skills of Stress Management

Treatment is based on managing stress.

—Patrick McGorry, M.D., Ph.D.
Royal Melbourne Hospital

There's good stress, there's tolerable stress, and there's toxic stress.

—Bruce McEwen
Rockefeller Institute

Regardless of your approach to interviewing and counseling, you will constantly be working with clients who have some form of stress. This is logical when you think about it: How often do you go through a stress-free day? This section defines stress more precisely and discusses the impact of both positive and negative stressors on key parts of the brain.

As emphasized in Chapter 1, we need some level of stress to get ready for an exam, an athletic event, or a job interview. Stress is necessary for learning. Positive stressors make us happy and joyful in many ways. Examples are planning for a big date or marriage, being deeply involved at an opera or a baseball game, rock climbing, running, or driving fast on a racetrack. Positive stress can provide fulfillment, such as the satisfaction of a job well done, being able to help another person, graduating from a master's degree program, or helping build a Habitat for Humanity house.

But continuous, severe, day-to-day stress can be seriously damaging to physical and mental health. Or a single traumatic incident can accomplish this in a few seconds or minutes. Severe stress leads to many mental and physical health issues. It will be helpful if you return to Chapter 1, page 17, and view brain activation under severe stress. Also recall that the body responds to stress in ways that are physically damaging.

All of us may experience one or more common life stressors, such as unemployment, divorce, illness or the illness of a close family member, death of a parent, financial reversals, a sibling with a mental illness, credit card debt, and many others. All of these stressors have produced the following results, potentially even trauma:

- Forty-three percent of all adults suffer adverse health effects from stress.
- From 75% to 90% of all doctor's office visits are for stress-related ailments and complaints.
- Stress can play a part in problems such as headaches, high blood pressure, heart problems, diabetes, skin conditions, asthma, arthritis, depression, and anxiety.
- The Occupational Safety and Health Administration (OSHA) has declared stress a hazard of the workplace. Stress costs American industry more than $300 billion annually.
- The lifetime prevalence of an emotional disorder is more than 50%, often due to chronic, untreated stress reactions. (Goldberg, 2012)

Figure 12.1 shows how stress can be either growth and resilience producing or, at high levels, destructive. Appropriate levels of stress can "pump us up" to prepare for that exam or other challenge. Normal stress helps the brain grow through neurogenesis, resulting in new neurons and neural connections. But severe stress results in negative neurogenesis, with neural loss and possible long-term damage to the brain. This loss can occur with one really traumatic event (war, rape, a severe accident) or a continued series of chronic damaging stressors, such as bullying, racism, poverty, abuse, neglect, and even so-called "normal" stress in the workplace.

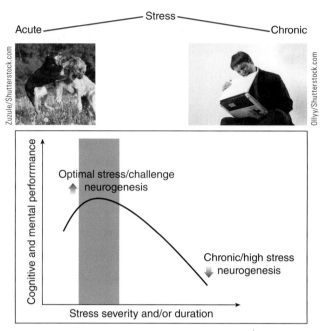

FIGURE 12.1 Optimal levels of stress contrasted with chronic stress.

Reflective Exercise Experiencing the impact of a simple stressor

With a partner: Try this exercise with a friend, taking turns as leader and follower. Have your partner close his/her eyes. Then say in a slow, even voice, "NO NO NO NO NO NO." The partner then opens his/her eyes and you debrief the experience. How was the word "NO" felt in the body? What thoughts and feelings occurred for your partner? You will find that repetition of the negative word "NO" has an almost immediate impact.

Now ask your partner to close his/her eyes again. In a slow, even voice, say, "YES YES YES YES YES YES." Open the eyes and debrief this experience, comparing it with "NO." Typically, it takes several "YES's" to relax a person from the multiple "NO's."

On your own: Close your eyes and visualize some error or difficult experience from the past, along with the word "NO." Take a moment to know both body and mind. Then follow this with a positive, joyful memory. Take some time to debrief.

This exercise provides some understanding of what even the simplest stressor can do to the body. Negative stressors over time produce negative neurogenesis.

A child or adult who receives a negative comment or goes through a difficult personal experience easily becomes stressed. These negative events have an impact on the brain and imprint memory at a deeper level than most positives. Some say that it takes 5 to 10 positives to counteract one negative comment. If there is a trauma, even 10 will not be enough, and the negative memory may take over one's life. A single negative—to a child "You're fat and a freak" or "We don't want to play with you"; or something as simple as a friend saying with a quizzical expression, "Your hair looks different today"—can ruin the whole day or more, no matter what else happens. Any number of direct or subtle put-downs that we all experience build a negative self-concept.

Our brains are organized for safety, and stressful negative events can become dominant. Clients who have a negative view of themselves can be encouraged to think about positive events and experiences, perhaps even making a journal of strength and resilience. Usually, they are surprised at how many positives have occurred without their full attention. Similarly, when we anticipate a negative experience, we can protect ourselves partially by deliberately planning at least three positives during the day before it happens. If the negative experience is anticipated, rather than an unpleasant surprise, the impact will be less powerful and damaging.

Those who find themselves coping with trauma, bullying, and the many forms of microaggressions will require special help in building resilience. The influencing action skills of this chapter are all oriented toward coping with both normal daily stressors and those that have been deeply embedded in one's view of self and the world. Whether the stress is positive or negative, its impact on the body and learning is impressive. Box 12.1 shows how the body reacts to stress.

BOX 12.1   The Stress Response System

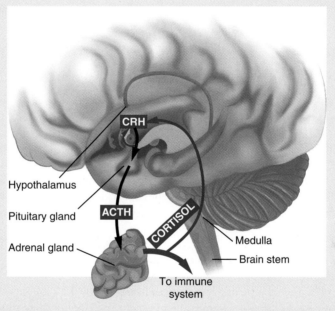

Critical for us to understand in stress management is the role of hormones in the stress response system associated with the emotional limbic system: the hypothalamus, pituitary, and adrenals (HPA) (see also Appendix IV). HPA hormones are activated by the energizing amygdala, and in times of emergency, fear, or anger, they may bypass prefrontal executive functioning. This protects us when we have no time to think and make decisions—when we see a snake, a fist coming at us, or a car swerving directly into our lane. We need the fast-acting limbic and stress systems, or we as a human species would not be here.

The *hypothalamus*, the H of HPA, is the master gland controlling hormones that affect heart rate, biological body responses to emotions such as hunger, sleep, aggression, and other biological factors.* The *pituitary* is another control gland that receives messages from the hypothalamus and influences growth, blood pressure, sexual functioning, the thyroid, and metabolism. The *adrenal glands* produce *corticosteroids*, including *cortisol*.†

Under significant negative emotional stress of many kinds, cortisol is potentially damaging. High cortisol levels from prolonged or chronic stress produce side effects such as cognitive and memory impairment, increased blood pressure, blood sugar imbalance, lowered immunity, and inflammatory responses. However, the normal stressors of a challenge, an examination, or a race produce appropriate levels of cortisol, which result in positive effects like improved memory, reduced sensitivity to pain, and increased sustained energy (Lee & Hopkins, 2009).

---

* *Corticotropin-releasing hormone (CRH)*, released from the hypothalamus, is a neurotransmitter involved in the stress response. Major depression and Alzheimer's disease are noteworthy for increased CRH production.

---

† *Adrenocorticotropic hormone (ACTH)* is produced by stress in the adrenal glands producing cortisol. Overproduction weakens the immune system and can affect gene functioning.

BOX 12.2	Example Stress Management Strategies

Theory and Methods Associated with Stress Management	Associated Primarily with Therapeutic Lifestyle Changes	
Cognitive behavioral and other theories of counseling Multicultural approaches Psychoeducation Socials skills training Assertiveness training Conflict resolution Gestalt exercises Biofeedback Neurofeedback Positive reframing Thought stopping Imagery, guided imagery Time management Relaxation training Action influencing skills Action planning	**The Big Seven TLCs** 1. Physical exercise 2. Nutrition 3. Sleep 4. Social relations 5. Cognitive challenge 6. Meditation 7. Cultural health **Other highly useful TLCs** Control screen time for TV, games, and computers Prayer Positive thinking/optimism No drugs/limited alcohol	Religion/spirituality/strong value system Taking a nature break rather than a coffee break Enjoyable hobbies of any type: art, music, collecting, cards, reading, etc. Helping others, social justice action Careful use of medications and supplements No smoking

Stress management traditionally has served as a *remedial* treatment for already stressed and needy clients. Stress management strategies such as therapeutic lifestyle changes (TLCs) and many other therapeutic techniques are typically focused on treatment and prevention of stress. Using the skills and strategies of counseling and therapy, we resolve issues and build resilience.

Box 12.2 presents a list of common stress management and TLC instructional strategies that can be used in counseling and clinical practice. Some of these will be elaborated briefly in this chapter.

The therapeutic lifestyle changes listed in Box 12.2 are, of course, central in Chapter 2 (pages 45–51). Being aware and competent in bringing these to the therapy session with clients—and the seriously distressed—is now a necessary set of skills for all of us. They are basic to mental and physical health. They are all effective, well-documented and researched stress management strategies.

Exercise is generally regarded as the number one TLC and is discussed in detail below. The section on psychoeducation at this end of this chapter further discusses TLCs. You will want to become skilled and competent on their use, both for yourself and for your clients.

Physical Exercise: Perhaps the Most Important Stress Management/Therapeutic Lifestyle Change for Client Mental and Physical Health

Evidence is mounting for the benefits of exercise, yet psychologists (and counselors and physicians) don't often use exercise as part of their treatment arsenal.

—Kirsten Weir

Just do it.™

—Nike

A central goal of stress management is to get blood flowing to your brain and body. Exercise increases brain volume, has been found as effective as meds for mild depression, may prevent cancer, and may slow the development of Alzheimer's disease. The best summary of exercise research and the implications for our practice may be found in John Ratey's (2008b) *Spark: The Revolutionary New Science of Exercise and the Brain.*

We love and work more effectively if we are comfortable in our bodies. A sound body is fundamental to mental health. It is recommended that adults get at least 150 minutes of exercise each week—20 minutes a day, which is often 20 minutes more than people claim that they do. Just 15 minutes of daily exercise resulted in a 10% decrease in cancer death and 14% decrease in death overall. This means that exercise can give you, on average, an additional three years of life (Weir, 2011).

Exercise and mental health was the focus of the cover story of the American Psychological Association's *Monitor on Psychology* for December 2011 (Weir, 2011). The most powerful study on exercise to date found that exercise can overcome genetic issues that lead to brain atrophy and depression. See Figure 12.2 for an illustration of the effects of exercise on the brain. Moreover, aerobic exercise can increase beneficial BNDF (brain-derived neurotrophic factor). BDNF has been referred to as "Miracle-Gro for the brain." It is critical for the growth of new neurons and synapses, especially in the hippocampus, cortex, and basal forebrain, which are all central to learning, memory, and higher thinking. This increase in BDNF leads to neurogenesis, a larger hippocampus, and less depression. This has been shown to be true in both mice and humans (Erickson, Miller, & Roecklin, 2012). The authors summarize their research findings with the following conclusions:

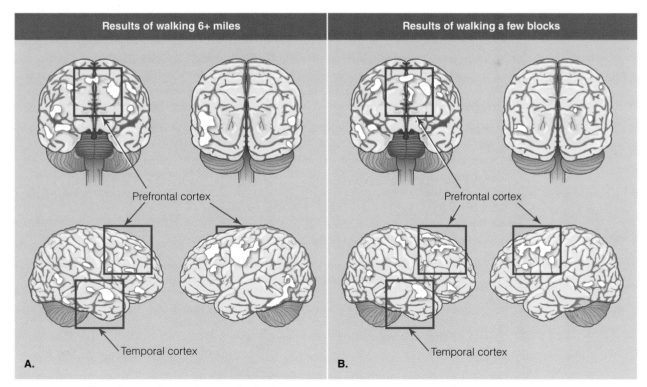

FIGURE 12.2 The effects of exercise on the brain

- Aging and lower BDNF produce lower hippocampal volume and thus poorer memory function. This, in turns, leads to increased likelihood of depression and possibly Alzheimer's.

- Exercise increases BDNF, serotonin, and hippocampus volume through neurogenesis. This results in improved memory and reduction or elimination of depression.

- Walking a few blocks weekly is useful, but 72 blocks (6–9 miles) is more effective.

- About 25% of the U.S. population gets no exercise at all.

- Exercise enhances mood. In some studies, clients with depression often do better with exercise than they do with medication. Those who continue to exercise do better than those on medication.

- Exercise is beneficial to people with cancer, diabetes, multiple sclerosis, and other physical challenges.

- Exercise is useful in the treatment of anxiety and panic disorders.

- Smoking cessation can be more successful with exercise.

- Mice and humans share similar reactions to stress. Researchers found that "bullied" mice without the opportunity to exercise in enriched cages and hassled by dominant alpha mice hid in shadows and showed signs of anxiety and depression. However, mice in enriched cages with considerable exercise were able to "shrug" off bullying and social defeat, did not show the negative signs, and handled mazes competently.

Other research studies with older adults clearly show the benefits of exercise. One found increased plasticity in older adults who exercise (aerobic training and walking). They were examined using fMRI and were found to have increased executive functioning and better connections in higher-level networks (Voss et al., 2010, 2013). A study of African American women at risk for cognitive loss found the same results, with increased activity in the prefrontal cortex (Carlson et al., 2009). These findings have relevance for your clients, regardless of age.

First, let us examine self-disclosure and feedback, which are drawn from your ability to listen to client stories and understand the meaning of their thoughts, feelings, and behavior. Next follows natural and logical consequences, and then the influencing skills of directives, instruction, and psychoeducation.

Awareness, Knowledge, and Skills of Empathic Self-Disclosure and Feedback

Feedback is the breakfast of champions.

—Kenneth Blanchard

Before I open up to you (self-disclose), I want to know where you are coming from . . . . In other words, a culturally different client may not open up (self-disclose) until you, the helping professional, self-disclose first. Thus, to many minority clients, a therapist who expresses his/her thoughts and feelings may be better received in a counseling situation.

—Derald Wing Sue and Stanley Sue

The skills of self-disclosure and feedback are close to the listening skill of summarization in that all require the counselor and therapist to seek holistic understanding of what may be going on in the client's mind—*empathic mentalizing*, integrated cognitive and affective empathy. However, self-disclosure and feedback move beyond the listening summary to add to client awareness. They are all whole brain activities that require special sensitivity and awareness of the client's world in both here-and-now and there-and-then conversation.

Before counselors share a personal self-disclosure or provide feedback, they need to have a solid understanding of where they are in the relationship that they are building with clients. Well-done self-disclosure early in the interview can smooth the process of relationship building. Later in the session, if you have similar experience to what the client has said, self-disclosure may make clients more comfortable. But if your life experience is distant from that of your clients, it is best to hold back, as you can lose trust. Timely feedback on what the client has said in the interview or has done in the recent or distant past can be helpful, but only if the client is ready. Mentalizing, seeing the interview and the client holistically in a timely fashion, is basic.

Following are brief definitions of the two skills and how the client might respond when they are used appropriately.

Empathic Self-Disclosure and Feedback	Anticipated Client Response
Both skills are used sparingly and only when the client appears to need more counselor involvement and support. They require careful listening and understanding before sharing your thoughts.	
Self-disclosure is sharing your own personal experience related to what the client has said and often starts with an "I" statement. It can also be sharing your own thoughts and feelings concerning what the client is experiencing in the immediate moment, in the here and now.	Clients respond well to carefully said self-disclosure, especially at the beginning of a session. They are often pleased to know more about you at that point. Later in the session, sharing your thoughts and feelings about the client can enable them to talk more openly about their issues. Self-disclosure almost always needs to be positive and supportive.
Feedback presents clients with clear, nonjudgmental information (and sometimes even opinions) on client thoughts, feelings, and behaviors, either in the past or in the here and now.	Feedback can be supportive or challenging. Supportive feedback searches for positives and strengths, while challenges ask clients to think more carefully about themselves and what they are saying.

The specific skills of self-disclosure and feedback are presented below.

- *Listen first.* Be fully empathically aware so that you can determine whether the client is ready for self-disclosure or feedback. If the clients solicit either by asking questions, they are more likely to respond positively. (Listen—provide self-disclosure or feedback—use checkout.)

- *Be brief and concrete.* Self-disclosure and feedback need to be short and specific. Then immediately returning focus to the client (for example, "How does that relate to you?"). Share and describe thoughts, feelings, or behaviors *briefly*. With self-disclosure: "I can imagine how much pain you feel." "I also grew up in an alcoholic family and understand some of the confusion you feel."

- *Use "I" statements.* Both skills frequently use "I" statements, including the pronouns *I, me,* and *my,* or the self-reference may be implied. With feedback, the focus is on the client rather than on yourself, and "you" statements focusing on the client are central. However, "I sense that you . . .," "Could it be that you . . .," and similar others make the same point more tentatively.

- *Be authentic and nonjudgmental.* Are your self-disclosures and feedback real? Don't make up things. Feedback needs to be nonjudgmental and come from the heart, with full understanding of where the client is coming from. It needs to be authentically honest, but also intentionally flexible so that the client can hear you and, hopefully, use the feedback for better self-understanding. With feedback, we are asking clients to think about self more fully (develop their own personal theory of mind and ability to mentalize about themselves).

- *Use appropriate immediacy and tense.* The most beneficial self-disclosures and feedback are usually made in the here and now, the present tense. "Right now I feel . . . ." "I am hurting for you at this moment—I care." "Let me offer some feedback—I hear you as hurting, but I sense your developing strength to cope with the issues." Focusing on the here and now brings immediacy. When speaking to the there and then of the past or future, seek to relate these to the moment when you are talking with the client. "Right now, I hear you saying you are hurting over what your partner said." "I get a feeling in my own stomach you are hurting." "We've talked about this before, and you clearly can live through a lot of stress and then bounce back." Immediacy is a vital part of the successful use of other skills and theories—being in the here and now often makes for a richer, more involving interview.

- *Consider cultural implications and explore differences.* You cannot expect to have experienced all that your clients bring to you. After an appropriate time for introducing the session, when a man works with a woman, for example, it may be useful to say, "Men don't always understand women's issues. The things you are talking about clearly relate to gender experience. I'll do my best, but if I miss something, let me know. Do you have any questions for me?" If you are White or African American working with a person of the other race, frank disclosure that you recognize the differences in cultures at the initiation of the session can be helpful in developing trust. Encourage the client to ask you further questions at any point in the interview.

Observe: Self-Disclosure and Feedback

Transcripts of interviews with a client named Alicia will be presented throughout this chapter as we follow her growth in self-confidence and significant moves to change and action. Alicia works as an assistant administrator in a community agency, funded by the local government and a few grants. She comes to the counselor upset and discouraged by a difficult situation at work.

The presenting issue concerns Alicia's increasing inability to speak up and share her ideas with her boss, particularly if she disagrees with him. As often happens, out of this come other and more challenging issues that Alicia is dealing with, including other difficulties in speaking up and, eventually, the real issue of dealing with her partner, Simon—whether she should stay with him or not. Each microskill will be discussed in relation to this client.

The following is an excerpt from the later stages of the second interview, in which the counselor, Onawumi, uses self-disclosure and feedback. Alicia is speaking of new difficulties with her manager, who has started "hitting" on her.

Counselor and Client Conversation	Process Comments
Stage 4: Restory	
53. *Onawumi:* Alicia, as we've talked these last two sessions, you come across as very able and sure of yourself. In this situation, you are not so sure. Now your boss, Jackson, is hitting on you, and you're scared for your job but still angry at what he is doing. The anger makes a lot of sense.	Onawumi starts here with brief feedback on Alicia's strengths, followed by contrasting her abilities with her current fears and anger. Two major discrepancies are confronted here: (1) internal strengths shown in the past, as contrasted with current fears and indecision; (2) the conflict with her boss, including her present silence and the need to keep her job. Onawumi also provides feedback on the validity of Alicia's anger, giving less attention to the fear. (Additive)
54. *Alicia:* That's right, what am I doing? What can I do?	As Alicia talked earlier about this issue, it was clear that she was moving between Levels 1 and 2 on the Client Change Scale (CCS). At times she denied that anything was happening, but then was able to examine the issues partially. Finally, she has moved to Level 3, showing clear awareness of what was happening. The restorying process continues below.
55. *Onawumi:* You've got considerable strength, and women go through this too often, but something can be done. What occurs to you?	Feedback again with the cognitive/emotional word *strength*, but this also provides feedback on other women. She then tosses the decision back to Alicia, thus providing both respect and potential empowerment. (Additive)
56. *Alicia:* Well, talking with you is helping. I think I see what's going on more clearly. I think I can pull myself together after all. I know that Jackson does like the work I am doing. I think I got that raise because my work is good, not because he had his eye on me.	Alicia is drawing on the positive feedback and showing more certainty about herself and her abilities. She has moved to Level 3 on the CCS, and her self-view and her story are changing.
57. *Onawumi:* So you know he respects your work; and I gather from what you say, you are holding an important project together for him and the company. Clearly you have some leverage in that. Go on, I think you are on the right track.	Cognitive paraphrase, feedback. This helps cement and firm up the changes we are seeing in cognition. (Additive)
58. *Alicia:* I wonder what would happen if I sat down with him and reviewed the present status of the project and how well things are going. He has never directly really forced or embarrassed me, but the hints are really there. Perhaps, after we review the project, I could simply say that I like working with him but feel that we must keep our relationship on a professional level. I've not been comfortable with a few things he has said, but I still respect him. Would he mind if we kept our focus on the project?	Recall that this did not happen that fast; only in the last 15 minutes of the session did Alicia show real movement, thanks to primary use of listening skills plus feedback and self-disclosures. Here she is at CCS Level 3, but moving toward Level 4.
59. *Onawumi:* I'm impressed with your view of the situation. Well done. If he is a reasonable person, that likely could work. You are giving him respect, but still standing up for yourself. Let's try your idea in a role-play. I'll be Jackson, and you go through and practice what you might say.	Positive and appropriate self-disclosure followed by feedback to Alicia on her ideas, followed by a directive for testing them out in a role-play.

The role-play then follows, and Alicia clearly summarizes the project and her feelings about the sometimes tense relationship. As she presents the situation, she firmly maintains her "cool," but also shows respect for the boss. She feels good about the way she handled the role-play. (See Box 12.4 later in this chapter for further information on role-plays.)

(continued)

Counselor and Client Conversation	Process Comments
Stage 4: Restory	
60. *Onawumi:* That was great, Alicia. I think that is about what you can try. Let's hope that it works. On the other hand, I can say that I've been in a similar situation, and I tried to do what we just role-played. He did not say a word, but he did start leaving me alone. Gradually things got better—the same can happen for you. Nonetheless, I wonder if I lost some power. You've got some ideas that might well work—but what if they don't?	Feedback followed by a self-disclosure. The self-disclosure of a parallel situation in Onawumi's life is helpful. Nonetheless, it is also useful for Alicia to explore the possible negative logical consequences of speaking up. (Additive)
61. *Alicia:* Oh, I know that he might become angry, but it's good to know that you had to deal with the same thing.	More indication of Level 4 responding. Note that the feedback and self-disclosures have encouraged Alicia to take action.
62. *Onawumi:* Alicia , women run into this all too often, and we have to start standing up for our rights. I think this boss can take it and may have to. So, what if it doesn't go well?	Focus on cultural/environmental context (CEC). Educational information giving, followed by an open question asking Alicia how she would react. (Potentially additive)
63. *Alicia:* Actually, I'm not as scared for my job as I was. I think I'll get through this. I have to stand up. He needs me for this project. And if it doesn't work out, I'm ready to leave as I know that I can find something else.	Level 4 on CCS. (The recent interventions have been additive.)
64. *Onawumi:* That makes me feel good that you are able to take that risk. And my sense is that you have the power and wisdom to make it work, no matter what the result is . I'll be waiting here to learn what happened. You are doing the right thing for you. Now, as we wind up, could we set up an action plan for you to take home and test out what we have discussed today?	Self-disclosure and feedback. The session is closing, and shortly Onawumi will work with Alicia on an action plan.

Developing an Action Plan with Alicia

We suggest that you continually recognize client uniqueness as you work collaboratively to outline action plans or homework. Box 12.3 presents the action plan that was decided as Onawumi worked collaboratively with Alicia.

BOX 12.3 Alicia's Action Plan

Past Thoughts, Feelings, and Behavior (Story)	Goal	Action Plan/Homework
Hesitant to speak to boss about unfair behavior. Typical of my behavior.	Speak up and change behavior to get positive results.	Do what we practiced. Stand up straight, change vocal tone, don't hurry. Hang in, don't give up immediately when challenged—be assertive.
Internal thoughts/cognitions focused on myself and my past failures.	Focus cognitions on my strengths and resources. Focus more on boss and his reactions.	Work on positive thoughts, and past successes; change from "I can't" to "I can."
Feelings in such situations are fear, but underlying anger at the way I am treated. My body even feels tense and awkward.	Move from negative thinking and nervous reactions to being more relaxed, more sure of myself, and in control of my body and enjoying it.	Remember my positive feelings of accomplishment, ratification, and pride in my job. Focus on my body—slow and relax, breathe normally.

In the next session with Alicia, Onawumi finds that her client was able to speak up and begin to resolve her issues. She was able to change her behavior and achieve her main goal. However, despite her focus on strengths and resources, many of her cognitions and affects were still not fully formed in a more positive fashion. In the next session, you will see that she has similar difficulties with a difficult, perhaps nasty, mechanic working on her car.

Awareness, Knowledge, and Skills of Natural and Logical Consequences

Not to decide is to decide.

—Harvey Cox, Theologian

Those in difficult situations can gain from understanding the consequences of their behavior and decisions—whether college decisions, work issues, or family relations. Clients facing possible changes in life direction will often profit from exploring the logical consequences, positive and negative, of change. In virtually all counseling and therapy, you will be facilitating the quality of client decisions. This includes not only decision counseling, problem-solving therapy, and motivational interviewing, but also many other approaches such as client-centered counseling, cognitive behavioral therapy, brief solution-oriented counseling, and feminist and multicultural counseling and therapy.

If you use natural and logical consequences, you can anticipate how clients will respond. If they do not response as expected, turn next to your ability to be intentional and move to other skills.

Natural and Logical Consequences	Anticipated Client Response
Explore with the client specific alternatives and the logical positive and negative concrete consequence of each decision possibility. "If you do _____ , then _____."	Clients will change thoughts, feelings, and behaviors through better anticipation of the consequences of their actions. When you explore the positives and negatives of each possibility, clients will be more involved in the process of making their creative new decision.

This strategy of logical consequences is most often used to help people sort systemically through issues when a decision needs to be made. With a complex decision, many clients find it useful to rank alternatives. The strategy of logical consequences was developed first by Alfred Adler in 1924 and has been continually emphasized by key Adlerian writers (e.g., Cox, 2015; Dreikurs & Gray, 1968; Sweeney, 1998).

The interviewer or counselor facilitates awareness of potential logical consequences of actions. Some examples include the client who is thinking of dropping out of school, the pregnant client who has not stopped smoking, or the client who wants to "tell off" a boss. Dropping out of school has serious consequences for now and later life. The baby could very likely be born less healthy. The client who talks back to the boss may lose the job.

It is equally important to help clients anticipate the results and rewards of good decisions. The pregnant woman's baby is likely to be healthier if she stops smoking; the client who graduates from school will probably find a better job. We likely will do better in the long run if we do not tell off the boss.

Clients can make better decisions when they can envision the likely consequences of any given action. Note in the following examples that adding the words "you decide" gives power to clients, thus showing them that they can take charge rather than letting others rule what they decide to do. Adler was concerned with helping individuals grow, rather than just resolving issues. Think of encouraging clients to take their own independent actions via "you decide."

> *Counselor:* What is likely to happen if *you decide* to continue smoking while you are pregnant?
>
> *Client:* I know that it isn't good, but I can't stop and I really don't want to.
>
> *Counselor:* Again, what are the possible negative consequences of your *deciding* to continue smoking?
>
> *Client:* (pause) I've been told that the baby could be harmed.
>
> *Counselor:* Right; is that something you want? What is the benefit of *deciding* to stop smoking for the baby?
>
> *Client:* No, I don't want to do harm. I'd be so guilty. But how can I stop smoking?
>
> *Counselor:* If *you decide* to stop smoking, the positive consequences are that your baby can grow in better health. Let's explore that decision. And to make this happen, to stop smoking, the *your next decision* is to *decide* which system works best for you. None of them will be easy. But let us consider . . . .

Sometimes in school and in the criminal justice system, the client is forced to come to you. In such situations, more power rests with you than in other interviews, but it can also build client resistance. The court may ask the interviewer to recommend actions that the legal system could take. Warnings are a form of logical consequences and may center on *anticipation of punishment*; if used effectively, warnings may reduce dangerous risk taking and produce desired behavior. At the same time, these clients need to be fully aware of the many positive consequences of changes in behavior. Help the client see that "*It is your decision you decide* what you are going to do."

Consider the following suggestions for using the strategy of logical consequences.

- Through listening skills, make sure you understand the situation and the way your client understands it. Draw out the story, summarize, and encourage the client to summarize what is happening.

- Use questions and brainstorming to generate alternatives for resolution. If necessary, provide additional ideas.

- Nonjudgmentally, outline with the client the positive and negative consequences of any potential decision. In addition, thinking ahead to the long term may be helpful. "Imagine two years from now. What will your life be like if *you decide* to choose and act on what we have discovered."

- Encourage client decision making as much as possible.

Observe: Case Study Applications of Natural and Logical Consequences

In the first two interviews, Alicia discussed general feelings of distress because of her inaction. In the third interview, Onawumi was pleased to see that the second interview homework/action was helpful and that Alicia had worked out most of the issues with her manager. Alicia wanted to go further with her difficulties with self-expression. It was decided to look at her interactions with her garage mechanic, Jon, who is ignoring her requests to fix her car—constantly putting her off.

Onawumi, using primarily self-disclosure and feedback skills, had already helped Alicia see and understand the need to change her thoughts and behaviors. Onawumi saw that counseling was building resilience in her client, but realized that further efforts were needed to help her along the road to resilience. Onawumi sensed that something else was still going on. We now come into the middle of the third session, and Alicia wants to talk about difficulties with a garage mechanic, Jon, who does not listen to her. As a result, she has real trouble getting her car repaired.

Alicia has talked about her difficulties with her boss, now the mechanic. Later she turns to even more complex issues with her partner, Simon.

32. *Onawumi:* To sum up so far, we see that you have a clear idea that you often *decide* to allow others to run over you and it makes you feel bad about yourself—frustrated and discouraged—but now you feel better, stronger, more sure of yourself, and confident. You look relaxed and happier. But the mechanic you are struggling with is another hassle. Have I heard you correctly? You are beginning to sense your power and strengths, but you want to do more—and mechanic Jon is the next place to go.	Summary of both cognition and affect. Onawumi is also demonstrating her ability to *mentalize* empathically—see the world as Alicia experiences it. Sometimes it is wise to develop trust as you counsel clients on smaller issues, but others will want "to get right to it." From the summary, we see the Alicia has reached Level 4 with her manager, as things are now smoother, but she also realizes that more needs to be done. (Significant and timely summaries such as this are usually additive empathy.)
33. *Alicia:* Exactly, you've got it perfectly. I think we should continue with my "friend" Jon, the mechanic (stops and smiles, recognizing that this is a small joke, as she feels just the opposite). I need help on following through on this decision.	Alicia has been influenced by Onawumi's warmth and listening skills. She is beginning to become empowered and aware that she is the one who makes the decisions, moving her from "other-centered and fearful" to increased strength. That Alicia can joke a little bit is an indication of increasing trust and self-confidence. (CCS Level 3)
34. *Onawumi:* Alicia, what are the consequences for mechanic Jon if *you decide* to speak up more forcefully? What is likely to happen?	Logical consequence. Two questions with focus on the concern and the key other person. (Potentially additive)
35. *Alicia:* Hmmm. Well, I imagine he would do one of three things. First, he might ignore me and continue, but I wouldn't allow that, as I want him to deal with me. Second, I bet he'll do a better job, and perhaps he will respect other women as well. The third possibility is that he will talk back to me rudely. But if so, I'm going to talk to the manager of the garage. I'm fed up. It is time that I took charge of my own decisions.	Here Alicia is looking at the logical consequences for the mechanic Jon if she becomes more assertive. (It was additive. Alicia seems now to be at Level 3 on the CCS, but will she be able to change her behavior and move to Level 4?)

Then the positive and negative possible consequences were role-played several times until Alicia was clear about what she might do to obtain a positive consequence. This included the decision to speak up without alienating mechanic Jon so that he would not stop work on her car. Basically, it was another application of the self-disclosure and feedback session above, but this time Onawumi focused more clearly on the skill of logical consequences.

The interview continues for several minutes, but then shifts to her relationship with her partner, Simon—perhaps the most important issue of all.

51. *Onawumi:* And what are the consequences if *you decide* to say that you want counseling because you want more equality in the relationship with Simon?	The two have started talking about the relationship. This brief paraphrase is almost a summary, as it focuses on the central issue.
52. *Alicia:* I think Simon will be put off; he is not very verbal and fears counseling. But I also know that he would like us to get along better.	Alicia can make a better decision if she anticipates what her decisions mean for the future. She is hoping for a positive consequence.
53. *Onawumi:* So, Simon might accept it. It does sound as if you want to be a stronger woman and *more decisive*. Your grandmother was a powerful role model, and you did say that talking back to Jon at the garage might be a strike for women in general, as well as for yourself.	Exploration of broader consequences of real change. Focus on Simon, then on family, and finally, the cultural/environmental context (CEC). Onawumi's counseling lead here is oriented to both Alicia and the issue of women in society.

(continued)

54. *Alicia:* One thing that I'm learning here is that *every woman has a responsibility to decide* to speak up. I need to be part of that.	This is a real move toward Level 4, but still not there until implementation of new meanings. We are seeing cognitive and emotional change, generating new solutions. But Alicia has to implement her thoughts and feelings in behavior to reach this level fully.
55. *Onawumi:* It makes me feel good to hear you say that you see your responsibility for others. That may help you "hang in" when the going gets tough as you change your style.	Self-disclosure followed by a reframe on the logical consequences for women in general as Alicia changes her style. (Additive)
56. *Alicia:* Thanks, I feel better. I wonder if should spend the rest of the time talking about Simon, my partner.	As she gains some strength from the session, Alicia loosens up further and now trusts Onawumi enough to talk about the real issue that brought her to counseling.
57. *Onawumi:* Tell me something about what is going on with you and Simon.	Implied open question that is slightly more directive, usually achieves the same result—"Tell me more . . . ."

Despite the earlier interviews, Onawumi had not realized previously that Alicia and Simon had been together for three years and now were contemplating marriage. She realized that the valid concerns with the boss and mechanic Jon were ways for Alicia to see if she could really trust the relationship and discuss deeper, more serious concerns. The fact that we finally get to Simon illustrates that what we consider a trusting relationship actually may be a time of testing us to see if it is safe to discuss the real issues.

Onawumi now suspects that the concerns with mechanic Jon and the manager may have started with difficulties at home. Onawumi realizes that it may take several interviews before the main issue appears.

When Alicia came back, Onawumi was pleased to know that learning in the interview had been transferred to mechanic Jon with some success. Alicia said that the third effort to speak up and reason with him finally reached him and they came to an agreement about the car, which now is running perfectly (Level 4 on the CCS).

However, Alicia soon wanted to talk about what had happened recently, and she reviewed the homework/action plan. Things had actually gotten worse during the week. Simon became upset and angry when she spoke more assertively and almost hit her, telling her to "shut up and sit down." After that, there was little conversation for the rest of the day. Simon sincerely apologized the next morning. Alicia said, "That is what always happens, he didn't hit me this time, but he pushed me down, he shouted. I got terribly frightened, and I wonder if it is only going to get worse."

With this, Onawumi brought out more stories from Alicia about the relationship as it became more traumatic through the months and years. Alicia still had deep positive feelings for Simon and was frightened to leave. This was obviously going to be a difficult and emotional decision. She wondered if it indeed was worthwhile to confront him and ask him to go to counseling. Onawumi's response was to determine the level of danger. If it was high, she knew how to refer Alicia to the safe house in town. As it turned out, Alicia still felt relatively safe, but she decided that she needed to explore whether or not she should leave.

Later in the session, Onawumi introduced the Cognitive/Emotional Balancing Sheet, a systematic review of the logical positive and negative consequences of any decision, short or long term. This could range from a behavior change to a decision about a new job, coping with bullying, or how to deal with racial, gender, or sexual orientation harassment.

Natural and Logical Consequences and the Cognitive/Emotional Decision Balance Sheet

At first glance, we think that decisions are primarily cognitive. But very few of us will be satisfied if our decisions reflect only rational cognitive processes. Decisions require emotional energy, and this is a critical part of establishing newly created decisions in our long-term memory. Eliciting and reflecting feelings throughout the balance sheet is vital.

The idea of a balance sheet was first developed by Yale University scholar Irving Janis in 1983 and has since become a formal and regular feature of both decision counseling and, more recently, motivational interviewing (Miller & Rollnick, 2013). Each alternative decision is written down with a list of gains and losses—the logical consequences of each action.

Table 12.1 shows Alicia's Cognitive/Emotional Decision Balance Sheet as she explores what to do about Simon and considers the logical consequences of her decisions. Her we see in a major way that Alicia now has full trust in the Onawumi and is sharing at a deeper level.

TABLE 12.1   The Cognitive and Emotional Decision Balance Sheet*

List below the logical positive and negative factual and emotional results for each of the possible alternatives. When there is more than one alternative, make a separate Cognitive and Emotional Balance Sheet for each one. However, when the key decision is clear, the balance sheet can serve as basis for future decisions.
The decision: *What happens if I leave my abusing partner?*

What are the possible gains for me?	What are the emotional gains for me?	What are the possible gains for others?	What are the emotional gains for others?
Abuse will stop and I won't get hurt.	I won't be so scared.	My mom won't have to worry and talk to me on the phone constantly.	Mom will be so relieved that it's over.
I'll be able to move on with my life.	Perhaps I can return to feeling good about myself.	My mom would like to help.	She'd feel that she is important to me again.
I can be myself.	I would feel OK again and that would be a relief.	When I feel better about myself, others will develop respect for me.	They may listen to me. They likely will enjoy me more and be happy that I am feeling better.
What are the possible losses for me?	**What are the emotional losses I might face?**	**What are the possible losses for others?**	**What are the emotional losses for others?**
I'll be on my own.	This frightens me as much as staying.	They may feel they need to help me.	They'll be happy to see him gone.
How can I finance things by myself?	This terrifies me.	My parents may have to support me for a while.	They aren't that well off, and they told me not to go out with him. They may be angry, even though they'll help.
I still love that man, despite it all.	I'll be lonely.	My friends will be there for me.	They'll be glad for me and listen.
He might follow me, and that might make it worse.	I'll have no future and be totally alone.	My counselor is there to advise and support me.	I can sense that I'm not as alone as I might think I am. I feel supported and cared for.

*Adapted from Leon Mann (Mann, 2001; Mann, Beswick, Allouche, & Ivey, 1989); also see Miller and Rollnick (2002).

The arguments for staying in an abusive relationship are emotionally powerful, despite Alicia' anxiety and fear. This is why your support in such cases—and encouraging clients to move toward safety—is particularly significant. The earlier sessions helped make Alicia become more resilient, self-confident, and strong. This new power will be vital for coping with this more serious and life-changing issue.

Alicia ended the session saying that she still cared for Simon and that she was fearful of leaving. Ultimately, she still wanted the relationship to work out. As she explored the negative consequences of speaking more forcefully, she realized the first negative consequence was that her partner might really hurt her. The financial challenge and the likely need to find a new and much less expensive apartment emerged as additional negative consequences. Alicia also feared being alone, as she has had other bad experiences with loss. These are common experiences around separation and unfortunately often result in an abused woman's returning to her abusing partner or spouse.

On the more positive side, Alicia realized how good it would feel if *she decided* to speak up for herself, and she hoped that it eventually might be possible. The success with mechanic Jon and her boss helped to realize that she was not as helpless as she feared. She would feel better about herself if she could learn to take decisive positions. As she balanced the positives and negatives, *she decided* she would wait for a better time to talk with Simon, one of the times when they were getting on well, which still did happen. She could not leave the relationship now, although she was fully aware of the possible consequences. A homework/action plan was generated. Alicia decided to write a positive balance sheet on what could happen if she stayed with Simon. She felt that this would be helpful in her final decision making. As part of the action plan, she also decided to share her concerns with a good friend, but not yet with her mother. Again, it was agreed that observation of the behavior between the two of them was important. Finally, Onawumi asked Alicia to write a strength inventory of past and present successes.

Incidentally, Alicia had developed sufficient relationship and trust to go on changing and growing, particularly as she was now becoming aware that "Alicia is the *decider*." That word coined by President George W. Bush is a good one to use from time to time. *Deciders* are empowered; becoming a *decider* is one worthy goal for clients.

Awareness, Knowledge, and Skills of Directives, Instruction, and Psychoeducation

The microskills presented here are the most active and influencing. It is here that the counselor or therapist is in danger of taking too much power. At the same time, you as counselor or therapist have the opportunity and responsibility to provide useful information, new ideas, and meaningful strategies that can lead to significant change. At times, even advice may be appropriate, for your comments can be insightful and make a difference. But think of how you respond to advice. Sometimes you listen; perhaps more often it offends you, and thus even good advice is ignored.

The point is obvious and we will not repeat it again: directives, instruction, and psychoeducation are best received in a good relationship with a solid working alliance. It will be received most often when significant decisions are made by the client, with collaborative support from you, the counselor or therapist.

If you use the action influencing skills, you can anticipate how clients will respond. If they do not respond as expected, your ability to be intentional, listen, and move to other skills is essential. The ability to be flexible and change direction is particularly necessary, as clients may respond differently than when you are listening.

Directives, Instruction, Psychoeducational Strategies	Anticipated Client Response
Clear directions, encouraging clients to do what you suggest, underlies instruction and psychological education. These offer specifics for daily life to help change thoughts, feelings, and behaviors. Providing useful instruction and referral sources can be helpful. Psychoeducational strategies include systematic educational methods such as therapeutic lifestyle changes. With all these, a collaboration approach is essential.	Clients will make positive progress when they listen to and follow the directives, use the information that you provide for them, consider your advice, and engage in new, more positive thinking, feeling, or behaving. Psychoeducation can lead to major life changes for physical and mental health.

Directives, instruction, and psychoeducational strategies are valuable in encouraging clients move to change and action. They are particularly useful in the interviews' fourth and fifth stages—restory and action. A positive new story may be sufficient for some clients, but many will profit from directive strategies outlining specific behaviors and actions they can use immediately. One directive strategy emphasized is the homework/action plan, which has been shown to be central in producing results.

The basic underlying skills of directives, instruction, and psychoeducational strategies are as follows:

- *Involve clients as co-participants.* Rather than telling clients what you want them to do, be sure that you have heard their story and the relationship is solid. Involve clients in directive strategies beforehand so that they understand what will happen. Encourage them to respond. However, some practitioners use surprises (e.g., Gestalt theory), and this sometimes is helpful.

- *Use appropriate visuals, vocal tone, verbal following, and body language.* Your attending behaviors need to provide empathic support as you intentionally flex in response to client needs. When challenging an acting-out teen or clients diagnosed as narcissistic or antisocial, you may need a stronger, more active persona with even clearer verbal and nonverbal behavior. With a quieter and more hesitant client, appropriate attending may require being more tentative as you share new ways of thinking about issues. Directives given softly can be very effective.

- *Be clear and concrete in your verbal expression, and time the information to meet client needs.* Directives need to be authoritative and clear but also stated in such a way that they are in tune with the unique client. Give advice with caution, but it can often help with the right client. Know what you are going to say, and say it clearly and explicitly. Relevant information that the client needs to know—medical referrals, how to find housing, the local gym, where to find child care—is usually received positively.

- *Check out if you were heard and understood.* Just because you think you are clear doesn't mean the client understands or remembers, especially if the ideas are complex. An anxious client often has trouble hearing exactly what you said. For example, "Could you repeat back to me how to get to Social Security?" or "I suggested three things for you to do for homework this coming week. Would you summarize them to make sure I've been clear?"

Awareness, Knowledge, and Skills: Making Action Skills Work

> Directives, instruction, and stress management psychoeducational strategies are preferred modes of treatment, useful for the majority of your clients. They also serve an important *preventive* function, thus enabling the client to make better decisions and cope more effectively with present and future stressors.
>
> —Mary Bradford Ivey

As you start using directives, instruction, and psychoeducation, remember they can come across as "telling clients what to do." It is very important not to get too enthusiastic with these strategies. Few of us like to be *told* what to do. Furthermore, many of us say we want advice and then don't listen. Or we may simply ignore suggestions from the counselor. Always remember to empower your clients so as to make them copartners in the here and now of the interview as you together select the skills and strategies that might help produce growth and the development of the *New*.

The homework/action plan presented in detail earlier in this chapter (see Box 12.3) illustrates how we can provide direction, information, and even advice if we work *with* the client. Directives developed *with* the client are most likely to become part of the client's cognitive, emotional, and behavioral style. Interventions developed with the client have more meaning.

The psychoeducational instructive strategies summarized in Box 12.4 can be used in several different approaches to counseling and therapy. In fact, with some clients, listening and working with them on the TLCs may be sufficient for behavior change. Test these strategies first by trying them on yourself. Later work with a friend or classmate, and then have that person test the same directive strategy with you. If you practice the details of directives, you will have a better idea of their potential and how to pace and time the strategy,

At this point we suggest that you return to a review of Chapter 2, where a great many therapeutic lifestyle changes are introduced. All of this can and will be useful—and important—to many of your clients, as all are key aspects of a healthier physical and mental lifestyle.

BOX 12.4   **Directives, Instruction, and Psychoeducational Strategies**

These example strategies are presented in very brief form. With further study and some imagination and practice—and client participation in the process—you can successfully use many of them. For more detailed presentations of these and other strategies in highly concrete form, see Ivey, D'Andrea, and Ivey's *Theories of Counseling and Psychotherapy: A Multicultural Perspective* (2012).

Therapeutic lifestyle changes (see pp. 288)

All of the TLCs are oriented to positive management of stress. Help clients learn and improve mental and physical health through TLCs such as exercise, nutrition, sleep, meditation, social relations, cognitive challenge, social justice action, joy, humor, and zest for living. While sharing the many possibilities, it is best that the client select one or two for action. KEEP IT SIMPLE!

The interview in the following section, "Observe: Integrating Therapeutic Lifestyle Changes into the Session," illustrates the use of TLCs within a decisional counseling framework.

Directives, sharing information, and advice

Detail and concreteness are very important when providing a directive.

> "I suggest you try . . ."
>
> "Vanessa, the next time you go to the garage and they start giving you a bad time, I'd like you to stand at the counter, make direct eye contact, and clearly and firmly tell the manager that you have a meeting at 10:00 and you need prompt service—now! If he says there will be a delay, get him to

make a firm time commitment. Then follow up 15 minutes later."

Spiritual images (useful as a TLC)

In recent years the counseling and interviewing field has recognized the strengths and power in spirituality and religion. Many clients benefit from spiritual imagery and often find peace in their inner body and strength to move on. Forgiveness of the transgressions and omissions of others can come from spiritual imagery, or your client may find new strengths to deal with a difficult illness or serious loss. A spiritual orientation even helps some clients recover from operations or serious illness.

"You say you gain strength from your spirituality and religion. Could you tell me about an image that comes to your mind related to a spiritual strength?" (Listen to the story and the feelings that go with it.)

"Now, close your eyes and visualize (that symbol, person, experience) and allow it to enfold you completely. Just relax, focus on that image, and note what occurs in your body."

Role-play enactment

Role-playing is an especially effective technique to make the abstract concrete. It makes the client's behavior clear and specific. This is one of the most basic techniques used in assertiveness training.

"Now return to that situation, and let's play it out."

"Let's role-play it again, only change the one behavior we agreed to."

Positive reframing (Chapter 11) combined with a directive

Taking real positives to attack problems through the body can be effective.

"We've identified the problem and how it feels. Now feel that wellness strength in your body. Do it fully, magnify it, and take it to the problem."

If the strength is not able to meet and counteract the negative, add another resource—or just have the strength approach one part of the negative at a time.

Mindfulness meditation

Mindfulness meditation is derived primarily from Buddhist practice through Kabat-Zinn's (2005, 2009) contacts with the Dalai Lama. There is no "goal" except to live as much as possible in the immediate here and now.

Practitioners usually lie comfortably on the floor or sit in a suitable chair, then close their eyes from 10 minutes to an hour. The focus becomes the Now and paying special attention to breathing, noting the breath coming in and out. Thoughts and feelings will likely start wandering through your mind. Do not fight them; let them come, but let them drift off.

After practice, usually for several weeks, you may find a near perfect "stillness" and awareness of the present moment. There is clear evidence that this state alone allows new neural connections to develop in positive areas of the brain, as well as increase brain gray matter (Hölzel et al., 2011). If you keep this up, you will eventually notice the here and now more fully throughout the day. You'll notice the beauty of the world in new ways.

A safe, but secondary, alternative to formal training is to refer clients to the website www.mindfulnesstapes.com, where they can purchase excellent audio mindfulness training and learn on their own. In addition, many YouTube videos are available under "meditation." Apps are also widely available for your cell phone.

Relaxation (like meditation, an important TLC)

The simplest way to learn the relaxation response is to (1) notice body tension, (2) take a deep breath, and (3) hold it for just a moment and let it go as the body gradually relaxes.

In teaching clients, the following may be useful:

"Close your eyes and focus on the moment."

"Tighten your forearm, very tight, now let it go."

(Directions for relaxation continue throughout the body parts, ending with full body relaxation.)

Clients may be tight and tense, but once they are able to relax and gain control of their body, they are better able to cope with stressful daily encounters. Teaching the relaxation response is an essential skill for all interviewers and counselors (Benson & Proctor, 2010).

Encouraging physical exercise and related therapeutic lifestyle changes

A past president of the American Psychiatric Association has stated that any physician who does not recommend exercise to patients is unethical. Interviewing, counseling, and therapy have been very weak in this area. Exercise is a preventive health activity that needs to become part of everyone's practice. Moreover, research is now showing that exercise helps clients deal with stress, which in turn helps with many difficult issues ranging from depression to Alzheimer's (Ratey & Manning, 2014).

A sound body is fundamental to mental health. Beyond encouraging clients to exercise regularly, remind them that proper eating habits and a regime of stretching and meditation can make a significant difference in their lives. Teaching clients how to nourish their bodies is becoming a standard part of counseling. We love and work more effectively if we are comfortable in our bodies.

(continued)

BOX 12.4 (continued)

Imagery focusing on a relaxing scene (useful as a TLC)

This directive may be used with any positive image, person, or situation. All of us have past positive experiences that are important for us—maybe a lakeside or mountain scene, or a snowy setting, or a special quiet place. The image can become a positive resource to use when you feel challenged or tense—for example, you feel tension in your body when anticipating making a presentation, being interviewed by an employer, or having difficulty falling asleep.

When giving a guided imagery directive, time your presentation to your observations of the client.

> Close your eyes and relax. [Pause] Notice your breathing and the general feelings in your body. Focus on that place where you felt safe, comfortable, and relaxed. Allow yourself to enter that scene. What are you seeing? . . . Hearing? . . . Feeling? . . . Allow yourself to enjoy that scene in full relaxation. Notice the good feelings in your body. Enjoy it now for a moment before coming back to this room. Now, as you come back, notice your breathing [pause], and as you open your eyes, notice the room, the colors, and your surroundings. How was this experience for you?

Thought stopping

"Wind back the tidal wave of the frenetic motion or the mind." —Joseph Ting

This simple strategy has consistently been found to be one of the most effective interventions we can use. It is useful for all kinds of client problems: perfectionism, excessive culture-based guilt or shame, shyness, and mild depression. Almost everyone engages in internalized negative self-talk—stressful thoughts you say to yourself, perhaps several times a day. For example:

> "Why did I do that?"
> "I'm always too shy."
> "Can't I stop making mistakes?"
> "I should have done better."
> "Life is so discouraging for me."
> "Nobody will listen to me."

The following is the basic process for learning and using thought stopping.

Step 1. Learn the basic process. Relax, close your eyes, and imagine a situation in which you make the negative self-statement. Take time and let the situation evolve. When the thought comes, observe what happens and how you feel after the negative self-talk. Then tell yourself silently "STOP." If you are alone, say it loudly and firmly.

Step 2. Transfer thought stopping to your daily life. Place a rubber band around your wrist and every time during the day that you find yourself thinking negatively, snap the rubber band and say "STOP!" This simple step almost sounds silly, but it works. (Snapping the rubber band is not a form of punishment but a way to interrupt or interfere with negative thinking. Be kind to yourself; use it with this purpose only.) The client can, of course, just say "Stop," but the rubber band adds extra reinforcement to the effort to change thought patterns.

Step 3. Add positive imaging. Once you have developed some understanding of how often you use negative self-talk, and after you say "STOP" or snap the rubber band, immediately substitute a more positive statement about yourself. You may use positive imagery, or think about an example when you had a positive experience, or use a brief broader statement emphasizing general strengths.

> "I can do lots of things right."
> "I am lovable and capable."
> "I sometimes mess up—no one's perfect."
> "I did the best I could."

Journaling

Keeping a journal is helpful to many clients. This helps them reflect on the interview and its impact on them during the week.

> "Alicia, you like to write and think about things. How would it be if you started a journal of your work with me? You might want to reflect on each interview and its impact and how what we discuss relates to what you see happening in your daily life. You can share this with me or not, as you choose."

Observe: Integrating Therapeutic Lifestyle Changes into the Session

Exercise is 30% of preventing cancer, another 20% is not becoming obese.

—James Watson, codiscoverer of DNA,
on his 10 years of research on cancer

As described in detail in Chapter 2, therapeutic lifestyle changes are a "different" way to approach interviewing and counseling, but they are well supported by psychological, neuroscience, and neurobiology research. With Dr. Watson's surprisingly strong support, it means that we need to take TLCs seriously.

Several sources have described exercise as the number one TLC (Ratey & Hagerman, 2013). Exercise not only builds the body, but the increased blood flow increases the possibility of neurogenesis and developing new neural nets. Once a client gets the body moving, it is more difficult to feel down and depressed. In fact, 25 research studies have found that exercise can prevent depression (Bergland, 2013).

We enter the session after 20 minutes of reviewing how Alicia is doing. She has now resolved the relationship with mechanic Jon, but after several sessions discussing Simon, she now finds herself sad and discouraged. She not sleeping well and shows signs of a mild depression. Otherwise, she is functioning well, but she still has not fully decided what to do about the relationship. Decision making that is "on hold" is stressful. The waiting time can sometimes be as emotionally troublesome here and now as the issues that brought about the thoughts of separation or even divorce. This is a good example of how decisions are involved throughout virtually all counseling and therapy issues.

Physical exercise obviously will not resolve Alicia's concerns, but it can help her build resilience and better control of her own body when she faces stress. This and other TLCs discussed in Chapter 2 build strength and self-confidence.

The following is excerpted from a 50-minute session.

Interviewer and Client Conversation	Process Comments
35. *Alicia:* Thanks, Onawumi, I think I'm feeling better through the imagery exercises, but lately I've not been feeling too good because things are not changing at home as I hope. Earlier you mentioned exercise as a good stress reliever. Let's talk about how I can get exercising again. I recall that you said exercise could change and perhaps even strengthen the brain. I used to run a lot, but stopped about a year after I moved in with Simon.	Her nonverbals are not as positive and strong as they were in the last session, and it becomes clear that more is needed. At the same time, we see evidence that Onawumi's ability to listen and, at times, confront Alicia's issues has been successful.
36. *Onawumi:* It's good that you have exercised in the past. Exercise is a real brain builder, not just the body. But first, I'd like to hear more about your past exercise. I hear you saying that you stopped about a year ago after moving in with Simon. How was running for you?	Onawumi starts the process by learning about her past experience with exercise. The entire session can be classified as "potentially additive," as we won't know how valuable this is until she reports back in later interviews whether or not she has started and is likely to continue exercising.
37. *Alicia:* The first half-mile always was difficult and sometimes even painful. But then that "runner's high" came on and the endorphins were great. I miss that; I was happier when I was running.	
38. *Onawumi:* I hear that. Sounds like you are ready to start again. You already know how good those endorphins feel, but did you know that 25 research studies have shown that exercise can prevent depression? Just like counseling, exercise changes the brain.	Providing information.
39. *Alicia:* No, I thought depression was just in the mind. Tell me more.	

Now we see a longer explanation, which is often part of psychoeducation. Psychoeducation in the early stages requires more talk time. For a superb presentation on depression as a physical issue, search for the national expert on YouTube: "Robert Sapolsky, depression."

(continued)

Interviewer and Client Conversation	Process Comments
40. *Onawumi:* Here I have a picture of the brain. There is too much detail here, but let me share some highlights and how the brain reacts to exercise. (See Appendix IV; consider photocopying for your practice sessions.)	First, notice that word *amygdala* (points). The amygdala is our energizer, but it is also the seat of those negative emotions you experience. When we are depressed, that area is working overtime. Now look at the prefrontal cortex (points). Easier to say PFC. That's where our thinking and decision making are primarily located. It is also where we" manage and control our emotions. Exercise increases blood flow to the brain and body. When people are depressed, they often stop exercising, they sleep less well, and eat less carefully. This actually increases the depression. What we want to do is increase the power of the PFC, the top part of the brain, to regulate and control those negative emotions and relieve depression. How does that sound? (Psychoeducation followed by checkout—potentially additive)
41. *Alicia:* (Interrupts) I get it, and the endorphins come out of that and I feel better. I used to love my runner's high. (Pauses and points) So, exercise endorphins likely hit the PFC here and allow it to better regulate or control the negative emotions. Is that right?	Counseling changes not only the brain, but also the body. Mentioning mental and physical benefits can help make more decisions to take the interview back home. Action beyond the session is the most important part of our work with clients.
42. *Onawumi:* Yes, exercise enlivens our whole brain. It brings about a better mood. Not only will you feel better with exercise, your brain will operate more effectively.	Summary of the possibilities within the TLC of exercise.
43. *Alicia:* Wow, makes exercise a must.	So far, Onawumi's psychoeducational efforts are additive, but that always depends on whether or not something happens after the session.

At this point, we leave the story of Alicia. Onawumi introduced meditation and encouraged Alicia to return to her church, which she had stopped after moving in with Simon. Friends and the church provided critical support during this difficult time. Over the next five sessions, she finally made the choice to leave. While this was extremely difficult for her, she now had the strength to move on. Her role-played practice sessions helped her speak more effectively with Simon, who it turned out was frightened of being alone. There were a few arguments, but finally he accepted what she said. They agreed to stay in touch and see what happened. She did not have to go to a safe house.

Action: Key Points of Influencing Skills and Stress Management

Stress management. Sustained, chronic, or extreme stress accelerates the normal wearing and tearing of our body and mind. Changing the stressors or changing our reactions to them are key goals of stress management. Psychoeducation, stress management, and TLC strategies are key to protecting our physical and mental health. Several strategies can help you achieve these goals, including relaxation, meditation, disputing irrational beliefs, thought stopping, time management, and many other techniques mentioned in this chapter.

Self-Disclosure. Indicating your thoughts and feelings to a client constitutes self-disclosure, which necessitates the following:

1. Use personal pronouns ("I" statements).

2. Use a verb for content or feeling ("I feel . . ." "I think . . .").

3. Use an object coupled with adverb and adjective descriptors ("I feel happy about your being able to assert yourself . . .").

4. Express your feelings appropriately.

Self-disclosure tends to be most effective if it is genuine, timely, and phrased in the present tense. Keep your self-disclosure brief. At times, consider sharing short stories from your own life.

Feedback. Feed back accurate data on how you or others view the client. Remember the following:

1. The client should be in charge.

2. Focus on strengths.

3. Be concrete and specific.

4. Be nonjudgmental.

5. As appropriate, provide here-and-now feedback.

6. Keep feedback lean and precise.

7. Check out how your feedback was received.

These guidelines are useful for all influencing skills.

Logical Consequences. This is a gentle skill used to help people sort through issues when a decision needs to be made. Decisions can have both negative and positive consequences. The focus is on potential outcomes, and the task is to assist clients to foresee consequences as they review alternatives for action. A common statement used here is "If you do _____, then _____ will possibly result."

   This skill predicts the probable results of a client's action, in five steps:

1. Listen to make sure you understand the situation and how the client understands what is occurring and its implications.

2. Encourage the client to think about possible positive and negative consequences of a decision.

3. If necessary, comment on the positive and negative consequences of a decision in a non-judgmental manner.

4. Summarize the positives and negatives.

5. Let the client decide what action to take.

Instruction and Psychoeducation. Instruction and psychoeducation are closely related. Instruction, providing information or advice, is brief, consisting of relatively short comments to facilitate action in the real world. Psychoeducation is more comprehensive. Many times clients need the counselor's knowledge and expertise around key life issues. The counselor knows the community and the resources available. He or she also knows the likely pattern and key issues of a divorce, the death of a family member, or other life changes. Psychoeducation is a more systematic way of teaching clients about new life possibilities; this may range from training in communication skills to developing a successful wellness plan.

Before you give instruction or engage in psychoeducation:

1. A solid working relationship is essential.
2. Hear the client's story and identify strengths.
3. Check out your client's interest and readiness for receiving information.
4. Be clear and concise, and encourage client participation and feedback.

Practice and Feedback: Individual, Group, and Microsupervision

Additional resources can be found by going to CengageBrain.com and logging into the MindTap course created by your professor. There you will find a variety of study tools and useful resources that include quizzes, videos, interactive counseling and psychotherapy exercises, case studies, the Portfolio of Competencies, and more.

We strongly suggest that you practice all of the skills presented in this chapter.

Individual Practice

Exercise 12.1 Self-Disclosure

Find a classmate or friend, get the person's permission, and the two of you try the strategy of self-disclosure on each other. What happens? What occurs for you? What did you learn? Would you like to continue and practice this skill further?

Exercise 12.2 Feedback

Again, find a classmate or friend, get the person's permission, and the two of you engage in the strategy of feedback. What happens? What occurs for you? What did you learn? Would you like to continue and practice this skill further?

Exercise 12.3 Logical Consequences

Using the five steps of the logical consequences strategy, briefly indicate to a client what might be the logical consequences of one of the following: staying in an abusive relationship; smoking while pregnant; moving from marijuana to cocaine.

1. Summarize the client's concern in your own words, using "if, then" language.
2. Ask specific questions about the positive and negative consequences of continuing the behavior.
3. Provide the client with your own feedback on the probable consequences of continuing the behavior. Use "if, then" language.
4. Summarize the differences between the feedback just given and the client's view when the client says she/he doesn't want to change (this implies the use of confrontation).
5. Encourage the client to make her/his own decision.

Exercise 12.4 Writing Logical Consequence Statements

By using questioning skills, you can encourage clients to think through the possible consequences of their actions. ("What result might you anticipate if you did that?" "What results are you obtaining right now while you continue to engage in that behavior?") However, questioning and paraphrasing the situation may not always be enough to make clients fully aware of the logical consequences of their actions. For each of the following clients and situations, write logical consequences statements that can help the client understand the situation more fully.

1. A student who is contemplating taking drugs for the first time.

2. A young woman contemplating an abortion.

3. A student considering taking out a loan for college.

4. An executive in danger of being fired because of poor interpersonal relationships.

5. A client who is consistently late in meeting you and is often uncooperative.

Group Practice and Microsupervision

Exercise 12.5 Practicing Strategies

This chapter includes many different possibilities for practice. Try out each skill or strategy as time permits. Work through each of the strategies before using them with a client.

Group work with these influencing skills requires practice with each if you are to develop competence. The general model of group work is suggested, but only one strategy should be used at a time.

Remember to include the Client Feedback Form from Chapter 1 as part of the practice session. This is a particularly important place to practice group supervision, sharing, and feedback.

Step 1: Divide into practice groups.

Step 2: Select a leader for the group.

Step 3: Assign roles for each practice session.

❑ Client

❑ Counselor, who will begin by drawing out the client story or issue using listening skills and then attempt one of the influencing skills and strategies from this chapter

❑ Observer 1, who will observe the client and complete the CCS Rating Form (see Chapter 10), deciding how much of an impact the counselor's influencing skills have made

❑ Observer 2, who will complete the Feedback Form in Box 12.5 or 12.6

Step 4: Plan. In using influencing skills, the acid test of mastery is whether the client actually does what is expected (for example, does the client follow the directive given?) or responds to the feedback, self-disclosure, and so on, in a positive way. For each skill, different topics are likely to be most useful. State goals you want to accomplish in each instance. Some ideas follow:

❑ *Logical consequences.* A member of the group may present a decision he or she is about to make. The counselor can explore the negative and positive consequences of that decision. In the process, the counselor may wish to make one or two self-disclosures and provide feedback to the client at the end of the session.

❑ *Instruction/psychoeducation.* The counselor may provide instruction (information) or psychoeducation about a particular issue to the individual or group, such as the value of a wellness plan or dealing with a death in the family. The group gives feedback on whether the counselor was able to give information in a way that was clear, specific, interesting, and helpful. We suggest that you consider teaching microskills as communication skills to your individual or group.

❑ *Stress management and TLCs.* Select one of the strategies presented in this chapter and work through the specific steps. Involve your client in the process; the two of you together can select the strategy that you would like to try. As part of the practice session, be sure to tell the client what to expect and the likely results.

❑ *Feedback and self-disclosure.* Ask the client to describe an experience and provide feedback and self-disclosure.

Step 5: Conduct a 5- to 15-minute practice session using the strategy. Use listening skills along with the selected strategy. Is the client connected and involved?

Step 6: Review the practice session and provide feedback for 10 to 12 minutes. Remember to stop the recording to provide adequate feedback for the counselor.

Step 7: Rotate roles.

BOX 12.5   Feedback Form: Self-Disclosure and Feedback

_____
(DATE)

_____   _____
(NAME OF COUNSELOR)   (NAME OF PERSON COMPLETING FORM)

Instructions: The two raters will complete the form and then discuss their observations with the practicing counselor and the volunteer client.

1. Did the counselor use basic listening sequence to draw out and clarify the client's story or concern? How effectively?

2. Provide nonjudgmental, factual, and specific feedback for the counselor on the use of the specific influencing skill or directive strategy. How empathic was the feedback?

3. As you view the totality of the session, where was the client at the beginning on the Client Change Scale? Where was he or she at the conclusion? What aspects of the skill or strategy impressed you as most useful and effective?

4. Evaluate the effectiveness and empathic level of the use of self-disclosure.

BOX 12.6 Feedback Form: Logical Consequences, Instruction/Psychoeducation, Stress Management, and TLCs

_____

(DATE)

_____   _____

(NAME OF COUNSELOR)   (NAME OF PERSON COMPLETING FORM)

Instructions: Observer 2 will complete the form and then discuss observations with the practicing counselor and the volunteer client.

1. Did the counselor use the basic listening sequence to draw out and clarify the client's story or concern? How effectively?

2. Provide nonjudgmental, factual, and specific feedback for the counselor on the use of the specific influencing skill (logical consequences, instruction/psychoeducation, stress management, or TLC).

3. As you view the totality of the session, where was the client at the beginning on the Client Change Scale? Where was he or she at the conclusion? What aspects of the skill or strategy impressed you as most useful and effective?

Portfolio of Competencies and Personal Reflection

This chapter is about multiple interpersonal influence strategies, and it covers considerable material. You cannot be expected to master these concepts until you have a fair amount of practice and experience. At this point, however, it will be helpful if you think about the major ideas presented in this chapter and where you stand currently. Also, where would you like to go in terms of next steps?

Assessing Your Level of Competence: Awareness, Knowledge, Skills, and Action

Use the following table as a checklist to evaluate your present level of mastery. As you review the items below, ask yourself, "Can I do this?" Check those dimensions that you currently feel able to do. Those that remain unchecked can serve as future goals. Do not expect to attain intentional competence on every dimension as you work through this book. You will find, however, that you will improve your competencies with repetition and practice.

	Awareness and Knowledge	Basic Competence	Intentional Competence	Pychoeducational Teaching Competence
Self-disclosure				
Feedback				
Logical consequences				
Providing information				
Microskills instruction				
Assertiveness training				
Thought stopping				
Positive Imagery				
Stress management & TLCs				
Exercise				
Nutrition				
Sleep				
Social relations				
Cognitive challenge				
Meditation				
Cultural health				
Spirituality/prayer				
Social justice action/ helping others				

Personal Reflection on Influencing Skills

You have encountered the most active set of microskills and strategies in Section IV and have had the opportunity for at least a brief introduction to each.

With which of these skills and strategies do you feel most comfortable? Which might you seek to use? Which might you avoid?

How do you feel about the idea of consciously influencing the direction of the session?

What single idea stands out for you among all those presented in this chapter, in class, or through informal learning? What stands out for you is likely to be important as a guide toward your next steps.

How would you use these skills to help clients (and maybe yourself) manage their stress?

What are your thoughts on multicultural issues and the use of this skill?

What other points in this chapter strike you as important?

How might you use ideas in this chapter to begin the process of establishing your own style and theory?

Given the complexity of this chapter and the many possible goals you might set for yourself, list three specific goals you would like to attain in the use of influencing skills and strategies within the next month.

Integrating Skill into Theory for Effective Practice, Personal Style, and Transcendence

What is your preferred style for counseling and psychotherapy? How would you use the ideas of this book in your own practice? This section provides a framework to help you integrate the many skills and concepts of this book. Central to this process is Chapter 13, where we examine in detail two counseling sessions demonstrating the application of crisis counseling and cognitive behavioral therapy. Then, in Chapter 14, we recommend completing a full interview with a volunteer and analyzing how your personal integration of skills affects the client. Competence and mastery of counseling begin to show when you can anticipate and evaluate the impact of your style on client change, growth, and development.

Chapter 13. Counseling Theory and Practice: How to Integrate the Microskills with Multiple Approaches Here you will see a brief summary of various theories' use of the microskills. This is followed by a more detailed presentation of crisis counseling and cognitive behavioral therapy. In addition, special attention is given to suicide prevention and culturally sensitive interventions.

Chapter 14. Skill Integration, Determining Personal Style, and Transcendence This final chapter outlines specifics for the recommended final interview and analysis. In addition, you will find useful information for practice, such as planning for the interview and a checklist of specifics that need to be considered as you meet each new client. Information about treatment planning, case management, and referral are important in this chapter. Also, you will be asked to think about transcendence. How will you go about helping others achieve change and create the New?

We are nearing the end of our journey through the basics of counseling and psychotherapy. You now have competencies that can be used in many settings, as the microskills and five stages are foundational units of all communication—in counseling, psychotherapy, business, sales, law, medicine, peer helping, and even working on issues within your own family.

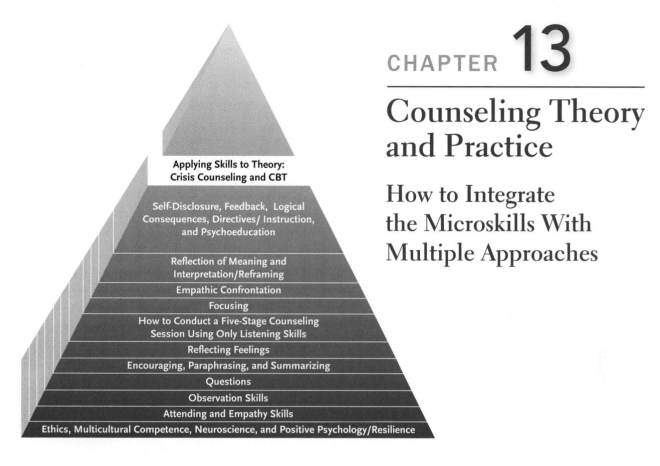

Applying Skills to Theory:
Crisis Counseling and CBT

Self-Disclosure, Feedback, Logical
Consequences, Directives/ Instruction,
and Psychoeducation

Reflection of Meaning and
Interpretation/Reframing

Empathic Confrontation

Focusing

How to Conduct a Five-Stage Counseling
Session Using Only Listening Skills

Reflecting Feelings

Encouraging, Paraphrasing, and Summarizing

Questions

Observation Skills

Attending and Empathy Skills

Ethics, Multicultural Competence, Neuroscience, and Positive Psychology/Resilience

Counseling Theory and Practice

How to Integrate the Microskills With Multiple Approaches

There is nothing so practical as a good theory.

—Kurt Lewin

Chapter Goals and Competency Objectives

Awareness and Knowledge

▲ Review how the microskills framework is used across multiple theories of counseling and psychotherapy.

▲ Examine the background and practice of crisis counseling and cognitive behavioral therapy.

▲ Become aware of the realities of suicide.

Skills and Action

▲ Read a transcript, practice, and engage in some of the basics of crisis counseling.

▲ Read a transcript, practice, and engage in some of the basics of a major approach to counseling and psychotherapy: cognitive behavioral therapy.

Introduction: Microskills, Five Stages, and Theory

The competencies you have developed with this book will enable you to understand and work with most counseling theories more quickly and competently. Although each theory offers a conceptual framework that differs from the others, virtually all theoretical approaches use microskills and the five stages presented in this book.

Table 13.1 reveals that each theory of counseling has a distinct pattern of microskill use. All use the basic listening sequence, but counselors using each system listen to stories from a different frame of reference. The person-centered system tends to use the listening skills most frequently and focuses most on the individual client, giving less attention to the cultural/environmental context. Brief counseling and counseling/coaching use many questions; CBT and Gestalt use many directives; and multicultural counseling and therapy (MCT) and feminist therapy pay more attention to the cultural/environmental context. Decisional counseling, MCT, and feminist theories are the most eclectic of the theories as they tend to use a variety of skills and focus dimensions.

Each theory, because of its particular philosophic orientation, emphasizes different aspects of life experience and gives different emphasis to each of the five stages. Pragmatic decision counseling, for example, tends to work in the here and now on immediate life issues or facilitate life planning. In contrast, the person-centered counselor emphasizes relationship and self-actualization, while logotherapy focuses on life's meaning and CBT on cognitions and behavior. MCT and feminist theory draw on all of the above, but always seek to situate the individual in the cultural/environmental context so that clients are well aware of the impact the surrounding world has on their cognitions and emotions.

Each theoretical system has considerable merit. This may help you to understand why theories have proliferated over the years. Perhaps Kurt Lewin's quotation could be rephrased: *There is nothing so practical as becoming competent in several theoretical approaches.*

We have chosen crisis counseling and cognitive behavioral therapy to illustrate how the five stages and microskills can be applied to theories other than those described earlier in this book. We begin with crisis counseling.

Awareness, Knowledge, and Skills of Crisis Counseling

> We want people to know that their emotions and reactions are completely normal.
>
> —Anonymous crisis helper

Philosophy of Crisis Counseling

Crisis counseling is the most pragmatic and action-oriented form of helping. The word *pragmatism* comes from the Greek word for deed, act, to practice, and to achieve πράγμα (*pragma*). Even more than decisional counseling, crisis counseling is concerned with action and useful, pragmatic results for the client. However, pragmatism is embedded in a caring attitude that is fully aware that most responses to a crisis can be considered completely normal.

What is a crisis? *Crisis* is closely related to *trauma*; it means stress, stress hormones, and a rise in cortisol to the brain. It has been pointed out that virtually all the world's

TABLE 13.1   Microskills Patterns of Differing Approaches to the Interview

Legend: ● = Frequent use of skill; ◒ = Common use of skill; ○ = Occasional use of skill

	Decisional counseling	Person-centered	Logotherapy	Multicultural and feminist therapy	Crisis counseling	Cognitive behavioral therapy	Brief counseling	Motivational interviewing	Counseling/coaching	Psychodynamic	Gestalt	Business problem solving	Medical diagnostic interview
Open question	●	○	●	◒	◒	◒	●	●	●●	◒	●	◒	◒
Closed question	◒	○	●	◒	◒	◒	◒	◒	◒	○	◒	◒	◒
Encourager	●	◒	●	◒	◒	◒	◒	◒	●	◒	◒	◒	◒
Paraphrase	●	●	●	◒	◒	◒	●	●	●	◒	○	◒	◒
Reflection of feeling	●	●	●	◒	◒	◒	●	●	◒	◒	○	◒	◒
Summarization	◒	◒	●	◒	◒	◒	●	◒	●	◒	○	◒	◒
Reflection of meaning	◒	●	●	●	○	○	○	◒	◒	◒	○	○	○
Interpretation/reframe	◒	○	◒	◒	◒	○	◒	●	○	●	●	◒	◒
Logical consequences	◒	○	○	●	○	◒	○	●	◒	○	○	◒	◒
Self-disclosure	◒	◒	◒	◒	○	○	◒	●	●	○	◒	◒	○
Feedback	◒	◒	◒	◒	○	◒	●	●	●	○	◒	◒	○
Instruction/ psychoeducation	●	○	○	●	◒	●	○	◒	●	○	○	●	◒
Directive	◒	○	◒	◒	◒	●	○	◒	◒	○	●	●	●
CONFRONTATION (Combined skill)	◒	◒	◒	●	○	◒	◒	●	◒	◒	●	◒	◒
Client	●	●●	●	◒	◒	●	●	●	●	●	●	◒	◒
Main theme/issue	●	○	◒	◒	●	●	●	●	◒	○	●	●	●
Others	◒	○	◒	◒	◒	◒	◒	◒	◒	◒	○	○	○
Family	◒	○	◒	◒	◒	◒	◒	○	◒	○	○	○	○
Mutuality	◒	◒	◒	◒	○	◒	◒	◒	●	○	○	○	○
Counselor/therapist	○	◒	◒	◒	○	○	○	○	○	○	○	○	○
Cultural/environmental/ contextual	◒	○	◒	●●	●	◒	◒	◒	◒	○	○	◒	○
ISSUE OF MEANING (Topics, key words likely to be attended to and reinforced)	Problem solving	Self-actualization, relationship	Values, meaning vision for life	How CEC impacts client	Immediate action, meeting challenges	Thoughts, behavior	Problem solving	Change	Strengths and goals	Unconscious motivation	Here-and-now behavior	Problem solving	Diagnosis of illness
	Medium	Low	Medium	Medium	High	High	Medium	Medium	Medium	Low	High	High	High

LEGEND

● Frequent use of skill   ◒ Common use of skill   ○ Occasional use of skill population experiences one or more crises/traumas in their lifetime. In that sense, crisis is a normal life event, and "normalizing" the crisis is one foundational idea to keep in mind.

You may encounter many different types of crisis. The first type of crises, immediate here-and-now crises, demand rapid practical action. These include flood, fire, earthquake, rape, war, refugee status, a school or community shooting, personal assault (including abuse), being held hostage, a serious accident, and the sudden discovery or diagnosis of a major medical challenge (e.g., heart attack, cancer, multiple sclerosis). These call for immediate supportive action. But many crisis/trauma survivors will benefit from further support and counseling.

The second type of crisis is a more "normal" part of life. Many of your clients will come to you to talk about divorce or the breakup of a long-term relationship, foreclosure of their home, job and income loss, a home break-in and burglary, or the death of a loved one. Examples of children's crises include bullying, dealing with their own or a parent's illness, leaving friends and moving to a new location, or parental divorce and remarriage. For some, not getting into a desired college or failing an exam will become a crisis situation.

Crisis counseling involves two major phases: (1) working through the initial trauma and (2) appropriate follow-up and further counseling. Counseling skills are obviously needed in both phases, but the immediate crisis will demand more cognitive and emotional flexibility— and the ability to join the client in the here and now. The second phase usually gives you more time and elements to work with and will begin to look like more typical counseling.

A more comprehensive team approach is required in serious crises, such as when a person with a gun approaches and attacks a school, university, bank, or office. The aftermath of a bombing, fire, earthquake, or other disaster will typically need follow-up group and individual work. See Box 13.1 for an example of how New Zealand planned and worked with one of the world's most damaging earthquakes.

| BOX 13.1 | Organizing a Crisis Team in a Major Earthquake |

Plan for Long-Term Involvement

In 2011, a magnitude 6.3 earthquake struck Christchurch, a seaside city of 350,000 on the South Island of New Zealand. A total of 185 people were killed, and the central city was decimated; about 1,000 buildings had to be demolished because of structural or land damage, and 180,000 homes were damaged or destroyed. Serious aftershocks continued for months, adding considerably to the stress and uncertainty in people's lives.

Civil Defense met this challenge by immediately assembling 530 people into teams of four that visited every home in the city over a period of two weeks. The teams checked the safety of the damaged homes, arranged for immediate temporary housing and other physical needs of residents, and checked each family for possibly needed crisis counseling.

Working with another psychologist, Robert Manthei, a professor of counseling at the University of Canterbury, helped organize, brief, and debrief the mental health members of the home visit teams. While the families welcomed the care, there appeared to be no large-scale immediate

trauma among the people visited; instead, the vast majority was remarkably resilient and coping effectively.

Nevertheless, many families did need what we would term coaching support as they dealt with immediate problems of reality and key decisions that they had to make. It is possible, perhaps even likely, that this practical approach to community crisis counseling contributed to the building of wider community resiliency.

Interestingly, over the next two or three years, the major stressors that many Christchurch residents experienced centered around challenges posed by the government earthquake commission and individuals' insurance companies, which argued about how much damage existed, who should pay for it, and whether the land was safe to still live on or to rebuild on. Some residents had to wait three years or longer for these decisions, leaving them in uncomfortable limbo with increasing anger and stress.

It is here that the coaching orientation could be of most use, but only if it combined empathic listening to each family's issues with effective suggestions for resolution, including referral to appropriate sources and services. Practical answers and action were needed at this stage, as residents were often caught up in ongoing disputes between funding organizations.

Community resilience in crisis is clearly widespread, but over time the continuing stressors of moving, financial uncertainty, community changes, and slow-moving government/insurance processes are often the most serious. Think of the New Orleans flood, the Haitian hurricane, or the tsunami in Japan. There is the immediate crisis, and then there is the long-term aftermath.

All too often, we think crisis is "managed" when it is immediately over and we have settled people for a day or two. Many possible long-term issues remain, however, and targeted, practical-oriented counseling with a coaching orientation still needs to be available.

As another example of the team approach, Mary Bradford Ivey, as a counselor in an elementary school, helped usher the excited elementary students into the school auditorium where they were to watch the liftoff of the NASA spaceship *Challenger*. Their excitement turned to fear and tears when the ship blew up in front of them. Mary and the teaching staff faced a major here-and-now crisis.

The teachers had worked with Mary to plan ahead if a school crisis should occur. It had been agreed that the teachers would take the children to their homerooms and encourage them to talk and ask questions, with awareness that each child would have her or his own unique reactions. The teacher needs to maintain composure and reassure children that they are safe. Encouraging children to ask questions helps give them control. Mary went from classroom to classroom, supporting each teacher and the students. If one or more children were particularly upset, Mary took them with her to the school counselor's office.

The next day and later that week, the debriefing continued. Some children wanted to express their feelings through art. Mary had had extensive experience in group work and set up several sessions for students who wanted or needed to explore issues in more detail.

The message here: *Prepare for crisis.* We never know when one will occur and what the crisis will be. In these days of school, theater, and other shootings, preparation to help is all the more important.

Key Strategies for Working with Those Who Experience Crisis

Although we are speaking here of the immediate survivors of a major crisis such as a fire or flood, the suggestions offered also hold for those you may meet after the event. Once again, you will find the five stages and the microskills a useful framework for thinking about how to help these clients. The transcript example in the next section, involving a woman the day after a frightening fire and the loss of her home, illustrates the five stages.

Normalizing. Do we call those who experience trauma "victims" or "survivors"? The second term is more empowering for clients and puts them more in control. Thinking of people as victims tends to depersonalize them and put them in a helpless position, controlled by external forces. Research shows that the majority of those who experience crisis have sufficient resilience and cognitive reserve to continue on, but the pain remains.

Many crisis workers object to the term *posttraumatic stress disorder* (PTSD), pointing out that virtually any serious encounter with crisis will produce extreme stress and challenges to the whole physical and mental system. *Disorder* is an inappropriate term because the client has actually responded in a normal fashion to an insane situation. Many prefer to refer to a normal *posttraumatic stress reaction* (PTSR). Your clients have gone through a far from normal experience. Helping them see that their "problem" is not inside them, but the logical result of external stressors, is one step toward normalizing the situation.

Furthermore, labeling the client in any way, even with a term such as *stress reaction*, should not be central—all these people are facing serious issues and need individual support and respect. Thus, the cultural/environmental/contextual focus remains central as we want to avoid attributing any client response as being solely "in the person." All survivors of a crisis need to know that however they responded to severe challenges is OK and to be expected.

Calming and Caring. A second major concept is to "normalize" the situation for the client and provide some sense of calm and possibility. This means establishing an empathic relationship by indicating that you care and will listen. Often the first thing in a crisis is that you must be calm and certain. Your personal bearing will do much to meet this first criterion of effective crisis work: *calming the client.*

Don't say "Calm down, it will be OK" or " You're lucky you survived." Better calming language includes such comments as "It's safe now" (if that is true), "We will see that this situation is taken care of," "I feel bad myself; that was a terrible thing to go through," "Your reaction and what you did are common and make sense," and the critical "What would help you right now?" For those who are having flashbacks (and this occurs with all types of trauma), calming and normalizing what they are experiencing are essential.

In particular, do not minimize the crisis. Think about the survivors of Hurricane Katrina. Some 20,000 people were housed in the New Orleans Superdome starting August 4, 2005; on September 4, the last group was evacuated to new shelters, where their crises continued. While they were at the Superdome, portions of the roof were blown away and water flooded in. Food and water were in short supply, and toilet facilities failed. Sleeping was more than a challenge.

Louisiana governor Kathleen Blanco termed the use of the Superdome an "experiment." The following words were attributed to Barbara Bush, mother of then-President George W. Bush, after her visit to the survivors of Katrina held at Houston's Astrodome:

> Everyone is so overwhelmed by the hospitality. And so many of the people in the arena here, you know, were underprivileged anyway, so this is working very well for them. ("Barbara Bush," 2005)

Consider these two attitudes and comments as examples of the privilege of the entitled, described in Chapter 2's multicultural section. These same attitudes toward minority and less privileged groups are often seen in the media.

In some situations, you will think and even know that the client is overreacting. Be aware of your thoughts and feelings, which may be valid, but join the client where he or she is. "Enter the client's shoes," as Carl Rogers might say.

Safety. Crisis and trauma survivors need to know that they are safe from the danger they have gone through. With soldiers and many others suffering from serious posttraumatic stress (not PTSD), creating a sense of safety and calm may not be accomplished immediately. Offer verbal reassurance that the crisis is over and they are now safe—again, if they are indeed safe. However, more than words may be needed. A woman who has experienced spousal abuse or a homeless and hungry person needs to find a safety house or a place to stay and eat immediately. To some counseling and therapist supervisors, this is "violating boundaries." This type of thinking is a relic of the field's past, but still there are some with these attitudes and beliefs. Stand up for what is right, help clients find what they need, connect them with resources. And consider the statement, with appropriate timing, "I'll be there with you to help."

Action. A good place to start is "What do you need now?" "What help do you want?" For yourself, what can you do that is *possible* in the here and now, and in the future? Do not overpromise. As noted above, some clients need a place to stay that night. Others need to know facts immediately. "Will I have to go through a vaginal exam after the rape?" "Are we going to be taken away by a bus?" "Where is the high ground in case the water comes again?" Answering these and other questions calmly and clearly will do much to alleviate anxiety.

The next step is to stay with clients and ensure that their needs are met. Crisis situations are often confusing. Following Katrina, many clients lost their helpers and thus experienced even more anxiety and tension. Volunteers in the Haiti earthquake went to help with good intentions and, indeed, did provide valuable assistance. But soon they had to return home, and often people were left "up in the air" with no knowledge of what to do next.

Debriefing the Story. Have you ever talked to family members or friends who have had a difficult and traumatic hospital operation? Have you noted that they often give you detail after detail? And then, the next time you see them, they tell you the same painful story . . . and perhaps even a third or fourth time. Freud called this "wearing away the trauma." People need to tell their stories, and many need to tell them again and again. Here the basic listening sequence becomes the treatment of choice. If you paraphrase, reflect emotions, and summarize what they have said authentically and accurately, they will know that someone has finally heard them.

Follow-up. Concrete action in the immediacy of crisis is essential. Where possible, you want to arrange to meet the client again for debriefing and planning in more detail for the future. In some cases, longer-term counseling and therapy will be needed.

Watch for strengths and resilience. If given sufficient early support, most people work through their crises. They have internal strengths that will carry them through. Look for these strengths and external resources that will enable them to recover. At the same time, even the most resilient survivors need to debrief what has happened.

Implications for Your Practice

There is much more to crisis counseling than what is said here, but you will find that competence and expertise in the basic listening sequence and the five-stage structure will provide a map that will help carry you through some challenging situations. Box 13.2 illustrates the importance of storytelling in the treatment of clients experiencing a traumatic event.

All of us need to be ready to help in crisis situations. We may have to deal with immediate crises such as the ones we have focused on here. But we also need to understand the concepts underlying crisis counseling because so many clients will have experienced, or be experiencing right now, extremely difficult situations.

BOX 13.2	**Research Evidence That You Can Use**

Systematic Emergency Therapy for Sexual Assault, Personal Assault, and Accident Survivors

Ideas from this research program should be used only under appropriate supervision and after you have acquired sufficient knowledge and practice. There are important implications for your practice here, but considerable experience with severe distress is necessary before exploring this experimental treatment.

In a study conducted in Atlanta by Rothbaum and Keane (2012), 137 trauma survivors (about one-third had experienced rape, one-third assault, and one-third an automobile accident) were divided into two groups. The first group received standard trauma assessment; the second group received the assessment plus the experimental systematic treatment, which consisted of an initial treatment followed by two additional sessions one week apart.

Dr. Rothbaum summarized the first session:

> We asked people to go back to the traumatic event, to go through it in their mind's eye and recount it out aloud over and over. We tape-recorded it, and we gave them that tape to listen to. All of this happened very quickly, in about an hour, because they had already been in the ER [emergency room] for a long time and just wanted to go home.

The patients were given cognitive behavioral therapy in which client negative thoughts were identified (e.g., "I never will feel safe again," "I'll never drive a car after that accident"). Strategies such as thought stopping and cognitive reframing were taught as ways to avoid harmful cognitions and emotions. Sessions 2 and 3 continued the process of debriefing, with continued emphasis on homework. Follow-up at 12 weeks revealed that the sexual assault victims had substantially reduced their amount of posttraumatic stress; the personal assault and accident survivors had also improved, but not as much. "More are going to end up with PTSD at week 4 and week 12 if they don't get the intervention," said Dr. Rothbaum.

This study obviously endorses the importance of storytelling to an empathic listener and the common need to repeat a traumatic story many times, whether it involves sexual assault, flood, or even a traumatic divorce.

Think of the need for counselors to debrief what people have witnessed at the scene of an accident, perhaps seeing a dead child with a bloody mother stuck in a seat belt and the father stunned and speechless. EMTs (and police and firefighters) don't forget experiences like this; they wear on them emotionally and frequently lead to depression. There is a real need to provide counseling and support after such traumatic experiences.

Counselors and therapists also experience trauma burnout. Counselors often suffer burnout when they work several days with a major disaster, listen to endless sad stories on a crisis line, or just do daily intervention work at a mental health center. The continual load of people in crisis wears on helpers, who may become traumatized themselves as they listen to horrific stories. Counselors need support when they work with these difficult situations. Counseling and therapy for the counselor needs to be considered as part of this support process.

Microskills and the five stages give you a start in understanding crisis work, but you have much more to learn to be fully helpful in such situations. At the same time, some crisis situations require many helpers and counselors. Seek some training and offer yourself to others.

Observe: Crisis Counseling First Session Transcript

Each type of crisis is different; adapt your approach accordingly. Establish trust and the working relationship as quickly as possible. You will often have to act swiftly and sometimes decisively to help your clients reach the next stage beyond that first session.

Following is a sample transcript involving a family in a big city dealing with the loss of their apartment after a fire. The counselor, Angelina Knox, meets the mother, Dalisay

Arroyo, in the office of one of the managers at the community center the day after the fire. Dalisay, 31, is employed as an aide in a nursing home and has two children. The father has only occasionally been involved since the children's birth. The fire department took Dalisay and her children to her parents' small apartment where they spent the night, but they obviously can't stay there more than a few days. Thus, as the counselor prepares to meet with the mother, first responders have already worked with early crisis safety and basic needs. The counselor may need to focus more on emotional reactions and planning for the future.

This transcript is an edited and condensed version of a half-hour session. This large city has a history of preparation for crisis; it is in a flood zone and subject to summer fires. In addition, Homeland Security has strengthened existing resources. Thus, the session occurs within an ideal support system, something that is not available in all settings. Therefore, we will review the crisis situation again after the transcript, outlining what can be done in more difficult situations with inadequate support systems.

Counselor and Client Conversation	Process Comments
1. *Angelina:* (Walks to the secretary's office, smiles warmly, and invites Dalisay in.) Hello Ms. Arroyo, I'm Angelina Knox. You and your children have had a terrible night. I'm a community counselor here in town and want to see how things are going and how I might be helpful. But, before we start, are there any questions that you might want to ask me?	Angelina is ready to spend time on developing an empathic relationship, but like many trauma survivors, the client wants to start immediately.
2. *Dalisay:* Angelina, I'm not sure where I should go or what I should do. I don't want to stay with my parents. They are good people, but don't have room for us and they get impatient with the kids. All my furniture is gone. I don't know what to do. (Starts crying softly.)	This is a common type of statement in crisis. Clients are "all over the place," topic-jumping. Other clients may be unable to talk coherently; still others angrily demand that action be taken immediately. Be ready for almost any reaction, and remember they are all normal and to be expected.
3. *Angelina:* It's really hard . . . really hard. (She sits in silence for a moment until Dalisay looks up.) Your reactions make sense and are totally normal. It will take some time to sort things out, but we have some resources here that will help. But, before going on, could you tell what happened?	Angelina acknowledges Dalisay's feelings and encourages her to tell her story, while seeking to normalize her thoughts and emotions. (Interchangeable empathy)

Clients reacting to trauma need to tell their stories. Some will tell them at length, while others may simply describe the bare facts. Emotions will vary from a loss of control to numbness without much feeling expressed. The following is a much shortened version of what was said over 5 minutes of interaction. More tears flowed, but Dalisay also felt relief that no one was hurt. Before calming can occur, we need to be with the client wherever he or she is.

4. *Dalisay:* I was almost asleep and then I smelled something strange in the kitchen. I went in and there was a small fire in the wastebasket. I must not have put the cigarette out. Then, all of sudden, it went "poof" and I ran to get the children out. . . . It spread so fast, but the neighbors called the fire department right away and only our apartment is gone. But then we got out in the cold. Firemen wrapped their blankets around us, asked if we had any help, and then they took us to my parents, 10 blocks away. But during all that time, the children were frantic and I couldn't quiet them. Their dolls and toys are gone. They couldn't stop crying until Grandma held them.	Throughout the longer story, previous 5 minutes, Angelina offered solid attention and a fair amount of natural spontaneous body mirroring. Her comments were short and usually took the form of encouragers and restatement. She did acknowledge Dalisay's emotions, but did not reflect them, believing that would be more appropriate later. Angelina's session behavior represented interchangeable empathy.

(continued)

Counselor and Client Conversation	Process Comments
The fire chief called this morning and said that you could likely help me figure out what to do next. My father drove me here on the way to work, the children are with their grandmother, but I guess I'll have to walk back to them, but all my warm clothes are gone. I called the nursing home today and the shift supervisor said that I could have the rest of the week off, but that likely means no pay and I can hardly pay bills now. (Serious crying) I don't have enough money to rent a new place, but I can't stay with my parents.	
5. *Angelina:* Dalisay, you and your daughters have had a terrible experience, but it is good that you got out in time and had your parents to stay with. I can sense the horror you must have experienced and felt, even though I wasn't there. And . . . then . . . the children. I can see that you worry about them. It's great that your parents are close, even though they can only do so much. As I listen to you, I get the feeling that you already have useful strengths and some clear ideas about what needs to be done. That contact with your shift supervisor was wise and will be helpful in the long run. Not everyone . . .	What we see here is a brief summary of Dalisay's situation and recognizing her emotions without pressing issues. Angelina then brings in a family focus, along with feedback supporting what Dalisay has done already to remedy her situation. Dalisay, like almost all people in crisis, has assets, strengths, and resources. (Interchangeable empathy with some additive dimensions)
6. *Dalisay:* (Interrupts anxiously) Thanks, but I can only be gone so long. As soon as everyone was safe, I started thinking how things could work out and I realized that I must hang onto my job. It's really scary. How am I to manage?	"What am I to do next?" While listening skills remain central, this is the time for Angelina to move to more direct influencing in the session. Dalisay needs listening and emotional support, but the real issues require action.
7. *Angelina:* I hear your worry. There are some things that our office can offer. We are lucky here in that we have a trauma relief center that provide much of what you need, including some limited financial help for a few days. You won't have to stay with your parents long, as I think we can arrange for a temporary furnished apartment for you. I've already contacted the Women's Center and they have clothes and some kitchen essentials that will help. If you are interested, I'll call and arrange for a time for you to meet with them. So, you see that there are several possibilities, but we don't want to do anything until it makes sense to you.	Angelina again acknowledges emotions and comes up with very specific directives and suggestions as to how she and her agency can help. Is Angelina offering too much material aid so soon? Certainly it would not have been too soon to offer hurricane or flood survivors clear statements of what actually could be done for them. Many felt lost in the vagueness of helping efforts, and sometimes more was promised than would ever be delivered. Here we see a large city well prepared for crisis—Homeland Security has done its job here. (Potential additive empathy with action)
8. *Dalisay:* (Seeming relieved) Wow, that is more than I expected. I thought I'd be hanging like Katrina survivors. It's terrific that I can get an apartment, and I'm amazed at the possibility of financial help. This will enable me to keep my job and take care of my children. But the next thing is, what about the children and school?	We see positive movement on the Client Change Scale. Dalisay is moving from "I can't" to "I think I can," with Angelina's help.
9. *Angelina:* Well, we need to talk about that. We will try to find an apartment near your old place so that they don't have to change, but that might not happen. We will do the best we can. I know that it isn't easy for you or them.	Try not to overpromise, but to say what you and your agency really can do in a straightforward fashion. Following information giving, we see a brief acknowledgment of feeling. (Lower-level interchangeable empathy)
10. *Dalisay:* I feel a little better, but still a lot anxious and worried.	More movement on the CCS.

Counselor and Client Conversation	Process Comments
11. *Angelina:* Clearly we need to talk over the fire in more detail, the fright you experienced, what it did to the children, and how you handled it. And then there is a lot of worry over what will happen next. If we can get together tomorrow or the next day, we can do some more serious debriefing of what happened. Would you like to do that?	As noted, this is an edited version of the longer session. The session has now gone on for about 20 minutes and provided the client with security about what will happen next and how Angelina will follow up. Debriefing of the story and the trauma needs to start as soon as possible. Note the open invitation to talk, rather than telling Dalisay that she has to return. This moves toward a more egalitarian relationship.
12. *Dalisay:* Angelina, you have been so much help and so understanding. Yes, I'd like to talk about what happened. All of us, the children and me, had nightmares last night. It wasn't good. But I know that we have to get settled right now, and I'll meet and talk with you later on this.	Dalisay is much calmer than she was at the beginning of the session. She has moved from a Level 1+ on the Client Change Scale to a beginning Level 3. Emotionally, at least in the moment, she may have reached Level 4, but don't expect this to hold unless further counseling and support are provided.
Like many crisis sessions, this one moved from topic to topic and stage to stage with a flow that did not follow the typical pattern. Those who have gone through trauma need (1) personal supportive contact; (2) understanding and clarification of the crisis trauma; (3) awareness of their own personal strengths, as well as what external resources are available to them; (4) some short-term achievable goals; and (5) an immediate, clear, concrete action plan, with arrangements for later personal follow-up and further discussion and debriefing as soon as possible.	
13. *Angelina:* Let's write down together where we are and what we can and need to do before we meet again. Where shall we start?	This directive brings in Dalisay as an egalitarian partner in finding solutions. Angelina could tell her client what to do, but success is much more likely if Dalisay is respected and fully involved. (Additive empathy)
14. *Dalisay:* I really appreciate that you could help us find housing. Could we begin with that? (Angelina brings out paper and pen for both of them, and they start to work.)	It takes 10 minutes for Angelina and Dalisay to write down the action plan—who will do what and when. Dalisay occasionally starts to cry, but more easily regains self-control. Out of this come workable alternatives that can be implemented in stages. Follow up on actions and debriefing of the trauma, but this has to be something that Dalisay wants. The children also need to tell their stories, and the school counselor needs to be consulted. At some point, it may be useful to bring the grandparents in.

Crisis counseling typically does not include formal action plans. What is needed, as illustrated above, is to write down clear action steps that the client can take today, tomorrow, and next week. If possible, arrange to follow up with clients about plan completion. In some situations where you are working with emergencies such as a flood, tornado, or hurricane, your major task is immediate support and helping them find a place to stay.

Crisis counseling demands much from you, the counselor, but it also provides many rewards when you can provide concrete help and see relief start to come in for the client and the family. But imagine what it feels like in a major crisis when you don't have the resources described above and you meet with 10 or more people who have just gone through a fire, flood, or earthquake the middle of the night in the rain. These clients have even more needs, and they could be hungry. You may only have 15 minutes and never see the person again. Thus, the calming and caring are needed continually, both for the client and yourself.

Again, burnout can be a problem for the crisis counselor. There is also your own emotional involvement. You may care for clients and their future, but follow up to make sure that they have followed an action plan may not be possible, leaving you wondering how helpful you (and the crisis team) really were. Thus, crisis counseling can often turn into a developing crisis for the helper. This means that you need to take care of yourself throughout each day. Take breaks, seek to get enough sleep, try to get a little exercise, and make sure that you debrief your experiences with understanding colleagues and/or supervisors.

Suicide Watch: Awareness and Knowledge[1]

> 94% of those who had tried to commit suicide on (San Francisco's) Golden Gate Bridge were still alive or had died of natural causes. . . .
>
> People who attempt suicide are always subject to sociological risk factors, but they need an idea or story to bring them to the edge and justify their act. If you want to prevent suicide, you want to reduce unemployment and isolation, but you also want to attack the ideas and stories that seem to justify it.
>
> —David Brooks

Suicide rates vary widely around the world, and even within states and provinces, with no clear patterns. We suggest that you search for data on your home country and region. Data for the United States vary, but mortality rates for suicide are high (Rockett et al., 2012). Internationally, we see the same trend. Youth suicide is of particular concern, as suicide is now the third leading cause of death in those 15–24 (Anika Foundation, 2013). More specific data for the United States may be seen in Table 13.2.

Background That Might Lead to a Suicide Attempt. A review of research literature lists 11 key factors to consider as indicating the possibility of a suicide attempt: severe anxiety, panic attacks, depression and the inability to experience pleasure, alcohol, difficulty in concentration, sleeplessness, hopelessness, employment problems, relationship loss, a history of physical/sexual abuse, and especially a history of past suicide attempts or deliberate self-harm (Sommers-Flannagan & Sommers-Flannagan, p. 248). To this list we would add the dangers of drug abuse, serious health issues, and serious interpersonal conflict such as bullying or harassment. Bad economic times, such as the recent long depression with difficulty in finding work to match talents (or any work), can also be a suicide cause.

TABLE 13.2 U.S. Suicide Data, 2014

	Number	Per Day	Rate	% of Deaths	Group (Number of Suicides)	Rate
Nation	42,773	117.2	13.4	1.6	White Male (29,971)	24.1
Males	33,113	90.7	21.1	2.5	White Female (8,704)	6.9
Females	9,660	26.5	6.0	0.7	Nonwhite Male (3,136)	9.6
Whites	38,675	106.0	15.4	1.7	Nonwhite Female (956)	2.7
Nonwhites	4,098	11.2	6.0	1.1	Black Male (1,946)	9.2
Blacks	2,421	6.6	5.5	0.8	Black Female (475)	2.1
Elderly (65+ yrs.)	7,693	21.1	16.6	0.4	Hispanic (3,244)	5.9
Young (15-24 yrs.)	5,079	13.9	11.6	17.6	Native Americans (489)	10.8
Middle-aged (45-64 yrs.)	16,294	44.6	19.5	3.1	Asian/Pacific Islanders (1,188)	6.1

American Association of Suicidology, http://www.suicidology.org/Portals/14/docs/Resources/FactSheets/2014/2014datapgsv1b.pdf.

[1]John Westefeld, University of Iowa, a nationally recognized expert on suicide, reviewed this section. We thank him for his comments.

The availability of guns has become an important element in suicide. About half of suicide deaths in the United States occur this way (Westefeld, Richards, & Levy, 2011). Needless to say, availability of weapons for depressed and suicidal clients is an important issue (Westefeld et al., 2012).

Risk Assessment. Understanding and working with suicide is not the province of this book, but it seems important that a few basics and suggestions for further reading and follow-up be provided. The most central of these issues in maintaining a watchful eye for suicide potential, providing immediate crisis support, and ensuring a careful referral with client follow-up to ensure that he or she actually appears for sessions. An excellent next step beyond this brief section is to download the *Suicide Risk Assessment Guide* at www.mentalhealth.va.gov/docs /suicide_risk_assessment_guide.doc. Several of their useful guidelines are summarized below.

Strengths and Resources. The background that may lead to suicide was outlined above, but the *Risk Assessment Guide* suggests looking for strengths and resources to build on, both for the here and now of the interview and for long-term safety of the client. The guide points out the following, which will be familiar to you from our emphasis on positive psychology and strength-based approaches. Use all of these as you seek to support your client while you plan for appropriate referral.

- Positive social support
- Spirituality
- Sense of responsibility to family
- Children in the home, pregnancy
- Life satisfaction
- Reality testing ability
- Positive coping skills
- Positive problem-solving skills
- Positive therapeutic relationship

As in decisional counseling, it can be helpful to discuss alternatives for resolution of issues, but the client does need to be able to listen to you. We would recommend keeping the counseling simple rather than theoretically complex. Be with the client empathically in the here and now.

Warning Signs of Impending Suicide. The three key warning signs are (1) actual threat to hurt or kill oneself; (2) seeking access to pills, guns, or other routes; (3) talking or writing about death, dying, or suicide (including giving away valued objects or pets to friends or family). In these cases, take immediate action. The *Risk Assessment Guide* reminds us to remove anything lethal and keep the client safe and with some caring person available. Depending on the level of risk, if necessary get immediate help and facilitate moving to a hospital.

The basic principles of crisis counseling remain. Your calmness, empathic caring, and ability to listen are central, but you are also required to make decisions. As much as possible share the decisions with the client. Often it will be important to bring in the family or friends to help provide further support and help to implement any plan.

Asking Key Questions. The *Risk Assessment Guide* includes a pocket card that summarizes key issues and recommended questions. Being direct is something that interviewers and counselors may have trouble with, but it is essential here. Be matter of fact, show concern, but not shock or worry. One way to start is:

> I appreciate how difficult this problem must be for you at this time. Some of my patients with similar problems/symptoms have told me that they have thought about ending their life. I wonder if you have had similar thoughts?

Then go on to ask:

> Are you feeling hopeless about the present or future?

If yes, ask:

> Have you had thoughts about taking your life?

If yes, ask:

> When did you have these thoughts and do you have a plan to take your life?
>
> Have you ever made a suicide attempt?

In this process, listening, nonjudgmental warmth, respect, and caring will facilitate openness and trust. Show that you are present in the here and now and available with understanding and support.

Avoiding the "Why" Question. This is not a time for focusing on rational explanations. The client may want to swear you to secrecy. This is not possible as it could lock out key safety procedures. Thus, be respectfully honest and open about the nature of the relationship.

Other Recommended Resources. YouTube has a number of role-plays of various lengths on crisis counseling. *Teen Suicide Crisis Counseling*, a 10-minute presentation on YouTube, covers many issues related to youth; it is also designed to be shared with teens. *Why We Choose Suicide: Mark Henick at TEDx Toronto*, a 15-minute presentation also on YouTube, discusses Henick's own attempted suicide.

Awareness, Knowledge, and Skills of Cognitive Behavioral Therapy

> [People] are disturbed not by things, but by the view that they take of them.
>
> —Epictetus

> The key to change . . . is doing.
>
> —Carlos Zalaquett

Philosophy of Cognitive Behavioral Therapy

Cognitive behavioral therapy (CBT) originated in two different philosophic traditions. The cognitive portion of CBT is rooted in Epictetus' famous Stoic statement above. Victor Frankl's logotherapy is often seen as the first cognitive theory because of his emphasis on reframing cognitions to more positive thought patterns, though Frankl also stressed the importance of taking thought into action. But it took Albert Ellis to bring cognitive work to center stage with what he first called rational emotive therapy (RET). He soon changed the name of his theory to rational emotive behavioral therapy (REBT), emphasizing the importance of making ideas and cognitions real through behavioral change and homework. Aaron and Judy Beck have since become central to cognitive therapy, and Donald Meichenbaum is known for his integrative CBT model, with a strong emphasis on behavioral change and stress management. CBT is now integrative and eclectic and includes many strategies advocated by other theories. Ultimately, we can view CBT as pragmatic and practical, searching constantly for "what works" with each client.

Key Methods of Cognitive Behavioral Therapy

The National Association of Cognitive-Behavioral Therapists defines CBT as follows (www .nacbt.org/basics-of-cbt.aspx):

> Cognitive-behavioral therapy does not exist as a distinct therapeutic technique. The term "cognitive-behavioral therapy (CBT)" is a very general term for a classification of therapies with similarities. There are several approaches to cognitive-behavioral therapy, including Rational Emotive Behavior Therapy, Rational Behavior Therapy, Rational Living Therapy, Cognitive Therapy, and Dialectic Behavior Therapy.

Frame of Reference. CBT is an information processing system in which thoughts influence our feelings and actions. The purpose of CBT is to explore thought patterns, help the client see that they are ineffective or irrational, and enable the client to "think different." Some specifics of CBT, as outlined by the national association, include the following:

1. A base of cognitive response as being key to change
2. Time limited with specific goals
3. A sound relationship is needed, but not central
4. A collaborative venture between counselor and client that uses a Socratic question-and-answer style
5. Based on aspects of Stoic philosophy ("It is not things, but what one thinks of things that counts.")
6. Structured and directive
7. Based on an educational model (psychoeducation)
8. Relies on induction (encourages clients to look at their thoughts and draw their own conclusions, although the counselor will use serious challenges and confrontation)
9. Homework is essential. Recently, Beck and Broder (2016) have substituted the term *action plan* for *homework*.

CBT also encourages self-healing and aims to increase clients' competency, providing coping skills they can use when facing new concerns and challenges.

Key CBT Propositions

- Cognitive activity affects behavior.
- Cognitive activity may be monitored and altered.
- Desired behavior change may be effected through cognitive change. (Dobson, 2009, p. 4)

As you can see, much of what we have emphasized in this book is in accord with CBT tenets. However, we believe that attention to relationship, feelings, and meanings is more essential than CBT typically suggests. An excellent illustration of why CBT also needs to consider emotional experience can be seen in the session transcript that follows. In fact, many CBT specialists agree with neuroscience findings that thoughts are often based on feelings and emotions. Box 13.3 highlights neuroscience findings related to CBT. Albert Ellis's rational emotive behavioral therapy (REBT), of course, includes emotional issues as a basic factor.

BOX 13.3	**Research Evidence That You Can Use**

How Neuroscience Research May Affect Our Daily Counseling and Therapy Practice

Functional magnetic resonance imaging (fMRI) of battered women has revealed that certain patterns of brain activity predict better response to cognitive behavioral therapy (Aupperle & Hunt, 2012). More specifically, greater anterior cingulate activity and less posterior insula activity were found to be critical in the different responses to treatment. The anterior cingulate monitors conflicts in information processing and guides decision making, while the insula has been found to relate to pain and the six basic feelings of Chapter 7.

With some studies showing only a 50% response rate to CBT in the domestic violence population, "there is room for improvement," noted lead investigator Robin Aupperle. "If we can find techniques to target these areas specifically to enhance CBT and future treatments, that is important."

This study, and more like it to follow, may well lead to more precise and specific counseling and therapeutic treatments. We can anticipate further research in neuroscience that will guide our practice and enable us to engage in more meaningful and successful counseling and therapy. We suggest that you "stay tuned" and maintain awareness of the avalanche of new findings that appear almost daily.

We have said that "counseling changes the brain." We now can help clients change memories, and thus their thinking, feeling, and behavior. With this new knowledge, clients can find new and more powerful meanings for their lives. We are likely nearing the time where we can anticipate with increasing precision the power of effective counseling and therapy.

Observe: Cognitive Behavioral Session Transcript

Following is a transcript of Carlos Zalaquett using basic cognitive therapy techniques. He focuses on identifying and changing negative automatic thoughts. The client, Renée, has been referred by her practicum site supervisor for specific work on her self-defeating thoughts producing anxiety and affecting her ability to perform as a counselor. She has completed her paperwork and discussed issues regarding confidentiality and differences in gender and culture.

We join the conversation about 15 minutes into the session after Carlos has structured the interview and established the relationship. Carlos has brought out the client's concerns, thoughts, and feelings around her work as a counselor. He has also searched for signs of resilience and brought out stories of strength. Specific goals have been defined. In this early phase of a CBT interview, the session will look much like other sessions you have read in this text, in that the emphasis is on listening and joining the client's world. The client has more talk time than the counselor.

As we turn to Stage 4 of the session, notice the use of the following three strategic questions to promote cognitive restructuring: What is the evidence supporting the conclusion currently held by the client? What is another way of looking at the same situation but reaching a different conclusion? What will happen if, indeed, the current belief is correct? Carlos also provides considerable support and encouragement, promotes active participation in the process, and encourages the client to take credit for her positive assets and achievements. Again, this transcript has been edited and shortened from an hour for clarity and space considerations. The focus here is on restorying, and Carlos uses many influencing skills. His talk time in the final two stages increases, but the focus remains on the client.

Counselor and Client Conversation *Stage 4. Restorying*	Process Comments
1. *Carlos:* Well, to review—this is an interesting situation because your internship site supervisor referred you. I understand that you did well in the program, but perhaps could use some help as you prepare for sessions. Is that right? Is that your goal?	Carlos briefly summarizes Stages 1, 2, and 3, and we see the reason for Renée's consultation. Notice how he uses a questioning tone at the end to check out the accuracy of his statement.
2. *Renée:* I've been feeling some anxiety with my new job as a counselor. A client that I really like didn't come back. I just recently graduated, so every day I'm feeling insecure about seeing clients. I just feel nervous all the time.	Renée's body language indicates tension. She is sitting back in her chair and her legs are jiggling. Renée seems to start the session at Level 2 on the Client Change Scale.
3. *Carlos:* I can see some discomfort as you speak about it. Are you hesitant about what the client's going to do, or you worry because you feel that you may not be as competent with them?	Acknowledgment of emotions followed by an interesting question on her nervousness coupled with a mild interpretation. Using a question for an interpretation at times softens the impact and allows the client to reflect on what you say. (Potentially additive)
4. *Renée:* That's exactly it. I feel that I'm second guessing myself. Am I doing the right thing? Is this what's best for them? You know, I just graduated. So I just have all this insecurity that I'm trying to deal with every day. It's becoming increasingly difficult.	The additive questioning interpretation brings out both cognitions and the emotions associated with her difficulties.
5. *Carlos:* I see. So, share with me some situations in which you felt that way.	This could be best described as a question formed as an encouraging statement. The focus is on the main theme or issue. (Potentially additive)
6. *Renée:* Before I see a client, I always feel this way.	Renée sees her behavior as a pattern (abstract formal operational thought).
7. *Carlos:* So, as you are going to see a client, you feel anxious about how will you do, how you will perform?	Reflection of feeling, combined with a question. Carlos seeks to confirm the experience of the client. (Interchangeable)
8. *Renée:* Am I good enough?	Here we see one of Renée's central cognitions, which in turn leads to feelings of anxiety.
9. *Carlos:* How good you are and how competent, I see. Well, I think I understand the reason for the referral. I'm going to focus on what we call cognitive behavior therapy, also known as cognitive therapy, to help you cope with your worrying thoughts. In cognitive therapy we believe that your thoughts affect your behavior and your mood. For example, a person may say "I'm going to see a client and I feel anxious," suggesting that the situation is what triggers the anxiety. However, what we have learned in CT is that it's not the situation that triggers the reaction. There is something in between. And this is your thoughts or images. The question is, do you understand what happens between seeing a client and your emotional response?	Carlos restates the central cognition and lets the client know he has understood her concern. He then structures using instruction/psychoeducation to introduce the basics of cognitive behavioral therapy. The question at the end seeks to check out if Renée has an understanding of what he just said.
10. *Renée:* I haven't really thought about it because I just feel so overwhelmed with emotions that I can't really put the thing into perspective.	Renée's emotions are leading her cognitions. Reversing this pattern is a major goal of CBT. The cognitive way we think about things affects the nature of things and how we feel about them.

(continued)

Counselor and Client Conversation *Stage 4. Restorying*	Process Comments
11. *Carlos:* So let me help you by engaging in a process of discovery of "what is in between." What's the first thing that happens when you picture in your mind when you are going to see a client?	Cognitive therapists see clients as practical scientists and help them to engage in self-discovery. Imagery is used to help the client gain a different understanding of her issue. (A key additive strategy)
12. *Renée:* I just get nervous.	
13. *Carlos:* You get nervous. Exactly. It's such an automatic connection. It is hard to ask what else could be there.	Restatement of feelings followed by information/psychoeducation to explain that immediacy of emotional reactions precludes clear thinking. (Continues the additive strategy)
14. *Renée:* Right.	
15. *Carlos:* Now, let me ask you then to stop for a second and notice what happens when you are the brink of seeing a client. . . . Now, let me ask what crossed your mind?	Imagery directive to help Renée observe her thoughts. One key to CBT is helping clients observe their own thoughts, feelings, and behaviors. The question is similar to free association strategies and is a useful directive in CBT. (Potentially additive)
16. *Renée:* Um, am I gonna use the appropriate counseling techniques for this client?	Renée thinks back and identifies the negative cognition. She is starting self-observation. This represents the awareness and the beginning of Level 3 on the CCS.
17. *Carlos:* Um-hum. And what do we call this thought that goes through your mind?	Carlos uses a minimal encourager and questions to draw out the client's label for the cognition. Use the client's key words and language from everyday vocabulary.
18. *Renée:* Negative thinking.	Voila! Renée shows clear understanding and thus is now at Level 3 on the CCS.
19. *Carlos:* Negative thinking. Very well, this is exactly how we call these in our work, negative thinking or negative thoughts. And the reason, as you see, is that they are negative in nature. Okay.	Restatement and psychoeducation/information helps Renée feel understood and provides her with additional CBT information. (Interchangeable and somewhat additive)
20. *Renée:* Okay.	
21. *Carlos:* So the sequence in our view is that when you are facing an event, something goes through your mind, and that is really what triggers your reaction, your emotional reaction. Are you with me?	Carlos provides further psychoeducation, but checks out to see if Renée understands. (Additive)
22. *Renée:* Yeah. Right now, I'm thinking about other things that go through my mind before I get ready to see a client. Even just talking about it now makes me feel a little tense.	Renée "gets it" and demonstrates a good level of understanding by providing further examples of thoughts and physical reaction. Clear Level 3 on the CCS.
23. *Carlos:* Very good. It's interesting you say this because I was at the brink of asking you to go back to the original situation to see if you could discover some other thoughts. So let's look at your situation again. Very good, educate me. Help me know what else goes through your mind as you are facing these types of situations?	Notice the frequent use of encouragers. The counselor engages the client as a co-collaborator, as she is the expert on her own experiences. Renée can help Carlos understand her better, promoting an active process of self-discovery. (Additive)
24. *Renée:* Is the client gonna want to come back?	Our greatest fear! And 25% of clients don't come back, more if the client is culturally different from you. There is always some truth behind cognitions that lead us to fear and inaction.
25. *Carlos:* Um-hum . . .	Use of encouragers to further promote self-discovery.

Counselor and Client Conversation	Process Comments
Stage 4. Restorying	

26. *Renée:* Am I good enough for that client? Am I gonna be successful? These types of thoughts.

> Notice how Renée's responses address previous questions raised in the session. She is engaged.

27. *Carlos:* Good, sometimes these thoughts, that we call negative thoughts, are part of what we call a core belief. These are long-held thoughts that affect our behavior and emotions. It's not easy to identify core beliefs, but they are central to whatever we do. I mention this because automatic thoughts spring from core beliefs, which should be the ultimate focus of our attention.

> But for the time being, we can focus on the negative thoughts. So this is a two-step process. Step one is working with the current thoughts. Step two is discovering the core beliefs and then dealing with these over time.

> But let's go back to your current situation because I understand you want to do something about the negative thoughts. Since you understand this very well, we will continue to analyze your situations from a CBT therapy point of view. Is this okay?

> Instruction/psychoeducation to further advance the CBT model. Carlos ends with a brief summary of Renée's concerns and, in the spirit of collaboration, checks to see if he is heading in the right direction and she concurs.
>
> What you see here is a brief summary of an hour session. Remember that CBT does not typically move this fast, nor do we always have a client who grasps the purpose of counseling this quickly.

28. *Renée:* Okay.

> Psychoeducation will be ineffective unless the client is ready, willing, and able.

29. *Carlos:* I have a chart that I share with my clients to record their automatic thoughts. As you can see, it's a chart with three columns. The first is used to report a specific event; okay, then here in the third column we look at the emotional or behavioral reaction; and then we spend time in the second column identifying the thought or image that may be involved in the situation.

> Counselor introduces a CBT chart to record automatic thoughts. This chart in its simplest form displays three columns: Event—Thought—Response. See Figure 13.1.

30. *Renée:* Okay.

31. *Carlos:* What we did before actually followed these three columns (points to chart), so let's look at it from this chart's point of view. Can you see it well?

> Carlos has laid the foundation for the work with this chart and builds on what has transpired in the session so far.

32. *Renée:* Yes, I see it.

33. *Carlos:* Think about your situation right now; you are going to see a client. Okay. And you're feeling anxious. As you represent this in your mind, pay close attention to your thoughts. What thoughts did you identify?

> Carlos instructs Renée in how to use the three-column chart. The event (seeing a client) is first. Her anxiety about the session is in the third column.

THOUGHT RECORD		
Event	**Thought**	**Response** (Emotions/Behaviors)
Meeting a client	*I am not good enough at counseling.*	*Anxious, worry, and feel less effective in the meeting, although somehow I survive. But, I wonder about the client.*

FIGURE 13.1 Basic thought record sheet.

(continued)

Counselor and Client Conversation	Process Comments
Stage 4. Restorying	
34. *Renée:* There were several of them: "I'm not sure if I'm going to be proficient." "I'm not sure I'm going to do well." "I'm not sure if the client will like to come back or if the client will be successful."	Here Renée has been asked to explore the cognitions that will be recorded in the second column. Her responses are Level 3 on the CCS scale as Renée shows awareness of what is happening to her, but already in a more optimistic framework.
35. *Carlos:* Good job, I'm going to put these thoughts down here. Now, we have identified the specific parts of this chart, including the negative thoughts that are creating the anxiety in your relationships with clients. The situation, the thoughts, and then the emotional reaction. This column representing the thoughts is essential. In our view, it's not the situation itself that upsets you. It's what you think or how you perceive the situation that affects you. That's the reason why the focus is on the thoughts. Sounds reasonable?	Carlos demonstrates the use of the chart and helps Renée familiarize with CBT's view of the importance of thought processes.
36. *Renée:* Yes, it does.	
37. *Carlos:* Now we are going to take one step forward. Think about past experiences. How many times have you been, for a lack of a better word, incompetent?	Carlos challenges the automatic thought of incompetence by asking for evidence in its support. CBT practitioners usually ask, "What's the evidence that the thought is true? That it's not true?" Notice the client's response. (Potentially additive)
38. *Renée:* I don't think I've ever been incompetent. I always try my best.	Anxiety rides on the shoulders of expectations, not of actual experiences.
39. *Carlos:* Um-hum. So let me ask you the opposite then. What evidence do you have that suggests that you may be competent?	Positive assets search. Counselor begins to search for strengths and positive experiences and thoughts. (Additive)
40. *Renée:* Well, in my work throughout practicum and internship, I did well. I completed all my counseling courses successfully. I graduated, so that must mean that I've received the proper training and that I should be competent to perform these new skills that I've been given.	A further demonstration of CBT: Negative thoughts have less actual support than positive thoughts, but the negatives influence behavior and feeling more. This is a truism reinforced by neuroscience research.
41. *Carlos:* Help me understand this. So you don't think that you may be very competent, but then there are these facts that, if I heard you correctly, demonstrate that you have been competent. How do you reconcile these?	Counselor uses gentle confrontation to help client restory.
42. *Renée:* How am I competent?	Notice how Renée's response demonstrates a shift to a positive self-perception. Early step to Level 4 of the CCS and real change.
43. *Carlos:* Yeah, how do you know that you have been competent?	Open question to promote further restorying and drawing of competence evidence.
44. *Renée:* Like I said, probably my graduate coursework. That to me is evidence that I've been competent. Perhaps—I guess I learned in my classes that my first sessions would not be perfect and that some clients do not return regardless.	

Counselor and Client Conversation	Process Comments
Stage 4. Restorying	
45. *Carlos:* How fascinating. You are right about what happens to all of us. I was curious about how you knew that you have been competent in the past. Now let me ask you a different question. Let's say that you see a client and that you don't do your very best. What's the worst that could happen?	The counselor continues to work with the client to challenge her negative belief. He introduces a CBT technique called *worst case scenario.*
46. *Renée:* Maybe the client won't come back. One didn't.	We've heard that before, but now we are working on the negative condition on a basis of positive assets and strengths.
47. *Carlos:* Maybe they won't come back. Uh-huh. And how bad would that be?	Restatement followed by question to further explore client's worst expectations. (Potentially additive)
48. *Renée:* I guess it isn't so terrible. Would just give me a chance to maybe practice more, work on my skills more. I guess it wouldn't be the end of the world or anything.	Client reveals the catastrophic thought underlying her fears but demonstrates she can challenge that thought.
49. *Carlos:* It wouldn't be . . .?	Minimal encourager is offered to get Renée to repeat her statement and reinforce thought change.
50. *Renée:* It wouldn't be the end of the world.	Another step to Level 4 change in cognitions.
51. *Carlos:* Oh, it wouldn't be the end of the world. I see. Sometimes when my clients say something like this, I ask them if they could restate their thoughts in positive terms. Can you do this?	Reframing. Carlos assists Renée in transforming her thought into a positive statement.
52. *Renée:* Absolutely. It will be an opportunity to improve.	
53. *Carlos:* I see. Is there other evidence that suggests you have the competence to do your work?	Positive asset search. Carlos continues searching for strengths and positives. (Additive)
54. *Renée:* I have a few clients that I've seen that always come back, and they look forward to scheduling the appointments with me and seeing me again, so I guess that's good news. I must be doing something right.	Note that a negative experience can lead to fears and ineffective cognitions, even when the evidence suggests otherwise.
55. *Carlos:* Uh-hum. You know, I always wonder when people say I'm doing well, because doing well for one person may have a different meaning than it has for you, so when you say doing well, in your case that clients are coming back, what helps you to do well?	Open question to help Renée own her skills.
56. *Renée:* Um, I guess being confident in the techniques that I use with my clients. They even want to come back and see me, and it seems like I have been using the appropriate ones and that they've been working well with the client, because they want to come back and continue to work on their issues with me.	Notice Renée's active involvement in cognitive restructuring and restorying.

(continued)

Counselor and Client Conversation	Process Comments
Stage 5. Action: Generalizing new cognitions and behaviors to the real world	
57. *Carlos:* Very well. So let me go back to our initial situation. We were talking about seeing clients and the fact that you were feeling anxious about it, and we made a connection that demonstrates that negative thoughts trigger our negative emotions. I would like you to think about seeing a client, a new client right now, what will go through your mind as you work to prepare to do that?	Counselor moves to prepare client for the action phase of counseling.
58. *Renée:* I will think about the positive things that I'm doing and probably how I've been successful so far, so there's no reason for me to think that I'm going to fail or am incompetent.	The basis for change and eventual maintenance of new patterns of thought.
59. *Carlos:* Very good. In the past, when you were at the brink of seeing a new client, you felt anxious. Concern was about not doing it well, not having the client returning, not having all the confidence to do your best work, and all that was defined as the words or thoughts that triggered your emotional reaction. Now, I see that when you look at these situations, you have more positive thoughts.	Carlos summarizes CBT model and session work.
60. *Renée:* Yes, I would be more confident and enjoy my work even if the client doesn't want to come back; I realize that this it's not the end of the world; it will give me a chance to improve.	Level 4 cognitive change has been achieved, but this is still not the real world. Follow-up action will be necessary to achieve lasting change.
61. *Carlos:* Good. As you can see, with this cognitive approach we not only begin to address your situation, but also learn about its practice. What I would like you to do is to use this very same chart to monitor your thoughts. Monitor what happens when you are going to see a client. So that will be the event, and then we will see how you feel emotionally about it. Then spend some time paying attention to identify the thoughts and the dreaded consequences or results that crossed your mind in that situation. Okay. And I have a date here because I'm going to ask you to do this throughout the week. This will give us a chance to identify more clearly negative thoughts and to use those as a foundation to look for core beliefs; and I will explain more of that with more experiences, but for now this will give us a way to work using a model similar to the one we used in this session.	Carlos adds more information about the model, assigns homework, and encourages action.
62. *Renée:* Okay.	
63. *Carlos:* Any questions, any comments about following up from what we discussed just now? Anything else?	Carlos uses questions to find out if Renée has doubts or concerns.
64. *Renée:* No. It is something I can do. It will be very helpful.	
65. *Carlos:* Very good. One last question. How did it go? How do you feel?	Final checkouts to determine client's satisfaction and feelings.
66. *Renée:* Good. I feel better.	

As demonstrated in Renée's experience, automatic thoughts can trigger feelings and affect behavior. Discovering negative automatic thoughts helped the client replace them with more appropriate thoughts and improve her situation. Renée has learned a new cognitive technique in this session. Clients like her can learn to detect automatic thoughts by

- Learning about the CBT conceptualization
- Discovering negative automatic thoughts using thought recording charts
- Finding ways to replace these thoughts with more appropriate ones
- Applying these techniques in their everyday situations to effect positive changes in their lives

Action Plan and Homework

Change doesn't come from the sky. It comes from human action.

—Tenzin Gyatso, The 14th Dalai Lama

The fifth stage of the microskills interview has long been *action*, and many of your clients will be comfortable with the action plan. The final part of the session needs to include a formulation of a concrete action plan based on what was learned on the session. Renée's action plan would be to continue using the Thought Record sheets to monitor her thoughts on a daily basis in between sessions. CBT practitioners have typically used the word *homework*, but clients who hear "homework" may think that suggestions are superficial; *action plan* seems to be more acceptable to some (Beck & Broder, 2016). Action planning and follow-up are essential parts of any treatment plan because they encourage clients to take home and act on what was learned in the session. Renée needs to engage in the action plan for at least 30 days to achieve success (Ellis & Ellis, 2011).

As counseling continues, it is highly likely that Carlos will discover other areas where Renée has difficulty expressing herself and exhibits a general lack of self-confidence. The topic and goals for further sessions will then change, as will the weekly action plan. More emphasis on behavior change is likely to be a focus. As thoughts change, so do behaviors.

Treatment with CBT in later sessions potentially will include other interventions such as assertiveness training, meditation and relaxation for Renée's general anxiety, and selected therapeutic lifestyle changes to build resilience. Exercise, of course, will be central in treatment planning, assuming that Renée is interested and motivated.

Action: Key Points of Counseling Theory and Practice

Microskills and Multiple Approaches. Table 13.1 summarizes microskill use in many different counseling approaches, including decisional counseling, person-centered counseling, logotherapy, multicultural counseling and therapy, feminist therapy, crisis counseling, cognitive behavioral therapy, brief counseling, motivational interviewing, and counseling/coaching. Though all these approaches may be explained and understood in terms of their use of microskills and how the session is structured, note that their emphases are quite different.

Consider just the two examples presented in this chapter. Can you see the differences in the theoretical approaches of crisis counseling and CBT? Crisis counseling, which emphasizes careful listening to the client's story/concern/challenge, seeks to work on the issues as quickly as possible and provide appropriate support. CBT actively seeks to encourage the client to change and adopt new thoughts and behaviors. The CBT process may last several sessions, and follow-up interviews to ensure that the client actually takes action are important.

Suicide. Monitor suicide. Provide immediate crisis support if needed, offer an appropriate referral, and confirm that the client attends that meeting. Look for strengths and resources to secure the client's safety. Depending on the level of risk, if necessary get immediate help. Ask direct questions regarding suicidal thoughts, risk, plan, and corresponding behaviors. Your calmness, empathic caring, and ability to listen are central. Bring in the family or friends to help provide further support and to implement any prevention plan when possible.

Multicultural Issues. Each theory requires different adaptations to be meaningful in multicultural situations. Particularly helpful in this regard is the concept of focus (Chapter 9). By focusing on the cultural/environmental/contextual dimensions, you can bring in these issues fairly easily to all helping approaches. However, you still must recognize that the aims of each approach may not be fully compatible with varying cultures. This same point, of course, should be made with regard to the client regardless of cultural background. Some clients may prefer the Rogerian person-centered approach; others may want solutions and cognitive behavioral action. Avoid stereotyping any client with prior expectations.

Practice and Feedback: Individual, Group, and Microsupervision

Additional resources can be found by going to CengageBrain.com and logging into the MindTap course created by your professor. There you will find a variety of study tools and useful resources that include quizzes, videos, interactive counseling and psychotherapy exercises, case studies, the Portfolio of Competencies, and more.

Individual and Group Practice and Microsupervision

Exercise 13.1 **Practice with Crisis Counseling and CBT**
Select one theory and build from there.

❏ Work with a partner, switching the roles of client and counselor. Plan for a minimum session of 15 minutes, as this is likely enough to cover basics. But be flexible, as more time is often needed.

❏ Select a concern for the role-play. This time the issues need to be very specific and meaningful to you, such as a past or present conflict on the job, in the family, or with a friend or partner. Consider issues of life goals and vision. Aim for concreteness and clarity throughout the storytelling.

❑ Record each of the sessions on audio or video, perhaps using your computer, cell phone, or video-equipped digital camera to provide some instant feedback.

❑ With your partner, after completing the role-play, review the discussion of each theory and create a tentative treatment plan in accord with the basic tenet of the theory, structure, and microskill usage.

❑ Review with your partner each of your sessions using the feedback form in Box 13.4. Determine strengths and areas for improvement, and offer one another suggestions for achieving desired outcomes.

BOX 13.4   **Feedback Form: Counseling Theories**

THEORETICAL SYSTEM SELECTED FOR PRACTICE: _____

_____

(DATE)

_____   _____
(NAME OF COUNSELOR)   (NAME OF PERSON COMPLETING FORM)

Empathic relationship: Initiating the session, rapport and structuring ("Hello; this is what might happen in this session"). How well did the counselor establish rapport, and how did he or she accomplish this objective? Were preliminary goals identified? According to the theory, was goal setting carried to more specificity?

Story and strengths: Gathering data, drawing out stories, concerns, problems, or issues ("What's your concern? What are your strengths or resources?"). Was at least one positive asset or strength of the client identified? How completely did the counselor draw out the story and/or issues? Were the strengths and resources adequately explored?

Goals: Mutual goal setting ("What do you want to happen?"). Was it effective? Were the original goals of the session reviewed, and were the client's desired outcomes really clear? With brief counseling, review of goals can also be helpful at this point.

Restory: Working. How was this approached? Were thoughts, feelings, behaviors, or meanings a primary focus? What specifics did the counselor use to encourage creation of new ways of thinking and being?

BOX 13.4 **(continued)**

Action: Did the counselor help the client plan specifics for generalization to daily life? How did the counselor go about helping the client create a concrete plan for action? Was follow-up agreed to by counselor and client?

General comments on the counselor and skill usage:

Portfolio of Competencies and Personal Reflection

Assessing Your Level of Competence: Awareness, Knowledge, Skills, and Action

Developing and evaluating your skills and competence using each of the theories practiced should be included in your Portfolio of Competencies.

We will not ask you to assess your competence in any of these approaches at this point, as it is far too early and you will want to work further with each one. Rather, please focus your attention on your early impressions and where you think you might go next in building competence in these or other theoretical orientations.

Personal Reflection on Microcounseling in Crisis Intervention and CBT

How does the concept of theoretical orientation relate to your own developing style and theory? Which of the approaches presented most appeals to you? What do you think about crisis counseling and cognitive behavioral therapy?

What single idea stands out for you among all those presented in this chapter, in class, or through informal learning? What stands out for you can be a guide toward your next step.

What are your thoughts on multicultural issues and crisis counseling and cognitive behavioral approaches?

What other points in this chapter strike you as useful?

How might you use ideas in this chapter to begin the process of establishing your own style and theory?

CHAPTER 14

Skill Integration, Determining Personal Style, and Transcendence

The pyramid (bottom to top):

- Ethics, Multicultural Competence, Neuroscience, and Positive Psychology/Resilience
- Attending and Empathy Skills
- Observation Skills
- Questions
- Encouraging, Paraphrasing, and Summarizing
- Reflecting Feelings
- How to Conduct a Five-Stage Counseling Session Using Only Listening Skills
- Focusing
- Empathic Confrontation
- Reflection of Meaning and Interpretation/Reframing
- Self-Disclosure, Feedback, Logical Consequences, Directives/Instruction, and Psychoeducation
- Applying Skills to Theory: Crisis Counseling and CBT
- Transcendence, Determining Personal Style, Skill Integration

I have learned that people will forget what you said, people will forget what you did, but people will never forget how you made them feel.

—Maya Angelou

Chapter Goals and Competency Objectives

Awareness and Knowledge

▲ Review theories emphasized throughout the book. (Note that detailed reviews of crisis counseling and cognitive behavior therapy were presented in Chapter 13.)

▲ Review and integrate concepts, skills, and strategies learned in previous chapters.

▲ Explore a pre-interview checklist to help ensure that critical points are covered in the first session.

▲ Understand the place of case conceptualization, treatment planning, referral, and relapse prevention, closely related to action planning.

Skills and Action

▲ Create long-term treatment plans for a client, and keep systematic records.

▲ Increase client take-home of thoughts, feelings, meanings, and behavior in the fifth stage of the interview through action planning and relapse prevention.

▲ *The central skill and action:* Record and analyze your own session, and compare your counseling style with earlier recorded interviews. Work toward defining your unique personal style and what you wish to see happening as you work with others.

Introduction: Defining Skill Integration

You have now reached the apex of the microskills hierarchy. You and your clients will greatly benefit from a naturally flowing session and treatment plan using a smooth integration of the skills, strategies, and concepts of intentional counseling and psychotherapy. The central aim of this chapter is for you to focus on integrating your thoughts and feelings as you examine your personal style of interviewing, counseling, and psychotherapy.

The aim of skill integration is to take the microskills, stages of the interview, and your natural expertise and then examine where you stand now. As you increase interviewing competence, the result will be increasing intentionality and the ability to flow naturally with your clients.

Skill Integration	Anticipated Client Response
Integrate the microskills into a well-formed interview and generalize the skills to situations beyond the training session or classroom.	Developing interviewers and counselors will integrate skills as part of their natural style. Each of us will vary in our choices, but increasingly we will know what we are doing, how to flex when what we are doing is ineffective, and what to expect in the interview as a result of our efforts.

Take a careful look at your interview behavior, as well as your competence in analyzing the session, as that is the central focus of this chapter. Here you will find guidelines for recording a session with a volunteer client. In preparation for this, several key issues are discussed—among them are an interviewing checklist, treatment planning, case management, referral, and action planning for relapse prevention.

Awareness, Knowledge, and Skills: Review of Theories of Counseling and Psychotherapy

There is nothing so practical as becoming competent in several theoretical approaches.

—Carlos Zalaquett

You will find that your competencies developed in this book will enable you to understand and work with each theoretical system more quickly and competently. Although each of them has a conceptual framework and a worldview that differ from the others, virtually all theoretical approaches use microskills and implicitly the five stages. However, many of them historically have not given much attention to the fifth stage and action planning.

You have seen that theories of counseling all now recognize the importance of relationship and the working alliance. You may wish to turn to Chapter 1, page 15, for a review of differential use of microskills. Each theory listens to and defines the meaning of stories from a different frame of reference. The person-centered system tends to use the listening skills most frequently and focuses most on the client, with historically less attention to the cultural/environmental context. Brief counseling and counseling/coaching use many questions, and CBT uses many directives, while multicultural counseling and therapy (MCT) and feminist therapy pay more attention to the cultural/environmental context. Counseling, MCT, and feminist theories are the most eclectic of the theories, as they tend to use a variety of skills drawn freely from the other theoretical approaches.

Crisis counseling and cognitive behavioral therapy were reviewed in Chapter 13. Four other approaches to counseling and psychotherapy are summarized here.

Decision Counseling

> The doors we open and close each day decide the lives we live.
> —Flora Whittemore

Philosophy. Decision counseling is representative of an underlying philosophy in the United States—pragmatism: Abstract theory is helpful, but "let's be practical and find something that works."

In the 1890s, C. S. Peirce and William James established the philosophy of pragmatism, which is an extension of Benjamin Franklin's three-stage decision model, created around 1750: define the problem, generate alternatives, and decide for action. Peirce and James's theory is that what we think needs to show itself useful in practical matters, the purpose of thought is to guide action, and "truth" shows itself by results. An updated version of pragmatism is "Walk the talk."

Key Methods and Strategies. The five-stage decision structure plus microskills is actually the basic strategy. You have seen decision counseling in practice throughout this book. Feeling and emotion are given central attention, and the neuroscience base is critical. Unless the client is emotionally satisfied with the decision, it is less likely to be practical or beneficial. Moreover, the balance sheet is a helpful strategy to organize the decision process. In addition, decision counseling does not hesitate to draw on other theoretical approaches as appropriate.

Implications for Your Practice. By this time you have likely mastered decision counseling and are ready to move on to other theoretical systems. Nonetheless, your clients will always need to make pragmatic decisions, particularly when they face crisis. Moreover, whether you favor person-centered, cognitive behavioral therapy, or some other system, all require decisions and ultimate client action in the real world.

Person-Centered Counseling

> When I look at the world I'm pessimistic, but when I look at people I am optimistic.
> In my early professional years I was asking the question: How can I treat, or cure, or change this person? Now I would phrase the question in this way: How can I provide a relationship that this person may use for his (or her) own personal growth?
> —Carl Rogers

Philosophy. Carl Rogers believed in a humanistic, person-centered approach. Clients are the experts in their own life history and direction. They need a counselor primarily as a facilitator. Out of this philosophy comes a belief in human dignity and personal self-actualization to reach one's full potential, and thus a very clear goal of individual decision making and determining one's own life path.

Key Methods. Rogers was the ultimate listener. View his famous films or look for him on YouTube, and you will see that he has perfect attending behavior and is superb with encouragements, paraphrasing, reflecting feelings (and meaning), and summarization. While the concept was not there in his time, he was superb at mentalizing and being with the client's experience.

Person-centered theorists typically have strongly opposed the use of questions, considering them intrusive and limiting the client's self-discovery process. However, careful examination of even Rogers's work reveals an occasional nondirective question, usually associated with some form of "What do you want?" Rogers would not use directives, as used in many theories, and often spoke strongly against them.

In summary, some key aspects of Rogers's lasting legacy are the importance of relationship and empathic understanding, the centrality of emotion, and his emphasis on listening (although he did not label the skills). In addition, he showed us the importance of looking at what actually happens in the session, through his courage in being the first to record his sessions and share them, and he was ahead of his time in encouraging and working for world peace and multicultural understanding through his group work and international presentations.

Implications for Your Practice. You have seen that Carl Rogers's influence is present throughout this book, even when we don't speak of him. Regardless of your chosen theory, listening will always be central in helping your client. Continue listening, seek empathic understanding, and keep your eye on client desires and goals, not your own wishes. Be patient and keep honing those attending skills. Make your own judgments on the value of questions and influencing skills.

Logotherapy

> Once at a therapy conference (a general, not logotherapy conference) someone from the audience asked one of the speakers: "What do you call the spiritual encounter between two people?" The speaker answered "Viktor Frankl."
>
> —Anonymous

Philosophy. There is always meaning to life. Logotherapy aims to help us become our own best person through discerning meaning and purpose. Even in the most miserable circumstances, life still has meaning and can provide support. We can choose to be unhappy and live life without meaning or choose life with meaning. Paraphrasing Nietzsche's philosophy: Those who have a *why* can find a *how* and bear any situation.

Frankl's survival in the Nazi death camps and his influential life serve as a model for us all. If we can discern a meaning for our lives, we can ourselves "survive" the many challenges and difficult decisions we face.

Key Methods. That word *meaning* and its relationship to counseling and psychotherapy are perhaps Frankl's most significant contribution. No other theory speaks so well and completely to the meaning of life, but you will find logotherapy discussed in very few textbooks. Yet we have found that his thinking is so powerful that it permeates our being. Discerning what we care for, what we are about, and our life's meaning and purpose seems to transcend all other issues. Whether in a prison or a palace, what meaning does the person have to live?

Many see Frankl as the first and original cognitive behavioral theorist, and he identified himself as such. How can that be? Simply put, skilled use of logotherapy is a powerful way to enable clients to *reframe* cognitions and emotions and think in new ways, thus leading to behavior change that is suggested by the client rather than the therapist or theory.

Implications for You. Look for meaning issues as you listen to the client's story. Working in depth with meaning and discernment of life's direction takes a bit more time than brief therapy, CBT, and others, but it can be life changing. Exploring meaning can be particularly useful in crisis situations. Once clients reframe the trauma or tragedy from the meaning frame of reference, they can more readily calm down and think through concrete action plans with your help.

Beyond that, Frankl emphasized that we must live and act on our meaning. He once commented, "It's not hard to make decisions when you know what your values are." You can easily bring basic concepts of logotherapy into other theories by making meaning a focus of some of your sessions.

Multicultural Counseling and Therapy (MCT) and Feminist Therapy

All counseling and psychotherapy are multicultural.

—Paul Pedersen

Women belong in the house . . . and the Senate.

—Author unknown

Philosophy. Multicultural counseling and therapy (MCT) and women's issues have both been discussed in this book. But why are we presenting the two together? Both rest on a foundation of the cultural/environmental context. Making CEC central in the session is seen as essential for effective counseling and therapy. Without this dimension, counseling and therapeutic work is incomplete.

Feminist therapy, also known as relational cultural therapy (RCT), is usually considered separate from MCT. It has created its own separate body of literature focusing on women's issues. At the same time, a commitment to feminist theory includes a respect for and inclusion of MCT. Similarly, MCT cannot truly exist without feminist concepts.

Key Methods. Both focus on raising personality consciousness of how the individual develops in a cultural/environmental context. They draw on all the other theories discussed in this book as they assist the individual increase competence and consciousness. In that sense, they are eclectic and integrative.

Implications for You. Remember that clients, regardless of race, ethnicity, gender, or socioeconomic background, come from a cultural background and history that affects their identity and all the issues that they face. We have spoken primarily of People of Color and women, but White people and those of all ethnicities and races, and men too, come from a CEC background with varying levels of identity development. White people in general (not all) are often unaware that being White represents a culture that often brings privileges denied to others because of the color of their skin.

Perhaps the central implication is that "there is a lot to learn and experience." We cannot know all dimensions of culture, but we can develop an appreciation and learn to use these concepts in the session. Give special attention to the most prominent cultural groups in your community.

Your Personal Style and Future Personal Theory

You have reviewed your personal style and current theories of counseling throughout this book. The theories presented here represent the different ways in which counselors and therapists understand human behavior and challenges. There are many more available. Learn from as many as possible, but most probably you will establish your own personal style and theory over time. This is a lifelong process, as explained below.

Determining Personal Style and Theory	Anticipated Client Response
As you work with clients, identify your natural style, add to it, and think through your approach to interviewing and counseling. Examine your own preferred skill usage and what you do in the session. Integrate learning from theory and practice in interviewing, counseling, and psychotherapy into your own skill set.	You, as a developing interviewer or counselor, will identify and build on your natural style. You will commit to a lifelong process of constantly learning about theory and practice while evaluating and examining your behavior, thoughts, feelings, and deeply held meanings.

Three major factors to consider as you move toward identifying your own personal style and integrating the many available theories are (1) your own personal authenticity; (2) the needs and style of the client; and (3) your own life goals, values, worldview, vision, and wisdom. Unless a skill or theory harmonizes with who you are and your sense of meaning, it will be false and less effective. Competence, caring, and a sense of direction and purpose are essential.

Remember that you are one of a kind, as are those whom you would serve. We all come from unique life experiences, varying families, differing communities, and distinct views of gender, ethnic/racial, spiritual, and other multicultural issues. It is obvious that modifying natural style and theoretical orientation will be necessary if you are to be helpful to the endlessly varying, challenging, and interesting clients that you will meet.

Awareness, Knowledge, and Skills: Case Conceptualization, the Interview Checklist, Treatment Planning, and Action Planning to Prevention Relapse

Case Conceptualization

Your understanding of your clients, their issues, and the decisions they want or need to make is fundamental. This involves cognitive and affective empathy, as well as your ability to mentalize and be with your client holistically. A case conceptualization is an individualized application of your theoretical model that takes into consideration the antecedents of the case and your observations and inferences.

The data to be considered include what you may find in the intake file, such as demographics, ethnic/cultural dimensions, personal history, and presenting issues or concerns. These are integrated with your interview observations as you listen and learn from and with the client. In the ongoing sessions, many other factors will result in constant changes in your case formulation. Some of these factors are changing client issues, nonverbals, emotional experience, your experience of the client, what happens during your interactions, test results, diagnoses, strengths and assets, weaknesses and gaps, supporting materials, inferences and assumptions, working hypotheses, goals of treatment, possible barriers, interventions, and evaluation of the impact and effectiveness of action plans.

Based on all this information, you produce a case formulation that becomes more useful and precise as the relationship develops. This may or may not include a diagnosis. Many counselors and psychotherapists suggest sharing your formulation with the client; certainly clients need to know specifics if you provide a specific diagnosis.

This formulation will be the working hypothesis used to guide treatment and change as interviews continue. It will evolve over sessions with the client and will help in making decisions on treatment plans, selecting interventions, assigning homework, coping with setbacks, assessing outcome, determining paths to action, and completing treatment.

The Interview Checklist: Planning the First Session

Given the complexity of relationships, particularly professional relationships, Atul Gawande has written *The Checklist Manifesto* (2009). Focusing first on medicine, he found that a surgical checklist of basic and often obvious factors significantly reduced dangerous errors during operations. He goes on to point out that thinking ahead about what one is going to do improves performance regardless of the field.

Box 14.1 provides a checklist for the first session, developed by the authors. Even the most experienced and confident counselor or therapist is likely to forget some of the items in the checklist. Review the list before you talk with your client; then review it again afterwards, checking to see what might have been missed. There are items here that need to be considered in every session. For you, personally, what might you add to or delete from this checklist? Adapting the checklist so that it fits you and your agency is essential. This is particularly important for the first session, as it is possible to overlook essential issues.

BOX 14.1 **Checklist for the First Session***

Before the Session

- Are you familiar with HIPAA (see Appendix II), the policies of your agency, and key state laws? Are key policies posted in the agency waiting area? These need to be shared early with the client.

- Is there a file, and have you read it? Do you need notes from the file as a refresher?

- Do the room and setup ensure confidentiality? If you are working in an open setting, how will you maintain privacy?

- Does the room provide adequate silence? Do you need a sound machine working outside your door? How will you handle these issues if you meet with the client in the home or community?

- Is there an inviting atmosphere where you will meet the client? Is it neutral, or do you have interesting and culturally appropriate art and objects relevant to those who may come to this office? Are chairs placed in a position where the power is relatively equalized?

- If the setting is informal, such as on the street or in a gym, again is the situation as comfortable as possible?

Stage 1: Empathic Relationship

Initiate the session. Develop rapport and structuring. "Hello, what would you like to talk about today?" Did you:

- Plan ahead flexibly to ensure an empathic relationship and connect with this unique client?

---

*This checklist was inspired by the book by A. Gawande, *The Checklist Manifesto* (New York: Holt, 2009). Gawande talks specifically about the importance of a checklist for successful surgery and suggests that the idea be taken into other areas as well. This counseling checklist was authored by Allen Ivey. However, he gives permission for anyone to photocopy this checklist with the request that he be given appropriate credit. Copyright © 2014 Allen Ivey.

(continued)

BOX 14.1 (continued)

- Discuss the client's rights and responsibilities? The counseling and therapy relationship works in part because of clearly defined rights and responsibilities of each person involved.

- Provide an explanation of what might happen in the session and/or how the conversation is likely to be structured?

- Review HIPAA, agency policies, and key legal issues? If your agency requires diagnostic labels, did you explain that to the client and offer to share that diagnosis if he or she wishes?

- Discuss confidentiality and its limits?

- Obtain the client's permission to take notes and/or record the session? Was the client informed that these notes and the recording are available if he or she wishes to review them?

- If working with an underage client, obtain the appropriate parental permission as required by your agency and/or state law?

- Provide an opportunity for the client to ask you questions before you started? Were issues of multicultural differences addressed?

- Work with the client to establish an early preliminary goal or objective for the session?

- Come prepared if the client immediately started talking about issues and concerns? Did you listen carefully and return later to cover those matters that you may not have had time to attend to?

Stage 2: Story and Strengths

Gather data. Use the BLS to draw out client stories, concerns, problems, or issues. "What's your concern?" "What are your strengths and resources?" Did you:

- Allow and encourage the client to present the story fully? Did you reframe the word *problem* into a more positive, change-oriented perspective using words such as *issue*, *challenge*, *concern*, or *opening for change*?

- Bring out the key facts, thoughts, feelings, and behaviors related to the story? Did you also look for underlying deeper meanings behind the story?

- Avoid becoming enmeshed in the client's story by becoming a voyeur (endless fascination and searching for details about the client's interesting issues) or by unconsciously putting a "negative spin" on what the client said? To paraphrase author and activist Eldridge Cleaver, "Is the counselor part of the problem or part of the solution?"

- Bring out stories and concrete examples of the client's personal strengths and external resources? Did you search for

specific images within these stories and perhaps anchor these positive images in specific areas of the body?

- Ask the critical questions "What else relates to what we've talked about so far?" "What else is going on in your life?" and "Is there anything else I should have asked you but didn't?"

Stage 3: Goals

Set goals mutually. The BLS will help define goals. "What do you want to happen?" "How would you feel emotionally if you achieved this goal?" Did you:

- Review the early goals set by the client and revise them in accordance with new information about the story and strengths?

- Jointly make these goals as specific and observable as possible?

- When necessary, break down large goals into manageable step-by-step objectives that can be reached over time? Did you prioritize these goals?

- Remind the client of the strengths and resources that he or she brings to achieve these goals?

Stage 4: Restory

Explore alternatives via the BLS. Confront client incongruities and conflict, restory. "What are we going to do about it?" "Can we generate new ways of thinking, feeling, and behaving?" Did you:

- Include brainstorming without a theoretical orientation? Confront with a supportive challenge, summarizing the goal and the issue? ("On one hand, the goal is ____, but on the other hand, the main challenges you face are ____. Now what occurs to you as a solution?") Often clients with your support will come up with their own unique and workable answers, often ones that you did not think of.

- Use appropriate theories and strategies with this client?

- Use a variety of listening and influencing strategies to facilitate client reworking and restorying of issues? What were they?

- Use identified positive strengths and resources to remind clients during low points of their own capabilities?

- Agree on homework or personal experiments to be completed after the session?

- Develop a clear definition of a more workable story that can lead to action and transfer to the real world?

Stage 5: Action
Plan for generalizing session learning to "real life." "Will you do it?" Use BLS to assess client commitment to taking action after the session. Did you:

• Build on the new story, or start of a new story, and work with the client to take specific action and learning to the "real world"?

• Agree on a plan for transfer of learning that is clear and doable?

• Work with the client to develop a systematic relapse prevention plan, if the sessions are more long term and/or complex.

• Contract with the client to do at least one thing differently during the week, or even tomorrow?

• Agree to plans to look at the results of the action plan during the next session?

• Check how it was for the client? Does the client think he or she could work with you? Did you agree to work together?

• Set a date for the next session and/or follow-up sessions?

• Write interview notes as soon as possible and seek consultation from supervisors or colleagues as necessary?

Your first counseling session with a client is always unique and gives you an opportunity to learn about diversity and the complexity of the world. Box 14.2 presents an international view of the work with clients and recommendation for practice.

BOX 14.2   National and International Perspectives on Counseling Skills

What's Happening with Your Client While You Are Counseling?
Robert Manthei, Christchurch University, New Zealand

There is more going on in counseling beyond what we see happening during the session. Clients are good observers of what you are doing, and they may not always tell you what they think and feel. Research shows that clients expect counseling to be shorter than do most counselors and therapists. Clients see counselors as more directive than counselors see themselves. And what the counselor sees as a good session may be seen otherwise by clients, and vice versa. Counselors and clients may vary in their perceptions of the effectiveness of their work together.

I conducted a study of client and counselor experience of counseling. Among the major findings are the following:

Clients Often Have Sought Help Before
Most people don't come for counseling immediately. Talking with friends and family and trying to work it out on their own were usually tried first. Reading self-help books, prayer, and alcohol and drugs are among other things tried. Some deny that they have difficulties until these become more serious.

Implications for Practice Ask clients what they have tried before they came to you, and find out what aspects of prior efforts seemed to have helped. You may want to build on past successes. This is an axiom of brief solution-oriented counseling.

First Impressions Make a Difference
That first meeting sets the stage for the future, and the familiar words "relationship and rapport" are central. I found that clients generally had favorable impressions of the first session and viewed what happened even more positively than counselors. Sometimes sharing experience helps. One client who did not feel positive about the first session commented, "Maybe if the counselor had gone through a similar experience of divorce and children, it would have helped."

Implications for Practice Obviously, be ready for that first session. Cover the critical issues of confidentiality and legal issues in a comfortable way. Structure and let the client know what to expect. Some personal sharing, used carefully, can help. And empathic listening always remains central.

Counseling Helps, but So Do Events Outside of the Session
Resolution of their issues was attributed to counseling by 69% of clients, while 31% believed events outside the session made the difference. Among things that helped were talking and socializing more with family and friends, taking up new activities, learning relaxation, and involvement with church.

Implications for Practice What you do in the session needs to be supplemented with plans for generalization of behavior and thought to daily life. Homework and specific

(continued)

BOX 14.2 **(continued)**

---

ideas for using what is learned in the session help ensure that action follows the session.

Things That Clients Liked

Relationship variables such as warmth, understanding, and trust are essential. Clients liked being listened to and being involved in making decisions about the course of counseling. Reframes and interpretations helped them see their

situations in a new way; also valued were new skills such as imagery, relaxation training, and thought stopping to eliminate negative self-talk.

All of the above speaks to respecting the client's ability to participate in the change process. We need to disclose with clients the rationale for what we are doing, but also ask them to share their perceptions of the session(s) with us. We can learn much from the client by adopting an egalitarian approach.

Multiple Applications of Skill Integration: Referral, Treatment Planning, Case Management, and Relapse Prevention

What treatment, by whom, is most effective for this individual with that specific problem, and under what set of circumstances?

—Gordon Paul

The best way to find yourself is to lose yourself in the service of others.

—Mahatma Gandhi

These two quotations provide a background for integrating and thinking through the many ideas, concepts, and theories discussed in this book. While we have broken down the counseling and therapy process into clear steps, it is the totality of the many basics of helping that enable us to realize the real complexity of this interactive process—we and the client learn and change with each other.

Whether you are using decision counseling, person-centered, crisis, or cognitive behavioral therapy, you need to consider referral, treatment planning, case management, and the action plan. The following sections address these key issues in planning for the future.

Referral

No counselor has all the answers. An important part of individual counseling is helping your clients find community resources that may facilitate their growth and development. The community genogram (presented in Chapter 9) helps counselors and clients think more broadly and consider appropriate referral sources.

Sometimes the counselor/client relationship simply doesn't work as well as we all would like. When you sense the relationship isn't doing well, avoid blaming either the client or yourself. Focus on clients' goals, and seek to hear their story completely and accurately. Ask clients for feedback on how you might be more helpful. Seek consultation and supervision, and most often these "difficult patches" can be resolved to the benefit of all. When an appropriate referral needs to be arranged, you do not want to leave clients "hanging" with no sense of direction or fearful that their problems are too difficult. Maintain contact with the client as the referral process evolves, sometimes even continuing for a session or two until arrangements are complete. It is critical that the client never feel rejected by you. Your understanding, empathic support during the referral process is essential.

Another key referral issue is whether counselor expertise and experience are sufficient to help the client. Even if you think that you are working effectively, this may not be enough; it may be a case in which supervision and case conferences can be helpful. Opening up your work to others' opinions is an important part of professional practice. Clients, of course, should be made aware that you as counselor or therapist are being supervised.

Treatment Planning

First Think Prevention. It is best to work with concerns and life challenges before they become serious issues. Positive psychology and developing respect and understanding for the culturally different and the underserved are underlying themes of this book. As we search for strength and build resilience, we are reducing the number of mentally ill over the lifespan. Stress management and therapeutic lifestyle changes are key in our prevention efforts. And, of course, they are also central treatment strategies.

If we are to build mental health and prevent mental illness, we need to look to children and youth. Developmental services can make a huge difference in developing resilience and encouraging self-actualization. The importance of youth counseling is provided in Figure 14.1, which shows the prevalence of disorders among children. The National Institute of Mental Health estimates that 26% of the U.S. population has the possibility of a diagnosis during a given year, while 6% face diagnosis of serious mental illness. Children and adolescents are increasingly being diagnosed, along with overuse of sometimes dangerous medications.

With prevention as our ultimate goal, let us turn to treatment planning itself.

Treatment planning. With specific written goals and objectives, treatment planning is becoming increasingly standardized and often required by agencies and insurance companies. When possible, negotiate specific goals with the client and write them down for joint evaluation. They should be as concrete and clear as possible, including specific indicators of behavioral change and emotional satisfaction. The more structured counseling theories, such as cognitive behavioral, strongly urge counseling and treatment plans

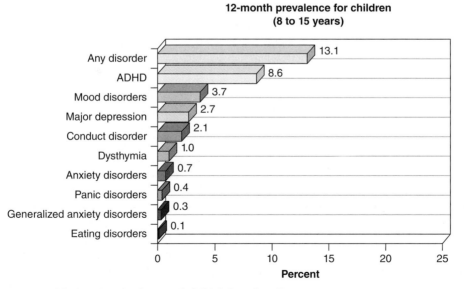

FIGURE 14.1 12-month prevalence of child "disorders."

with specific action goals developed for each issue. Their counseling and treatment plans are often more specific than the one presented in the example that follows (see relapse prevention, below).

Less structured counseling theories (psychodynamic, person-centered) tend to give less emphasis to treatment plans, preferring to work in the moment with the client. In short-term counseling and interviewing, the interview plan serves as the treatment plan. As you move toward longer-term counseling (5 to 10 sessions), a more detailed treatment plan with specific goals is often required. Many agencies also use case management as part of the treatment plan and bring in other professionals from social work, education, and medicine.

Counselors and therapists working in community and hospital clinics are often required to have precise goals in their treatment plans, both for the good of the client and to fulfill the requirements of many insurance companies. A brief outline of what many agencies consider important in a treatment plan is shown in Figure 14.2. Note the emphasis on concrete goals, specificity of interventions, and a planned date for evaluation of goal achievement. Increasingly, you will find yourself working with some variation of this goal-oriented form, regardless of setting.

Case Management

Although this book focuses on the session and counseling skills, treatment planning often needs to be extended to *case management*. For human service professionals, social workers, and school counselors, case management will be as important as or more important than treatment planning. Case management requires the professional helper to coordinate community services for the benefit of the client, and very often the client's family as well. Let's look at the complexity of case management with an example.

> A single mother and her 10-year-old boy are referred to a social worker in a family services agency. The family physician thinks that the child's social interaction issues may be a result of Asperger's syndrome. The child has few friends, but is doing satisfactorily in school. The social worker interviews the mother and reports the child's social and academic situation at school. The mother says that she has financial problems and difficulty in finding work. Meeting with the child a few days later, the worker notes good cognitive and language competence, but that the child is unhappy and demonstrates some repetitive, almost compulsive behavioral patterns.

The agency staff meets and starts to initiate a case management treatment plan. It is clear that the mother needs counseling and that the child needs psychological evaluation and likely treatment as well. At the staff meeting, the following plan is developed: The child is to be referred to a psychologist for evaluation; based on that evaluation, recommendations for treatment are likely to be made. The social worker is assigned to do supportive counseling with the mother and to take overall responsibility for case management. The social worker will eventually need a treatment plan for the mother and a case management plan for the family.

School performance is good, but the social worker contacts the elementary school counselor and finds that the counselor already has the child in counseling. In fact, the counselor has been working with the teacher to help the client develop better classroom and playground relationships. The school counselor and social worker discuss the situation and realize that the boy is often left alone without much to do. Staying after school for an extended day would likely be helpful. The school counselor, with the social worker's backing, contacts a source for funding through a local men's club.

COMMUNITY CLINIC
Behavioral Treatment Plan

This form will be reviewed again in no more than two months, and progress toward goals will be noted. Changes in interventions or goals should be noted immediately.

Patient's Name, Address, Phone, Email: _____

_____

Clinic Record Number _____   Insurance _____

Diagnosis: Summary of Patient's Original Concerns:

_____   _____

_____   _____

_____   _____

_____   _____

_____   _____

Identified Patient Strengths and Resources (to be added to throughout therapy):

_____

_____

_____

_____

Interview Progress Narrative

Problem/Concern #1		
Goal	Interventions	Progress Toward Goal

Problem/Concern #2		

Problem/Concern #3		

Signature _____   Date _____

Patient signature _____   Date _____

If patient is a child:

Name of child _____   Age _____

Parent signature _____   Date _____

FIGURE 14.2 Treatment plan example.

This is but a beginning for many cases—case management involves multiple dimensions beyond this basic scenario. Counseling and therapy are an important part of case management, but only a part. The Division of Youth Services or Family and Children Services may need to be called to review the situation if the counselor or another mandated reporter suspects signs of abuse or neglect; the mother may need short-term financial assistance and career counseling. If the father has not been making child support payments, legal services may have to be called in.

Through all of the above, the social worker maintains awareness of all aspects of the case. Through individual counseling with the mother, the worker is in constant touch with all that is going on and remains in contact with key figures working toward the positive treatment of the child. But the child can only be successfully treated in the family, school, and community situation.

Advocacy action is important throughout, and it does not happen by chance. Throughout this process, the social worker will constantly advocate for the mother and the child by establishing connections with various agencies. The school counselor who goes to the local men's club for funding has to advocate for the child; in addition, encouraging a teacher to change teaching style to meet individual child needs requires advocacy. Elementary counselors spent a good deal of time in various activities advocating for students individually, and often advocating for fair school policies as well.

Social justice issues may appear. In many communities and agencies, all of the above services may not exist—and, equally or more likely, they may not work together effectively. This may require organizing to produce change. The child may be bullied, but the school has no policy to prevent bullying. Social justice and fairness demand that each child be safe. Discrimination against those who are poor or are from minority backgrounds may exist. Individual and group education or actual forced change may be necessary.

Relapse Prevention: Maintaining Planned Change for Long-Term, Complex Issues

Change that is not planned is likely to be lamented.

—Allen Ivey

An action plan is essential for counseling. Contracted general change goals are complemented more concretely with a clear action plan. This is to ensure success in reaching short-term and long-term outcomes. The action plan is reinforced with behavioral and cognitive specifics to work on at home, which can enable emotional regulation and help the client feeling more positive and optimistic.

Some clients need relapse prevention to supplement the action plan. A portion of the clients who respond well to counseling and therapy will relapse and return to former behavior, thoughts, and feelings. This is particularly so with alcohol and drug clients, those with severe depression, and trauma survivors. Relapse prevention (RP), originally developed for use in alcohol or drug abuse treatment, has become common practice for many counseling interventions.

Change that is not planned and contracted with the client is less likely to occur, and even less likely to happen in the long term. The action plan helps ensure that clients have specific action goals and that decisions from the interview are taken home and practiced during the next day and week. Furthermore, many action plans are for the longer term. These goals for more significant change need to be recorded as part of the treatment plan.

This approach, originally developed for use in alcohol or drug abuse treatment, is used in motivational interviewing, cognitive behavioral therapy, and other theories as well (Marlatt, Larimer, & Witkiewitz, 2011). It may be particularly helpful in a variety of behavioral issues ranging from anger management to getting exercise, complying with medical advice, and many others.

The Maintaining Change Worksheet: Self-Management Strategies for Skill Retention (Box 14.3) helps the client plan to avoid relapse or slips into the old behavior. The counselor hands the client the worksheet and they work through it together, giving special emphasis to things that may come up to prevent treatment success. The decision process is joint, but ultimately the client will decide. If the client is not "fully on board," the chances for relapse are increased. Research and clinical experience in counseling reveal that RP can be very effective with clients, helping to ensure that they actually *do* something different as a result of their experience in the session or treatment series. Remember that change takes time and effort. Maintaining an intentional effort to change will pay off at the end.

BOX 14.3 Maintaining Change Worksheet: Self-Management Strategies for Skill Retention

I. Choose an Appropriate Behavior, Thought, Feeling, or Skill to Increase or Change

Describe in detail what you intend to increase or change.

_____

_____

Why is it important for you to reach this goal?

_____

_____

What will you do specifically to make it happen?

_____

_____

II. Relapse Prevention Strategies

A. Strategies to help you anticipate and monitor potential difficulties: regulating stimuli

Strategy	*Assessing Your Situation*
Do you understand that a temporary slip may occur but it need not mean total failure?	_____
Support network: Who can help you maintain the skill?	_____
Triggers and high-risk situations: What kinds of people will tempt you? What places should your avoid? How will you handle alcohol at a party—or even keeping ice cream in the fridge?	_____

Permission to use this adaptation of the Relapse Prevention Worksheet was given by Robert Marx, University of Massachusetts, Amherst.

(continued)

BOX 14.3	(continued)

B. Strategies to increase emotional regulation of thoughts and feelings

*Where can you turn for cognitive and emotional direction
and support? Who will help?*

What can you do when alone?

What might be an unreasonable emotional response to a
temporary slip or relapse?

How can you think more effectively in tempting situations or after a relapse?

How can you forgive yourself for a lapse and move on?

_____

C. Strategies to diagnose and practice related support skills: regulating behaviors

What additional support skills do you need to retain the
skill? Exercise? Assertiveness? Relaxation? Communication
microskills?

D. Strategies to provide appropriate outcomes for behaviors: regulating consequences

Can you identify some probable outcomes of succeeding
with your new behavior?

How can you reward yourself for a job well done? Generate specific rewards and satisfactions.

III. Predicting the Circumstance of the First Possible Failure (Lapse)
Describe the details of how the first lapse might occur; include people, places, times, and emotional states.

_____

_____

_____

Action: Key Points and Practice for Skill Integration and Determining Personal Style

Skill Integration. Skill integration combines the microskills, stages of the interview, and your natural expertise into the interview or counseling session. As you engage in the integration process, examine where you stand now. Have you increased your interviewing competence? Are you more intentional and able to flow naturally with your clients? Where do you need to grow or continue practicing?

Multiple Counseling Approaches. The chapter presents a brief review of several counseling approaches, including decisional counseling, person-centered counseling, logotherapy, multicultural counseling and therapy, and feminist therapy. Note that their emphases are quite different. For example, consider the differences between decisional counseling, which emphasizes careful listening to the story/concern/challenge of the client before acting, and person-centered counseling, which stresses providing core attitudes so the client can explore their feelings in detail. On the other hand, multicultural counseling and therapy (MCT) and feminist therapy pay more attention to the cultural/environmental/context.

Case Conceptualization. A case conceptualization is an individualized application of your theoretical model that takes into consideration the antecedents of the case and your observations and inferences. A good conceptualization demonstrates your understanding of your clients, their issues, and the decisions they want or need to make. It provides the foundations for your work.

Planning the First Session and Using the Checklist. The first session presents many challenges. Using the checklist helps you systematically plan for the initial session using the five-stage structure and reduces the chances of missing the basics. The five stages provide a useful model to ensure that you cover all points, even if the session does not go as expected. Be intentionally flexible and ready to change your plan if events in the session suggest that another approach is needed.

Referral. Know your community. What medical offices are available? What social services are available for additional treatments, financial assistance, and advice? Being able to suggest places for help will be one of your major functions. Of course, when the client you are working with does not seem appropriate for your background and/or expertise, know which colleagues or services might be more helpful. Do not just send clients off, but stay in touch until they have found something that fits their needs.

Treatment Plan. This is a long-term plan for conducting a course of counseling sessions. Work with your client to establish mutually agreed upon goals and planned treatment. Clients need to be part of this plan; be sure to discuss benchmarks for change and the achievement of clients' goals.

Case Management. Case management is an important part of the work done by human service professionals, social workers, and school counselors. Case management requires the professional helper to coordinate treatment plans with community services and family support systems for the benefit of the client.

Maintaining Change. Maintaining change can be achieved by implementing an agreed upon action plan and a set of practice homework assignments. These activities extend the work done in the session to the time between sessions. The use of homework is no longer optional; it is an integral component of successful counseling and therapy, one that ensures that the fifth stage of the interview is indeed accomplished.

Preventing Relapse. A relapse prevention plan also helps accomplish the fifth stage of the counseling interview as it reduces slips or relapses. The Maintaining Change Worksheet: Self-Management Strategies for Skill Retention (Box 14.3) helps clients plan to avoid relapses or slips.

Developing and Reviewing Your Final Interviewing Transcript. Using the constructs of this book, examine your own style of microskill usage, focus, structure of the session, and the resultant effect on a client's cognitive and emotional development style. Your transcript will help you conduct a systematic analysis of your work. Also, it will be useful to evaluate client movement on the Client Change Scale.

Your Personal Theory and Transcendence. Identify your natural style, add to it, and think through your approach to interviewing and counseling as you grow in the field. Examine your own preferred theories and expand them as you learn from your work with diverse clients. As you develop your personal theory, remember to remain client focused. Your ultimate goal is to help others improve their well-being, achieve happiness, and succeed in their positive goals. A respectful and client-centric approach is basic to your effectiveness in helping others.

Practice and Feedback: Individual, Group, and Microsupervision

Additional resources can be found by going to CengageBrain.com and logging into the MindTap course created by your professor. There you will find a variety of study tools and useful resources that include quizzes, videos, interactive counseling and psychotherapy exercises, case studies, the Portfolio of Competencies, and more.

You have engaged in the systematic study of the counseling process and have experienced many ideas for analyzing your style and skill usage. These responses must be genuinely your own. If you use a skill or strategy simply because it is recommended, it could be ineffective for both you and your client. We hope that you will draw on the ideas presented here, but ultimately *you* will put the science together in your own art form.

You have practiced varying patterns of helping skills with diverse clientele. We hope that you have developed awareness and knowledge of individual and multicultural differences. Study and learn how to "flex" and be intentional when you encounter varied clients with differing needs. For example, you may be more comfortable with teenagers than with children or adults, or you may have special abilities with elders.

Conducting and Transcribing a Full Session

Now that you are finishing this chapter, to help ensure your understanding of your own style, complete another audio or video session. Use the following guidelines.

1. *Find a volunteer client* willing to role-play a concern, issue, or opportunity.
2. *Counsel the volunteer client* for at least 15 minutes. Avoid sensitive topics. Feel free to go further to gain a sense of completion.
3. *Use your own natural communication style.*
4. *Ask the volunteer client,* "May I record this session?"
5. *Inform the client* that the recording device may be turned off any time he or she wishes.
6. *Select a topic.* You and the client may choose interpersonal conflict, a specific issue, or one of the elements from the RESPECTFUL model.

7. *Follow the ethical guidelines* from Chapter 2. Common sense demands ethical practice and respect for the client.

8. *Obtain feedback.* You will find it very helpful to get immediate feedback from your client. As you practice the microskills, use the Client Feedback Form in Chapter 1 (Box 1.4).

9. *Also obtain feedback from another student.* We have found that it is very helpful and clarifying if you and a student partner exchange transcripts and comment fully on each. This gives you important additional feedback.

10. *Develop an action plan with the client.*

Additional guidelines for transcribing the session are presented in Box 14.4. Transcribe the session and use what you have learned so far to fully analyze your work. A careful analysis of your behavior in the session will aid in identifying your natural style and its special qualities as well as your skill level. Use the ideas presented in this chapter to further examine and analyze your work in the session.

Remember that *you* are the person who will integrate what you have learned here into your own practice. Identifying the nature of your personal style and current skill level will set you on the road to continued growth and competence as a professional counselor or therapist. Please give special attention to your understanding and use of cultural/environmental/contextual issues. Look at your natural style of counseling, and evaluate your multicultural expertise.

Finally, go back to the transcript or recording of the session you recorded earlier, in Chapter 1, and note how your style has changed and evolved since then. What particular strengths do you note in your own work?

BOX 14.4 Transcribing Sessions

Organize the transcript in a format similar to the transcripts presented throughout this book. The session transcript may be useful in demonstrating your competence and obtaining an internship or job. Consider the transcript a permanent part of your developing professional life and your portfolio.

You don't have to transcribe a full session, although a full transcript likely will be most beneficial to you. A 20-minute transcript from within a longer session is enough. But if you do such an excerpt, be sure to indicate what happened in the rest of the counseling session so that the context of the transcript is clear.

Check off the following points to make sure you have included all the necessary information in the transcript:

- Describe the client briefly. Do not use the client's real name.

- Outline your session plan *before* the session begins.

- Be sure you obtain the client's permission before recording the session, and include a summary of this agreement in the transcript. The client should be free to withdraw at any time. Ethically, we protect the rights of the client.

- Number all interactions, and be sure to indicate who is speaking at the beginning of each interaction.

- Mark the focus of each interaction, and note your use of attending and influencing skills.

- If you confront, note this in the Process Comments; also note how the client responds, using the Client Change Scale. Pay special attention to any bullying or microaggressions that you may encounter.

- Comment on your interactions as appropriate to the situation. Discuss what you feel was good or needs improvement in your skills. If you feel that you used a skill inappropriately, describe what you believe would have been a better approach. Note also what skills worked well!

- Indicate when you think you have reached the end of a stage. Do not feel that you must cover all stages; in some cases, you may cycle back to an earlier stage or forward to a later stage.

- Write a commentary that summarizes what happened.

- Summarize your use of skills through a skill count.

- Assess your competence levels. What skills have you mastered, and what do you need to do next? This is also a summary of your strengths and the areas that need further development. What did you like and not like

(continued)

BOX 14.4 **(continued)**

about your work? Your ability to understand and process "where you are" and discuss yourself is necessary for personal and professional growth.

• End the transcript with an action plan. As appropriate, write a possible treatment plan as if you were to continue for a series of interviews.

• Work with one of your classmates to obtain feedback about your work. Share the audio, video, or transcript with others. Take turns providing feedback and enriching each other's skills. Don't forget to emphasize the positive strengths observed in your practice.

Portfolio of Competencies and Personal Reflection

Assessing Your Level of Competence: Awareness, Knowledge, Skills, and Action

The transcript and your Portfolio of Competencies can provide a permanent course summary for you. In addition, when you apply for practicum or internship positions, it may be useful as a demonstration of what you can do as a therapist or counselor. At times, the portfolio will even be helpful as you search for a position.

At the level of personal skills, reflect on your competence level in each of the major microskill areas. Use the self-assessment tool provided in Table 14.1 to guide your reflection. You may use this as a summary and as a plan for future growth. Can you demonstrate awareness and knowledge by identifying each skill or concept and classifying its place in the interview? Can you demonstrate basic competence by using the skill in the session? Most important, can you identify specific things that happen with your clients as a result of your interventions and success in the interview?

TABLE 14.1   Self-Assessment Summary

Skills and Concepts	Awareness and Knowledge	Basic Competence	Intentional Competence	Evidence of Achieving Competence Level
1. Attending behavior				
2. Questioning				
3. Observation skills				
4. Encouraging				
5. Paraphrasing				
6. Summarizing				
7. Reflecting feelings				

8. Basic listening sequence				
9. Positive asset search				
10. Empathy---Cognitive, Affective, and Mentalizing				
11. Five stages of the interview				
12. Focusing				
13. Confrontation				
14. Client Change Scale				
15. Eliciting and reflection of meaning				
16. Interpretation/reframe				
17. Self-disclosure				
18. Feedback				
19. Logical consequences				
20. Information psychoeducation				
21. Directives				
Key Issues and Practical Applications of Microskills and the Five-Stage Interview				
22. Stress and stress management				
23. Ethics				
24. Multicultural competence				
25. Social justice				
26. Positive psychology				
27. Therapeutic Lifestyle Changes				
28. Homework/action plan				
29. Community genogram				
30. Client Change Scale				
31. Discernment				
32. Decision counseling				
33. Person-centered counseling				
34. Crisis counseling				
35. Cognitive Behvioral Therapy				
36. Relapse Prevention				
37. Defining personal style and self-assessment via your own final interview.				

Concluding Personal Reflection and Review of Intentionality

We have come to the end of this phase of your counseling and psychotherapy journey. You have had the chance to learn the foundational skills and how they are structured in a variety of theoretical and practical approaches. Skills that may have seemed awkward and unfamiliar are often now automatic and natural. As in the Samurai effect, you now do not need to think of them constantly. The basic listening sequence is likely part of your being at this point. Moreover, expect that the *empathic relationship—story and strengths—goals—restory—action* framework will become a very useful part of your practice, enabling you to adapt and work with many different theories of counseling and psychotherapy.

The fact that you took this course in counseling and psychotherapy skills and read this book suggests that you have a strong interest in working with and serving others. The helping fields are rich in opportunity for personal joy and satisfaction. You can make a difference in other people's lives. As you think about your personal style and future theoretical/practical orientation, what do you want to give to your clients and the world?

There are four major factors to consider as you move toward identifying your own personal style and integrating the many available theories: *authenticity, competence, caring, and a sense of purpose and the meaning of life are essential.* All this needs to be done with awareness of the cultural/environmental context. Multicultural understanding is now essential for effective practice.

Consider these issues as you continue the process of identifying your natural style and future theoretical/practical integration of skills and theory.

Goals: What do you want to happen for your clients as a result of their working with you? What would you *desire* for them? How do you plan to help them achieve their own goals? What else?

Skills and strategies: You have identified your competence levels. What do you see as your special strengths? What are some of your needs for further development in the future? What else?

Cultural intentionality and cultural health: Can you focus on the positive resilience factors within yourself and within your clients as you consider the RESPECTFUL model? Are you able to help clients work through issues of bullying and microaggressions toward cultural identity and cultural health? Where are you on social action beyond the interview to the classroom and community?

What is your assessment of your present level of intentionality? We have included Table 14.2 to help you reflect on these major factors and further advance your review. The table illustrates the differences between intentional counselors and ineffective ones. Where do you see yourself on each one of these attributes?

TABLE 14.2   The Qualities, Skills, and Strategies of the Intentional and Ineffective Counselor and Therapist

Quality/Skill/Strategy	Intentional Counselor and Therapist	Possible Behaviors of the Ineffective
Relationship/working alliance	Center of attention throughout, especially at the beginning.	Starts questioning and solving "problems" immediately.
Ethics and multiculturalism/social justice	Knows and follows ethical standards. Aware and skilled in the multicultural and social justice issues, but constantly seeking new knowledge. Becoming active in school and community outside the office.	Somewhat aware of importance, but not sure how to implement multicultural issues. Likely to see social justice as irrelevant to individual counseling and therapy.

Quality/Skill/Strategy	Intentional Counselor and Therapist	Possible Behaviors of the Ineffective
Positive psychology, wellness, therapeutic lifestyle changes (TLCs)	Listens fully to client concerns, but always seeks to focus on strengths and resources to build resilience. Uses TLCs as a route toward prevention and personal growth.	Stays with the traditional individualistic problem-oriented approach, with little attention to strengths. Unaware of positive psychology and TLCs.
Definition of concern(s)	Listens to multiple issues and then prioritizes them with the client, saving others for later. Aware that the key issues change during counseling. Sees the client in total cultural/environmental context and how it relates to personal issues.	Wanders from topic to topic, or fixates on one single definition of what "should" be discussed. Feels that context and resources are at best secondary issues.
Listening and observation	Skilled in attending to both verbal and nonverbal communication.	Not effective, may focus too much on self, single issues, or avoid client's main concern.
Basic listening sequence	Able to use all skills with anticipated results, but ready to flex intentionality in the moment as client comments change.	Not aware, random, may miss emotion or key cognitive issues.
Influencing skills	Works collaboratively with the client. Can empathically confront, reframe, supply direction and instructions as needed.	Decides what the client needs and freely uses directives and advice.
Psychoeducation	Knowledgeable in alternative forms of treatment that may be key to prevention and developing resilience.	Unaware or does not consider them relevant.
Planning for action and homework generalization after the session	Fully aware that the time between sessions is when clients can reflect and implement a jointly prepared action plan.	Terminates the interview, usually without clear plans for client follow-up.
Neuroscience and neurobiology	Becoming a neurocounselor. Aware that the National Institute of Mental Health's Brain-Based Assessment and Treatment System may be ready for practical application in 2025, thus changing the Diagnostic and Statistical Manual as well as conceptions of counseling and therapy practice. Seeks to learn constantly about this rapidly evolving science that will have increasing influence on our practice over time.	Not interested, prefers to stay where he or she is now. May be upset when clients ask about recent neuroscience findings.
Future plans	Realizes that collaborative supervision with colleagues is essential. Joins appropriate national, state, and local profession associations. Takes meaningful continuing education courses to improve knowledge and skills. Searches the media and Internet for new information and explores YouTube for visual examples of interviews and neuroscience/neurobiology.	Avoids professional associations as too expensive and time-consuming. Relies on those who do to maintain state licensing requirements for them. Does the minimum of new learning. Finds an easy way to obtain CE credits, often just through short tests in journals.

What Is Your Story of Interviewing and Counseling?

> Trust yourself. You know more than you think you do.
>
> —Benjamin Spock

Summarize your own story of interviewing and counseling. You may want to base it on the interview transcript you completed in this chapter as well as on your success in previous microskill practice exercises. What have you learned by completing these exercises? What can you say about your natural style of interviewing and counseling? Where would you like to go next?

Counseling and psychotherapy are noble and altruistic professions. Our ultimate goal is the betterment and well-being of individuals, groups, and societies. Transcendence is beyond ourselves. Or, as the theologian Paul Tillich might say, "Love is listening."

Transcendence	Anticipated Client Response
Transcendence speaks about your capacity to go beyond yourself and successfully apply your newly mastered skills to help others. The ultimate test of your capacities is the benefit they can afford others. Your growth is wonderful; helping others growth is even better!	Clients will perceive your genuine interest and your intentional and collaborative work with them. Trust in the relationship will increase, and clients will benefit more from working with you. Their meaning and life vision will become clearer.

Becoming an effective counselor or therapist is a wonderful goal, but we believe the ultimate measure of your success is the success of others.

Goodbye and Thanks for Being Here

We have enjoyed sharing this time with you. Many of the ideas in this book come from interaction with students. We hope you will take a moment to provide us with your feedback and suggestions for the future. Please see the evaluation forms at the end of this book. This book will be constantly updated with new ideas and information. You have joined a never-ending time of growth and development. Welcome to the field of interviewing, counseling, and psychotherapy!

Allen, Mary, and Carlos

Suggested Supplementary Readings

The literature of our field is extensive, and you will want to sample it on your own. We would like to share some books that we find helpful as next steps to follow up ideas presented here. All of these build on the concepts of this book, but we have recommended several books that take different perspectives from our own.

Microskills

Evans, D., Hearn, M., Uhlemann, M., & Ivey, A. (2016). *Essential interviewing* (9th ed.). Belmont, CA: Brooks/Cole.
Microskills in a programmed text format.

Ivey, A., Gluckstern, N., & Ivey, M. (2015). *Basic attending skills* (5th ed.). Alexandria, VA: Alexander Street Press/Microtraining Associates.

Brief and perhaps the most suitable book for beginners and those who would teach others microskills. Supporting videotapes are available (www.academicvideostore.com /microtraining).

Zalaquett, C., Ivey, A., Gluckstern, N., & Ivey, M. (2008). *Las habilidades atencionales básicas: Pilares fundamentales de la comunicación efectiva.* [Book and Training Videos]. Alexandria, VA: Alexander Street Press/Microtraining Associates.

An introduction to the microskills and the five steps of the interview in Spanish. Una introducción a las microhabilidades y la entrevista de cinco etapas en Español. [Libro y Videos de entrenamiento.] www.academicvideostore.com/microtraining

Visit this website for videos on the listening and influencing skills, as well as many theoretical orientation demonstrations. Special attention is given to multicultural counseling and therapy.

Theories of Interviewing and Counseling with an Orientation to Diversity

Ivey, A., D'Andrea, M., & Ivey, M. (2012). *Theories of counseling and psychotherapy: A multicultural perspective* (7th ed.). Thousand Oaks, CA: Sage.

The major theories are reviewed, with special attention to multicultural issues. Includes many applied exercises to take theory into practice.

Sue, D., & Sue, D. M. (2007). *Foundations of counseling and psychotherapy: Evidence-based practices for a diverse society.* New York: Wiley.

An excellent text, focusing on evidence-based approaches on a multicultural base.

Multicultural Counseling and Therapy

Cheek, D. (2010). *Assertive Black . . . puzzled White: A Black perspective on assertive behavior.* Parker, CO: Outskirts Press.

This book needs to be on every interviewer's and counselor's shelf.

Chung, R. C., & Bemak, F. P. (2000). *Social justice counseling: The next steps beyond multiculturalism.* Thousand Oaks, CA: Sage.

This book advances awareness and knowledge of multicultural progress and reviews the central concepts of social justice counseling. Authors see social justice counseling as a natural extension of the multicultural movement.

Sue, D. W. (2010). *Microaggressions in everyday life: Race, gender, and sexual orientation.* New York: Wiley.

This book caused a sensation in multicultural circles. It shows how everyday events shape our worldview and the harm and cumulative trauma that microagressions can cause. Out of awareness one can build a stronger cultural identity. View Dr. Sue at www.youtube.com/watch?v=4pZy7JaO3FE.

Sue, D. W., Carter, R., Casas, M., Fouad, N., Ivey, A., Jensen, M., et al. (1998). *Multicultural counseling competencies.* Beverly Hills, CA: Sage.

The original and most comprehensive coverage of the necessary skills and competencies in the multicultural area.

Sue, D. W., Ivey, A., & Pedersen, P. (1999). *A theory of multicultural counseling and therapy.* Pacific Grove, CA: Brooks/Cole.

A general theory of multicultural counseling and therapy, with many implications for practice.

Sue, D. W., & Sue, D. (2012). *Counseling the culturally diverse* (6th ed.). New York: Wiley.

The classic of the field, this book helped launch a movement.

Integrative/Eclectic Orientations

Beck, J. (2011). *Cognitive behavior therapy: Basics and beyond* (2nd ed.). New York: Guilford Press.

The popular CBT, along with extensive research, has become an integrative theory. It includes many techniques and strategies drawn from many sources. Originally a derivative of Albert Ellis's REBT framework, it has redefined his work. It has added meditation, relaxation training, and a host of strategies, many of them original and others useful adaptations. Visit the Beck Institute online (www.beckinstitute.org) for extensive information and many new planning and feedback forms.

Dobson, K. S. (2011). *Handbook of cognitive-behavioral therapies* (3rd ed.). New York: Guilford Press.

A thorough review of the basic tenets, models, and applications of CBT, complemented with a chapter of applications on diverse populations.

Ivey, A., Ivey, M., Myers, J., & Sweeney, T. (2005). *Developmental counseling and therapy: Promoting wellness over the lifespan.* Belmont, CA: Wadsworth.

This book elaborates the developmental stages discussed in connection with discernment (Chapter 11). It shows practice specifics for elaborating cognitive and emotional experience, how to use a positive developmental model with severely distressed clients, and applications in multicultural counseling and therapy with emphasis on cultural identity theory, spirituality, and family therapy.

Lazarus, A. A. (2006). *Brief but comprehensive psychotherapy: The multimodal way.* New York: Springer.

The basic book for multimodal therapy. You will find the BASIC-ID model useful in conceptualizing broad treatment plans.

Books for Follow-up on Neuroscience

Grawe, K. (2007). *Neuropsychotherapy: How the neurosciences inform effective psychotherapy.* Mahway, NJ: Erlbaum.

This is quite challenging reading, but it was the first to show how neuroscience relates to our practice. It remains innovative and important. Grawe, sadly, died before his book was published, but you can get a sense of his major contribution at www.psychotherapyresearch.org/displaycommon.cfm?an=1&subarticlenbr=50.

Hoffman, M. (2015). *Brain beat: Scientific foundations and evolutionary perspectives on brain health.* New York: Page.

Despite the title, this is a clear, accurate, fascinating, brief overview of neuroscience and neurobiology. Neuroarcheology outlines the history of the brain and body from the very beginning. You will discover how brain structures evolved over time and their relationship to the natural environment—and now the changing environments of culture. This is a fascinating read that will excite and interest you to think about neuroscience and neurobiology in a new way.

Siegel, D. (2015). *The developing mind* (2nd ed.). New York: Guilford Press.

Dr. Siegel has gained the most prominence for beginning readers. Watch for his continuing series.

APPENDIX I

The Ivey Taxonomy
Definitions of the Microskills and Strategies with Anticipated Client Response

Skill, Concept, or Strategy	Anticipated Client Response
Ethics and Morals Ethics are rules, typically prescribed by social systems and, in counseling, as professional standards. They define how things are to be done. Morals are individual principles we live by that define our beliefs about right and wrong. A moral approach to interviewing and counseling allows us to apply ethical principles respectfully to our clients and ourselves.	Following professional ethics results in client trust and provides us with guidelines for action in complex situations. Morals represent our individual efforts and actions to follow ethical principles. A moral approach to interviewing and counseling helps us to remember that our personal actions count, both inside and outside the session. Furthermore, a moral approach to the session may ask you to help clients examine their own moral and ethical decisions.
Multicultural Competence Your competence in multiculturalism is based on your level of *awareness, knowledge, skills,* and *action.* Self and other awareness and knowledge are critical, but one must also have the skills and the ability to act.	Anticipate that both you and your clients will appreciate, gain respect, and learn from increasing knowledge in ethics and multicultural competence. You, the interviewer, will have a solid foundation for a lifetime of personal and professional growth. You will be challenged to consider implications of social justice for your practice.
Positive Psychology and Resilience Help clients discover and rediscover their strengths. Find strengths and positive assets in clients and in their support system. Identify multiple dimensions of wellness. In addition to listening, actively encourage clients to learn new actions that will increase their resilience.	Clients who are aware of their strengths and resources can face their difficulties and resolve issues from a positive foundation. They become resilient and can bounce back from obstacles and defeat.
Therapeutic Lifestyle Changes (TLCs) Therapeutic lifestyle changes facilitate development of new connections, and brain health can be increased. The TLCs positively affect mental and physical health, self-esteem, cognitive reserve, happiness, and length of life.	Positive approaches to mental and physical health recognize the role of TLCs in interviewing and counseling. TLCs are cost effective and are supported by research, but a team approach may be required. You cannot be an expert in all therapeutic lifestyle changes; as needed, refer clients to medical personnel, nutritionists, physical therapists, personal trainers, and other behavioral health professionals.

(continued)

Skill, Concept, or Strategy	Anticipated Client Response
Attending Behavior Support your client with individually and culturally appropriate visuals, vocal quality, verbal tracking, and body language, including facial expressions.	Clients will talk more freely and respond openly, particularly about topics to which attention is given. Depending on the individual client and culture, anticipate fewer breaks in eye contact, a smoother vocal tone, a more complete story (with fewer topic jumps), and a more comfortable body language.
Empathy Experiencing the client's world and story as if you were that client; understanding his or her key issues and saying them back accurately, without adding your own thoughts, feelings, or meanings. This requires attending and observation skills plus using the important key words of the client, but distilling and shortening the main ideas.	Clients will feel understood and engage in more depth in exploring their issues. Empathy is best assessed by clients' reaction to a statement and their ability to continue discussion in more depth, and eventually with better self-understanding.
Basic Empathy Interviewer responses are roughly interchangeable with those of the client. The counselor is able to say back accurately what the client has said.	This the most common level of empathy provided in counseling and therapy. According to Rogers, listening in itself is necessary and sufficient to produce client change.
Additive Empathy Interviewer adds meaning and feelings beyond those originally expressed by the client.	Clients reach a better understanding of their own issues and engage in more depth in exploring of these issues.
Subtractive Empathy Interviewer responses give back to the client less than what the client said and perhaps even distort what has been said. In this case, the listening or influencing skills are used inappropriately.	Skill is used inappropriately and subtracts from client's experience. Client doesn't feel understood.
Client Observation Skills Observe your own and the client's verbal and non-verbal behavior. Anticipate individual and multicultural differences in nonverbal and verbal behavior. Carefully and selectively feed back some here-and-now observations to the client as topics for exploration.	Observations provide specific data validating or invalidating what is happening in the session. Also, they provide guidance for the use of various microskills and strategies. The smoothly flowing session will often demonstrate movement symmetry or complementarity. Movement dyssynchrony provides a clear clue that you are not "in tune" with the client.
Open Questions Begin open questions with the often useful *who, what, when, where,* and *why. Could, can,* or *would* questions are considered open but have the additional advantage of being somewhat closed, thus giving more power to the client, who can more easily say that he or she doesn't want to respond.	Clients will give more detail and talk more in response to open questions. *Could, would,* and *can* questions are often the most open of all, because they give clients the choice to respond briefly ("No, I can't") or, much more likely, explore their issues in an open fashion.
Closed Questions Closed questions may start with *do, is,* or *are.*	Closed questions may provide specific information but may close off client talk. As such, they need to be asked carefully. But if the relationship is solid and the topic important, the client may talk as much as if given an open question.

Skill, Concept, or Strategy	Anticipated Client Response
Encouraging Encourage with short responses that help the client keep talking. These responses may be verbal (repeating key words and short statements) or nonverbal (head nods and smiling).	Clients elaborate on the topic, particularly when encouragers and restatements are used in a questioning, supportive tone of voice.
Paraphrasing Shorten or clarify the essence of what has just been said, but be sure to use the client's main words when you paraphrase. Paraphrases are often fed back to the client in a questioning tone of voice.	Clients will feel heard. They tend to give more detail without repeating the exact same story. They also become clearer and more organized in their thinking. If a paraphrase is inaccurate, the client has an opportunity to correct the interviewer. Paraphrasing of client statements is important in cognitive empathy.
Summarizing Summarize client comments and integrate thoughts, emotions, and behaviors. Summarizing is similar to paraphrasing but used over a longer time span.	Clients will feel heard and discover how their complex and even fragmented stories are integrated. The summary helps clients make sense of their lives and will facilitate a more centered and focused discussion. Secondarily, the summary also provides a more coherent transition from one topic to the next or a way to begin and end a full session. As a client organizes the story more effectively, we are seeing growth in brain's executive functioning and better decision making.
Checkout/Perception Check Periodically, check with your client to discover how your interviewing lead or skill was received. "Is that right?" "Did I hear you correctly?" "What might I have missed?"	Interviewing leads such as these give clients a chance to pause and reflect on what they have said. If you indeed have missed something important or have distorted their story and meaning, they have the opportunity to correct you. Without an occasional checkout, it is possible to lead clients away from what they really want to talk about.
Reflection of Feelings Identify the key emotions and feed them back to clarify affective experience. With some clients, a brief acknowledgment of feelings may be more appropriate. Reflection of feelings is often combined with paraphrasing and summarizing. Include a search for positive feelings and strengths.	Clients will experience and understand their emotional states more fully and talk in more depth about feelings. They may correct the counselor's reflection with a more accurate descriptor. In addition, client understanding of underlying feelings leads to emotional regulation with clearer cognitive understanding and behavioral action. Critical to lasting change is a more positive emotional outlook.
Basic Listening Sequence (BLS) The basic listening sequence (BLS), based on attending and observing, consists of these microskills: using open and closed questions, encouraging, paraphrasing, reflecting feelings, and summarizing.	Clients will discuss their stories, issues, or concerns, including the key facts, thoughts, feelings, and behaviors. Clients will feel that their stories have been heard. In addition, these same skills will help friends, family members, and others to be clearer with you and facilitate better interpersonal relationships.

(continued)

The Five Stages/Dimensions of the Well-Formed Interview

1. Empathic Relationship *Initiate the session.* Develop rapport and structuring. "Hello, what would you like to talk about today?" "What might you like to see as a result of our talking today?"	The client feels at ease with an understanding of the key ethical issues and the purpose of the interview. The client may also know you more completely as a person and a professional—and has a sense that you are interested in his or her concerns.
2. Story and Strengths *Gather data.* Use the BLS to draw out client stories, concerns, problems, or issues. "I'd like to hear your story." "What are your strengths and resources?"	The client shares thoughts, feelings, and behaviors; tells the story in detail; presents strengths and resources.
3. Goals *Set goals mutually.* The BLS will help define goals. "What do you want to happen?" "How would you feel emotionally if you achieved this goal?" One possible goal is exploration of possibilities, rather than focusing immediately.	The client will discuss directions in which he or she might want to go, new ways of thinking, desired feeling states, and behaviors that might be changed. The client might also seek to learn how to live more effectively with stressful situations or events that cannot be changed at this point (rape, death, an accident, an illness). A more ideal story might be defined.
4. Restory *Explore alternatives.* Explore alternatives via the BLS. Confront client incongruities and conflict. "What are we going to do about it?" "Can we generate new ways of thinking, feeling, and behaving?"	The client may reexamine individual goals in new ways, solve problems from at least those alternatives, and start the move toward new stories and actions.
5. Action *Conclude.* Plan for generalizing session learning to "real life." "Will you use what you decided to do today, tomorrow, or this coming week?"	The client demonstrates changes in behavior, thoughts, and feelings in daily life outside of the interview conversation. Or the client explores new alternatives and reports back discoveries.
Focusing Intentionally focus the counseling session on the client, theme/concern/issue, significant others (partner/spouse, family, friends), a mutual "we" focus, the counselor, or the cultural/environmental context as necessary to gain a broader understanding of client and issue. You may also focus on what is going on in the here and now of the interview.	As the counselor brings in new focuses, the story is elaborated from multiple perspectives. If you selectively attend only to the individual, the broader dimensions of the social context are likely to be missed and counseling and therapy may fail in the long run.
Empathic Confrontation Supportively challenge the client to address observed discrepancies and conflicts: 1. Listen, observe, and note client conflict, mixed messages, and discrepancies in verbal and nonverbal behavior. Give attention to both cognitive and emotional dimensions. 2. Paraphrase and reflect feelings to clarify internal and external discrepancies. As the issues become clarified, empathically summarize what has been said—for example, "on one hand you feel _____, but on the other hand you feel _____." Bring both cognition and emotions into most summaries.	Clients will respond to effective confrontation of discrepancies and conflict by creating new ideas, thoughts, feelings, and behaviors, and these will be measurable on the five-point Client Change Scale. Again, if no change occurs, listen. Then try an alternative style of confrontation.

3. Evaluate how the client responds and whether the confrontation leads to client movement or change. If the client does not change, flex intentionally, try another skill, and approach the conflict from another direction.

Client Change Scale (CCS)
The CCS helps you evaluate where the client is in the change process.
Level 1. Denial
Level 2. Partial examination
Level 3. Acceptance and recognition, but no change
Level 4. Generation of a new solution
Level 5. Transcendence

The CCS can help you determine the impact of your use of skills. This assessment may suggest other skills and strategies that you can use to clarify and support the change process. You will find it invaluable to have a system that enables you to (1) assess the value and impact of what you just said; (2) observe whether the client is changing in response to a single intervention; or (3) use the CCS as a method for examining behavior change over a series of sessions.

Reflection of Meaning
Meanings are close to core experiencing. Encourage clients to explore their own meanings and values in more depth from their own perspective, but also the perspectives of others. Questions eliciting meaning are often a vital first step. A reflection of meaning looks very much like a paraphrase, but focuses beyond what the client says. Appearing often are the words *meaning, values, vision,* and *goals.*

The client discusses stories, issues, and concerns in more depth, with a special emphasis on deeper meanings, values, and understandings. Clients may be enabled to discern their life goals and vision for the future.

Interpretation/Reframing
Provide the client with a new perspective, frame of reference, or way of thinking about issues. Interpretations/reframes may come from your observations; they may be based on varying theoretical orientations to the helping field; or they may link critical ideas together.

The client may find another perspective or way of thinking about a story, issue, or concern. New perspective could have been generated by a theory used by the interviewer, from linking ideas or information, or by simply looking at the situation afresh.

Empathic Self-Disclosure
Self-disclosure is sharing your own personal experience related to what the client has said and often starts with an "I" statement. Or it may be sharing your own thoughts and feelings concerning what clients are experiencing in the immediate moment, the here and now of the interview.

Clients respond well to self-disclosure, carefully put, especially at the beginning of a session. They are often pleased to know more about you at that point. Later in the session, sharing your thoughts and feelings about clients can enable them to talk more openly about their issues. Self-disclosure almost always needs to be positive and supportive.

Empathic Feedback
Feedback presents clients with clear, nonjudgmental information (and sometimes even opinions) on client thoughts, feelings, and behaviors, either in the past or here and now.

Feedback can be supportive or challenging. Supportive feedback searches for positives and strengths, while challenges ask clients to think more carefully about themselves and what they are saying.

Natural and Logical Consequences
Explore specific alternatives and the logical positive and negative consequences of each decision possibility with the client. "If you do _____, then _____."

Clients will change thoughts, feelings, and behaviors through better anticipation of the consequences of their actions. When you explore the positives and negatives of each possibility, clients will be more involved in the process of making creative new decisions.

(continued)

Directives, Instruction, Psychoeducational Strategies

Clear directions, encouraging clients to do what you suggest, underlie instruction and psychological education. These offer specifics for daily life to help change thoughts, feelings, and behaviors. Providing useful instruction and referral sources can be helpful. Psychoeducational strategies include systematic educational methods such as therapeutic lifestyle changes. With all these, a collaboration approach is essential.

Clients will make positive progress when they listen to and follow the directives, use the information that you provide for them, consider your advice, and engage in new, more positive thinking, feeling, or behaving. Psychoeducation can lead to major life changes for physical and mental health.

Skill Integration

Integrate the microskills into a well-formed interview, and generalize the skills to situations beyond the training session or classroom.

Developing interviewers and counselors will integrate skills as part of their natural style. Each of us will vary in our choices, but increasingly we will know what we are doing, how to flex when what we are doing is ineffective, and what to expect in the interview as a result of our efforts.

Determining Personal Style and Theory

As you work with clients, identify your natural style, add to it, and think through your approach to interviewing and counseling. Examine your own preferred skill usage and what you do in the session. Integrate learning from theory and practice in interviewing, counseling, and psychotherapy into your own skill set.

You, as a developing interviewer or counselor, will identify and build on your natural style. You will commit to a lifelong process of constantly learning about theory and practice while evaluating and examining your behavior, thoughts, feelings, and deeply held meanings.

Transcendence

Transcendence speaks about your capacity to go beyond yourself and successfully apply your newly mastered skills to help others. The ultimate test of your capacities is the benefit they can afford others. Your growth is wonderful; helping others' growth is even better!

Clients will perceive your genuine interest and your intentional and collaborative work with them. Trust in the relationship will increase, and clients will benefit more from working with you.

APPENDIX II

Ethics

Ethics and Morals: Professional and Personal

Interviewers, counselors, therapists, and other professionals observe and practice ethically. Their actions and interventions follow aspirations and professional standards offered by their professional organization's code of ethics and their licensing institution's standards of professional conduct.

If you behave ethically and intentionally, you can anticipate that the relationship with your interviewee or client will be enhanced, will proceed more smoothly, and will build trust as your client will feel protected.

Professional ethics codes provide you with guidelines for action in complex situations. Morals represent your individual efforts and actions to follow ethical principles.

A moral approach to interviewing and counseling helps us to remember that our personal actions count, both inside and outside the session. The same approach may lead you to help clients examine their own moral and ethical decisions.

A Brief History of the Multicultural Foundation of Ethics

A major conference on the future of professional psychology stated that

> the provision of professional services to persons of culturally diverse backgrounds by persons not competent in understanding and providing professional services to such group shall be considered unethical; . . . it shall be equally unethical to deny such persons professional services because the present staff is inadequately prepared; . . . it shall be the obligation of all service agencies to employ competent persons or to provide continuing education for the present staff to meet the service needs of the culturally diverse population it serves. (Korman, 1973, p. 105)

Despite the clarity of this statement, the field moved slowly to implementing these recommendations. Pedersen and Marsella (1982) commented, "A serious moral vacuum exists in that the delivery of cross-cultural counseling and therapy because the values of a dominant culture have been imposed on a culturally different consumer" (p. 498).

Finally, in 1986, the *General Guidelines for Providers of Psychological Services* provided the first formal recognition of the importance of multicultural issues in psychology. The *Guidelines* of the American Psychological Association are oriented to defining good practice for the field. As part of the preamble, we find:

> These *General Guidelines* have been developed with the understanding that psychological services must be planned and implemented so that they are sensitive to factors related to life in a pluralistic society such as age, gender, sexual orientation, culture, and ethnicity.

The definition and language of the key multicultural issues has changed, and now the helping fields give central attention to this in ethical practice. But it remains important to remember that this was not always so. The view of counseling and psychotherapy has progressed, but there is still much work to do as we move toward an increasingly aware and active multicultural understanding and its relationship to ethical practice.

Ethical Codes

Ethical codes can be summarized with the following statement: "Promote the well-being of your clients; treat them responsibly with full awareness of the social context of helping; do no harm to your clients." As interviewers and counselors, we are morally responsible for our clients and for society as well. At times these responsibilities conflict, and you may need to seek detailed guidance from documented ethical codes and standards of professional conduct, as well as from your supervisor, colleagues, or other professionals or professional organizations' ethics committees.

The following sections review essential ethical guidelines and aims of various professional codes of ethics. This information expands what was presented in Chapter 2, as we believe that ethics is at the foundation of interviewing, counseling, and therapy.

Confidentiality: Our Moral Foundation

Confidentiality is the cornerstone of our tool kits.

—Robert Blum

We have said that the empathic relationship is central to developing a working alliance with clients. But without confidentiality and trust as our basis, we will have no relationship. Thus, from the very beginning, the amount of confidentiality you can provide your client needs to be crystal clear. Your clients need to know that absolute confidentiality is legally impossible, and thus it is essential to spell out the limits of confidentiality in your setting at the beginning. Box II.1 provides additional information regarding confidentiality and its limits from an international perspective.

BOX II.1   Confidentiality and Its Limits

The following provisions regarding confidentiality are from the Code of Ethics and Practice of the Australian Counselling Association (2015).

3.4 Confidentiality

(a) Confidentiality is a means of providing the client with safety and privacy and thus protects client autonomy. For this reason any limitation on the degree of confidentiality is likely to diminish the effectiveness of counselling.

(b) The counselling contract will include any agreement about the level and limits of the confidentiality offered. This agreement can be reviewed and changed by negotiation between the counsellor and the client. Agreements about confidentiality continue after the client's death unless there are overriding legal or ethical considerations. In cases where the client's safety is in jeopardy any confidentially agreements that may interfere with this safety are to be considered void (see 3.6 'Exceptional circumstances').

3.6 Exceptional Circumstances

(a) Exceptional circumstances may arise which give the counsellor good grounds for believing that serious harm may occur to the client or to other people. In such circumstance the client's consent to change in the agreement about confidentiality should be sought

BOX II.1   (continued)

whenever possible unless there are also good grounds for believing the client is no longer willing or able to take responsibility for his/her actions Normally, the decision to break confidentiality should be discussed with the client and should be made only after consultation with the counselling supervisor or if he/she is not available, an experienced counsellor.

(b) Any disclosure of confidential information should be restricted to relevant information, conveyed only to appropriate people and for appropriate reasons likely to alleviate

the exceptional circumstances. The ethical considerations include achieving a balance between acting in the best interests of the client and the counsellor's responsibilities under the law and to the wider community.

(c) While counsellors hold different views about grounds for breaking confidentiality, such as potential self-harm, suicide, and harm to others they must also consider those put forward in this Code, as they too should imbue their practice. These views should be communicated to both clients and significant others, e.g., supervisor, agency, etc.

As a student taking this course, you are a beginning professional; you usually do not have legal confidentiality. Nonetheless, you need to keep to yourself what you hear in class role-plays or practice sessions. Trust is built on your ability to keep confidences. Be aware that state laws on confidentiality vary. Informed consent is an ethical issue discussed later in the chapter, along with some ways to share the concept of confidentiality with clients.

Professionals encounter many challenges to confidentiality. Some states require you to inform parents before counseling a child and to share information from interviews with them if they ask. If issues of abuse should arise, you must report this to the authorities. If the client is a danger to self or others, then rules of confidentiality change; the issue of reporting such information needs to be discussed with your supervisor. As a beginning interviewer, you will likely have limited, if any, legal protection, so limits to confidentiality must be included in your approach to informed consent.

Dual relationships can present challenging ethical and moral issues. They occur when you have more than one relationship with a client. Another way to think of this is the concept of conflict of interest.

If your client is a classmate or friend, you are engaged in a dual relationship. If you live in a small town, you are likely to encounter some of your clients at the grocery store or elsewhere. These situations may also occur when you counsel a member of your church or school. Personal, economic, and other privacy matters can become complex issues. You can examine statements on dual relationships in more detail in professional ethical codes.

Diversity, Multiculturalism, Ethics, and Morality

We need to help students and parents cherish and preserve the ethnic and cultural diversity that nourishes and strengthens this community—and this nation.

—Cesar Chavez

The American Counseling Association (2014) focuses the preamble to its code of ethics on diversity as a central ethical issue:

The American Counseling Association (ACA) is an educational, scientific, and professional organization whose members work in a variety of settings and serve in multiple capacities. Counseling is a professional relationship that empowers diverse individuals, families, and groups to accomplish mental health, wellness, education, and career goals.

Professional values are an important way of living out an ethical commitment. The following are core professional values of the counseling profession:

1. enhancing human development throughout the lifespan;
2. honoring diversity and embracing a multicultural approach in support of the worth, dignity, potential, and uniqueness of people within their social and cultural contexts;
3. promoting social justice;
4. safeguarding the integrity of the counselor–client relationship; and
5. practicing in a competent and ethical manner.

Human services professionals have added a moral personal dimension to this ethical statement (National Organization for Human Services, 2015). They first speak of the need for advocacy for the rights of others, particularly groups that have been disadvantaged or oppressed. For us, this means that following ethical principles in the office often is not enough. Morally, we are asked to move to the community and work to prevent discrimination and oppression. It also means that we need to look at our own RESPECTFUL identity and consider the morality of how we have been treated in the past and how we might want to treat our clients with moral respect.

You will work with clients who have made mistaken moral judgments about themselves. Some clients may be judging themselves too harshly for what they have done or left undone. Other clients may fail to recognize their own moral failures. You will have an interesting challenge as you face the moral dilemmas of clients, both those of which they are aware and those that they deny, ignore, or may not even be aware of.

Ethics, Morality, and Competence

Don't judge each day by the harvest you reap but by the seeds that you plant.

—Robert Louis Stevenson

Awareness of what we can and cannot do is basic to moral competence.

> **C.2.a. Boundaries of Competence.** Counselors practice only within the boundaries of their competence, based on their education, training, supervised experience, state and national professional credentials, and appropriate professional experience. Whereas multicultural counseling competency is required across all counseling specialties, counselors gain knowledge, personal awareness, sensitivity, dispositions, and skills pertinent to being a culturally competent counselor in working with a diverse client population. (American Counseling Association, 2014)

We all need to constantly monitor whether we are competent to counsel clients around the issues that they present to us. For example, you may be able to help a client work out difficulties occurring at work, but you discover a more complex underlying issue of serious depression undercutting the client's ability to find or keep a job. You may be competent to help this client with career and vocational issues but may not have had enough experience with depression. While maintaining a supportive attitude, you refer the client to another counselor for therapy while you continuing to work with the job issues. Although it is essential that you not work beyond your competence, the morality of ethics demands that you continue studying to expand your competence through reading, inservice training, and supervision.

Informed Consent

Counseling is an international profession. The Canadian Counselling and Psychotherapy Association's (2007) approach to informed consent is particularly clear:

> **B4. Client's Rights and Informed Consent.** When counselling is initiated, and throughout the counselling process as necessary, counsellors inform clients of the purposes, goals, techniques, procedures, limitations, potential risks and benefits of services to be performed, and other such pertinent information. Counsellors make sure that clients understand the implications of diagnosis, fees and fee collection arrangements, record keeping, and limits of confidentiality. Clients have the right to participate in the ongoing counselling plans, to refuse any recommended services, and to be advised of the consequences of such refusal.

The American Psychological Association (2010) stresses that psychologists should inform clients if the interview is to be supervised and provides additional specifics:

> **Standard 10.01 (c)** When the therapist is a trainee and the legal responsibility for the treatment provided resides with the supervisor, the client/patient, as part of the informed consent procedure, is informed that the therapist is in training and is being supervised and is given the name of the supervisor.

> **Standard 4.03** Before recording the voices or images of individuals to whom they provide services, psychologists obtain permission from all such persons or their legal representatives.

When you work with children, the ethical issues around informed consent become especially important. Depending on state laws and practices, it is often necessary to obtain written parental permission before interviewing a child or before sharing information about the interview with others. The child and family should know exactly how any information is to be shared, and interviewing records should be available to them for their comments and evaluation. An essential part of informed consent is stating that both child and parents have the right to withdraw their permission at any point. Needless to say, these same principles apply to all clients—the main difference is parental awareness and consent.

When you enter into role-plays and practice sessions, inform your volunteer "clients" about their rights, your own background, and what clients can expect from the session. For example, you might say:

> I'm taking an interviewing course, and I appreciate your being willing to help me. I am a beginner, so only talk about things that you want to talk about. I would like to [audio or video] record the interview, but I'll stop immediately if you become uncomfortable and delete it as soon as possible. I may share the recording in a practicum class or I may produce a written transcript of this session, removing anything that could identify you personally. I'll share any written material with you before passing it in to the instructor. Remember, we can stop any time you wish. Do you have any questions?

You can use this statement as a starting point and eventually develop your own approach to this critical issue.

Privacy Rules

The Health Insurance Portability and Accountability Act (HIPAA) took effect in 1996. We include it here because, among other functions, it requires the protection and confidential handling of protected health information.

Following is a summary of some key elements of the Privacy Rule, including who is covered, what information is protected, and how protected health information may be used and disclosed. For a complete outline of HIPAA requirements, visit the website of the U.S. Department of Health and Human Services, Office for Civil Rights, at http://www.hhs.gov.

1. Protected Health Information. The Privacy Rule defines "protected health information (PHI)" as all individually identifiable health information held or transmitted by a covered entity or its business associate, in any form or media, including electronic, paper, or oral. "Individually identifiable health information" is information, including demographic data and personal identifiers such as name, address, birth date, and social security number, that identifies the individual, or could reasonably be used to identify the individual, and that relates to:

- The individual's past, present, or future physical or mental health or condition
- The provision of health care to the individual
- The past, present, or future payment for the provision of health care to the individual

The Privacy Rule does not include protected health information from employment records that a covered entity maintains in its capacity as an employer as well as education and certain other records subject to, or defined in, the Family Educational Rights and Privacy Act, 20 U.S.C. §1232g.

2. De-identified Health Information. This is information that makes it impossible for others to identify a client, and there are no restrictions on its use or disclosure. Information can be de-identified in two ways: (1) a formal determination by a qualified statistician or (2) the removal of specified identifiers of the individual and of the individual's relatives, household members, and employers. De-identification is adequate only if the covered entity has no actual knowledge that the remaining information could be used to identify the individual.

When you visit a physician, you are asked to sign a version of the privacy statement. Mental health agencies make their privacy statements clearly available to clients and often post them in the office.

3. Mental Health Information. One exception to the general Privacy Rule, which applies to PHI regardless of the type of information, is psychotherapy notes. Psychotherapy notes are defined as notes recorded by a mental health professional to document or analyze the contents of a conversation during a private counseling session (e.g., individual, couple, group, or family session). These notes are separate from the rest of the patient's medical record and are treated differently than other mental health information because they contain particularly sensitive information. Also, they are the personal notes of the counselor or therapist that usually are not required or useful for treatment, payment, or the operations of the clinic or hospital. Important to note is that these notes do not include any information about medication prescription and monitoring, counseling session start and stop times, the modalities and frequencies of treatment furnished, or results of clinical tests; nor do they include summaries of diagnosis, functional status, treatment plan, symptoms, prognosis, and progress to date. Psychotherapy notes also do not include any information that is maintained in a patient's medical record. See 45 CFR 164.501.

The Privacy Rule requires a covered entity to obtain clients' authorization prior to disclosure of psychotherapy notes for any reason, except when disclosure is required by another law, such as mandatory reporting of abuse or mandatory "duty to warn" situations involving threats of serious and imminent harm made by the patient (state laws vary as to whether such a warning is mandatory or permissible). See 45 CFR 164.508(a)(2).

Social Justice as Morality and Ethics in Action

I've become even more convinced that the type of stress that is toxic has more to do with social status, social isolation, and social rejection. It's not just having a hard life that seems to be toxic, but it's some of the social poisons that can go along with the stigma of poverty.

—Kelly McGonigal

Jane Addams, the founder of social work, has infused her thinking throughout the National Association of Social Workers. Their code of ethics (2008) is strongest on social justice and emphasizes that action beyond the interview in the community may be needed to address unfairness of many types.

> **Ethical Principle:** *Social workers challenge social injustice.* Social workers pursue social change, particularly with and on behalf of vulnerable and oppressed individuals and groups of people. Social workers' social change efforts are focused primarily on issues of poverty, unemployment, discrimination, and other forms of social injustice. These activities seek to promote sensitivity to and knowledge about oppression and cultural and ethnic diversity. Social workers strive to ensure access to needed information, services, and resources; equality of opportunity; and meaningful participation in decision making for all people.

We now know that childhood poverty, adversity, and stress produce lifelong damage to the brain. These changes are visible in cells and neurons and include permanent changes in DNA. Therefore, just treating children of poverty through supportive counseling is not enough; for significant change to occur, prevention and social justice action are critical.

There are two major types of social justice action. The first and most commonly discussed is action in the community to work against the destructive influences of poverty, racism, and all forms of discrimination. These preventive strategies are now considered a vital dimension of the "complete" counselor or therapist. Getting out of the office and understanding society's influence on client issues is central. Clients who have suffered social injustice of virtually any type (poverty, bullying, sexism, heterosexism) will also benefit from joining groups that work in some way to prevent or alleviate the impact of oppression. One route toward healing is working with others, or even by oneself, for those who have experienced injustice. This can range from work in soup kitchens to participating in a protest, joining a Take Back the Night walk, or simply writing letters or articles in the local paper.

The second type of social justice action occurs in the interview. When a female client discusses mistreatment and harassment by her supervisor, the issue of oppression of women should be named as such. The social justice perspective requires you to help her understand that the problem is not caused by her behavior or how she dresses. By naming the problem as sexism and harassment, you often free the client from self-blame and empower her for action. You can also support her in efforts to bring about change in the workplace.

You will sometimes be challenged in the session by clients who hold moral and ethical values that differ from yours. The question always comes up: Should you confront and challenge them? First and foremost, how fragile is the client, and will your comments be disrupting? If so, it is best to hold your tongue and seek to support the client where he or she needs help. When the difficult topics come up (religion, politics, and oppression of many types), be prepared with your own moral values, but apply them carefully. If it becomes too difficult, refer; do not impose your values.

Box II.2 lists websites of some key ethical codes in English-speaking areas of the globe. The last link offers access to codes in Spanish-speaking countries. All codes provide guidelines on competence, informed consent, confidentiality, and diversity. Issues of power and social justice are explicit in social work and human services and implicit in other codes.

BOX II.2   Professional Organizations with Ethical Codes

American Academy of Child and Adolescent Psychiatry (AACAP)	http://www.aacap.org
	https://www.aacap.org/App_Themes/AACAP/docs/about_us/transparency_portal/aacap_code_of_ethics_2012.pdf
American Association for Marriage and Family Therapy (AAMFT) Code of Ethics	http://www.aamft.org
	http://aamft.org/iMIS15/AAMFT/Content/Legal_Ethics/Code_of_Ethics.aspx
American Counseling Association (ACA) Code of Ethics	http://www.counseling.org
	https://www.counseling.org/resources/aca-code-of-ethics.pdf
American Psychological Association (APA) Ethical Principles of Psychologists and Code of Conduct	http://www.apa.org
	http://www.apa.org/ethics/code/index.aspx
American Psychiatric Association (APA)	http://www.psychiatry.org
	http://www.psychiatry.org/psychiatrists/practice/ethics
American School Counselor Association (ASCA)	http://www.schoolcounselor.org
	http://www.schoolcounselor.org/asca/media/asca/Resource%20Center/Legal%20and%20Ethical%20Issues/Sample%20Documents/EthicalStandards2010.pdf
Australian Psychological Society (APS) Code of Ethics	http://www.psychology.org.au
	http://www.psychology.org.au/about/ethics
British Association for Counselling and Psychotherapy (BACP) Ethical Framework	http://www.bacp.co.uk
	http://www.bacp.co.uk/ethical_framework

BOX II.2   **(continued)**

The Canadian Counselling and Psychotherapy Association Code of Ethics	https://www.ccpa-accp.ca
	https://www.ccpa-accp.ca/wp-content/uploads/2014/10/CodeofEthics_en.pdf
Commission on Rehabilitation Counselor Certification (CRCC) Code of Professional Ethics for Rehabilitation Counselors	https://www.crccertification.com
	https://www.crccertification.com/code-of-ethics-3
International Union of Psychological Science (IUPsyS) Universal Declaration of Ethical Principles for Psychologists	http://www.iupsys.net/about/governance/universal-declaration-of-ethical-principles-for-psychologists.html
National Association of School Nurses (NASN)	http://www.nasn.org
	http://www.nasn.org/RoleCareer/CodeofEthics
National Association of School Psychologists (NASP)	http://www.nasponline.org
	http://www.nasponline.org/standards-and-certification/professional-ethics
National Association of Social Workers (NASW) Code of Ethics	http://www.naswdc.org http://www.socialworkers.org/pubs/Code/code.asp
National Career Development Association (NCDA)	http://www.ncda.org
	http://www.ncda.org/aws/NCDA/asset_manager/get_file/3395/ncda_code_of_ethics_2-24-15.pdf
New Zealand Association of Counsellors (NZAC) Code of Ethics	http://www.nzac.org.nz
	http://www.nzac.org.nz/code_of_ethics.cfm
School Social Work Association of America (SSWAA)	http://www.sswaa.org
	http://www.sswaa.org/?page=459&hhSearchTerms=%22code+and+ethics%22
International Coach Federation	www.coachfederation.org/ethics
Ethical codes of Latin American countries	http://sipsych.org/index.php/es/grupos-de-trabajo-e-iniciativas/grupo-de-trabajo-de-etica-y-dentologia/codigos-de-etica-por-paises http://sipsych.org/index.php/es/grupos-de-trabajo-e-iniciativas/grupo-de-trabajo-de-etica-y-dentologia

APPENDIX III

The Family Genogram

The Individual Develops in a Family Within a Culture

You and your clients will more easily understand the self-in-relation concept if you help them draw a family genogram. We suggest that you consider developing both family and community genograms with many of your clients (see Chapter 9 on focusing). If you keep the genograms displayed during the session, they will remind you and your clients of the cultural/environmental context in which we all live. Moreover, some clients find them comforting, as the genograms bring their family history to the interview. In a sense, we are never alone; our family and community histories are always with us.

Much important information can be collected in a family genogram. Many of us have family stories that are passed down through the generations. These can be sources of strength (such as a favorite grandparent or ancestor who endured hardship successfully). These family stories are real sources of pride and can be central in the positive asset search. There is a tendency to look for problems in the family history, and of course this is appropriate. But use this important strategy positively whenever possible. Be sure to search for positive family stories as well as problems. How can family strengths help your client?

Children often enjoy the family genogram, and a simple adaptation called the "family tree" makes it work for them. The children are encouraged to draw a tree and put their family members on the branches, wherever they wish. This strategy has the advantage of allowing children to present the family as they see it, permitting easy placement of extended family and important support figures as well as immediate family members. Many adolescents and adults may also respond better to this more individualized and less formal approach to the family.

Box III.1 illustrates the major "hows" of developing a family genogram. The classic source for family genogram information is McGoldrick and Gerson (1985). Specific symbols and conventions have been developed that are widely accepted and help professionals communicate information to each other. There is a convention of placing an "X" over departed family members. Once we were demonstrating the family genogram strategy with a client and she commented, "I don't want to cross out my family members—they are still here inside me all the time." We believe that it is important to be flexible and work with the clients' view of family and their choice of symbols. The family genogram is one of the most fascinating exercises that you can undertake. You and your clients can learn much about how family history affects the way individuals behave in the here and now.

We have found family genograms helpful and use them frequently; however, there are situations in which some clients find them less satisfying than the community genogram. There is a Western, linear perspective to the family genogram that does not fit all individuals and cultural backgrounds. It is important to adapt the family genogram to meet individual and cultural differences. You will find *Ethnicity and Family Therapy* (McGoldrick, Giordano, & Garcia-Preto, 2005) a most valuable and enjoyable tool to expand your awareness of racial/ethnic issues.

BOX III.1 Drawing a Family Genogram

This brief overview will not make you an expert in developing or working with genograms, but it will provide a useful beginning with a helpful assessment and treatment technique. First go through this exercise using your own family; then you may want to interview another individual for practice.

1. List the names of family members for at least three generations (four is preferred), with ages and dates of birth and death. List occupations, significant illnesses, and cause of death, as appropriate. Note any issues with alcoholism or drugs.
2. List important cultural/environmental/contextual issues. These may include ethnic identity, religion, economic, and social class considerations. In addition, pay special attention to significant life events such as trauma or environmental issues (e.g., divorce,

economic depression, major illness).
3. Basic relationship symbols for a genogram are shown in Figure III.1; an example of a genogram is shown in Figure III.2.
4. As you develop the genogram with a client, use the basic listening sequence to draw out information, thoughts, and feelings. You will find that considerable insight into one's personal life issues may be generated in this way.

Close	════
Enmeshed	≡≡≡
Estranged	─//─
Distant	------
Conflictual	WWW
Separated	─/─

FIGURE III.1 Basic relationship symbols.

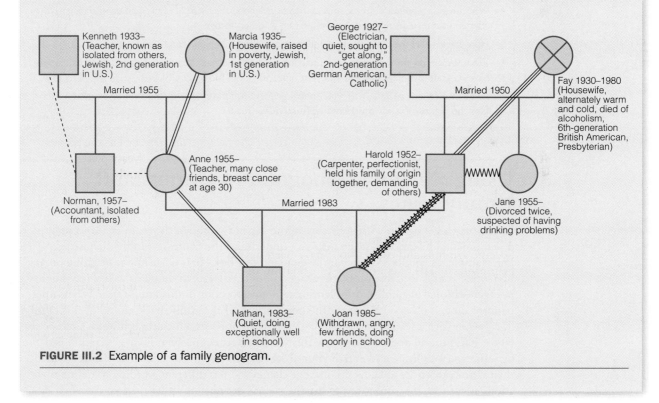

FIGURE III.2 Example of a family genogram.

The family genogram is most effective with a client who has a nuclear family and can actually trace the family over time. We developed the community genogram because some of our clients were uncomfortable with the family genogram. Clients who have been adopted sometimes find the genogram inappropriate. Single-parent families may also feel "different," particularly when important caregivers such as extended family and close community friends are not included. We have talked with gay and lesbian clients who have very differing views of the nature of their family. Family-oriented genograms can help clarify complex family and psychological issues and patterns. The Transgenerational Trauma and Resilience Genogram (TTRG; Goodman, 2013) can help in compressive trauma assessment and intervention from a transgenerational trauma and resilience framework. The TTRG emphasizes an ecosystemic view of trauma and focuses on culturally relevant, social justice–oriented, and strength-based interventions. Also attend to sociopolitical concerns that may affect trauma and recovery.

Exercise III.1 Developing a Family Genogram

Develop a family genogram with a volunteer client or classmate. After the two of you have created the genogram, ask the client the following questions and note the impact of each question. Change the wording and the sequence to fit the needs and interests of the volunteer.

- What does this genogram mean to you? (individual focus)

- As you view your family genogram, what main theme, problem, or set of issues stands out? (main theme, problem focus)

- Who are some significant others, such as friends, neighbors, teachers, or even enemies, who may have affected your own development and your family's? (others focus)

- How would other members of your family interpret this genogram? (family, others focus)

- What impact do your ethnicity, race, religion, and other cultural/environmental/contextual factors have on your own development and your family's? (CEC focus)

- As an interviewer working with you on this genogram, I have learned [state your own observations]. How do you react to my observations? (interviewer focus)

Using a Family Genogram to Understand Family Issues

Developing a genogram with your clients and learning some of the main facts of family developmental history will often help you understand the context of individual issues. For example, as you look at the family genogram in Box III.1, what might be going on at home that results in Joan's problems at school? Why is Nathan doing so well? How might intergenerational alcoholism problems play themselves out in this family tree? What other patterns do you observe? What are the implications of the ethnic background of this family? The person with a Jewish and Anglo background represents a bicultural history. Change the ethnic background and consider how this would affect counseling. Four-generation genograms can complicate and enrich your observations. (*Note:* The clients here have defined their ethnic identities as shown. Different clients will use different wording to define their ethnic identities. It is important to use the client's definitions rather than your own.)

APPENDIX IV

Counseling, Neuroscience/Neurobiology, and Microskills[1]

Experiences, thoughts, actions, and emotions actually change the structure of our brains. . . . Indeed, once we understand how the brain develops, we can train our brains for health, vibrancy, and longevity.

—John Ratey
Harvard Medical School

Counseling and psychotherapy change the brain. You are entering our field at what likely will be its most exciting and productive time. The bridge between biological and psychological processes is erasing the old distinction between mind and body, between mind and brain— *the mind is the brain.* We believe it is time to embrace a broader view that integrates counseling and psychotherapy, neuroscience, neuroimaging, molecular biology, and the medical and cognitive sciences. This is a time of the "mind/brain/body," as we now are aware of the intimate connections and how change in one affects the others.

Neuroscience and neurobiology research and theory lend strong scientific support to what we have long been doing in counseling and psychotherapy. Furthermore, developing knowledge of the brain will continually enable us to become more precise and effective in our work with clients. As noted throughout this book, each microskill, used effectively, makes a difference. Add to this positive psychology, therapeutic lifestyle changes (TLCs), and the multiple strategies of varying theoretical approaches; all of these together will give you an increasingly effective approach to counseling and psychotherapy.

As early as 1989, Eric Kandel argued that because learning produces structural changes in the brain, and because psychotherapy involves learning new ways of functioning, structural changes occurring in client brains would soon be detectable by neuroimaging machines that identify specifically what is going on inside the brain. Helping prove that prediction today are positron emission tomography (PET) scans, functional magnetic resonance imaging (fMRI), diffusion tensor imaging (DTI), and others. Since then, extensive research has succeeded in relating neuroscience and neurobiology to counseling and psychotherapy (Dean, 2014). Following are a few illustrative findings from Dean's summary that tell us what to expect in the future.

1. Using interpersonal therapy (IPT) with depressed clients was found to "normalize" and decrease overactive functioning in the prefrontal cortex.
2. Cognitive behavioral therapy with obsessive compulsive disorder (OCD) clients decreased the hyperactivity of the caudate nucleus. The caudate nucleus is involved with threshold control, a key factor in cognitive and emotional regulation. It is also associated with goal directed action, memory and learning, emotion, and language. Clients do not respond equally to CBT, but the effect was most evident in people who had a good response to CBT.
3. The relationship of medication and psychotherapy has been studied frequently. In the depression study mentioned above, both IPT and the antidepressant Paxil decreased

---

[1] © 2013, 2017 Allen E. Ivey. Released to Cengage, Inc. for this ninth edition of *Intentional Interviewing and Counseling.* Information and further permissions may be obtained from the author (allenivey@gmail.com).

prefrontal activity. Prozac and psychotherapy have been found useful with OCD patients. The parallels do not always hold. Dean observes that the antidepressant Effexor changes different structures in the brain than interpersonal therapy. Research on differential effects of psychotherapy or medication alone or paired is a critical research area now and in the future.

4. CBT has been found to be effective with some forms of physical disease. A study in the Netherlands focused on people with chronic fatigue syndrome (CFS), who suffered from debilitating fatigue. CFS has been found to reduce gray matter volume in the brain. After 16 sessions of CBT, the researchers found significant increases in gray matter volume in the prefrontal cortex. There is other evidence that effective counseling, as well as mindfulness meditation, increases gray matter volume.

With the help of counseling (or medication at times), clients are capable of functionally "rewiring" the brain. In turn, positive brain changes can lead to better mental and physical health. The following section shows us the future impact of neuroscience and neurobiology on counseling and psychotherapy. Expect to see these changes in the near future, and new frameworks for therapy within a decade.

The National Institute of Mental Health Brain-Based Initiative: Is Neurocounseling Our New Direction?

Neuroscience and neurobiology are changing the face of counseling and psychotherapy. The National Institute of Mental Health (NIMH) has ceased its strong support for traditional diagnostic procedures. The new model is designed to replace the current *Diagnostic and Statistical Manual (DSM-5)* (American Psychiatric Association, 2013) with a framework that will change the way we think and practice. NIMH's Thomas Insel (2013) writes:

> The goal of [the DSM], as with all previous editions, is to provide a common language for describing psychopathology. While DSM has been described as a "Bible" for the field, it is, at best, a dictionary, creating a set of labels and defining each. . . . The weakness is its lack of validity.

> Patients with mental disorders deserve better. NIMH has launched the Research Domain Criteria (RDoC) project to transform diagnosis by incorporating genetics, imaging, cognitive science, and other levels of information to lay the foundation for a new classification system.

Counselors and therapists need not be dismayed at Insel's strong words. First, the RDoC is being designed to include treatment recommendations, whereas the old DSM merely focused on its controversial approach to diagnosis. While we question the use or that unfortunate word *disorder*, we find that our areas of human development, multicultural issues, and behavior are central aspects of the new model. What RDoC adds to assessment and treatment includes the following (Insel, 2013):

- A diagnostic approach based on the biology as well as symptoms must not be constrained by the current DSM categories.

- Mental disorders are biological disorders involving brain circuits that implicate specific domains of cognition, emotion, or behavior.

- Mapping the cognitive, circuit, and genetic aspects of mental disorders will yield new and better targets for treatment.

With this book, you already have a good outline of the importance of neuroscience and how you can use these concepts in the session. You have seen learning change in client neural networks in the session transcripts of Chapters 9, 10, and 12. Neuroscience research on stress, listening skills, and cognitive and affective empathy illustrate that the vast majority of what traditional counseling and psychotherapy has done over the years is correct.

Neuroscience is bringing new terminology to the counseling and psychotherapy field. Three of these terms are *neurocounseling*, *neurotherapy*, and *neuroeducation*. Each of these describes a new and fascinating future as we encounter the NIMH brain-based initiative. Russell-Chapin and Jones (2015) have commented:

> The goal of counseling has been to change behaviors and thoughts and help clients feel healthier. Today a major goal of neurocounseling is to additionally help clients develop and enhance necessary skills for emotional and self-regulation. The brain and the body assess and decide what neurological, somatic, autonomic systems and internal and external stressors are needed to regain or maintain a sense of balance or regulation. What goes on in the mind goes on in the body—and vice versa in a self-repeating system.

Dr. Lori Russell-Chapin has an excellent 6-minute video on neurocounseling that you can view by inserting the word *neurocounseling* in the search function of YouTube and viewing the one that has Bradley University written underneath. She has a shorter YouTube video on the same page, also titled *Neurocounseling*.

The Holistic Mind/Brain/Body and the Possibility of Change

The more we pulverize matter, the more it insists on its fundamental unity.

—Teilhard de Chardin

Stories appear to be a fundamental way in which the brain organizes information in a practical and memorable manner.

—Antonio Damasio

The whole brain is greater than the sum of its parts, and the brain is a constantly interacting system within itself and in relation to input from the cultural/environmental context (CEC). Each component, even small structures, affect the total system of the interacting brain and body. Of necessity, the following discussion breaks down the brain into specific structures that are critical for you to know if you are to communicate with other professionals in the near future. In addition, there is an ongoing focus on how the structures of the brain interact through "hubs," which in turn interact with other hubs (e.g., the attentional, auditory, and memory systems, as well as control hubs integrating various systems).

The brain is, simultaneously, a localized and a distributed system. While some of its functions are associated with specific brain structures and regions, these regions act in concert with other, sometimes distant, brain regions. What we experience as "mind" is the result of this intense connectivity. The idea that one brain structure has one purpose is long gone, as we have discovered that as each connects with other structures, new possibilities for action evolve. Each of our 100 billion neurons connects through even more synapses with an almost infinite number of receptors. Early on, brain interactivity was highlighted by Freed and Mann (2007), who reviewed 22 studies examining sadness and the brain. There is evidence that sadness causes reactions in at least 77 different brain regions.

The Human Connectome Project.   Scientists are currently developing a detailed map of all the neural connection paths within the brain (Seung, 2012). Each of us, as a result of genetics and environmental experience, will have unique connections and pathways—in effect, *you are your connectome.* The connectome studies will ultimately provide us with a clear map of how distinct parts of our brains are joined via neural networks. Already, scientists are discovering important new and very basic connections. For example, cells are now examined for their own way of connecting throughout the brain and body (Seung & Sümbül, 2014).

Neuroplasticity.   The key term for this new future is *neuroplasticity*—the brain's ability to change and reorganize itself throughout life. For counseling and therapy, this means the brain can change—it is not fixed, but responds to external environmental events and actions or initiations by the individual. The old idea that the brain does not change is simply wrong. Neuroplasticity means that even in old age, new neurons, new connections, and new neural networks are born and can continue development—a brain can rewire itself.

Particularly fascinating is *neurogenesis*, the development of completely new neurons, even in the aged. Neurogenesis occurs primarily in the *hippocampus*, the main seat of memory (see Figure IV.1). The conversation between Allen and Nelida in Chapters 9 and 10 is illustrative of how counseling can affect the generation of new neural connections, leading to client change. The TLCs are critical for neurogenesis. Exercise is particularly relevant as a lifetime process to ensure brain and physical health (Ratey, 2008, 2012). Exercise increases blood flow and the release of positive neurotransmitters such as serotonin. Exercise is particularly helpful for depression because of serotonin release. If you are sad— walk or run! If you can't run, meditate and use relaxation training.

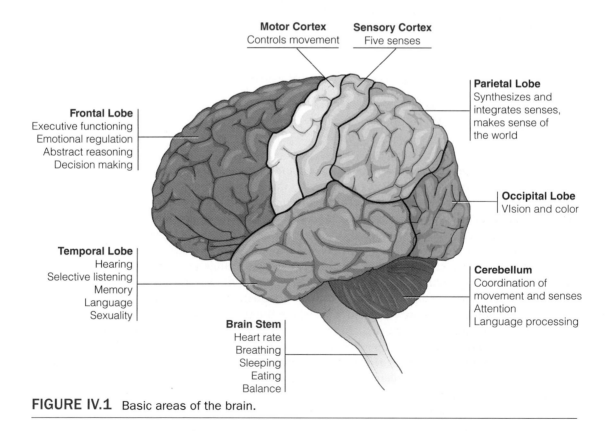

FIGURE IV.1 Basic areas of the brain.

Many of you reading this have experienced the serotonin "high" of a beautiful sunset, the here-and-now immediacy of a close relationship, prayer and meditation, exercise, and other therapeutic lifestyle changes. All these can facilitate physical and mental health.

The Brain Lobes and Their Implications for Counseling and Psychotherapy

Familiarity with some key aspects of the brain will enable you to understand and converse with physicians, neuropsychologists, and others who will be important for your career. Counselors and therapists need to be aware of these areas and potential serious concerns that necessitate referral to physicians. This is an introductory summary of some key structures and issues.

The *frontal lobe* is our chief operating officer (CEO); it is associated with executive functioning, abstract reasoning, and decision making. Critical for long-term memory, it is also the focus for attentional processes and much of motor behavior so that we are effective in social systems. Emotional regulation is located here through connections with the limbic system. However, in dangerous or emergency situations, mental distress, or through the influence of drugs or alcohol, the limbic system may take over. Counselors need to be aware that clients with frontal lobe issues may show attentional problems, poor emotional control, language problems, personality changes, apathy, or inability to plan. In addition, moral and value decisions (good/bad) rest here.

The *parietal lobe* gives us our spatial sense, but it also serves as a critical integrating force between the senses (see/hear/feel/taste/touch) and our motor abilities. Synthesizing, putting things together, is a function of the parietal. Problems in the parietal lobe may show in personality change, lack of self-care or dressing, or difficulty in making things or drawing. Any failure to integrate may involve the parietal lobe. Difficulties with these functions are often associated with Alzheimer's disease.

The *temporal lobe* is concerned with auditory processing, language and speech production, aspects of sexuality, and memory. The hippocampus, center of long-term memory, is located in the temporal lobe. Issues with the temporal lobe may include difficulty in recognizing faces, aphasia, attention, short-term memory loss and interference with long-memory, increased aggression, and sexual changes; it is also associated with Alzheimer's.

The *motor cortex* integrates information from the senses and controls movement. The *sensory cortex* processes and integrates sight, hearing, touch, taste, and touch. The *occipital lobe* is for visual processing and color recognition.

The *cerebellum* is approximately 10% of the brain's volume, but contains more than 50% of the total number of neurons. Not so long ago, it was ignored as a vestige of the past, but recent research has revealed its centrality and significance in brain functioning—with more research to come. It is a vital part of smooth, coordinated body movement and balance. It also has a role in several cognitive functions, including attention, language processing, and the sensory modalities. A common test for healthy cerebellum motor control is to ask the person to move the fingertip in a rapid straight trajectory; a person with damage will move slowly and erratically.

The *brain stem* connects the brain to the spinal cord and the rest of the body. It is a conduit for integrating the whole brain and is critical for central nervous system functioning such as heart rate, respiration, attention, and consciousness. It also regulates the sleep cycle. Balance, dizziness, and nausea may also be related to the brain stem.

Relevant to counseling and therapy is the autonomic nervous system (ANS), which includes the sympathetic and parasympathetic nervous systems. The sympathetic nervous system (SNS) operates through a series of interconnected neurons and activates the fight-or-flight response. However, more important than fight or fight, it takes in stimuli and engages the total body in appropriate action. Effects of overactivation include pupil dilation, increased sweating, increased heart rate, and increased blood pressure. But sufficient activation leads to ability to learn and work and relate to others. The calming parasympathetic nervous system, also called the rest-and-digest system, serves to conserve energy. Effects of its activation include slowing the heart rate, increasing intestinal and gland activity, and relaxing sphincter muscles in the gastrointestinal tract. Both systems play a pivotal role in the stress response. The issue is finding an allostatic balance between the two.

Executive Functioning, Emotional Regulation, Hormones and Other Structures

No cognition without emotion. No emotion without cognition.

—Jean Piaget

The term *executive functioning* has been emphasized in neuroscience research, but now is recognized as central in counseling and psychotherapy—in fact, improved executive functioning is what we have been seeking for our clients over the years. Neuroscience provides us with a more specific and useful point of view for immediate counseling and clinical practice. Executive functions are a set of processes that enable us and our clients to manage ourselves and our resources in order to achieve a goal and live in relative harmony with the world.

We have made a strong case in this book for understanding limbic HPA emotion in counseling and therapy. At the same time, we also emphasize the TAP as our executive control station and the prefrontal cortex as the CEO making decisions and monitoring the more capricious HPA. Figure IV.2 shows the main structures of the HPA and TAP.

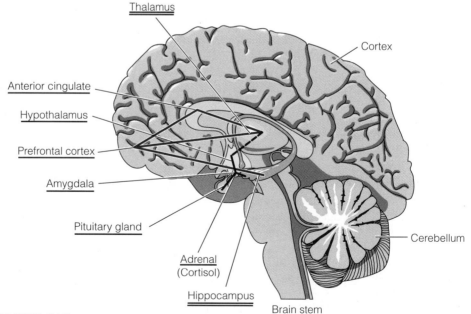

FIGURE IV.2 The limbic system and the HPA (hypothalamus, pituitary, adrenals) and TAP (thalamus, anterior cingulate cortex, prefrontal cortex) axes.

Think of the *hypothalamus* as a "switching station" for the HPA in which messages from inside and outside are transferred; it controls hormones that affect sex, hunger, sleep, aggression, and other biological factors. The *pituitary* is a "control" gland that relates to the hypothalamus. It also influences growth and blood pressure, sexual functioning, the thyroid, and metabolism. The *adrenal glands* produce all-important *corticosteroids*, including *cortisol* (potentially damaging with too much stress, but also necessary for stimulating memory), as well as the neurotransmitter *epinephrine* (also known as adrenaline or norepinephrine), which regulates heart rate and the fight-or-flight response.

The *prefrontal cortex* (PFC) is the "P" of the TAP. Its functions have already been defined, as it is a major lobe of the brain focused on executive functioning, emotional regulation, and decision making.

The *thalamus* is a switching station responsible for relaying sensorimotor signals to the cerebral cortex; it also regulates consciousness, sleep, and alertness. It is central in emotional regulation as well, because of its connections with both the PFC and the limbic HPA.

The *anterior cingulate cortex* (ACC) is a "collar" around the corpus callosum relaying neural signals between the left and right brain (see later discussion). It regulates important cognitive functions, including decision making, empathy, and emotion. In addition, it plays a part in blood pressure and heart rate. The ACC links the body, brain stem, limbic, cortical, and social processes into one functional whole and is central in emotional expression and regulation. You can get a better sense of how mirror neurons work if you notice what occurs in your body when you see an exciting ballgame or an involving movie. Many of us find ourselves tensing up and clenching our fists in close or exciting situations. We may even sway as the pass receiver grabs the ball and heads down the field. In good movies that touch you emotionally in some way, the same thing happens. You heart rate goes up and you may duck a swing from the villain as you sit on the edge of your seat.

The *nucleus accumbens* (not shown in Figure IV.2) is our pleasure center; it plays a part in reward, laughter, addiction, aggression, fear, and the placebo effect. GABA, the inhibitory neurotransmitter, is produced here, along with acetylcholine, which transmits information throughout the brain via the peripheral nervous system (PNS). The nucleus accumbens is significant in sexual functioning and the "high" from certain recreational drugs, which increase the supply of pleasurable dopamine. Key in understanding addiction, the nucleus accumbens is particularly responsive to marijuana, alcohol, and related chemicals. Cocaine, for example, offers more dopamine to the nucleus accumbens than does sexual experience.

This provides a partial explanation of what is involved in working with drug addicts, and with sex addiction as well. One of our great challenges is helping these clients examine and rewrite their stories and find new actions through healthy alternative highs to replace the strengths of addiction. When you find these clients developing new life satisfactions and interests (therapy, wellness education), you are influencing them toward behavior that can result in new positive responses in the nucleus accumbens and other parts of their brain.

You will find clients who tend to operate spontaneously, on the spur of the moment. They may be impulsive, and it gets them into trouble. They may be creative, but have difficult in organizing their many ideas. They may be out of control with overmedication or drugs. First it is important to join them in their stories and understand their emotions. Then more linear and cognitive theories, such as cognitive behavioral therapy or motivational interviewing, may be useful. For those caught in obsessive cognitive thinking and rumination, emotion-focused therapies may be helpful. The surprise of an effective confrontation or providing useful factual information via psychoeducation, stress management, and therapeutic lifestyle changes may make the difference.

Emotion underlies all cognition, but new cognitions and meanings change emotional experience. We believe that cognitive behavioral therapy will be most effective when relationship and emotions in the here and now, as well as there-and-then experience, are given special attention. Rogerian person-centered therapy needs a balance of feelings and cognitions, perhaps even a bit more emphasis on executive functioning. Decision counseling and problem-solving counseling at first glance are very cognitive, but Chapters 9 and 10 show that decisions that are made without consideration of emotion are likely to be unsatisfactory. There is a need for a balance between the limbic HPA and the executive TAP.

The Limbic System: Basics of Emotion

The *limbic system* is of prime importance for us as counselors and therapists, as it helps us to understand issues of emotion, feeling, and memory (see Figure IV.2). Through understanding emotion, we can help clients improve TAP executive emotional regulations. As stressed throughout the book, the *amygdala* is the energizer of emotive strength. It is the power of emotion that places information in memory in the hippocampus; thus the interrelationship of the amygdala and hippocampus is central. There is also an equally important direct connection between the amygdala and the prefrontal cortex. Drawing information from sensory information (what is seen, heard, felt, tasted, smelled) and other parts of the brain, the amygdala signals intensity.

The *hippocampus* is our memory "organ" and works closely with the amygdala and cerebral cortex, distributing information throughout the brain for storage. Energy from the amygdala tells the hippocampus which information should be remembered. When there is not enough interest or energy, no memory is produced. In contrast, a highly stressful event, such as war or rape, can overwhelm the whole system like a lightning bolt and result in destruction of neurons and distressed memory. New, negative neural networks take over. The research discussed above shows us the importance of wellness and positive assets as we seek to develop and strengthen positive memories. Again, *effective counseling can affect the brain in positive ways.* Positive psychology and wellness education build resilience.

The amygdala is recognized as the central area for emotions, particularly the four negative feelings of fear, anger, sadness, and disgust. Surprise, of course, can be negative or positive, depending on the situation and context. All these five feelings are protective, having evolved over time to keep us from danger. The sixth, gladness and its varieties, ranging from satiation through happiness, joy, and contentment, is believed to have evolved later, and thus the later-to-develop prefrontal cortex becomes central.

The social emotions, such as guilt and shame, are blends of feelings as we cognitively respond to what occurs in our social-emotional environment and relations with others. These require more sophisticated cognitive processes than the basic feelings.

It can be argued that the six basic feelings developed through evolution can be reduced to only two fundamental behavioral reactions: approach and avoidance. At all levels of development, from small organisms through snakes and mammals to humans, survival depends on knowing when and how to obtain food and reproductive opportunities and how to avoid danger of all types. Drawing from this, it is helpful to think again of the HPA and TAP axis (hypothalamus, pituitary, adrenals and thalamus, anterior cingulate cortex, and prefrontal cortex). The HPA, of course, is the location of the amygdala and the seat of protective feelings, while the evolutionarily more recent TAP is deeply involved in basic positives, as well as defining the social emotions, as outlined above.

Accepting this explanation, you can see the critical importance of a positive approach, stress management, and therapeutic lifestyle changes (TLCs) in the counseling and therapy process. Building on this foundation, neuroscience research has offered exciting findings. For example, we can now identify specific neurons in the amygdala that affect anxiety, depression, and posttraumatic stress. These harmful networks remain in place unless treated effectively. Using classic behavioral methods derived from Pavlov's work, evidence is that we can change the power of these neurons "through presenting the feared object in the absence of danger." Medications can do the same thing as counseling and therapy if targeted to specific "intercalated neurons" (Ekaterina, Popa, Apergis-Schoute, Fidacaro, & Paré, 2008). These are very clear examples of the approach/avoidance hypothesis. The practical application for counseling and building resilience, of course, is to provide a zone of safety where clients can discuss fearful issues. The relationship plus a strength-based approach are basic whether you use CBT, person-centered, decision counseling, or some other framework.

The amygdala has complex responses to our social environment, enlarging with some experiences and decreasing with others. For example, the broader your social environment, the larger your amygdala will be; on the other hand, trauma has the opposite effect. In one study (Bickart, Wright, Dautoff, Dickerson, & Barrett, 2011), 24 traumatized women diagnosed with borderline personality disorder were compared with 25 healthy controls. It was found that their amygdala was reduced in size by 2%, and the hippocampus by 11%. In addition, significantly impaired cognitive performance was noted.

The Nobel Prize–winning psychologist Daniel Kahneman (2011) carries this discussion a bit further. He comments that our emotional likes and dislikes determine what we believe about the world—politics, irradiated food, global warming, motorcycles, tattoos. Once we are settled emotionally, it is difficult for us or our clients to change. Many times, perhaps most of the time, *our emotions guide our cognitions*. Thus, again, we see the importance of exploring and reflecting feelings and emotions. When we see a picture of a scary spider or view blood and gore, the amygdala and negative feelings are activated—both verbally and nonverbally. With things that we like or have a deep interest in, our pupils dilate with positive feelings.

Left Brain Versus Right Brain, or an Integrative Team

Researchers have demonstrated that right-brain/left-brain theory is a myth, yet its popularity persists. Why? Unfortunately, many people are likely unaware that the theory is outdated. Specifically, left-brain-dominant people are said to be more logical, and right-brain-dominant people are more creative. This theory stems from an overextension of the lateralization of the brain—that each hemisphere has specific cognitive processes (Lilienfeld, Lynn, Ruscio, & Beyerstein, 2010). Searching online, one can find hundreds of quizzes that help determine which side of your brain is dominant; additionally, there are books, study aids, and curricula that are specific for right-brain- or left-brain-dominant learners. All these fail to see that the two hemispheres work together for creative living.

The *corpus callosum* connects the two hemispheres. Both sides work together, and their differences and similarities go beyond the common generalization of the linear (and somewhat boring) left brain and the intuitive (and more interesting and supposedly creative) right brain. The left hemisphere is primarily associated with positive emotions while the right is more associated with the less positive but protective emotions (fear, anger, disgust), but they work together on emotional experience—and even this is an oversimplification of their complex relationship. The cognitive CEO of the full prefrontal cortex (left and right) manages emotional regulation and decides what behavior is most appropriate.

Neurons, Neural Networks, and Neurotransmitters

Neurons that fire together wire together.

—Donald Hebb

Research estimates indicate that there are between 85 and 100 billion *neurons* or nerve cells in the brain, which have been generated by stem cells. At the center of the neuron is DNA. Neurons are connected through *synapses* to other neurons in *neural networks*. A neural network and its connections with others is shown in Figure IV.3, which includes an enlarged representation of the end of the neuron connecting to another neuron via the synapse and neurotransmitters. Neurotransmitters are chemical molecules that transmit signals from one neuron to another. Without neurotransmission, nothing happens in the brain or body—no movement, no learning. The interaction in counseling and therapy affects the transmission of neural impulses, as do medications such as Prozac or drugs such as alcohol, marijuana, and cocaine. Both counseling and medication can increase neurogenesis, while there is evidence that drugs destroy.[2]

"Neurons that fire together wire together." It takes more than new neurons to produce significant change. The *neural network* related to a single neuron is shown at the top of Figure IV.3. Neurons fire when we have any type of experience or stimulus, including the counseling interview. The neural net is where learning from counseling and therapy ultimately takes place through the transmission of signals by neurotransmitters. If strong or frequent enough, this information becomes part of memory in the hippocampus. You can have a large influence on the developing brain through neuroplasticity. Your counseling skills and strategies can facilitate the movement of neurotransmitters and encourage strengthened neural connections. In our language, we call that *learning* or *change*, and we can measure it through positive development on the Client Change Scale. As a person learns, we can also see such change in brain scans using techniques such as PET and fMRI. As research evolves and becomes more precise, scans may become key diagnostic instruments and even show that your work has actually affected specific areas of the brain.

Sigmund Freud was a young medical student at the University of Vienna when he realized that the brain was composed of cells. He then predicted the future and the reality of neurotransmitters when he called the gap between cells *contact barriers*. In his "Project for a Scientific Psychology," Freud presented his model of the brain and mind, even describing neurons responsible for consciousness, memory, and perception (Freud, 1895/1953).

---

[2] As in all our discussions of the brain, neurotransmission is more complex than suggested here. For a more elaborate presentation, we recommend Leslie Samuels's excellent 6-minute YouTube presentation, reached by inserting "016 The Release of Neurotransmitter" in the search box.

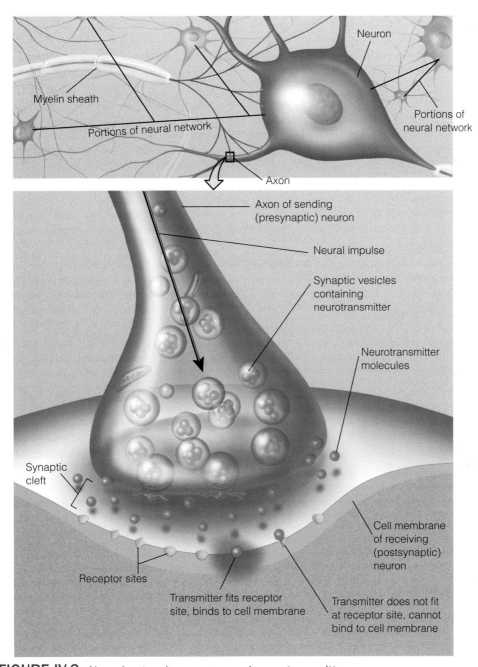

FIGURE IV.3 Neural network, neurons, and neurotransmitters.

The nerve impulse travels through the axon of the sending neuron and, if of sufficient strength, impels the *synaptic vesicles* to release the transmitter. The chemical neurotransmitter molecules then enter the synapse or synaptic cleft, where they seek to bind with their unique *receptor sites* in the next neuron. There are more than 100 identified neurotransmitters, each with its own unique set of receptors.

The receptor sites can be fooled if a foreign chemical enters the bloodstream. For example, alcohol influences several transmitters including pleasurable dopamine, enhances inhibitory GABA, inhibits the excitatory learning glutamate, and induces an endorphin high. Along with these good feelings come less control, less attention to consequences, and reduced effectiveness of motor control and cognition.

Psychiatric medications focus on and influence the action of transmitters, and thus neural nets, thereby affecting memory and behavior. As an example, consider selective serotonin reuptake inhibitors (SSRIs, such as Prozac) used for depression. This medication, which mimics serotonin neurotransmitters, can often alleviate major depression, although evidence for SSRI effectiveness with moderate or minor depression is more equivocal (Fournier et al., 2010). Medications are considered by many as a mixed blessing. For example, a study of depression among 7,696 pregnant women found that untreated women had babies with slower rates of head and body growth. Women treated with SSRIs had reduced depression, but head growth was again delayed and they were at higher risk for preterm babies (Marroun et al., 2012).

At another level, an increasing number of antipsychotics (first-generation Haldol, Thorazine, atypical second- and third-generation Abilify, Risperdal, Zyprexa) have targeted severe mental and emotional distress, such as schizophrenia, with varying success. The primary focus of these powerful antipsychotics is dopamine, but some also affect serotonin, noradrenaline, and acetylcholine. However, antipsychotics have been found to reduce gray matter and brain volume, and to have been used indiscriminately and dangerously with children (Ho, Andreasen, Ziebell, Pierson, & Magnotta, 2011; Lewis, 2011).

Given the mixed findings on medications, we need to consider whether effective therapy will influence the outcome. Cognitive behavioral therapy has often been found as effective or more effective with depression than medication, including more success at 6-month follow-up (e.g., Fava, Rafanelli, Grandi, Conti, & Belluardo, 1998). A study of posttraumatic stress survivors found that 12 weeks of CBT were more effective than medications, which had no impact (Shaley et al., 2012). Research with teens in danger of psychosis found that CBT coupled with a broad array of supplementary preventions (including nutrition, family counseling, and social skills education) significantly reduced the number who actually become psychotic.

Several studies of this type have been conducted by Patrick McGorry and his staff at Royal Melbourne Hospital in Australia and, by extension, others throughout the world. In these programs, an attempt is made to limit medications and avoid antipsychotics, if at all possible. Research consistently attests to effective improvement in at-risk teens (McGorry, 2012). On the other hand, the DSM-5 has chosen the words "attenuated psychosis syndrome" rather than high-risk (see, for example, Woods, Walsh, Saksa, & McGlashan, 2010). This nomenclature tends to pathologize teens and suggests that antipsychotics are much more likely to be used by naïve therapists than prevention strategies and counseling.

While this book has given attention to broad issues of neuroscience and the impact of counseling on the brain, we will take a risk and say that affecting neurotransmitters through *effective* and *quality* counseling and therapy is where the "rubber hits the road." And, as we have said, creating new neural networks through counseling is change—the creation of the *New*. Science and the art of counseling come together at this point.

Consider Table IV.1 as a beginning presentation showing how your practice can influence neurotransmitters, produce change, and create the *New*. Art becomes science, and science becomes art.

TABLE IV.1 Neurotransmitters and Possible Treatment Strategies

Neurotransmitter	Possible Impact of Counseling and Therapy
Glutamate. Most important brain excitatory neurotransmitter, vital for neuroplasticity, movement, memory, and learning. Moderates neural firing. Monosodium glutamate (MSG), chemically close to glutamate, and can cause problems for some.	Generally, we want to increase this central neurotransmitter. Exercise facilitates glutamate production. Stress management and wellness activities are useful for balancing. Preliminary evidence of glutamate abnormalities in depression and schizophrenia. (Medications: glutamate uptake inhibitors)
GABA (gamma-aminobutric acid). Inhibitory, prevents neurons from becoming too active and regulates neuron firing. Important in limbic system and amygdala. Alcohol and barbiturates increase GABA, which results in lowered sensitivity to stimuli, along with cognitive and sensorimotor issues.	Calming strategies of CBT stress counseling, meditation, and the here-and-now emphasis are likely to be useful and increase the release of GABA. The basic listening sequence will help clients as you listen to their stories. (Medications: minor tranquilizers, antianxiety medications, lithium for low GABA)
Dopamine. Attentional processes, pleasure, memory, reward system, fine motor movement. Addictive substances increase release. Low dopamine common in depression.	The relationship itself increases dopamine production. Therapeutic lifestyle changes and counseling focused on stories of strengths and positive narratives should help dopamine production. All effective restorying should improve dopamine release as we move away from depression and ineffective behavior. (Medications: dopamine reuptake inhibitors [NDRIs] as antidepressants)
Serotonin. Vital to mood, sleep, anxiety control, and self-esteem. Implicated in depression, impulsiveness, and anger/aggression.	Think of the serotonin "high" of running. Get clients moving. It is hard to be depressed when one is exercising. Wellness, meditation, cognitive behavioral counseling, and finding clear visions and meaning for life should be helpful. Positive restorying and action following the interview are important. (Medications: SSRIs, permitting more transmission. Ketamine has been shown experimentally to improve depression rapidly. However, ketamine is an ingredient of the dangerous hallucinogenic street drug Special K, also known as Spice.)
Norepinephrine (closely related to epinephrine, also known as adrenaline). Released immediately in stress, but also makes one more cognitively and physically aware and active. Involved in heart rate and helps new information transfer to long-term memory. With too much, damaging cortisol is released. Related to anxiety, depression, and bipolar diagnosis.	Again, get clients active and moving. But when needed, use stress management, decision counseling, CBT, or other approaches to lessen stress. As always, telling one's story in a relationship of caring is calming. People can become addicted to an adrenaline high—you may have seen this in runners and even people overinvolved and excited at work. Finding meaning should help clients meet the challenges of life more effectively. (Medications: SSRIs, sometimes coupled with dopamine as an antidepressant)
Anandamide. Affects cannabinoid receptors (yes, that is what they are called). Marijuana affects nucleus accumbens, the brain's pleasure center. Involved in addictive behavior. *Tetrahydrocannabinol (THC).* The active ingredient of marijuana activates receptors.	Key is finding pleasure and meaning in life. TLCs and work with meaning through purpose and visioning can be helpful. Motivational interviewing is likely the most effective theory, as it attacks addictive issues directly. The client has enjoyed the "highs" of drugs and needs alternative approaches to find positives and strengths in life. Referral to Alcoholics Anonymous or support groups focusing on other issues (e.g., sexual, drug, and other addictions) likely to be helpful. Marijuana appears to be helpful with many medical issues and Alzheimer's disease, but it has also been shown to increase teen suicide and potential for psychosis. (Medications: none available, but some in trials)
Acetylcholine (ACH). First neurotransmitter to be discovered. Affects memory, cognitive functioning, emotion, aggression, central nervous system. Loss of ACH is a central indicator of Alzheimer's disease.	Exercise, meditation, social relationships, positive activities can slow Alzheimer's. Your work with families will be central to help make decisions and support the client appropriately. You will work with clients to help them deal this increasingly common challenge of life. (Medication: Aricept ™, cholestine inhibitors, many new medications in advanced testing stages)
Enkephalins and endorphins. Endogenous morphine-like peptides such as enkephalins and Beta-endorphine are present within the central nervous system. Endorphins are released in response to pain or sustained exertion. They serve as internal analgesics and seem to have a role in appetite control.	Pain management has an increasing role in counseling and therapy through therapeutic lifestyle changes such as relaxation, exercise, and meditation. Modulate pain, reduce stress and produce a sensation of calm. Meditation, mindfulness training, and related counseling strategies have been found very useful for pain relief and are considered preferable to potentially addictive pain relievers, which usually have side effects. (Medications: an array of over-the-counter and prescription pain and headache relievers, with codeine and morphine as examples)

Microskills and Their Potential Impact on Change

The microskills of attending, observation, and the basic listening sequence are vital for the communication of empathy. We start with the biological possibility of "feeling the feelings" of others because of mirror neurons. Through our childhood and later developmental experiences, we become more or less attuned to others. Neuronal structures of empathic understanding can pass away if not nourished. In turn, the teaching of empathy, particularly through the listening skills, may be helpful in human change. Moreover, if you are empathic with a client, you are helping that person become more understanding of others.

A classic study by Restak (2003, p. 9) found that training volunteers in movement sequences produced sequential changes in activity patterns of the brain as the movements became more thoroughly learned and automatic. Systematic step-by-step learning, such as that emphasized in this book, is an efficient learning system also used in ballet, music, golf, and many other settings. If there is sufficient skill practice, changes in the brain may be expected, and increased ability in demonstrating these skills will appear in areas ranging from finger movements to dance—and from the golf swing to counseling skills. Following is a summary of how various microskills relate to the learning process involved in counseling and psychotherapy.

Attending Behavior.   Attention is measurable through brain imagining. When client and counselor attend to the story, the brain of both counselor and client become involved. Factors in attention are arousal and focus. Arousal involves the brain's core, which transmits stimuli to the cortex and activates neurons firing throughout many areas. Selective attention "is brought about by . . . a part of the thalamus, which operates rather like a spotlight, turning to shine on the stimulus" (Carter, 1999, p. 186). If you listen with energy and interest, and this is communicated effectively, expect your client to receive that affect as a positive resource in itself. Attention is central to functioning of the CEO prefrontal cortex. Where our attending is directed influences not only cognitions but also emotions and feelings. Attention to positives reduces the effect of negative issues in one's life.

Questions.   New histories and stories are written in the counseling session. The very asking of questions affects long-term memories stored in the hippocampus and throughout the brain. Creating "new history is influenced by current determinants of neural experience, and such factors are usually very different from those that affected the original experience a long time ago" (Grawe, 2007, p. 67).

Observation.   As you learn to observe your client more effectively, your brain is likely developing new connections. Expect your multicultural learning to become one of those new connections. Japanese have been found to be more holistic thinkers than Westerners. Expect different cognitive/emotional styles when you work with people who are culturally different from you—but never stereotype!

Blacks and Whites both exhibit greater brain activation when they view same-race faces and less when race is different. We tend to feel more comfortable when people are like us. This suggests that discussing racial and other cultural differences early in the session can be a helpful way to build trust. Interestingly, similar findings exist for political persuasions.

Expressions can transmit emotions to others. If you smile, the world does indeed smile with you (up to a point). Experiments in which tiny sensors were attached to the "smile" muscles of people looking at faces show that the sight of another person smiling triggers automatic mimicry—albeit so slight that it may not be visible. The brain concludes that something good is happening out there and creates a feeling of pleasure.

Encouraging, Paraphrasing, and Summarization.   Active listening with cognitive empathy is a key aspect of relationship. Consider the importance of listening to wellness strengths as well as client challenges; if you listen to problems only, expect the nerve cells to communicate that as well. Summaries can affect client mentalizing and thoughts and feelings about themselves (Theory of Mind).

Reflection of Feeling.   Based on emotion, affective empathy includes traditional categories of feeling (sad, mad, glad, fear) that appear in brain imaging. The limbic system organizes bodily emotions and includes the amygdala, hypothalamus, thalamus, hippocampal formation, and cortex. The cortex receives this information, determines how to name feelings, and regulates emotions and what can be done about them. The central feelings of fear are located in the amygdala, which also transmits the *intensity* of emotions. In times of emergency or impulse, the limbic system can and will overcome the judgments of the cortex.

Reflection of feeling is central to communicating empathy. The counselor's mirror neurons "light up" when hearing the emotions and stories of the client. We light up the client's brain with intentional counseling and therapy.

Confrontation.   Confrontation is about creating the New. All listening and influencing skills affect creativity, but confrontation appears to be a prime route for change, as resolving contradictions requires creativity. Early on, activity of the prefrontal cortex can slow the creative process, which partially explains how the "right brain" myth arose. Manish Saggar, a psychiatrist at Stanford, summarizes: "The more you think about it, the more you mess it up" (Saggar et al., 2015). There is a need to "let go" of too much thought.

The creative process is not a one-hemisphere process. The cerebellum appears to be central in the creative process because it coordinates data, while other structures in the brain take on other tasks. Creation of the New is ultimately a holistic brain function.

Focusing.   Client selective attention is guided by existing patterns in the mind; focusing is an intentional skill that can open up more possibilities for client thoughts, feelings, and actions. A number of regions of the prefrontal cortex are activated during attentional task preparation and execution. Executive functioning and cognitive empathy are central here in understanding the client so deeply affected by attentional systems. As clients become aware of their cultural/environmental context (or other focus dimension), they then have the knowledge to reframe their lives. Or the counselor may find useful information to confront the client.

Reflection of Meaning.   Ratey (2008a, p. 41) has commented:

> You have to find the right mission, you have to find something that's organic, that's growing, that keeps you focused on and continues to provide meaning and growth and development for yourself. . . . Spirituality even lights up key centers in the brain. Meaning drives the lower centers and is connected to emotions and motivational areas. . . . If you can get people into a situation where they have the meaning direction provided by their mission or their job or their goal, they don't need medicine.

The dorsal lateral prefront cortex (dlPFC) is involved in moral decision making and when individuals have to make choices involving ethics. This could include working with clients who may make unwise moral decisions or those who need to think about repairing broken relationships. It also evokes a preference toward the most moral and equitable choice and can work against selfish individual focus for personal gain.

Interpretation/Reframing, Logical Consequences, Information/Advice, and Directives.   Some clients enter therapy with negative emotions and an amygdala on overdrive, anxiously fearing what the therapy relationship will hold. In the language of neurobiology, the aim is to reduce this hyperactivity and bolster the activity of the nucleus accumbens, a brain area associated with pleasure.

We need to activate cortical functions where positive thoughts and feelings are generated so that we can deal effectively with issues and problems. For example, cognitive therapy can encourage activity to gain control over negative emotions. The influencing skills and strategies, used effectively, provide clients with specific things they can do to build more positive thoughts, feelings, and behaviors. In this way, clients can deal more effectively with their issues.

Under times of severe stress or panic, the amygdala can take over. Thus you will find many clients who fail to use more positive memories and personal skills to counteract negativity. We need to build positive emotions to cope with the negative. Building on wellness and strengths will enable clients to cope with major challenges. Some even speculate that practitioners will be able in the not too distant future to tailor specific treatments to modify brain circuits through counseling, medication, meditation, or other positive interventions

The Default Mode Brain Network: What's Happening When the Brain Is at Rest?

A great deal of meaningful activity is occurring in the brain when a person is sitting back and doing nothing at all.

—Marcus E. Riachle

The brain is always active, awake or sleeping. The active brain, also called the *task-positive brain* (TPB), has been the focus of this book and this brief summary of key brain structures. The TPB is concerned with daily life, doing and acting, thinking, feeling, and behavior. Surprisingly, the resting brain, the default mode network (DMN), is even more active. It was not until 2001 that the brain at rest was studied seriously (Raichle, 2015; Raichle & Snyder, 2007). The active TPB uses only 10%–20% of the brain's potential energy, while the DMN consumes many times more than that. Virtually anything we do, from swatting a fly to writing a complex paper or hitting a tennis ball, requires more effort from the brain than what occurs at the default level.

The default mode network (DMN) is closely related to long-term memory, past life events, and stories (episodic memory). It is also concerned with information about the autobiographical self (events, thoughts, and emotions). It is deeply involved in mentalizing/Theory of Mind (ToM), where we think about the thoughts and feelings others may have. The DMN, then, is a key aspect of empathy and creativity, and basic to the counseling process. It tends to be automatic and thus is described by some as unconscious thought processing.

Furthermore, it is involved with moral reasoning, attitudes toward social concepts (e.g., race, politics, abortion, religion), and thinking about characteristics of social groups (e.g., low income/high income, politics, multicultural issues). The DMN remembers the past and imagines the future. From all this, many think of the DMN as the site of unconscious experience.

The influence of race on the DMN is one example of how unconscious thought processing affects our perceptions and behavior. A widely quoted study by Mathur, Harada, and Chia (2011) found:

> Racial identification shapes self-concept and how people share in and respond
> to the emotional states of others around them. Prior neuroimaging studies
> have demonstrated the role of the neural default network in self-referential
> and empathic processing. . . . Our results demonstrate that degree of racial
> identification predicts activity within cortical midline structures of the default
> network in response to viewing racial ingroup, relative to outgroup members,
> and activity within the medial temporal lobe subsystem of the default network
> in response to viewing racial outgroup, relative to ingroup members.

Our view of others different from ourselves deeply affects our thoughts, feelings, and behavior. The default brain, associated with memories, thoughts, and feelings about the self and others, views the faces of African American and European American races differently. The background (unconscious) DMN interacts with the conscious task-positive brain, bringing varying meaning to interpersonal and intrapersonal contact. Values remembered in the DMN can lead to positive interracial contact or to overt acts of racism.

Mind wandering is an example we all experience—when we are driving, our minds often move from the task-positive brain's here-and-now focus on the road to the DMN and thinking about almost anything, ranging from what we had for breakfast to remembering a childhood fall and crying. If something happens that may cause an accident, the TPB takes over (based on amygdala stimulation) and saves us. When we focus intently on our work, the DMN is suppressed and very little mind wandering occurs.

Recently, it was believed that the TPB and the DMN were separate. However, new research is providing another answer in that the DMN contributes to TPB prefrontal activity when there is information from the past that helps work in the present (Hill, 2014). Specifically relevant to counseling, research examined responses to familiar and unfamiliar faces and found that the DMN became active (along with the TPB) with familiar but not with stranger faces (Spreng et al., 2014). In short, the DMN becomes more active if the task involves something in long-term memory that is relevant to the immediate context. This study shows that talking with clients about familiar issues will activate memories along with what they are saying. It is here that you may want to access these memories through gentle listening or suggesting free association.

The change in beliefs about the DMN's role is best exemplified in research by Elton and Gao (2015), who conclude:

> Indeed, our results demonstrate significant increases in DMN connectivity with
> task-promoting regions (e.g., anterior insula, inferior frontal gyrus, middle frontal
> gyrus) across all six tasks [including autobiographical memory, emotion, and
> inhibitory control]. Furthermore, canonical correlation analyses indicated that the
> observed task-related connectivity changes were significantly associated with
> individual differences in task performance. Our results indicate that the DMN may
> not only support a "default" mode but may play a greater role in both internal and
> external tasks through flexible coupling with task-relevant brain regions.

It is believed that our brain consolidates so much incoming information that we cannot deal consciously with it all. In actuality, we are taking in more than we are aware of. We suggest that you now view a 2-minute introduction to the default mode network by going to the YouTube search area and inserting "A Brief Introduction to the Default Mode Network." Viewing that video will enrich your understanding and ability to use your own DMN in practice.

In a successful counseling session, we first need to attend to clients and listen to what they are saying—this is the TPB in action. Mind wandering and the DMN enter in when you start thinking internally about what the client has said and its meaning, both to you and to what the client is thinking and feeling internally—empathy in action! You draw on the immediate experience in the session but couple it with memories from the past, which leads you to your next comment and also starts the process of mentalizing, empathically understanding the client more holistically. Then you may suppress these thoughts and focus intently on the here and now of a trauma the client shares with you. When your mind wanders to the DMN, you are using even more brain energy than when focusing on the here and now of the interview.

Gonçalves (2015) comments:

> It is not uncommon for counselors to feel apprehension and guilt when recognizing that their minds are wandering despite trying to stay focused on client experiences. . . . Nevertheless, mind wandering is an unavoidable consequence of focused attention. . . . The counseling relationship is a bidirectional stream . . . the mind wandering helps to guide our attentional focus.

What Dr. Gonçalves is saying here is allow yourself to let your mind wander in the interview, but the task-positive brain then needs to suppress most mind wandering, particularly as you listen to the next client comment. Your microskill of summarization takes your mind wandering to the TPB with how you have empathically mentalized client thought. This is the bidirectional stream.

Reflective Exercise Practice mind wandering and apply it to the interview via free association

First, when driving, notice how you are attending to the road with your task-positive brain (TPB), but shortly find that you have drifted off to thoughts about the work you have to do today—your DMN in operation. You return to TPB control, but then your DMN takes over and you start thinking of that kiss from your loved one this morning. A stoplight ahead turns red and, thankfully, your TPB returns to control. If not, your DMN could take you right through it for a ticket. This is the bidirectional stream that Gonçalves is speaking about.

It will be to your advantage if you practice deliberate mind wandering and then returning to task-positive thought. One way to do this is relax, take a deep breath, and with your TPB stare fixedly at an object in the room or a distant tree—continue breathing. Very likely your mind will start to wander; allow this to happen. Then use the TPB and deliberately stop and reflect on the meaning of what occurred while your mind was afloat and wandering. Further practice with this, particularly around a concern or a challenge you face. This process introduces you to free association, common in classical psychoanalysis. It can be a powerful and creatively freeing process. Continue moving back and forth bidirectionally.

Free association is a helpful strategy in reframes, interpretations, and meaning. With your clients, ask them to relax and focus wholeheartedly on a single word that they have just said—or an emotional event or a dream. Suggest that they relax, possibly even close their eyes, and tell you whatever comes into their mind. Over time, free association can lead to creative new insights that are part of successful resolutions of present concerns. Images from the present or past are another place to start with free association.

Figure IV.4 shows the inner complexity of the DMN—the multiple areas that influence how we think and feel, think about self and others, and organize our experience. The dorsal medial prefrontal subsystem (dMPFC) is oriented to mentalizing and provides much of the *remembered* moral and social reasoning. It also is related to cognitive processing. Memory is the foundation of the medial temporal lobe (MTL) system, both autobiographical and episodic, working with both past and present. Here lies much of our basis for imagery and creative thinking. The contextual associations are what we try to reach

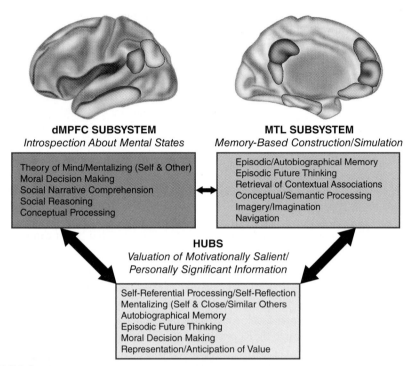

dMPFC SUBSYSTEM
Introspection About Mental States

Theory of Mind/Mentalizing (Self & Other)
Moral Decision Making
Social Narrative Comprehension
Social Reasoning
Conceptual Processing

MTL SUBSYSTEM
Memory-Based Construction/Simulation

Episodic/Autobiographical Memory
Episodic Future Thinking
Retrieval of Contextual Associations
Conceptual/Semantic Processing
Imagery/Imagination
Navigation

HUBS
*Valuation of Motivationally Salient/
Personally Significant Information*

Self-Referential Processing/Self-Reflection
Mentalizing (Self & Close/Similar Others
Autobiographical Memory
Episodic Future Thinking
Moral Decision Making
Representation/Anticipation of Value

FIGURE IV.4   The subsystem and hubs of the default brain network.

when we explore if a client can think systemically, particularly about multicultural issues. This is what we are talking about in the focus dimension of the cultural/environmental context (Andrews-Hanna, 2012; Andrews-Hanna, Smallwood, & Spreng, 2014; Spreng & Andrews-Hanna, 2015).

The hubs are value indicators connecting to key areas of the brain (dPFC and MTL). The hubs bring together relevant neural networks as indicated in Figure IV.4. Among the results of their interaction are mind wandering, mentalizing, and informing us about thinking and behaving, often at an unconscious level. This is why we automatically/ unconsciously swerve to miss on oncoming baseball bat, or a fist headed toward our face. The processes of the DMN are foundational to meaning and the mentalizing skills of interpretation/reframe and reflection of feeling.

Disruption of the DMN can occur as a result of external stressors or body concerns ranging from a cold or upset stomach to cancer and diabetes. Neuroimaging research reveals that psychiatric issues including depression, anxiety, posttraumatic stress, and ADHD are all related to change in the DMN. Similarly, primarily medical issues are also involved with the DMN—autism, chronic pain, amyotrophic lateral sclerosis (ALS), persistent vegetative state, and many others (Andrews-Hanna, 2012).

Allen likes to think that the DMN can be defined metaphorically as a ball with an almost infinite number of small balls or molecules within, representing memories, cognitions, feelings, and so on. These are active, interactive balls, and when they bounce against one another, they may ignore each other, synthesize to develop a new meaning, feeling, or understanding—and may even split, leaving the *New* synthesis, and continue on to bump against other molecules. This might explain why memory is often so imperfect and how false memories of abuse or distorted views of other people develop.

Again, when we are involved and focused on a task, or giving here-and-now fully focused attention to the client, DMN damps down, but at some point, mind wandering helps us continue. Being aware of bidirectionality enables us both to focus on the client more effectively *and* to profit from closely related mind wandering. This wandering could involve not just this one client, but also draw from your experience with other clients.

Many believe that the DMN represents much of unconscious thought (e.g., Viamontes & Beitman, 2007), validating much of what Freud said about the mind. A child's developmental experiences are imprinted in long-term memory and organized in the wandering mind of the default mode. Throughout our lives as adults, these and new experiences constantly change memories in our hippocampus and are part of the neurogenesis process. This complex array of sensory information is abstracted and forms the substrate of the unconscious. Only a small amount of subconscious data ever appears in conscious thought in the task-positive brain. Yet it is always there at some level in the DMN, ultimately influencing conscious decisions. Some go so far as to suggest that consciousness and individual agency may not exist (Lacan, 1977; Ivey, 1986/2000).

There is evidence that the default networks of those diagnosed with autism, schizophrenia, ADHD, and Alzheimer's are distinctly different from others. Each of these appears to have an overactive DMN. Some believe that assessment of the DMN can lead to therapeutic change, and neurofeedback is considered a promising therapeutic mode. For example, it was found that 40 sessions of neurofeedback served a calming function, smoothing out the brain waves of ADHD children through fMRI evaluation. ADHD children are constantly active and changing what they are doing. The calming from neurofeedback also resulted in significant differences in behavior at school and at home (Chapin & Russell-Chapin, 2014).

Research on the DMN is expanding rapidly, and it is anticipated that it will have a powerful impact on theories of personality and human development, as well as being useful for us in counseling and therapeutic practice. There is evidence through fMRI study that different personality types have varying patterns of activity in the DMN. The "Big Five" personality styles (extroversion, neuroticism, openness/intellect, agreeableness, and conscientious) have been identified (Sampaio, Soares, Coutinho, Sousa, & Gonçalves, 2014):

> Extraversion and agreeableness were positively correlated with activity in the midline core of the DMN, whereas neuroticism, openness, and conscientiousness were correlated with the parietal cortex system. Activity of the anterior cingulate cortex (ACC) was positively correlated with extraversion, but negatively with introversion. Regions of the parietal lobe were differentially associated with each personality dimension.

Social Stress and Its Impact on the Body

Stress impacts the total body, not just the mind.

—Allen Ivey

"Depression is as real a disease as diabetes." This statement by Stanford's famous Robert Sapolsky is based on considerable research showing that psychological depression has a deep impact on the body. In turn, dysfunction of the body through diet and obesity, infection/inflammation, and illness all lead to depression as well. Our cognitions, beliefs, emotions, and behavior can build bodily health, or they can be as toxic as illness or environmental pesticides. The bidirectional brain/body feedback loop can increase or decrease depression. Research reveals that positive attitudes and beliefs, exercise, and lifestyle affect the immune

system in healthy ways. Counseling can rewire the brain, with subsequent bodily changes as well. For a clear and practical background on depression and bidirectionality, search YouTube for "Robert Sapolsky depression."

The Autonomic Nervous System and Vagus Nerve: Connecting Brain and Body for Stress Resilience and Allostasis

One way to think about this book is in terms of listening to calm and support the client through attending, observation, and the basic listening sequence. Establishing the relationship and drawing out stories of strength and resilience, as well as family and friend resources, provide a foundation for growth. The influencing skills (confrontation, interpretation, directives, and so on) are activating in bringing new thoughts and feelings to the client. At the same time, paradoxically, the use of influencing skills, particularly therapeutic lifestyle changes, both calms and activates.

The *autonomic nervous system* (ANS) regulates the body's unconscious actions of heart, lungs, esophagus, stomach, and gastrointestinal system and consists of two divisions: the *sympathetic*, focused on response to stimuli and activation, and the *parasympathetic*, focused on calming and balance. The ANS is connected to the brain stem in a bidirectional pattern. What happens in the brain affects both sympathetic stimulation (e.g., stress) and parasympathetic calming. In turn, bidirectional crosstalk means that action in the ANS affects the brain. Another way to think about stress is in terms of *activation* that can be destructive or strength building (e.g., the influencing skills).

The "calming and activating" or "stop and go" actions of the parasympathetic system are repeated throughout the entire stress system through the vagus nerve, from our neurotransmitters to every region and cell of our body. For example, the neurotransmitter glutamate activates and makes learning possible, while GABA is necessary for balanced calming. Hormones in the brain and body interact with the immune system in positive and negative ways through cytokines. Even our gastrointestinal system with its microbiota, an interactive imbalance of highly diverse microorganisms, can lead to poor mental and physical health, but a healthy gut through diet, exercise, and a positive attitude improve our mental well-being. Through our listening, we seek to calm clients. Through reframing, confrontation, and influencing skills, we seek to activate change. Each of our counseling interventions affects the holistic body, as well as the mind.

The connections of the parasympathetic vagus nerve are basic to the calming process for our overstressed clients and a key to developing allostasis and resilience (see Figure IV.5). Polyvagal theory holds that physiological state dictates the range of behavior and psychological experience (Porges, 2011). Porges provides specific suggestions to help clients cope with flight-or-fight sympathetic overstimulation: deep breaths, attention to the here and now, eye contact, and social engagement. Interestingly, his suggestions overlap with this book's discussion of attending and observation. In this body-aware framework, we help clients become aware of the power of unconscious body processing and show them how they can calm the vagus nerve and themselves, through biofeedback and control of heart rate, breathing exercises and the relaxation response, as well as neurofeedback. Of course, our relationship and counseling with clients can be calming—our words and nonverbal behavior offer the promise of calming and appropriate activation. An oversimplification would be to think of the first half of this book on attending and listening as basic to social engagement and calming. But effective listening can also lead to action.

The influencing skills are activating and more closely related to the sympathetic nervous system (SNS), which takes input/stimulation and distributes it throughout the body. While a vast oversimplification, many relate the SNS to "fight or flight." However, we do not survive without an activated SNS, as it influences the amygdala, the HPA, and all parts of the body necessary for action. Thus, confrontation, focusing, directives, and psychoeducation are often association with the SNS. But, as said (perhaps too many times) throughout this book, *it is critical to listen before you seek to influence the client.* And, of course, active listening itself can be both calming and activating.

Illustrating the Vagal and the Microbiota–Gut–Brain Axis

Figure IV.5 presents vagal connections from the brain's perceptions, cognitions, and emotions to our HPA hormone production of cortisol (hypothalamus, pituitary, adrenals). Cortisol in an allostatic balance facilitates learning, while overabundance can be seriously damaging to the brain and body. This same stimulation reaches down to the heart, lungs, and onward to the gut microbiota flora with its 100 trillion microbes. Grenham, Clarke, Cryan, and Dinan (2011) summarize:

> A stable gut microbiota is essential for normal gut physiology and contributes to appropriate signaling along the brain–gut axis and to the healthy status of the individual, as shown on the left hand side of the diagram. Conversely, as shown on the right hand side of the diagram, intestinal dysbiosis can adversely influence gut physiology leading to inappropriate brain–gut axis signaling and associated consequences for CNS functions and disease states.

There really is a thing called "gut feelings" as the microbiota–gut–brain axis is disturbed by imbalances in the autonomic nervous system (ANS), the brain, the body, and by any external or internal stressor. Another example is that our gut produces more serotonin than

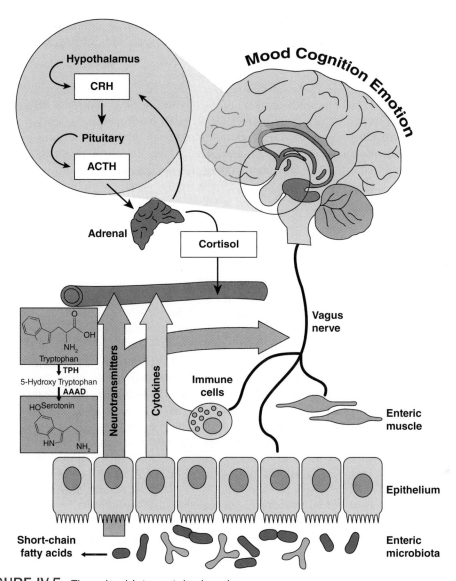

FIGURE IV.5 The microbiota–gut–brain axis.

Grenham, S., Clarke, G., Cryan, J. F., & Dinan, T. G. (2011). Brain-gut-microbe communication in health and disease. *Frontiers in Physiology, 2,* 94. By permission of the authors.

our brain. Our stress system is holistic, and the psychic distress reverberates throughout the body, just as illness does the same to our mind/brain/body.

In Figure IV.5, we see the reciprocal bidirectional crosstalk interconnections of the brain, the gastrointestinal system, and the immune system, all connected by the autonomic nervous system, particularly the vagus nerve. These bidirectional interconnections are also labeled as "the brain to body and the body to brain" or "top down to bottom up." The HPA axis (hypothalamus, pituitary, adrenals) generates and passes on hormones throughout the body. Important here is the production of cortisol, necessary for learning, but typically dysregulated in serious situations such as war, rape, or the repeated traumas of bullying, poverty, racism, and harassment. This disruption of cortisol can lead to damage to key brain structures, as well as increased heart rate, breathing rate changes, and disturbances to the gut or gastrointestinal system (Chrousos, 2015).

Nearly every chemical that controls the brain is also located in the stomach region, including hormones and neurotransmitters such as serotonin, dopamine, glutamate, GABA, and norepinephrine. The gut produces more of the neurotransmitter serotonin than the brain. No longer should we think of "gut feelings" as just a passing thought.

In turn, a recycling negative feedback loop can lead from the gut to the brain and to the immune system, with accompanying inflammation. Stress increases inflammation, and it has been found that bodily inflammation accompanies depression and other psychological diagnoses. Interacting with the HPA are the cytokines, produced in both brain and gut. The cytokines are the proteins and chemicals that are most central in producing inflammation. Inflammation is a central issue to which counseling gives virtually no attention. Yet depression and other distressing issues that we discuss with clients are usually accompanied by inflammation, which can be dangerous to physical health over time—even to the point of ultimately reaching epigenetic changes in DNA.

While stress (also known as oxidative stress) is often central in producing inflammation in brain and body, physical illness (cancer, diabetes, severe flu or cold) also is a cause. Chemicals, pesticides, gluten (for some), and other pathogens also produce inflammation. It is important to realize that oxidative stress also comes from interpersonal relations and self-talk. One's thoughts, beliefs, and behaviors produce inflammation.

Important in this process, and not receiving enough attention, is maintaining a balance of our trillions of gut microbiota. An imbalance of too many negative microbes can be the result of external stressors and emotional imbalance, a poor diet (particularly sugar), allergies, or environmental toxins, even genetically modified food for some. The imbalance is another route toward inflammation and has been proven to be an issue in depression and other diagnoses. The inflammatory actions are both caused and activated by cytokines. Stress and diet, of course, are a central cause of body inflammation, which in turn can lead to psychic distress. Interestingly, research shows that a change in lifestyle can move the balance of microbes from negative to positive.

At this point, we also need to consider the mitochondria, found in large numbers within the cells, which produce the energy that moves our brain and body. We now argue that it is basic for all counselors and therapists to be aware of the role of mitochondria in our lives. While mitochondria enable us to move our muscles and think clearly through the production of the fuel ATP, they also need strengthening themselves. ATP is the molecular unit that energizes our metabolism and enables our muscles to contract and us to move and breathe. Among other things, it also is important in nervous system and cell signaling, as well as DNA synthesis. Therefore, we help our mitochrondria through exercise, diet, and positive health habits, the very same treatment methods we have emphasized throughout this book—therapeutic lifestyle changes. Mitochondria are also in continuing bidirectional drama with cytokines, as each can destroy or enhance the other.

Cytokines are small, vital proteins released from cells that affect communication among cells and their behavior. There are more than 30, and possibly they are growing in numbers. They have been found to interact bidirectionally with multiple genes. For example, interferons are produced by T cells and regulate the immune system. Cytokines are closely related to depression. Research is under way examining cytokines in the blood of clients pre- and post-counseling. Another bidirectional aspect of cytokines is their relationship to hormones in the brain. Dysfunction here leads to inflammation in the brain, a factor that we have not yet considered in our practice.

Figure IV.6 is too detailed to discuss fully, but please note the nature of bidirectional crosstalk and how it relates to allostasis: (1) diet and exercise have a profound impact on the energy producing mitochondria; (2) mitochondria ATP energy produces BDNF (brain derived neurotrophic factor, "miracle grow" for the brain); (3) these lead to synaptic plasticity, brain growth, and sharper cognition. Also key is ROS, reactive oxygen species that speak to oxidation in the body and brain. Illness, poor diet, lack of exercise, and depression all can lead to oxidation, inflammation, apoptosis (death) of the mitochondria, and oxidation and cognitive issues. In addition, throughout the process, epigenetic change to genes can be positive or negative (there actually should be a bidirectional arrow in the epigenetic–cognitive relationship, as cognitions can possibly affect epigenetics).

Mitochondria contain more DNA than the cells within which they live. It is at this foundational level, through epigenetics, that counseling can even be part of enabling genes to turn on or off in ways that lead to healthier living and even a longer life with better health. Conversely, social oppression, trauma, negative experience, depression, and illness can all lead to the death of mitochondria and dangerous changes in DNA through epigenetics. A recent article carries this full cycle back to the CNS. Fisher and Maier (2015) comment:

> Neuroinflammation and mitochondrial dysfunction are common features of chronic neurodegenerative diseases of the central nervous system. Both conditions can lead to increased oxidative stress by excessive release of harmful reactive oxygen and nitrogen species (ROS and RNS), which further promote neuronal damage and subsequent inflammation resulting in a feed-forward loop of neurodegeneration.

In summary, the discussion in this section includes considerable data and some terms that may be unfamiliar at this point, but eventually the bidirectionality of the stress system will become a standard part of our training and thinking in the practice of counseling and psychotherapy.

1. External psychological stressors or pathogens from the environment or internal physical stressors from illness or the nature of one's inherited genes can lead to the six senses perceiving stress, threat, or challenge affecting the autonomic nervous system, the flow of neurotransmitters and hormones through the HPA axis.

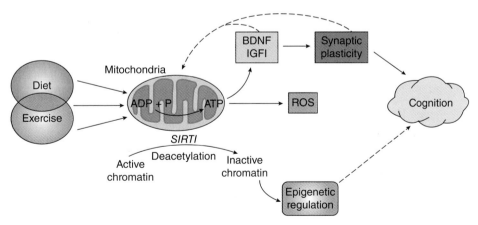

FIGURE IV.6 The effect of nutrients and exercise on cognition.

National Institute of Health.
Fernando Gómez-Pinilla. (2008). Brain foods: The effects of nutrients on brain function *Nature Reviews Neuroscience, 7*, 568–578.

2. Pro-inflammatory or anti-inflammatory cytokines interact throughout the body at all levels from the HPA axis to the gut microbiota. Particularly important is the impact on the immune system and inflammation.

3. The calming parasympathetic system and the vagus nerve are basic to physical and mental health. As counselors and therapist, we have considerable resources to help our clients build stress resilience and allostasis.

Thus we see that psychological distress, such as depression, has a profound impact on the body as well as the mind. In counseling, historically we have thought we were working only with the mind. We now realize that the mind/brain/body is one holistic enterprise.

An Optimistic View of Our Ability to Build Allostasis and Stress Resilience

Of course, caution is essential. We are not healers of the body; that is the role of the physician. With each client we are honored to work with, we need to be alert as possible to the reality of unseen illness as it manifests itself and have referral sources available. For example, Sapolsky has stated that when we see depression and anxiety, we also need to think of the possibility of thyroid problems and other possibilities. More and more, counselors and psychotherapists need to consider referral to physicians to ensure that severe, and potentially disabling, bodily issues are not involved in client concerns.

It is now virtually a truism: *Relationship and the working alliance are 30% of effective counseling and therapy.* Carl Rogers lives! Now to repeat ourselves, we highly recommend looking seriously at therapeutic lifestyle changes as a proven way to improve mental and physical health. John Ratey of Harvard Medical School stated that it is unethical for a physician not to prescribe appropriate exercise to all patients. The same holds for us as counselors. Are you also considering the importance of your clients' diet, their sleep patterns, and their willingness to take on cognitive challenge? We are rather good at helping clients with their social relations, so basic to calming or activating the autonomic nervous system. Cultural health and social justice action have positive mental and physical health benefits. Beyond these, other therapeutic lifestyle changes, all based in some research in neuroscience and neurobiology, are well worth considering as adding to your present skills in CBT, REBT, motivational interviewing, decision counseling, psychodynamic, or other therapeutic system.

Social Justice and Stress Management for Physical and Mental Health

Poverty in early childhood poisons the brain. . . . neuroscientists have found that many children growing up in very poor families with low social status experience unhealthy levels of stress hormones, which impair their neural development. The effect is to impair language development and memory—and hence the ability to escape poverty—for the rest of the child's life.

—Paul Krugman

If you are not part of the solution, you are part of the problem.

—Eldridge Clever

Stress is a factor in virtually all the issues clients bring to you. It will show in body tension and nonverbal behavior. Cognitive/emotional stress is demonstrated in vocal hesitations, emotional difficulties, and the conflicts/discrepancies clients face in their lives.

New discoveries in neuroscience reveal the negative impact that stressful, unjust, and oppressive systems, environments, and relationships have on individuals and societies. Poverty, discrimination, abuse, deprivations, and lack of freedom all affect the mind, brain, and body. These social injustices mean that many face discrimination in housing, often pay more for cars, are followed in stores, and receive less attention from medical services. Social justice counseling provides a viable route for action dedicated to eradicate, remediate, and advance a more just society (Zalaquett & Ivey, 2011, 2014).

First, let us look at the immediate practical implications of implementing social justice in the counseling and therapy session. In Chapter 2, Harvard's Jenny Galbraith tells us her story of alienation and loneliness that an African American experiences in a predominately White classroom. In Chapter 3, Mary's interview with Damaris, a low-income child, illustrates how teasing a child about less expensive shoes becomes a counseling issue. Shoes and clothes in themselves can lead to social exclusion, one of the most painful experiences that young girls experience. In Chapters 9 and 10, Nelida Zamora shares her story of a classroom microaggression that stayed with her throughout the term. In a sense, all these are minor, as least as seen by a majority person. However, repeated over time, these events produce damaging cortisol and become embedded in the hippocampus as permanent memories, with accompanying body changes that have potential long-term implications.

There are three main avenues in social justice counseling: the interview, seeking change in the social context, and broader social action in the community.

The Individual Interview. Nelida's session in Chapters 9 and 10 illustrates specifics of what a casual microaggression can do to self-confidence and comfort in the classroom, lasting over a full term and closing her off from some classmates. The memory of this "small" (to some) incident became part of her internalized permanent memory, which in turn, affected her behavior. Steps toward psychological liberation included:

1. *Relationship:* establishing a relationship in which gender and cultural background were briefly discussed.
2. *Story and strengths:* drawing out the story and the cognitive, emotional, and behavioral results of the incident, followed by focusing on individual and cultural strengths to build cultural health via the community genogram. Here memories of strength were brought to fore and were helpful in developing resilience and coping skills.
3. *Goals:* discussing behavioral change.
4. *Restory:* Nelida's memory began to be "rewired" with a new story, becoming part of hippocampus memory.
5. *Action:* directives encouraging her *fuerza* (strength) to build resilience and think differently about herself and challenges and other microaggressions she might encounter in the future—specifically, cultural health.

Individual Counseling with Social Justice Action in the School. The five-stage interview was used to enable Damaris to reframe and think about self and situation more positively. However, teasing easily slips into verbal bullying and exclusion via the power of "mean girls." This cultural/environmental context is challenging. Mary was a team leader establishing a school bullying policy with planned action steps. Teachers were informed of what was occurring in the classroom and playground and could take immediate steps to stop the behavior. Mary set up a friendship group that met for lunch for games and talk, and two of the members were leaders in the group that excluded Damaris. As part of the school bullying policy, further classroom instruction on the policy and potential consequences were made clear.

Larger Institutional and Community Change. This requires a long-term effort. Many minorities and those with lower incomes go through culture shock on arrival at college. Despite affirmative action, minorities often do not feel accepted, keep to themselves, and feel much the same alienation as Jenny Galbraith. Rules, punishments, and friendship groups cannot be implemented as in elementary school. One of the important benefits of attending college is meeting friends who may help you find opportunities throughout the lifespan. This is less available to minority students.

Given the size of these issues, what can you do? Here some example programs where you can be part of the solution. While not named as social justice actions, colleges have sought to make campuses more friendly and safe—counselors can be important in this process. Admitting minorities is not enough; groups have to work together to bring awareness to high school students that the college is truly interested in them and will support them once they get there. It is essential to increase the number of People of Color among both staff and faculty, plus invited campus speakers on cultural issues. Black and Chicano alliances have been effective in providing support, but keep their members separate from some opportunities and building interpersonal contacts. Campus activity groups and student government are another area that has helped integrate minorities into the community. Some campus programs focus on work in food pantries, tutoring, and other community service. An ongoing program of building awareness in administration, faculty, and students is required. Are courses taught that discuss these challenging issues in a direct and honest fashion? Are the textbooks biased? At another level is campus outreach to local schools. The list of actions to build change is infinite. You can never be the solution by yourself, but you can definitely be part of the solution by joining others in social justice action.

Toxic and long-term stress is damaging. *"Cortisol* is the long-acting stress hormone that helps to mobilize fuel, cue attention and memory, and prepare the body and brain to battle challenges to equilibrium. Cortisol oversees the stockpiling of fuel, in the form of fat, for future stresses. Its action is critical for our survival. At high or unrelenting concentrations such as post-traumatic stress, cortisol has a toxic effect on neurons, eroding their connections between them and breaking down muscles and nerve cells to provide an immediate fuel source" (Ratey, 2008b, p. 277).

However, many of the strategies of stress management, counseling, and psychotherapy may not be appropriate for many, as described in the RESPECTFUL model (Chapter 2). It takes money or insurance and verbal skills to go to a professional helper. Many of the less fortunate also find that the communication style of a counselor or therapist does not seem relevant. Therapeutic lifestyle changes have been recommended as key counseling strategies, but a gym for exercise, a good diet, time to sleep, and a safe community may not be available or too costly. It is here that counselors and therapists need to think seriously about their role in social change and enabling clients to find resources and supports in their home communities. Do not expect to do it all alone; seek others with the same purposes and values.

An article titled "Excessive Stress Disrupts the Architecture of the Developing Brain" (National Scientific Council on the Developing Child, 2005) offers many useful points, including the following:

1. In the uterus, the unborn child responds to stress in the mother, while alcohol, drugs, and other stimulants can be extremely damaging.
2. For the developing child, neural circuits are especially plastic and amenable to growth and change, but again excessive stress results in lesser brain development, and in adulthood that child is more likely to have depression, an anxiety disorder, alcoholism, cardiovascular problems, and diabetes.

3. Positive experiences in pregnancy seem to facilitate child development.
4. Caregivers are critical to the development of the healthy child.
5. Children of poverty or who have been neglected tend to have elevated cortisol levels.

If you review the sections above, particularly those talking about the frontal cortex and limbic system, you can obtain some sense of what poverty and challenges such as racism and oppression do to the brain. Incidents of racism place the brain on hypervigilance, thus producing significant stress, with accompanying hyperfunctioning of the amygdala and interference with memory and other areas of the brain. We need to be aware that many environmental issues, ranging from poverty to toxic environments to a dangerous community, all work against neurogenesis and the development of full potential. And let us expand this list to include trauma.

Let us recall that the infant, child, and adolescent brain can only pay attention to what is happening in the immediate environment. Again think of the varying positive and negative environments that your clients come from. One of the purposes of the community genogram is to help you and the client understand how we as individuals relate and have related to individuals, family, groups, and institutions around us. The church that welcomes you helps produce positive development, while the bank that refuses your parents a loan or peers that tease and harass you harm development.

Clients need to be informed about how social systems affect personal growth and individual development. Our work here is to help clients understand that the problem does not lie in them but in a social system or life experience that treated them unfairly and did not allow an opportunity for growth.

Finally, there is social action. What are you doing in your community and society to work against social forces that bring about poverty, war, and other types of oppression? Are you teaching your clients how they can work toward social justice themselves? A social justice approach includes helping clients find outlets to prevent oppression and work with schools, community action groups, and others for change.

YouTube Videos for Further Study

Here are some useful presentations that you can view to follow up on some key dimensions of neuroscience and neurobiology.

Robert Sapolsky: Depression

This is our recommended first choice. You will find an outstanding presentation that supplements this book and gives you a solid understanding of depression as a biological disease.

Robert Sapolsky: 41 Lectures at Stanford University

There is likely no better way to understand and work with neuroscience and biology than to commit yourself to time with this prize-winning and entertaining professor as he teaches.

Joan Chiao: Cultural Neuroscience

A leading expert on cultural neuroscience talks about neuroscience's powerful implications for multiculturalism.

Daniel Siegel: Interpersonal Connection, The Neurological Basis of Behavior, Mindsight and Neural Integration

Dr. Siegel is the person who has been the most effective in bringing neuroscience to counseling and therapy, medicine, and other fields.

Allen and Mary Ivey: Basic Lecture on Neuroscience and Counseling

Allen and Mary's neuroscience presentation.

More specific videos, full of useful information, can be found by searching for the following names: *Richard Davidson, Bruce McEwen, John Kabat-Zinn, Eric Kandel, John Ratey*. A search for *neuroscience* or *neurobiology* will produce many other interesting videos.

REFERENCES

Abe, N., Okuda, J., Suzuki, M., Matsuda, T., Mori, E., Minoru, T., et al. (2008). Neural correlates of true memory, false memory, and deception. *Cerebral Cortex, 18,* 2811–2819.

Adler, A. (1924). *The practice and theory of individual psychology.* Abington, UK: Routledge, Trench, Tubner.

Alim, T., Feder, A., Graves, R., Wang, Y., Weaver, J., Westphal, M., et al. (2008). Trauma, resilience, and recovery in a high-risk African-American population. *American Journal of Psychiatry, 165,* 1566–1575.

Armenta, B. E., Whitbeck, L. B., & Habecker, P. N. (2015). The Historical Loss Scale: Longitudinal measurement equivalence and prospective link to anxiety among North American indigenous adolescents. *Cultural Diversity and Ethnic Minority Psychology, 22,* 1–10.

American Counseling Association. (2014). *2014 ACA code of ethics.* Retrieved from http://counseling.org/docs/ethics/2014-aca-code-of -ethics.pdf?sfvrsn=4

American Psychiatric Association. (2013). *Diagnostic and statistical manual of mental disorders* (5th ed.). Washington, DC: Author.

American Psychological Association. (2010). *Ethical principles of psychologists and code of conduct.* Washington, DC: Author.

Andrews-Hanna, J. R. (2012). The brain's default network and its adaptive role in internal mentation. *The Neuroscientist, 18,* 251–270.

Andrews-Hanna, J. R., Smallwood, J., & Spreng, R. N. (2014). The default network and self-generated thought: Component processes, dynamic control, and clinical relevance. *Annals of the New York Academy of Sciences, 1316,* 29–52.

Anika Foundation. (2013). *Explaining the rise in youth suicide.* Retrieved from www.anikafoundation.com/rise_in_suicide.shtml.

Asbell, B., & Wynn, K. (1991). *Touching.* New York: Random House.

Aupperle, R., & Hunt, A. (2012, April). *fMRI may predict response to cognitive behavioral therapy.* Anxiety Disorders Association of America (ADAA) 32nd Annual Conference.

Australian Counselling Association. (2015). *Code of ethics and practice.* Retrieved from http://www.theaca.net.au/becoming-a-member.php

Barbara Bush calls evacuees better off. (2005, September 7). *New York Times.* Retrieved from http://www.nytimes.com/2005/09/07/us /nationalspecial/barbara-bush-calls-evacuees-better-off.html

Barrett-Lennard, G. (1962). Dimensions of therapist response as causal factors in therapeutic change. *Psychological Monographs, 76,* 43 (Ms. No. 562).

Baumeister, R. F., Bratslavsky, E., Finkenauer, C., & Vohs, K. D. (2001). *Review of General Psychology, 5,* 323–370.

Beck, J. (2011). *Cognitive behavior therapy: Basics and beyond* (2nd ed.). New York: Guilford Press.

Beck, J., & Broder, F. R. (2016). The new "homework" in cognitive behavior therapy. Retrieved from https://www.beckinstitute.org/blog /the-new-homework-in-cognitive-behavior-therapy

Bensing, J. (1999a). *Doctor–patient communication and the quality of care.* Utrecht, Netherlands: Nivel.

Bensing, J. (1999b). The role of affective behavior. *Communication,* 1188–1199.

Bensing, J., & Verheul, W. (2009). Towards a better understanding of the dynamics of patient provider interaction: The use of sequence analysis. *Patient Education and Counseling, 75*(2), 145–146.

Bensing, J., & Verheul, W. (2010). The silent healer: The role of communication in placebo effects. *Patient Education and Counseling, 80*(3), 293–299.

Benson, H., & Proctor, W. (2010). *Relaxation revolution: The science and genetics of mind body healing.* New York: Simon & Schuster.

Bergland, C. (2013, October 29). 25 studies confirm: Exercise prevents depression. *Psychology Today.* http://www.psychologytoday.com/blog /the-athletes-way/201310/25-studies-confirm-exercise-prevents -depression

Bickart, K., Wright, C., Dautoff, R., Dickerson, B., & Barrett, L. (2011). Amygdala volume and social network size in humans. *Nature Neuroscience, 14,* 163–164.

Binkley C., & Whack, E. (2015, November 13). Beyond Columbia: Black students around the US complain of everyday racism. *Herald Tribune,* p. A4.

Blackburn, E. H., Epel, E. S., & Lin, J. (2015). Human telomere biology: A contributory and interactive factor in aging, disease risks, and protection. *Science, 350,* 1193–1198.

Blonna, R., Loschiavo, J., & Watter, D. (2011). *Health counseling: A microskills approach for counselors, educators, and school nurses* (2nd ed.). Sudbury, MA: Jones & Bartlett.

Bodie, G., Vickery, A., Cannava, K., & Jones, S. (2015). The role of "active listening" in informal helping conversations: Impact on perceptions of listener helpfulness, sensitivity, and supportiveness and discloser emotional improvement. *Western Journal of Communication, 79*(2), 1–23.

Bourdieu, P., & Passeron, J. (1990). *Reproduction in education, society and culture.* London: Sage.

Boyle, P., Buchman, A., Barnes, L., & Bennett, D. (2010). Effect of a purpose in life on risk of incident Alzheimer disease and mild cognitive impairment in community-dwelling older persons. *Archives of General Psychiatry, 67,* 304–310.

Campó, R., & Carter, R. (2015). The appropriated racial oppression scale: Development and preliminary validation. *Cultural Diversity and Ethnic Minority Psychology, 21,* 497–506.

Canadian Counselling and Psychotherapy Association. (2007). *Ethics code.* Retrieved from https://www.ccpa-accp.ca/wp-content/uploads /2014/10/CodeofEthics_en.pdf

Carkhuff, R. (1969). *Helping and human relations: Practice and research.* New York: Holt, Rinehart, Winston.

Carkhuff, R. (2000). *The art of helping in the 21st century.* Amherst, MA: Human Resources Development Press.

Carl, J., Soskin, D., Kerns, C., & Barlow, D. (2013). Positive emotional regulation in emotional disorders: A theoretical review. *Clinical Psychology Review, 33,* 343–360.

Carlson, M., Erickson, K., Kramer, A., Voss, M., Bolea, N., Mielke, M., et al. (2009). Evidence for neurocognitive plasticity in at-risk older adults: The Experience Corps program. *Journals of Gerontology Series A: Biological Sciences and Medical Sciences, 64*(12), 1275–1282.

Carlstedt, R. (2011). *Handbook of integrative clinical psychology, psychiatry, and behavioral medicine: Perspectives, practices, and research.* New York: Springer.

Carstensen, L., Pasupathi, M., Mayr, U., & Nesselroade, J. (2000). Emotional experience in everyday life across the life span. *Journal of Personality and Social Psychology, 79,* 644–655.

Carter, R. (1999). *Mapping the mind.* Berkeley: University of California Press.

Chabris, C., & Simon, D. (2009). *The invisible gorilla and other ways our illusions deceive us.* New York: Broadway.

Chapin, T. J., & Russell-Chapin, L. A. (2014). *Neurotherapy and neurofeedback: Brain-based treatment for psychological and behavioral problems.* New York: Routledge.

Cheek, D. (2010). *Assertive Black . . . puzzled White: A Black perspective on assertive behavior.* Parker, CO: Outskirts Press.

Chung, R. C., & Bemak, F. P. (2000). *Social justice counseling: The next steps beyond multiculturalism.* Thousand Oaks, CA: Sage.

Chrousos, G. (2015). Stress and disorders of the stress system. *Medscape Psychiatry and Mental Health.* Retrieved from http://www.medscape .com/viewarticle/704866

Collura, T. F., Zalaquett, C. P., Bonnstetter, R. J., & Chatters, S. J. (2014). Toward an operational model of decision making, emotional regulation, and mental health impact. *Advances in Mind-Body Medicine, 28*(4), 18–33.

Contrada, R., Goyal, T., Cather, C., Rafalson, L., Idler, E., & Krause, T. (2004). Psychosocial factors in outcomes of heart surgery: The impact of religious involvement and depressive symptoms. *Health Psychology, 23,* 227–238.

Coventry, P. A., Small, N., Panagioti, M., Adeyemi, I., & Bee, P. (2015). Living with complexity, marshalling resources: A systematic review and qualitative meta-synthesis of lived experience of mental and physical multimorbidity. *BMC Family Practice, 16,* 171–182.

Cox, A. (2015). *Why should I care about Alfred Adler?* Chicago: Harrier.

Cross, W. (1971). The Negro-to-Black conversion experience. *Black World, 20,* 9, 13–27.

Cross, W. (1991). *Shades of black: Diversity in African-American identity.* Philadelphia: Temple University Press.

D'Andrea, M., & Daniels, J. (2001). RESPECTFUL counseling: An integrative model for counselors. In D. Pope-Davis & H. Coleman (Eds.), *The interface of class, culture and gender in counseling* (pp. 417–466). Thousand Oaks, CA: Sage.

D'Andrea, M., & Daniels, J. (2015, March). *Neuroscience and implementing the RESPECTFUL Counseling framework: A social justice advocacy framework.* Presentation at the annual meeting of the American Counseling Association, Orlando, FL.

Daniels, T. (2010). A review of research on microcounseling: 1967–present. In A. Ivey, M. Ivey, & C. Zalaquett, *Intentional interviewing and counseling: Your interactive resource* (CD-ROM) (7th ed.). Belmont, CA: Brooks/Cole.

Danner, D., Snowdon, D., & Friesen, W. (2001). Positive emotion in early life and longevity. *Journal of Personality and Social Psychology, 80,* 804–813.

Davidson, R. (2004). Well-being and affective style: Neural substrates and biobehavioral correlates. *The Philosophical Transactions of the Royal Society, 359,* 1395–1411.

Davis, M., Eshelman, E., & McKay, M. (2008). *The relaxation and stress management workbook* (6th ed.). Oakland, CA: New Harbinger.

Dean, S. (2014, March 12). How psychotherapy changes the brain. *Healthy minds, healthy lives.* Retrieved from http://apahealthyminds .blogspot.com/2014/03/how-psychotherapy-changes-brain.html

Decety, J., & Jackson, P. (2004). The functional architecture of human empathy. *Behavioral and Cognitive Neuroscience Reviews, 3,* 71–100.

Dietrich, A., & Kanso, R. (2010). A review of EEG, ERP, and neuroimaging studies of creativity and insight. *Psychological Bulletin, 136,* 822–848.

Dobson, K. (Ed.). (2009). *Handbook of cognitive behavioral therapies* (3rd ed.). New York: Guilford Press.

Donk, L. (1972). Attending behavior in mental patients. *Dissertation Abstracts International, 33* (Ord. No. 72-22 569).

Dreikurs, R., & Grey, L. (1968). *Logical consequences: A new approach to discipline.* New York: Dutton.

Duncan, B., Miller, S. D., Hubble, M., & Wampold, B. E. (Eds.). (2010). *The heart and soul of change: Delivering what works* (2nd ed.). Washington, DC: American Psychological Association.

Duncan, B., Miller, S., & Sparks, J. (2004). *The heroic client.* San Francisco: Jossey-Bass.

Duran, E. (2006). *Healing the soul wound: Counseling with American Indians and other Native peoples.* New York: Teachers College Press.

D'Zurilla, T., & Nezu, A. (2007). *Problem-solving therapy: A positive approach to clinical intervention* (3rd ed.). New York: Springer.

Egan, G. (2010). *The skilled helper* (9th ed.). Belmont, CA: Brooks/Cole.

Ekaterina, L., Popa, D., Apergis-Schoute, J., Fidacaro, G., & Paré, J. (2008). Amygdala intercalated neurons are required for expression of fear extinction. *Nature, 454,* 642–645.

Ekman, P. (1999). Basic emotions. In T. Dalgleish & M. Power (Eds.), *Handbook of cognition and emotion.* Sussex, UK: Wiley.

Ekman, P. (2007). *Emotions revealed* (2nd ed.). New York: Henry Holt.

Elliott, A. M., Alexander, S. C., Mescher, C. A., Mohan, D., & Barnato, A. E. (2016). Differences in physicians' verbal and nonverbal communication with black and white patients at the end of life. *Journal of Pain Management, 51*(1), 1–8.

Ellis, A., & Ellis, D. (2011). *Rational emotive behavior therapy.* Washington, DC: American Psychological Association.

Elton, A., & Gao, W. (2015). Task-positive functional connectivity of the default mode network transcends task domain. *Journal of Cognitive Neuroscience, 27,* 2369–2381.

Eres, R., Decety, J., Louis, W., & Molenberghs, P. (2015). Individual differences in local gray matter density are associated with differences in affective and cognitive empathy. *Neuroimage, 15,* 305–310.

Erickson, K., Miller, D., & Roecklein, K. (2011). The aging hippocampus: Interactions between exercise, depression, and BDNF. *The Neuroscientist, 18,* 82–97.

Ericsson, A. K., Charness, N., Feltovich, P., & Hoffman, R. R. (2006). *Cambridge handbook on expertise and expert performance.* Cambridge, UK: Cambridge University Press.

Fall, K., Fang, F., Mucci, L. A., Ye, W., Andrén, O., Johansson, J. E., et al. (2009). Immediate risk for cardiovascular events and suicide following a prostate cancer diagnosis: Prospective cohort study. *PLoS Medicine, 6*(12), e1000197.

Fan, J., Gu, X., Liu, X., Guise, K. G., Park, Y., Martin, L., et al. (2011). Involvement of the anterior cingulate and frontoinsular cortices in rapid processing of salient facial emotional information. *Neuroimage, 54,* 2539–2546.

Farnham, S., Gill, J., McLean, R., & Ward, S. (1991). *Listening hearts.* Harrisburg, PA: Morehouse.

Fava, G., Rafanelli, M., Grandi, S., Conti, S., & Belluardo, P. (1998). Prevention of recurrent depression with cognitive behavioral therapy. *Archives of General Psychiatry, 55,* 816–820.

Fiedler, F. E. (1950a). A comparison of therapeutic relationships in psychoanalytic, nondirective, and Adlerian therapy. *Journal of Consulting Psychology, 14,* 435–436.

Fiedler, F. E. (1950b). The concept of an ideal therapeutic relationship. *Journal of Consulting Psychology, 14,* 239–245.

Figley, C. R. (1995). *Compassion fatigue: Coping with secondary traumatic stress disorder in those who treat the traumatized.* New York: Brunner/Mazel.

Fischer, R., & Maier, O. (2015). Interrelation of oxidative stress and inflammation in neurodegenerative disease. *Oxidative Medicine and Cellular Longevity.* Retrieved from http://dx.doi.org/10.1155/2015/610813

Forsyth, J., & Carter, R. T. (2012). The influence of racial identity status attitudes and racism-related coping on mental health among Black Americans. *Cultural Diversity and Ethnic Minority Psychology, 18,* 128–140.

Fournier, J., DeRubeis, R., Hollon, S., Dimidjian, S., Amsterdam, J., Shelton, R., et al. (2010). Antidepressant drug effects and depression severity: A patient-level meta-analysis. *Journal of the American Medical Association, 303,* 47–53.

Fowler, J. H., & Christakis, N. A. (2010). Cooperative behavior cascades in human social networks. *Proceedings of the National Academy of Sciences of the United States of America, 107,* 5334–5338.

Frankl, V. (1959). *Man's search for meaning.* New York: Simon & Schuster.

Fredrickson, B., Tugade, M., Waugh, C., & Larkin, G. (2003). A prospective study of resilience and emotion following the terrorist attacks on the United States on September 11, 2001. *Journal of Personality and Social Psychology, 84,* 365–376.

Freed, P. J., & Mann, J. J. (2007). Sadness and loss: Toward a neurobiopsychosocial model. *American Journal of Psychiatry, 164,* 28–34.

Freire, P. (1970). *Pedagogy of the oppressed.* New York: Continuum.

Freud, S. (1953). Project for a scientific psychology. *Complete psychological works of Sigmund Freud* (Vol. 1, pp. 283–397). London: Hogarth Press. (Original work published 1895)

Fukuyama, M. (1990, March). *Multicultural and spiritual issues in counseling.* Workshop presentation for the American Counseling Association Convention, Cincinnati.

Galbraith, J. (2015, January/February). My Harvard education. *Harvard Magazine,* pp. 35–36.

Gallace, A., & Spence, C. (2010). The science of interpersonal touch: An overview. *Neuroscience and Biobehavioral Reviews, 34,* 246–259.

Gawande, A. (2009). *The checklist manifesto: How to get things right.* New York: Holt.

Gearhart, C., & Bodie, G. (2011). Active-empathic listening as a general social skill: Evidence from bivariate and canonical correlations. *Communication Reports, 24,* 86–98.

Gendlin, E., & Henricks, M. (n.d.). *Rap manual* [Mimeographed]. Cited in E. Gendlin, *Focusing*. New York: Everest House.

Gergen, K., & Gergen, M. (2005, February). The power of positive emotions. *The Positive Aging Newsletter*. Retrieved from www.healthandage.com

Goldberg, J. (2012) The effects of stress on your body. *WebMD*. http://www.webmd.com/mental-health/effects-of-stress-on-your-body

Goleman, D. (2013, October 5). Rich people just care less. *New York Times*. Retrieved from opinionator.blogs.nytimes.com/2013/10/05/rich-people-just-care-less/?_r=0

Gonçalves, Ó. (2015, April 13–15). The counselor's wandering mind: Being empathic by default. *Counseling Today*. Arlington, VA: American Counseling Association.

Goodman, R. D. (2013). The transgenerational trauma and resilience genogram. *Counselling Psychology Quarterly, 26*(3–4), 386–405.

Goodwin, L., Lee, S., Puig, A., & Sherrard, P. (2005). Guided imagery and relaxation for women with early stage breast cancer. *Journal of Creativity in Mental Health, 1*(2), 53–66.

Gottman, J. (2011). *The science of trust: Emotional attunement for couples*. New York: Norton.

Gould, E., Beylin, A., Tanapat, P., Reeves, A., & Shors, T. (1999). Learning enhances adult neurogenesis in the hippocampal formation. *Nature Neuroscience, 2*(3), 260–265.

Grawe, K. (2007). *Neuropsychotherapy: How the neurosciences inform psychotherapy*. London: Erlbaum.

Greene, D., & Stewart, F. (2011). African American students' reactions to Benjamin Cooke's "Nonverbal communication among Afro-Americans: An initial classification." *Journal of Black Studies, 42*, 389–401.

Greene, J. D. (2009). The cognitive neuroscience of moral judgment. In M. S. Gazzaniga (Ed.), *The cognitive neurosciences* (4th ed.). Cambridge, MA: MIT Press.

Greene, J. D., Nystrom, L. E., Engell, A. D., Darley, J. M., & Cohen, J. D. (2004). The neural bases of cognitive conflict and control in moral judgment. *Neuron, 44*, 389–400.

Grenham, S., Clarke, G., Cryan, J. F., & Dinan, T. G. (2011). Brain–gut–microbe communication in health and disease. *Frontiers in Physiology, 2*, 94. Available at http://journal.frontiersin.org/article/10.3389/fphys.2011.00094/full

Hall, E. (1959). *The silent language*. New York: Doubleday.

Hall, J., & Schmid Mast, M. (2007). Sources of accuracy in the empathic accuracy paradigm. *Emotion, 7*, 438–446.

Hanson, J. L., Nacewicz, B. M., Sutterer, M. J., Cayo, A. A., Schaeffer, S. M., Rudolph, K. D., et al. (2015). Behavioral problems after early life stress: Contributions of the hippocampus and amygdala. *Biological Psychiatry, 77*, 314–323.

Hargie, O., Dickson, D., & Tourish, D. (2004). *Communication skills for effective management*. New York: Palgrave Macmillan.

Haskard, K., Williams, S., DiMatteo, M., Heritage, J., & Rosenthal, R. (2008). The provider's voice: Patient satisfaction and the content-filtered speech of nurses and physicians in primary medical care. *Journal of Nonverbal Behavior, 32*, 1–20.

Herring, M. P., Jacob, M. L., Suveg, C., Dishman, R. K., & O'Connor, P. J. (2012). Feasibility of exercise training for the short-term treatment of generalized anxiety disorder: A randomized controlled trial. *Psychotherapy and Psychosomatics, 81*, 21–28.

Hoffman, M. (2015). *Brain beat: Scientific foundations and evolutionary perspectives on brain health*. New York: Page.

Hill, C. E. (2009). *Helping skills* (3rd ed.). Washington, DC: American Psychological Association.

Hill, C. E. (2014). *Helping skills: Facilitating exploration, insight, and action* (4th ed.). Washington, DC: American Psychological Association.

Hill, C. E., & O'Brien, K. (1999). *Helping skills*. Washington, DC: American Psychological Association.

Hill, C. E., & O'Brien, K. (2004). *Helping skills: Facilitating exploration, insight, and action*. Washington, DC: American Psychological Association.

Hill, S. (2014). The default network, task-positive network and goal-directed problem-solving. *Western Undergraduate Psychology Journal, 2*(1), Article 9. Retrieved from http://ir.lib.uwo.ca/wupj/vol2/iss1/9

Ho, B., Andreasen, N., Ziebell, S., Pierson, R., & Magnotta, V. (2011). Long-term antipsychotic treatment and brain volumes: A longitudinal study of first-episode schizophrenia. *Archives of General Psychiatry, 68*, 128–137.

Hölzel, B., Carmody, J., Vangel, M., Congletona, C., Yerramsetti, S., Gard, T., et al. (2011). Mindfulness practice leads to increases in regional brain gray matter density. *Psychiatry Research: Neuroimaging, 191*, 36–43.

Hunter, W. (1984). *Teaching schizophrenics communication skills: A comparative analysis of two microcounseling learning environments*. Unpublished doctoral dissertation, University of Massachusetts, Amherst.

Insel, T. (2013, April 29). *Transforming diagnosis*. National Institute of Mental Health. Retrieved from http://www.nimh.nih.gov/about/director/2013/transforming-diagnosis.shtml

International Coach Federation. (2015). *What is coaching?* Retrieved from http://coachfederation.org/need/landing.cfm?ItemNumber=978&navItemNumber=567

Ishiyama, I. (2006). *Anti-discrimination response training (A.R.T.) program*. Framingham, MA: Microtraining Associates.

Ivey, A. (1973). Media therapy: Educational change planning for psychiatric patients. *Journal of Counseling Psychology, 20*, 338–343.

Ivey, A. (2000). *Developmental therapy: Theory into practice*. North Amherst, MA: Microtraining Associates. (Originally published 1986)

Ivey, A., D'Andrea, M., & Ivey, M. (2012). *Theories of counseling and psychotherapy: A multicultural perspective* (7th ed.). Thousand Oaks, CA: Sage.

Ivey, A., & Daniels, T. (in press). Systematic interviewing microskills: Developing bridges between the fields of communication and counseling psychology. *International Journal of Listening*.

Ivey, A., Ivey, M., Gluckstern-Packard, N., Butler, K., & Zalaquett, C. (2012). *Basic influencing skills* (4th ed.) [DVD]. Alexandria, VA: Microtraining/Alexander Street Press.

Ivey, A., Ivey, M., Myers, J., & Sweeney, T. (2005). *Developmental counseling and therapy: Promoting wellness over the lifespan*. Belmont, CA: Wadsworth.

Ivey, A., Ivey, M., Zalaquett, C., & Daniels, T. (2014). *Intentional interviewing and counseling: CourseMate interactive website*. Belmont, CA: Brooks/Cole.

Ivey, A., Normington, C., Miller, C., Morrill, W., & Haase, R. (1968). Microcounseling and attending behavior: An approach to prepracticum counselor training [Monograph]. *Journal of Counseling Psychology, 15*, Part II, 1–12.

Ivey, A., Pedersen, P., & Ivey, M. (2001). *Intentional group counseling: A microskills approach*. Belmont, CA: Brooks/Cole.

Ivey, A., & Zalaquett, C. (2009). Psychotherapy as liberation: Multicultural counseling and psychotherapy (MCT) contributions to the promotion of psychological emancipation. In J. L. Chin (Ed.), *Diversity in mind and in action*. Santa Barbara, CA: Praeger.

Ivey, A., & Zalaquett, C. (2011). Neuroscience and counseling. *Journal for Social Action in Counseling and Psychology, 3*, 103–116. Retrieved from www.psysr.org/jsacp/ivey-v3n1-11_103-116.pdf

Janis, I. L. (1983). *Short-term counseling: Guidelines based on recent research*. New Haven, CT: Yale University Press.

Kabat-Zinn, J. (2005). *Coming to our senses: Healing ourselves and the world through mindfulness*. New York: Hyperion.

Kabat-Zinn, J. (2009). *Wherever you go, there you are: Mindfulness meditation in everyday life* (10th anniversary ed.). New York: Hyperion.

Kabat-Zinn, J., & Davidson, R. (2012). *The mind's own physician: A scientific dialogue with the Dalai Lama on the healing power of meditation*. New York: New Harbinger.

Kahnemann, D. (2011). *Thinking fast and slow*. New York: Farrar, Straus and Giroux.

Kaiser, C., Drury, B., Malahy, L., & King, K. (2011). Nonverbal asymmetry in interracial interactions: Strongly identified Blacks display friendliness, but Whites respond negatively. *Social Psychology and Personality Science, 2*, 554–559.

Kawamichi, H., Yoshihara, K., Sasaki, A., Sugawara, S., Tanabe, H., Shinohara, R., et al. (2014). Perceiving active listening activates the reward system and improves the impression of relevant experiences. *Social Neuroscience.* Online publication 4 September, http://www.tandfonline.com/doi/abs/10.1080/17470919.2014.954732#.VC-ha3VdWUk

Kellermann, N. (2013). Epigenetic transmission of Holocaust trauma. *Israel Journal of Psychiatry and Related Sciences, 50*(1), 33–39.

Kelley, J., Kraft-Todd, G., Schapira, L., Kossowsky, J., & Riess, H. (2014). The influence of the patient-clinician relationship on health-care outcomes: A systematic review and meta-analysis of randomized controlled trials. *PLOS One, 9*(4), e94207.

Kim, E., Park, N., & Peterson, C. (2011). Dispositional optimism protects older adults from stroke: The Health and Retirement Study. *Stroke.* Retrieved from stroke.ahajournals.org/content/early/2011/07/21/STROKEAHA.111.613448.full.pdf?ijkey=EgeC0lK195rBJDH&keytype=ref

Kolb, B., & Whishaw, I. (2009). *Fundamentals of human neuropsychology* (6th ed.). New York: Worth.

Korman, M. (1973). *Levels and patterns of training in psychology.* Washington, DC: American Psychological Association.

Lacan, J. (1977). *The four fundamental concepts of psychoanalysis: The seminar of Jacques Lacan, Book XI* (Jacques-Alain Miller, Ed.; Alan Sheridan, Trans.). New York: Norton.

LaFrance, M., & Woodzicka, J. (1998). No laughing matter: Women's verbal and nonverbal reactions to sexist humor. In J. Swim & C. Stangor (Eds.), *Prejudice: The target's perspective.* San Diego, CA: Academic Press.

Lambert, M. J. (2013). *Bergin and Garfield's handbook of psychotherapy and behavior change* (6th ed.). New York: Wiley.

Lazarus, A. A. (2006). *Brief but comprehensive psychotherapy: The multimodal way.* New York: Springer.

Lee, C. (*1992*). *Empowering young black males.* Ann Arbor, MI: ERIC.

Lee, J. R., & Hopkins, V. (2009). Cortisol and the stress connection. *Virginia Hopkins Test Kits.* http://www.virginiahopkinstestkits.com/cortisolstress.html

Lewis, D. (2011). Antipsychotic medications and brain volume: Do we have cause for concern? *Archives of General Psychiatry, 68,* 126–127.

Li, J., & Lambert, V. (2008). Job satisfaction among intensive care nurses from the People's Republic of China. *International Nursing Review, 55,* 34–39.

Lilienfeld, S. O., Lynn, S. J., Ruscio, J., & Beyerstein, B. L. (2010). *50 great myths of popular psychology.* Chichester, UK: Wiley-Blackwell.

Loftus, E. (1997, September). Creating false memories. *Scientific American,* pp. 51–55.

Loftus, E. (2003). Our changeable memories: Legal and practical implications. *Nature Reviews: Neuroscience, 4,* 31–34.

Loftus, E. (2011). We live in perilous times for science. *Skeptical Inquirer, 35,* 13.

Logothetis, N. (2008). What we can do and what we cannot do with fMRI. *Nature, 453,* 869–878.

Lucas, L. (2007). The pain of attachment—"You have to put a little wedge in there": How vicarious trauma affects child/teacher attachment. *Childhood Education, 84,* 85–91.

Mahoney, M., & Freeman, A. (Eds.). (1985). *Cognition and psychotherapy.* New York: Springer.

Mann, L. (2001). Naturalistic decision making. *Journal of Behavioural Decision Making, 14,* 375–377.

Mann, L., Beswick, G., Allouache, P., & Ivey, M. (1989). Decision workshops for the improvement of decision making skills. *Journal of Counseling and Development, 67,* 237–243.

Marci, C. D., Ham, J., Moran, E., & Orr, S. P. (2007). Physiologic correlates of perceived therapist empathy and social-emotional process during psychotherapy. *Journal of Nervous and Mental Disease, 195,* 103–111.

Marlatt, G., Larimer, M., & Witkiewitz, K. (2011). *Harm reduction: Pragmatic strategies for managing high-risk behaviors* (2nd ed.). New York: Guilford Press.

Marroun, H., Jaddoe, V., Hudziak, J., Roza, S., Steegers, E., Hofman, A., et al. (2012). Maternal use of selective serotonin reuptake inhibitors, fetal growth, and risk of adverse birth outcomes. *Archives of General Psychiatry, 69,* 706–714.

Martinez-Conde, S., & Macknik, S. L. (2013). The neuroscience of illusion: How tricking the eye reveals the inner workings of the brain. *Scientific American.* Retrieved from http://www.scientificamerican.com/article/the-neuroscience-of-illusion

Maslow, A., Frager, R., & Fadiman, J. (1987). *Motivation and personality.* New York: Harper.

Mathur, A., Harada, T., & Chia, J. (2011). Racial identification modulates default network activity for same and other races. *Human Brain Mapping, 33,* 1883–1893.

Matsumoto, D., Hwang, H., Skinner, L., & Frank, M. (2011, June). Evaluating truthfulness and detecting deception. *FBI Law Enforcement Bulletin.*

McGoldrick, M., & Gerson, R. (1985). *Genograms in family assessment.* New York: Norton.

McGoldrick, M., Giordano, J., & Garcia-Preto, N. (2005). *Ethnicity and family therapy* (3rd ed.). New York: Norton.

McGorry, P. (Ed.). (2012). *Early intervention in psychiatry.* Retrieved from onlinelibrary.wiley.com/journal/10.1111/(ISSN)1751-7893/issues

McIntosh, P. (1988). *White privilege and male privilege: A personal account of coming to see correspondences through work in women's studies.* Wellesley, MA: Wellesley College Center for Research on Women.

McMahon, H. G., Mason, E. C. M., Daluga-Guenther, N., & Ruiz, A. (2014). An ecological model of professional school counseling. *Journal of Counseling & Development, 92,* 459–471.

Meara, N., Pepinsky, H., Shannon, J., & Murray, W. (1981). Semantic communication and expectation for counseling across three theoretical orientations. *Journal of Counseling Psychology, 28,* 110–118.

Meara, N., Shannon, J., & Pepinsky, H. (1979). Comparisons of stylistic complexity of the language of counselor and client across three theoretical orientations. *Journal of Counseling Psychology, 26,* 181–189.

Mikecz, R. (2011). Interviewing elites: Addressing methodological issues. *Qualitative Inquiry, 18,* 482–493.

Miller, G. (1956). The magical number 7 plus or minus 2: Some limits on our ability for processing information. *Psychological Review, 63,* 81–87.

Miller, K. (2007). Compassionate communication in the workplace: Exploring processes of noticing, connecting, and responding. *Journal of Applied Communication Research, 35,* 223–245.

Miller, S., Duncan, B., & Hubble, M. (2005). Outcome-informed clinical work. In J. Norcross & M. Goldfried (Eds.), *Handbook of psychotherapy integration* (pp. 84–104). Oxford, UK: Oxford University Press.

Miller, W. R., & Rollnick, S. (2013). *Motivational interviewing: Helping people change* (3rd ed.). New York: Guilford Press.

Moore, M. (2015). *Coaching psychology manual.* Baltimore, MD: Lippincott, Williams, and Wilkens.

Moos, R. (2001, August). *The contextual framework.* Presentation at the American Psychological Association, San Francisco.

National Association of Social Workers. (2008). *Code of ethics.* Retrieved from http://socialworkers.org/pubs/code/default.asp

National Heart, Lung, and Blood Institute. (2013). *Reduce screen time.* http://www.nhlbi.nih.gov/health/educational/wecan/reduce-screen-time

National Organization for Human Services. (2015). *Ethical standards for human service professionals.* Retrieved from http://www.nationalhumanservices.org/ethical-standards-for-hs-professionals

National Scientific Council on the Developing Child. (2005). *Excessive stress disrupts the architecture of the developing brain* (Working Paper No. 3). Retrieved from www.developingchild.harvard.edu

Nes, L., & Segerstrom, S. (2006). Dispositional optimism and coping: A meta-analytic review. *Personality and Social Psychology Review, 10,* 235–251.

Nezu A., & Nezu, C. (2013). *Problem-solving therapy: A treatment manual.* New York: Springer.

Norcross, J. C. (Ed.). (2011). *Psychotherapy relationships that work* (2nd ed.). New York: Oxford University Press.

Nwachuku, U., & Ivey, A. (1992). Teaching culture-specific counseling use in microtraining technology. *International Journal for the Advancement of Counseling, 15,* 151–161.

Ogbonnaya, O. (1994). Person as community: An African understanding of the person as intrapsychic community. *Journal of Black Psychology, 20*, 75–87.

Oliveira-Silva, P., & Gonçalves, O. F. (2011). Responding empathically: A question of heart, not a question of skin. *Applied Psychophysiology and Biofeedback, 36*, 201–207.

Pedersen, P. B., & Marsella, A. J. (1982). The ethical crisis for cross-cultural counseling and therapy. *Professional Psychology, 13*(4), 492–500.

Pfiffner, L., & McBurnett, K. (1997). Social skills training with parent generalization: Treatment effects for children with attention deficit disorder. *Journal of Consulting and Clinical Psychology, 65*, 749–757.

Pierce, C. (1974). Psychiatric problems of the Black minority. In S. Arieti (Ed.), *American handbook of psychiatry* (pp. 512–523). New York: Basic Books.

Porges, S. (2011). *The polyvagal theory: Neurophysiological foundations of emotions, attachment, communication, and self-regulation.* New York: Norton.

Pos, A., Greenberg, L., Goldman, R., & Korman, L. (2003). Emotional processing during experiential treatment of depression. *Journal of Clinical and Consulting Psychology, 73*, 1007–1016.

Posner, M. (Ed.). (2004). *Cognitive neuropsychology of attention.* New York: Guilford Press.

Power, S., & Lopez, R. (1985). Perceptual, motor, and verbal skills of monolingual and bilingual Hispanic children: A discrimination analysis. *Perceptual and Motor Skills, 60*, 1001–1109.

Probst, R. (1996). Cognitive-behavioral therapy and the religious person. In E. Shafranski (Ed.), *Religion and the clinical practice of psychology* (pp. 391–408). Washington, DC: American Psychological Association.

Raichle, M. E. (2015, March 19). The brain's default mode network—what does it mean to us? *The Meditation Blog.* Retrieved from http://www.themeditationblog.com/the-brains-default-mode-network-what-does-it-mean-to-us

Raichle, M. E., & Snyder, A. Z. (2007). A default mode of brain function: A brief history of an evolving idea. *NeuroImage, 37*, 1083–1090.

Ratey, J. (2008a). *Neuroscience and the brain* [Transcript from video interview]. Framingham, MA: Microtraining Associates.

Ratey, J. (2008b). *Spark: The revolutionary new science of exercise and the brain.* New York: Little, Brown.

Ratey, J. (2012, April). *Exercise as a key to mental and physical health.* Presentation to Trends in Neuroscience Conference, Bradley University, Peoria, IL.

Ratey, J., & Hagerman, E. (2013). *Spark: The revolutionary new science of exercise and the brain.* New York: Little, Brown.

Ratey, J., & Manning, R. (2014). *Go wild: Free your body and mind from the afflictions of civilization.* New York: Little, Brown.

Ratts, M. J., Singh, A. A., Nassar-McMillan, S., Butler, S. K., & McCullough, J. R. (2015). *Multicultural and social justice counseling competencies.* Retrieved from http://www.multiculturalcounseling.org/index.php?option=com_content&view=article&id=205 :amcd-endorses-multicultural-and-social-justice-counseling-competencies&catid=1:latest&Itemid=123

Restak, R. (2003). *The new brain: How the modern age is rewiring your mind.* New York: Rodale Press.

Riess, H. (2015). Impact of clinical empathy on patients and clinicians. *AJOB Neuroscience, 6*(3), 51–53.

Rigazio-DiGilio, S., Ivey, A., Grady, L., & Kunkler-Peck, K. (2005). *The community genogram.* New York: Teachers College Press.

Rockett, I., Regier, D., Kapusta, N., Coben, J., Miller, T., Hanzlick, R., et al. (2012). Leading causes of unintentional and intentional injury mortality: United States, 2000–2009. *American Journal of Public Health,102*, e84–e92.

Rogers, C. (1951). *Client-centered therapy.* Boston: Houghton Mifflin.

Rogers, C. (1957). The necessary and sufficient conditions of therapeutic personality change. *Journal of Consulting Psychology, 21*, 95–103.

Rogers, C. (1961). *On becoming a person.* Boston: Houghton Mifflin.

Roskies, A. L. (2012). How does the neuroscience of decision making bear on our understanding of moral responsibility and free will? *Current Opinion in Neurobiology, 22*, 1022–1026.

Rothbaum, B., & Keane, T. (2012, April 13). *Emergency therapy may prevent PTSD in trauma victims.* Anxiety Disorders Association of America (ADAA) 32nd Annual Conference, Session 318R.

Russell-Chapin, L., & Jones, L. (2015) Neurocounseling: Bringing the brain into clinical practice. *Fact Based Health.* Retrieved from http://factbasedhealth.com/neurocounseling-bringing-brain-clinical-practice

Saggar, M., Quintin, E.-M., Kienitz, E., Bott, N. T., Sun, Z., Hong, W.-C., et al. (2015). Pictionary-based fMRI paradigm to study the neural correlates of spontaneous improvisation and figural creativity. *Scientific Reports, 5.* doi:10.1038/srep10894.

Sampaio, A., Soares, J. M., Coutinho, J., Sousa, M., & Gonçalves, Ó. F. (2014). The big five default brain: Functional evidence. *Brain Structure and Function, 219*, 1913–1922.

Scheier, M., Carver, C., & Bridges, M. (1994). Distinguishing optimism from neuroticism (and trait anxiety, self-mastery, and self-esteem): A reevaluation of the Life Orientation Test. *Journal of Personality and Social Psychology, 67*, 1063–1078.

Schlosser, L. Z. (2003). Christian privilege: Breaking a sacred taboo. *Journal of Multicultural Counseling and Development, 31*, 44–51.

Schwartz, J., & Begley, S. (2003). *The mind and the brain: Neuroplasticity and the power of mental force.* New York: Regan.

Seki, T., Sawamoto, K., Parent, J., & Alvarez-Buylla, A. (2011). *Neurogenesis in the adult brain I: Neurobiology.* New York: Springer.

Seligman, M. E. P. (2009). *Authentic happiness.* New York: Free Press.

Seligman, M. E. P. (2012). *Flourish: A visionary new understanding of happiness and well-being.* New York: Simon & Schuster.

Sepella. E. (2013). The compassionate mind: Science shows why it's healthy and how it spreads. *Observer.* Association of Psychological Science. Retrieved from http://www.psychologicalscience.org/index.php/publications/observer/2013/may-june-13/the-compassionate-mind.html

Seung, S. (2012). *Connectome: How the brain's wiring makes us who we are.* Boston: Houghton Mifflin Harcourt.

Seung, S., & Sümbül, U. (2014). Neuronal cell types and connectivity: Lessons from the retina. *Neuron, 83*, 1262–1272.

Shalev, A., Ankri, Y., Israeli-Shalev, Y., Peleg, M., Adessky, R., & Freedman, S. (2012). Prevention of posttraumatic stress disorder by early treatment: Results from the Jerusalem trauma outreach and prevention study, *Archives of General Psychiatry, 69*, 166–176.

Sharpley, C., & Guidara, D. (1993). Counselor verbal response mode usage and client-perceived rapport. *Counseling Psychology Quarterly, 6*, 131–142.

Sharpley, C., & Sagris, I. (1995). Does eye contact increase counselor-client rapport? *Counselling Psychology Quarterly, 8*, 145–155.

Shenk, D. (2010). *The genius in all of us.* New York: Doubleday.

Sherrard, P. (1973). *Predicting group leader/member interaction: The efficacy of the Ivey Taxonomy.* Unpublished doctoral dissertation, University of Massachusetts, Amherst.

Shoshan, T. (1989). Mourning and longing from generation to generation. *American Journal of Psychiatry, 43*, 193–207.

Shostrom, E. (1966). *Three approaches to psychotherapy* [Film]. Santa Ana, CA: Psychological Films.

Singer, T., Seymour, B., O'Dougherty, J., Kaube, H., Dolan, R., & Frith, C. (2004). Empathy for pain involves the affective but not sensory components of pain. *Science, 303*, 1157–1161.

Smith, B. W., Dalen, J., Wiggins, K., Tooley, E., Christopher, P., & Bernard, J. (2008). The brief resilience scale: Assessing the ability to bounce back. *International Journal of Behavioural Medicine, 15*, 194–200.

Soloman, Z., Kotter, M., & Mikulincer, M. (1988). Combat-related posttraumatic stress disorder among second-generation Holocaust survivors. *American Journal of Psychiatry, 145*, 865–868.

Somers, T. (2006). The sounds of silence: Brains are active in absence of sound. *Society for Neuroscience.* Retrieved from http://www.sfn.org/Press-Room/News-Release-Archives/2006/The-Sounds-of-Silence

Sommers-Flannagan, J., & Sommers-Flannagan, R. (2015). *Clinical interviewing* (5th ed.). New York: Wiley.

Spreng, R. N., & Andrews-Hanna, J. R. (2015). The default network and social cognition. *Brain Mapping: An Encyclopedic Reference.* Retrieved from http://dx.doi.org/10.1016/B978-0-12-397025-1.00173-1.

Spreng, R. N., DuPre, E., Selarka, D., Garcia, J., Gojkovic, S., Mildner, J., et al. (2014). Goal-congruent default network activity facilitates cognitive control. *Journal of Neuroscience, 34*, 14108–14114.

Stepanikova, I. (2012). Racial-ethnic biases, time pressure, and medical decisions. *Journal of Health and Social Behavior, 53*, 329–343.

Stephens, G. J., Silbert, L. J., & Hasson, U. (2010). Speaker–listener neural coupling underlies successful communication. *Proceedings of the National Academy of Sciences of the United States of America, 107*(32), 14425–14430.

Steward, R., Neil, D., Jo, H., Hill, M. & Baden, A. (1998). *White counselor trainees: Is there multicultural counseling competence without formal training?* Poster session presented at the Great Lakes Regional Conference of Division 17 of the American Psychological Association, Bloomington, IN.

Sue, D., & Sue, D. M. (2007). *Foundations of counseling and psychotherapy: Evidence-based practices for a diverse society.* New York: Wiley.

Sue, D. W. (2010). *Microaggressions in everyday life: Race, gender, and sexual orientation.* New York: Wiley.

Sue, D. W., Carter, R. T., Casas, J. M., Fouad, N. A., Ivey, A. E., Jensen, M., et al. (1998). *Multicultural counseling competencies.* Beverly Hills, CA: Sage.

Sue, D. W., Ivey, A., & Pedersen, P. (1999). *A theory of multicultural counseling and therapy.* Pacific Grove, CA: Brooks/Cole.

Sue, D. W., & Sue, D. (2015). *Counseling the culturally diverse: Theory and practice* (7th ed.). New York: Wiley.

Sweeney, T. J. (1998). *Adlerian counseling: A practitioner's approach* (4th ed.). Muncie, IN: Accelerated Development.

Swift, J. K., & Greenberg, R. P. (2012). Premature discontinuation in adult psychotherapy: A meta-analysis. *Journal of Consulting and Clinical Psychology, 80*, 547–559.

Szczygieł, D., Buczny, J., & Bazińska, R. (2012). Emotion regulation and emotional information processing: The moderating effect of emotional awareness. *Personality and Individual Differences, 52*(3), 433–437.

Tamase, K. (1991). Factors which influence the response to open and closed questions: Intimacy in dyad and listener's self-disclosure. *Japanese Journal of Counseling Science, 24*, 111–122.

Tamase, K., Otsuka, Y., & Otani, T. (1990). Reflection of feeling in microcounseling. *Bulletin of Institute for Educational Research* (Nara University of Education), *26*, 55–66.

Tamase, K., Torisu, K., & Ikawa, J. (1991). Effect of the questioning sequence on the response length in an experimental interview. *Bulletin of Nara University of Education, 40*, 199–211.

Torres-Rivera, E., Pyhan, L., Maddux, C., Wilbur, M., & Garrett, M. (2001). Process vs. content: Integrating personal awareness and counseling skills to meet the multicultural challenge of the twenty-first century. *Counselor Education and Supervision, 41*, 28–40.

Truax, C. (1961). *A tentative approach to the conceptualization and measurement of intensity and intimacy of interpersonal contact as a variable in psychotherapy.* Washington, DC: Eric Clearinghouse. Retrieved from www.eric.ed.gov/PDFS/ED133613.pdf

Uchino, B. N., Bowen, K., Carlisle, M., & Birmingham, W. (2012). What are the psychological pathways linking social support to health outcomes? A visit with the "ghosts" of research past, present, and future. *Social Science and Medicine, 74*, 949–957.

United Nations. (1989). *Convention on the rights of the child.* Retrieved from http://www.ohchr.org/Documents/ProfessionalInterest/crc.pdf

United Nations. (2015). *Convention on the rights of the child.* Retrieved from http://www.ohchr.org/Documents/ProfessionalInterest/crc.pdf

United Nations International Children's Emergency Fund (UNICEF). (2009). *About the convention.* Retrieved from http://www.unicef.org/rightsite/237_202.htm

United Nations International Children's Emergency Fund (UNICEF). (2014). *Convention on the Rights of the Child: Activating a promise.* Retrieved from http://www.unicef.org/crc/index_30160.html

U.S. Department of Labor. (2012). *Occupational outlook handbook 2010–2011.* Retrieved from www.bls.gov/oco/ooh_index.htm

Van der Molen, H. (1984). *Aan verlegenheid valt iets te doen: Een cursus in plaats van therapie* [How to deal with shyness: A course instead of therapy]. Deventer, Netherlands: Van Loghum Slaterus.

Van der Molen, H. (2006). Social skills training and shyness. In T. Daniels & A. Ivey (Eds.), *Microcounseling* (3rd ed.). Springfield, IL: Charles C Thomas.

Viamontes, G., & Beitman, B. (2007). Mapping the unconscious in the brain. *Psychiatric Annals, 37*, 243–258.

Voss, M., Heo, S., Prakash, R., Erickson, K., Alves, H., Chaddock, L., et al. (2013). The influence of aerobic fitness on cerebral white matter integrity and cognitive function in older adults: Results of a one-year exercise intervention. *Human Brain Mapping, 34*, 2972–2985.

Voss, M., Prakash, R., Erickson, K., Basak, C., Chaddock, L., Kim, J., et al. (2010). Plasticity of brain networks in a randomized intervention trial of exercise training in older adults. *Frontiers in Aging Neuroscience, 2*, 1–17.

Wade, S., Borawski, E., Taylor, H., Drotar, D., Yeates, K., & Stancin, T. (2001). The relationship of caregiver coping to family outcomes during the initial year following pediatric traumatic injury. *Journal of Consulting and Clinical Psychology, 69*, 406–415.

Wang, S. (2012, August 30). Coffee break? Walk in the park? Why unwinding is hard. *Wall Street Journal.* Retrieved from wsj.com/news/articles/SB10001424053111904199404576538260326965724

Weir, K. (2011). The exercise effect. *Monitor on Psychology, 42*(11), 48–52.

Welvaert, M., & Rosseel, Y. (2014). A review of fMRI simulation studies. *PLoS One, 9*(7). Retrieved from http://www.plosone.org/article/fetchObject.action?uri=info:doi/10.1371/journal.pone.0101953&representation=PDF

Westefeld, J. S., Casper, D., Lewis, A., Manlick, C., Rasmussen, W., Richards, A., et al. (2012). Physician-assisted death and its relationship to the human services professions. *Journal of Loss and Trauma, 18*, 539–555.

Westefeld, J. S., Richards, A., & Levy, L. (2011). Protective factors. In D. Lamis & D. Lester (Eds.), *Understanding and preventing college student suicide* (pp. 170–182). Springfield, IL: Charles C. Thomas.

Westen, D. (2007). *The political brain.* New York: Public Affairs.

Williams, K. D., & Nida, S. A. (2014). Ostracism and public policy. *Policy Insights from the Behavioral and Brain Sciences, 1*(1), 38–45.

Woods, S., Walsh, B., Saksa, J., & McGlashan, T. (2010). The case for including attenuated psychotic symptoms syndrome in DSM-5 as a psychosis risk syndrome. *Schizophrenia Research, 123*, 199–207.

Woodzicka, J., & LaFrance, M. (2002). Real versus imagined gender harassment. *Journal of Social Issues, 1*, 15–30.

Xu, P., Gu, R., Broster, L. S., Wu, R., Van Dam, N. T., Jiang, Y., et al. (2013). Neural basis of emotional decision making in trait anxiety. *Journal of Neuroscience, 33*, 18641–18653.

Yun, K., Watanabe, K. & Shimojo, S. (2012). Interpersonal body and neural synchronization as a marker of implicit social interaction. *Scientific Reports, 2*, Article 959. doi: 10.1038/srep00959

Zalaquett, C., & Chatters, S. J. (2014). Cyberbullying in college: Frequency, characteristics, and practical implications. *SAGE Open, 4*, 1–8.

Zalaquett, C., Foley, P., Tillotson, K., Hof, D., & Dinsmore, J. (2008). Multicultural and social justice training for counselor education programs and colleges of education: Rewards and challenges. *Journal of Counseling and Development, 86*, 323–329.

Zalaquett, C., & Ivey, A. (2014). Neuroscience and psychology: Central issues for social justice leaders. In H. Friedman & C. Johnson (Eds.), *The Praeger Handbook of Social Justice and Psychology.* Santa Barbara, CA: Praeger.

Zalaquett, C., Ivey, A., Gluckstern-Packard, N., & Ivey, M. (2008). *Las habilidades atencionales básicas: Pilares fundamentales de la comunicación efectiva.* Alexandria, VA: Microtraining Associates/Alexander Street Press.

Zur, O. (2015). *The HIPAA compliance kit.* Retrieved from http://www.zurinstitute.com/hipaakit.html

NAME INDEX

SUBJECT INDEX